DRILL HALL LIBRARY
MEDWAY

Block Copolymers in Nanoscience

Edited by
Massimo Lazzari, Guojun Liu,
and Sébastien Lecommandoux

Related Titles

M. Köhler, W. Fritzsche

Nanotechnology

An Introduction to Nanostructuring Techniques

Second Edition

2007. ISBN 3-527-31871-2

C. N. R. Rao, A. Müller, A. K. Cheetham (Eds.)

Nanomaterials Chemistry

Recent Developments and New Directions

2007. ISBN 3-527-31664-7

C. S. S. R. Kumar, J. Hormes, C. Leuschner (Eds.)

Nanofabrication Towards Biomedical Applications

Techniques, Tools, Applications, and Impact

2005. ISBN 3-527-31115-7

R. C. Advincula, W. J. Brittain, K. C. Caster, J. Rühe (Eds.)

Polymer Brushes

Synthesis, Characterization, Applications

2004. ISBN 3-527-31033-9

F. Caruso (Ed.)

Colloids and Colloid Assemblies

Synthesis, Modification, Organization and Utilization of Colloid Particles

2004. ISBN 3-527-30660-9

P. Gómez-Romero, C. Sanchez (Eds.)

Functional Hybrid Materials

2004. ISBN 3-527-30484-3

I. Manners

Synthetic Metal-Containing Polymers

2004. ISBN 3-527-29463-5

C. N. R. Rao, A. Müller, A. K. Cheetham (Eds.)

The Chemistry of Nanomaterials

Synthesis, Properties and Applications

2004. ISBN 3-527-30686-2

H.-G. Elias

An Introduction to Plastics

Second Edition

2003. ISBN 3-527-29602-6

Block Copolymers in Nanoscience

Edited by
Massimo Lazzari, Guojun Liu,
and Sébastien Lecommandoux

WILEY-
VCH

WILEY-VCH Verlag GmbH & Co. KGaA

The Editors

Prof. Dr. Massimo Lazzari
Institute of Technical Investigations
Department of Physical Chemistry
University of Santiago de Compostella
15782 Santiago de Compostela
Spain

Prof. Dr. Guojun Liu
Department of Chemistry
Queen's University
90 Bader Lane
Kingston, Ontario K7L 3N6
Canada

Prof. Dr. Sébastien Lecommandoux
Laboratory of Organic Polymer Chemistry
University of Bordeaux 1
16 Avenue Pey Berland
33607 Pessac
France

■ All books published by Wiley-VCH are carefully produced. Nevertheless, authors, editors, and publisher do not warrant the information contained in these books, including this book, to be free of errors. Readers are adviced to keep in mind that statements, data, illustrations, procedural details or other items may inadvertently be inaccurate.

Library of Congress Card No: applied for

British Library Cataloging-in-Publication Data:
A catalogue record for this book is available from the British Library

Bibliographic information published by the Deutsche Nationalbibliothek
The Deutsche Nationalbibliothek lists this publication in the Deutsche Nationalbibliografie; detailed bibliographic data are available in the Internet at http://dnb.d-nb.de

© 2006 WILEY-VCH Verlag GmbH & Co. KGaA, Weinheim

All rights reserved (including those of translation into other languages). No part of this book may be reproduced in any form — by photoprinting, microfilm, or any other means — nor transmitted or translated into machine language without written permission from the publishers. Registered names, trademarks, etc. used in this book, even when not specifically marked as such, are not to be considered unprotected by law.

Typesetting Hilmar Schlegel, Berlin
Printing and Bookbinding Markono Print Media Pte Ltd, Singapore

Printed in Singapore
Printed on acid-free paper

ISBN-13: 978-3-527-31309-9
ISBN-10: 3-527-31309-5

Contents

Preface *XIII*

List of Contributors *XV*

1 An Introduction to Block Copolymer Applications: State-of-the-Art and Future Developments *1*
Sébastien Lecommandoux, Massimo Lazzari, and Guojun Liu
References *6*

2 Guidelines for Synthesizing Block Copolymers *9*
Daniel Taton and Yves Gnanou
2.1 Introduction *9*
2.2 Free-radical Polymerization *13*
2.3 Coupling Reactions of Homopolymers *13*
2.4 Sequential Anionic Polymerization *14*
2.5 Sequential Group Transfer Polymerization *16*
2.6 Sequential Cationic Polymerization *17*
2.7 Non-radical Metal-catalyzed Polymerization *18*
2.8 Controlled Radical Polymerization *19*
2.8.1 Atom Transfer Radical Polymerization (ATRP) *20*
2.8.2 Nitroxide-mediated Polymerization (NMP) *23*
2.8.3 Reversible Addition Fragmentation Chain Transfer (the RAFT Process) *25*
2.9 Switching from One Polymerization Mechanism to Another *27*
2.10 Use of "Dual" Initiators in Concurrent Polymerization Mechanisms *29*
2.11 Chemical Modification of Pre-formed Block Copolymers *30*
2.12 Methods for the Synthesis of Block Copolymers with a Complex Architecture *31*
2.13 Conclusion *33*
References *35*

3	**Block Copolymer Vesicles** *39*
	Alessandro Napoli, Diana Sebök, Alex Senti, and Wolfgang Meier
3.1	Introduction *39*
3.2	Chemistry of Vesicle-forming Block Copolymers *41*
3.3	Block Copolymer Vesicle Formation in Water *46*
3.4	Block Copolymer Vesicle Formation in Organic Solvents *48*
3.5	Properties of Polymer Vesicles *51*
3.5.1	Morphology and Size of Polymer Vesicles *51*
3.5.2	Membrane Properties *53*
3.5.2.1	Polymer Membrane Thickness *53*
3.5.2.2	Mechanical Properties of Polymer Vesicles *54*
3.5.2.3	Adhesion of Polymer Vesicles *57*
3.5.2.4	Fusion and Fission of Polymer Vesicles *58*
3.6	Functional Polymer Vesicles *59*
3.7	Biohybrid Polymer Vesicles *60*
3.7.1	Polypeptide-based Copolymer Vesicles *60*
3.7.2	Protein Incorporation into Polymer Vesicles *62*
3.8	Potential Applications of Polymer Vesicles *64*
3.9	Concluding Remarks *66*
	References *66*
4	**Block Copolymer Micelles for Drug Delivery in Nanoscience** *73*
	Younsoo Bae, Horacio Cabral, and Kazunori Kataoka
	References *87*
5	**Stimuli-responsive Block Copolymer Assemblies** *91*
	Jean-François Gohy
5.1	Introduction *91*
5.2	Stimuli-sensitive Micellization *92*
5.2.1	Temperature-sensitive Micellization *93*
5.2.2	pH-sensitive Micellization *95*
5.2.3	Ionic Strength Sensitive Micellization *98*
5.3	Stimuli-responsive Micelles *100*
5.4	Multi-responsive Micellar Systems *103*
5.5	Stimuli-responsive Thin Films from Block Copolymers *106*
5.6	Stimuli-responsive Block Copolymers in the Bulk *109*
5.7	Conclusions and Outlook *112*
	References *114*
6	**Self-assembly of Linear Polypeptide-based Block Copolymers** *117*
	Sébastien Lecommandoux, Harm-Anton Klok, and Helmut Schlaad
6.1	Introduction *117*
6.2	Solution Self-assembly of Polypeptide-based Block Copolymers *119*

6.2.1	Aggregation of Polypeptide-based Block Copolymers	119
6.2.1.1	Polypeptide Hybrid Block Copolymers	119
6.2.1.2	Block Copolypeptides	123
6.2.2	Polypeptide-based Hydrogels	124
6.2.3	Organic/Inorganic Hybrid Structures	124
6.3	Solid-state Structures of Polypeptide-based Block Copolymers	126
6.3.1	Diblock Copolymers	126
6.3.1.1	Polydiene-based Diblock Copolymers	126
6.3.1.2	Polystyrene-based Diblock Copolymers	127
6.3.1.3	Polyether-based Diblock Copolymers	131
6.3.1.4	Polyester-based Diblock Copolymers	133
6.3.1.5	Diblock Copolypeptides	133
6.3.2	Triblock Copolymers	134
6.3.2.1	Polydiene-based Triblock Copolymers	134
6.3.2.2	Polystyrene-based Triblock Copolymers	138
6.3.2.3	Polysiloxane-based Triblock Copolymers	139
6.3.2.4	Polyether-based Triblock Copolymers	140
6.3.2.5	Miscellaneous	144
6.4	Summary and Outlook	146
	References	147

7 Synthesis, Self-assembly and Applications of Polyferrocenylsilane (PFS) Block Copolymers 151

Xiaosong Wang, Mitchell A. Winnik, and Ian Manners

7.1	Introduction	151
7.2	Synthesis of PFS Block Copolymers	152
7.3	Solution Self-assembly of PFS Block Copolymers	158
7.4	Shell Cross-linked Nanocylinders and Nanotubes	161
7.5	Self-assembly of PFS Block Copolymers in the Solid State	164
7.6	Summary	166
	References	167

8 Supramolecular Block Copolymers Containing Metal–Ligand Binding Sites: From Synthesis to Properties 169

Khaled A. Aamer, Raja Shunmugan, and Gregory N. Tew

8.1	Introduction	169
8.2	Block Copolymers with Chain-end Containing Metal Complexes	172
8.2.1	Metal Complexes in the Center of Star Polymers	172
8.2.2	Supramolecular Diblock Copolymers Connected by Metal Complexes	174
8.3	Block Copolymers with Side-chain Metal Complexes in One Block	178
8.3.1	Polymerizing Pre-formed Metal Complexes	178
8.3.2	Block Copolymers Containing Free Metal–Ligand Side-chains	180
8.3.2.1	Polymerization of Metal–Ligand Containing Monomers	180

8.3.2.2	Post-polymerization Attachment of the Metal–Ligand *184*
8.4	Conclusion *187*
	References *187*

9 Methods for the Alignment and the Large-scale Ordering of Block Copolymer Morphologies *191*
Massimo Lazzari and Claudio De Rosa

9.1	Introduction *191*
9.1.1	Motivation *191*
9.1.2	Organization of the Chapter *192*
9.2	How to Help Phase Separation *193*
9.3	Orientation by External Fields *195*
9.3.1	Mechanical Flow Fields *196*
9.3.2	Electric and Magnetic Fields *197*
9.3.3	Solvent Evaporation and Thermal Gradient *202*
9.4	Templated Self-assembly on Nanopatterned Surfaces *203*
9.5	Epitaxy and Surface Interactions *205*
9.5.1	Preferential Wetting and Homogeneous Surface Interactions *205*
9.5.2	Epitaxy *206*
9.5.3	Directional Crystallization *209*
9.5.4	Graphoepitaxy and Other Confining Geometries *212*
9.5.5	Combination of Directional Crystallization and Graphoepitaxy *214*
9.5.6	Combination of Epitaxy and Directional Crystallization *215*
9.6	Summary and Outlook *223*
	References *225*

10 Block Copolymer Nanofibers and Nanotubes *233*
Guojun Liu

10.1	Introduction *233*
10.2	Preparation *235*
10.2.1	Nanofiber Preparation *235*
10.2.2	Nanotube Preparation *238*
10.3	Solution Properties *240*
10.4	Chemical Reactions *247*
10.4.1	Backbone Modification *247*
10.4.2	End Functionalization *251*
10.5	Concluding Remarks *253*
	References *254*

11 Nanostructured Carbons from Block Copolymers *257*
Michal Kruk, Chuanbing Tang, Bruno Dufour, Krzysztof Matyjaszewski, and Tomasz Kowalewski

11.1	Introduction *257*
11.2	Discussion *259*

11.2.1	Well Defined PAN Polymers and Copolymers 259
11.2.2	Carbon Films from Phase-separated Block Copolymers 260
11.2.3	Carbon Nanoobjects from Water-soluble Precursors 265
11.2.4	Nanoporous Carbon from Block Copolymers Using Silica as an Auxiliary Component 266
11.2.5	Nanoporous Carbon from Phase-separated Block Copolymers 268
11.2.6	Carbons Synthesized Using Other Block Copolymers 270
11.3	Conclusion 271
	References 272

12 Block Copolymers at Interfaces 275
Mark Geoghegan and Richard A. L. Jones

12.1	Introduction 275
12.2	Block Copolymer Films 277
12.3	Block Copolymers on Heterogeneous Surfaces 278
12.4	Environmental Control of Block Copolymer Films 279
12.5	Block Copolymer Brushes 282
12.6	Surface Regeneration 286
12.7	Conclusions and Outlook 286
	References 287

13 Block Copolymers as Templates for the Generation of Mesostructured Inorganic Materials 291
Bernd Smarsly and Markus Antonietti

13.1	Introduction 291
13.2	General Mechanism 292
13.3	Details of the BC Templating Mechanism 298
13.4	Crystalline, Mesoporous Metal Oxides 299
13.5	Mesoporous Metals 304
13.6	Conclusion and Outlook 304
	References 305

14 Mesostructured Polymer–Inorganic Hybrid Materials from Blocked Macromolecular Architectures and Nanoparticles 309
Marleen Kamperman and Ulrich Wiesner

14.1	Introduction 309
14.2	AB Diblock Copolymers as Structure-directing Agents for Aluminosilicate Mesostructures 310
14.2.1	Formation Mechanisms 313
14.2.2	Flow-induced Alignment of Mesostructured Block Copolymer–Sol Nanoparticle Co-assemblies 316
14.3	Generalization to Other Blocked Macromolecular Amphiphiles as Structure-directing Agents for Mesostructured Materials 319

14.4	Generalization to Other Inorganic Materials Systems *322*	
14.4.1	Mesoporous Aluminosilicate Materials with Superparamagnetic γ-Fe$_2$O$_3$ Particles Embedded in the Walls *322*	
14.4.2	Ordered Mesoporous Ceramics Stable up to 1500 °C from Diblock Copolymer Mesophases *324*	
14.5	Generalization from Bulk Mesostructured Hybrids to Mesostructured Thin Films *328*	
14.6	Conclusions *331*	
	References *332*	

15 Block Ionomers for Fuel Cell Application *337*
Olivier Diat and Gérard Gebel

15.1	Introduction *337*
15.2	Definitions and Investigations *342*
15.3	Polymer Modification *344*
15.3.1	Post-sulfonation *344*
15.3.2	Grafting *345*
15.3.3	Blends *346*
15.4	Copolymerization of Functionalized Monomers *347*
15.4.1	Main-chain Type Co-ionomer *347*
15.4.1.1	Sulfonated Polyimides *347*
15.4.1.2	Polyarylene Systems *351*
15.4.2	"Side Chain" Co-ionomers *353*
15.5	Di- and Triblock Ionomers *357*
15.6	Conclusion *362*
	References *363*

16 Structure, Properties and Applications of ABA and ABC Triblock Copolymers with Hydrogenated Polybutadiene Blocks *367*
Vittoria Balsamo, Arnaldo Tomás Lorenzo, Alejandro J. Müller, Sergio Corona-Galván, Luisa M. Fraga Trillo, and Valentín Ruiz Santa Quiteria

16.1	Introduction *367*
16.2	Applications of SEBS Triblock Copolymers *371*
16.2.1	Adhesives, Sealants and Coatings *372*
16.2.2	Bitumen Modification *373*
16.2.3	Compounding and Plastic Modification *375*
16.2.4	Miscellaneous Applications *377*
16.2.4.1	Gels and Nanocomposites *377*
16.2.4.2	Medical Applications *378*
16.2.5	Future Trends *379*
16.3	Semicrystalline Triblock Copolymers with One or More HPB Blocks *379*
16.3.1	Semicrystalline ABA Triblock Copolymers *380*

16.3.2	Semicrystalline ABC Triblock Copolymers *381*	
16.4	Conclusions *387*	
	References *388*	

17	**Basic Understanding of Phase Behavior and Structure of Silicone Block Copolymers and Surfactant–Block Copolymer Mixtures** *391*	
	Carlos Rodríguez, Arturo López-Quintela, Md. Hemayet Uddin,	
	Kenji Aramaki, and Hironobu Kunieda[†]	
17.1	Introduction *391*	
17.2	General Aspects of Phase Behavior and Liquid Crystal Phases *393*	
17.3	Phase Behavior and Microstructure of $Si_mC_3EO_n$ Melts *395*	
17.4	Phase Behavior and Microstructure $Si_mC_3EO_n$ in Water *400*	
17.4.1	Phase Diagrams of Water–$Si_mC_3EO_n$ Systems as a Function of Temperature *400*	
17.4.2	Phase Diagrams of Water–$Si_mC_3EO_n$ Systemsas a Function of PEO Chain Length *402*	
17.4.3	Phase Diagram of Water–$Si_mC_3EO_{51.6}$ System as a Function of PDMS Chain Length *403*	
17.4.4	Effect of PEO and PDMS Chain Lengths on the Effective Cross-sectional Area per Copolymer Molecule, a_P *404*	
17.5	Phase Behavior of $Si_mC_3EO_n$ in Non-polar Oil *406*	
17.6	Phase Behavior of $Si_mC_3EO_n$ in Non-aqueous Polar Solvents *408*	
17.7	Mixing of Poly(oxyethylene)–Poly(dimethylsiloxane) Copolymer and Non-ionic Surfactant in Water *410*	
17.8	Conclusions and Outlook *415*	
	References *415*	

Subject Index *419*

Preface

Nanoscience and technology deal with the preparation, study, manipulation and application of nanometer-sized structures. "Nanoscience and technology have the potential for revolutionizing the ways in which materials and products are created and the range and nature of functionalities that can be accessed. It is an interdisciplinary area of research and development activity that has been growing explosively worldwide in the past decade".[1] Some example areas that nanotechnology may impact profoundly include the continuing miniaturization of electronic and memory devices, the development of more potent drugs that can recognize and attack only the diseased sites, the development of more accurate and effective diagnostic procedures and the design and synthesis of more robust catalysts.

Physicists and engineers prepare nanostructures by taking the *top-down* approach using techniques such as lithography, which "carves" a large piece of material into smaller pieces. They can now routinely make structures on an industrial scale with dimensions approaching around 100 nm by lithography. Chemists make nanostructures by assembling molecules together or taking the *bottom-up* approach. Structures made from the spontaneous assembly or self-assembly of organic and inorganic molecules are normally smaller than 10 nm. An imminent challenge facing the nanotechnology community is the development of methodologies for the preparation of materials with dimensions occupying the size range between 10 and 100 nm, a task that can be readily accomplished by block copolymer (BC) researchers.

This book gives an overview of recent developments in the nanoscience and technology of block polymers. Instead of a simple collection of review chapters, our objective was to compile a handbook that carefully evaluates all types of applications for block copolymers: as tools for fabricating other nanomaterials, as structural components in hybrid materials and nanocomposites, and as functional materials. Following a critical review of the state-of-the-art and future developments of BC applications (Chapter 1) and a general overview on the various synthetic approaches that can be used to make BCs (Chapter 2), through a multidisciplinary approach some of the most experienced specialists in the world cover all aspects of BC nanoscience and technology, ranging from chemical synthesis, characterization

---

[1] National Science and Technology Council (USA), *Nanostructure Science and Technology — A Worldwide Study*, **1999**.

Block Copolymers in Nanoscience. Massimo Lazzari, Guojun Liu, Sébastien Lecommandoux
Copyright © 2006 WILEY-VCH Verlag GmbH & Co. KGaA, Weinheim
ISBN: 3-527-31309-5

and large-scale ordering of morphologies to applications in nanotechnology and also in nanobiotechnology (Chapters 3–15). In addition, recent developments in industrial applications of BCs as thermoplastic elastomers and in the basic aspects of silicon surfactants are covered in the last two chapters of the book (Chapters 16 and 17, respectively).

While our understanding of BC self-assembly in bulk is relatively mature, a better comprehension of their self-assembly behaviour in block-selective solvents only began to emerge in 1995 with the seminal discovery by Zhang and Eisenberg of the multiple morphologies of block copolymers. In the subsequent 10 years, much effort focused on the study of amphiphilic copolymers with blocks ranging from organic polymers to metal-containing polymers (Chapter 8), polypeptides (Chapter 6) and stimuli-responsive polymers (Chapter 5). By using the appropriate copolymers, BC micelles can be made water-dispersible. These micelles should be useful in controlled drug delivery and diagnostics (Chapter 4). BCs with appropriate block lengths can self-assemble into vesicles, which are ideal nanocapsules that have potential for use in a large number of applications (Chapter 3).

Self-assembled BC nanostructures are held together by van der Waals forces and are stable only under a limited set of conditions. Permanent nanoobjects can be prepared by performing cross-linking chemistry to the self-assembled nanostructures. The physical understanding and fascinating chemistry of "permanent" BC nanostructures are the topics of Chapters 7 and 10. Also discussed in some depth is the use of self-assembled block copolymers as precursors to carbon nanoobjects (Chapter 11) and mesostructure materials (Chapters 13 and 14). Apart from the selective cross-linking of domains of a block-segregated copolymer, it is also possible to degrade the domains in a BC film selectively, thus yielding thin films with nanochannels of uniform size. More generally, the surface self-assembly and the alignment of the nanodomains formed can be used to design specific thin films (Chapters 9 and 12), which may find applications, for example, as fuel cell membranes (Chapter 15).

This book is dedicated to young research scientists who want to get into this field, and also to specialists who want to access an easy overview of the most recent developments.

We are grateful to Bettina Bems and her colleagues from Wiley-VCH for their help, and our families (Ana, Jiandong Wang — Karine, Charline — Maxence) for their encouragement, support and patience. Finally, we would like to thank all of the authors for their contributions and also the specialists who gave external support: they are all reasons for this book being a success.

Massimo Lazzari
Guojun Liu
Sébastien Lecommandoux

List of Contributors

Khaled A. Aamer
Department of Polymer Science and Engineering
University of Massachusetts, Amherst
120 Governors Drive
Amherst, MA 01003
USA

Markus Antonietti
Max-Planck Institute of Colloids and Interfaces
Department of Colloid Chemistry
Research Campus Golm
14424 Potsdam
Germany

Kenji Aramaki
Graduate School of Environment and Information Sciences
Yokohama National University
Tokiwadai 79-7
Hodogaya-Ku
Yokohama 240-8501
Japan

Younsoo Bae
Division of Clinical Biotechnology
Center for Biology and Integrative Medicine
Graduate School of Medicine
The University of Tokyo
7-3-1 Hongo, Bunkyo-ku
Tokyo 113-0033
Japan

Vittoria Balsamo
Polymer Group USB
Department of Science and Materials
University of Simón Bolívar
Apartado 89000
Caracas 1080-A
Venezuela

Horacio Cabral
Department of Materials Engineering
Graduate School of Engineering
The University of Tokyo
7-3-1 Hongo, Bunkyo-ku
Tokyo 113-8656
Japan

Sergio Corona-Galván
Center of Technology Repsol YPF
Carretera de Extremadura, Km. 18
28931 Móstoles
Madrid
Spain

Claudio De Rosa
Department of Chemistry
University of Napoli "Federico II"
Complesso Monte S. Angelo
Via Cintia
80126 Napoli
Italy

Block Copolymers in Nanoscience. Massimo Lazzari, Guojun Liu, Sébastien Lecommandoux
Copyright © 2006 WILEY-VCH Verlag GmbH & Co. KGaA, Weinheim
ISBN: 3-527-31309-5

Olivier Diat
Structures and Properties of
Molecular Architecture
UMR 5819 (CEA, CNRS, UJF)
Ion Conducting Polymer Group
DRFMC/SPrAM, CEA Grenoble
38054 Grenoble Cedex 9
France

Bruno Dufour
Department of Chemistry
Carnegie Mellon University
4400 Fifth Avenue
Pittsburgh, PA 15213
USA

Luisa M. Fraga Trillo
Center of Technology Repsol YPF
Carretera de Extremadura, Km. 18
28931 Móstoles
Madrid
Spain

Gérard Gebel
Structures and Properties of
Molecular Architecture
UMR 5819 (CEA, CNRS, UJF)
Ion Conducting Polymer Group
DRFMC/SPrAM, CEA Grenoble
38054 Grenoble Cedex 9
France

Mark Geoghegan
Department of Physics and
Astronomy
University of Sheffield
Hicks Building
Hounsfield Road
Sheffield S3 7RH
United Kingdom

Yves Gnanou
Laboratory of Organic Polymer
Chemistry, LCPO
CNRS, ENSCPB University of
Bordeaux-1
16 Avenue Pey Berland
33607 Pessac
France

Jean-François Gohy
Research Group in Macromolecular
Chemistry (MACRO)
Chemistry of Inorganic and Organic
Materials Unit (CMAT)
Department of Chemistry
Catholic University of Louvain
Place L. Pasteur 1
1348 Louvain-la-Neuve
Belgium

Richard A. L. Jones
Department of Physics and Astronomy
University of Sheffield
Hicks Building
Hounsfield Road
Sheffield S3 7RH
United Kingdom

Marleen Kamperman
Department of Materials Science
and Engineering
Cornell University
304 Thurston Hall
Ithaca, NY 14853
USA

Kazunori Kataoka
Department of Materials Engineering
Graduate School of Engineering
The University of Tokyo
7-3-1 Hongo, Bunkyo-ku
Tokyo 113-8656
Japan

and
Division of Clinical Biotechnology
Center for Biology and Integrative
Medicine
Graduate School of Medicine
The University of Tokyo
7-3-1 Hongo, Bunkyo-ku
Tokyo 113-0033
Japan

Harm-Anton Klok
Ecole Polytechnique Fédérale de
Lausanne (EPFL)
Material Sciences Department,
Polymer Laboratory
STI–IMX–LP, MXD 112
Bâtiment MXD
Station 12
1015 Lausanne
Switzerland

Tomasz Kowalewski
Department of Chemistry
Carnegie Mellon University
4400 Fifth Avenue
Pittsburgh, PA 15213
USA

Michal Kruk
Department of Chemistry
Carnegie Mellon University
4400 Fifth Avenue
Pittsburgh, PA 15213
USA

Hironobu Kunieda[†]
Graduate School of Environment
and Information Sciences
Yokohama National University
Tokiwadai 79-7
Hodogaya-Ku
Yokohama 240-8501
Japan

Massimo Lazzari
Institute of Technological
Investigations and Department
of Physical Chemistry
Faculty of Chemistry
University of Santiago de Compostela
15782 Santiago de Compostela
Spain
On leave from:
Department of Chemistry IPM and
Nanostructured Interfaces and
Surfaces Centre of Excellence
University of Torino
Via P. Giuria 7
10125 Torino
Italy

Sébastien Lecommandoux
Laboratory of Organic Polymer
Chemistry, LCPO (UMR5629)
CNRS, ENSCPB University of
Bordeaux-1
16 Avenue Pey Berland
33607 Pessac
France

Guojun Liu
Department of Chemistry
Queen's University
90 Bader Lane
Kingston, Ontario K7L 3N6
Canada

Arturo López-Quintela
Department of Physical Chemistry
Faculty of Chemistry
University of Santiago de Compostela
Avenida das Ciencas
15782 Santiago de Compostela
Spain

List of Contributors

Arnaldo Tomás Lorenzo
Grupo de Polímeros USB
Departamento de Ciencia de los Materiales
Universidad Simón Bolívar
Apartado 89000
Caracas 1080-A
Venezuela

Ian Manners
Department of Chemistry
University of Toronto
80 St. George Street
Toronto, Ontario M5S 3H6
Canada

Krzysztof Matyjaszewski
Department of Chemistry
Carnegie Mellon University
4400 Fifth Avenue
Pittsburgh, PA 15213
USA

Wolfgang Meier
Department of Chemistry
University of Basel
Klingelbergstrasse 80
4056 Basel
Switzerland

Alejandro J. Müller
Polymer Group USB
Department of Materials Science
University of Simón Bolívar
Apartado 89000
Caracas 1080-A
Venezuela

Alessandro Napoli
Department of Chemistry
University of Basel
Klingelbergstrasse 80
4056 Basel
Switzerland

Carlos Rodríguez
Department of Physical Chemistry
Faculty of Chemistry
University of Santiago de Compostela
Avenida das Ciencas
15782 Santiago de Compostela
Spain

Valentín Ruiz Santa Quiteria
Center of Technology Repsol YPF
Carretera de Extremadura, Km. 18
28931 Móstoles, Madrid
Spain

Helmut Schlaad
Max-Planck Institute of Colloids and Interfaces
Department of Colloid Chemistry
Research Campus Golm
14424 Potsdam
Germany

Diana Sebök
Department of Chemistry
University of Basel
Klingelbergstrasse 80
4056 Basel
Switzerland

Alex Senti
Department of Chemistry
University of Basel
Klingelbergstrasse 80
4056 Basel
Switzerland

Raja Shunmugan
Department of Polymer Science and Engineering
University of Massachusetts, Amherst
120 Governors Drive
Amherst, MA 01003
USA

Bernd Smarsly
Max-Planck Institute of Colloids
and Interfaces
Department of Colloid Chemistry
Research Campus Golm
14424 Potsdam
Germany

Chuanbing Tang
Department of Chemistry
Carnegie Mellon University
4400 Fifth Avenue
Pittsburgh, PA 15213
USA

Daniel Taton
Laboratory of Organic Polymer
Chemistry, LCPO
CNRS, ENSCPB University of
Bordeaux-1
16 Avenue Pey Berland
33607 Pessac
France

Gregory N. Tew
Department of Polymer Science
and Engineering
University of Massachusetts, Amherst
120 Governors Drive
Amherst, MA 01003
USA

Md. Hemayet Uddin
Department of Applied Chemistry
and Chemical Technology
Faculty of Science and Technology
Islamic University
Kushtia
Bangladesh

Xiaosong Wang
Department of Chemistry
University of Toronto
80 St. George Street
Toronto, Ontario M5S 3H6
Canada

Ulrich Wiesner
Department of Materials Science
and Engineering
Cornell University
330 Bard Hall
Ithaca, NY 14853
USA

Mitchell A. Winnik
Department of Chemistry
University of Toronto
80 St. George Street
Toronto, Ontario M5S 3H6
Canada

1
An Introduction to Block Copolymer Applications: State-of-the-art and Future Developments

Sébastien Lecommandoux, Massimo Lazzari, and Guojun Liu

Self-assembly of either molecular or non-molecular components by non-covalent interactions is an enormously powerful tool in modern material science, which enables the preparation of structures often not accessible by any other fabrication process [1–3]. This concept was initially associated with the use of synthetic strategies for the preparation of nanostructures around 1990 [4–6], and since then the interest in developing bottom-up approaches with the aim of offering an alternative to "traditional" top-down fabrications has grown dramatically. Albeit that such energetically convenient strategies may sometimes appear to be an excessively fashionable field of investigation, it is reasonable to assume that self-assembled materials in general, and those with a length scale smaller than the actual limits of conventional manufacturing in particular, are going to occupy a privileged position in the 21st century, as conventional plastics, alloys and semiconductors did in the 20th century. In this context, polymers are likely to play a key role, not just for the same reasons that have made them successful materials so far, i.e., ease of synthesis and processing, low cost, variability of chemical functionality and physical properties, but also because of their intrinsic dimensions (typically tens of nanometers) and, even more important, the peculiar mesophase segregation in the case of block copolymers (BCs) [7].

BCs are a particular class of polymers that belong to a wider family known as soft materials [8] that, independent of the procedure of synthesis, can simply be considered as being formed by two or more chemically homogeneous polymer fragments (blocks) joined together by covalent bonds. In the simplest case of two distinct monomers, conventionally termed A and B, linear diblock (AB), triblock (ABA), multiblock or star-block copolymers can be prepared. The phase behavior of such diblock copolymers has been the subject of numerous theoretical and experimental studies over recent decades, and is relatively well understood [9–12]. This self-assembly process is driven by an unfavorable mixing enthalpy and a small mixing entropy, while the covalent bond connecting the blocks prevents macroscopic phase separation. The microphase separation of diblock copolymers depends on the total degree of polymerization $N (= N_A + N_B)$, the Flory–Huggins χ-parameter (which is a measure of the incompatibility between the two blocks) and the volume fractions of the constituents blocks (f_A and f_B, $f_A = 1 - f_B$). The segregation prod-

Block Copolymers in Nanoscience. Massimo Lazzari, Guojun Liu, Sébastien Lecommandoux
Copyright © 2006 WILEY-VCH Verlag GmbH & Co. KGaA, Weinheim
ISBN: 3-527-31309-5

Fig. 1.1 Schematic representations of the morphologies obtained for diblock copolymer melts (reproduced from [13], copyright (1995) with permission from The American Chemical Society).

uct χN determines the degree of microphase separation. Depending on χN, three different regimes can be distinguished: (1) the weak-segregation limit (WSL) for $\chi N \leq 10$; (2) the intermediate segregation region (ISR) for $10 < \chi N \leq 50$ and (3) the strong segregation limit (SSL) for $\chi N \to \infty$.

In bulk, the minority block is segregated from the majority block forming regularly-shaped and uniformly-spaced nanodomains [12]. The shape of the segregated domains in a diblock is governed by the volume fraction of the minority block, f, and block incompatibility. Figure 1.1 shows the equilibrium morphologies documented for diblock copolymers [13]. At a volume fraction of $\approx 20\%$, the minority block forms a body-centred cubic spherical phase in the matrix of the majority block. It changes to hexagonally packed cylinders at a volume fraction $\approx 30\%$. Alternating lamellae are formed at approximately equal volume fractions for the two blocks. At a volume fraction of $\approx 38\%$, the minority block forms gyroid or perforated layers at moderate and high incompatibility, respectively. Furthermore, the smallest dimension of a segregated domain, e.g., the diameter of a cylinder, is proportional to the two-thirds power of the molar mass of the minority block and can typically be tuned from ≈ 5 to ≈ 100 nm by changing the molar mass of the block [14].

In analogy to their bulk behavior, diblock copolymers also self-assemble in block-selective solvents, which solubilize one but not the other block, forming micelles of various shapes [15, 16]. If the soluble block is predominant, the insoluble block aggregates to produce spherical micelles. As the length of the soluble block is decreased relative to the insoluble block, cylindrical micelles or vesicles are formed. For a given diblock, unusual micelles with shapes differing from spheres can sometimes be induced by the use of increasingly poor solvent for the insoluble block [15, 17]. Figure 1.2 illustrates schematically the structure and chain packing in four types of diblock copolymer micelles where the black block is insoluble.

ABC triblock copolymers or $(A)_n(B)_m(C)_l$ also undergo self-assembly in bulk or block-selective solvents. Triblocks have many more block segregation patterns than diblocks and some of the patterns are aesthetically appealing. Figure 1.3 illustrates triblock bulk segregation patterns that were already known by 1995 [18]. Over the next 10 years or so, approximately ten more structures were added to the triblock copolymer morphology list. Block segregation pattern complexity increases further for tetra- and pentablocks, thus making the list almost infinite.

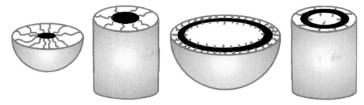

Fig. 1.2 Cross-sectional view of chain packing in diblock copolymer spherical, cylindrical, vesicular and tubular micelles.

Aside from the linear architecture, BCs can be made into many other fascinating star-like macromolecules, such as miktoarm star structures, i. e., BCs where arms with different chemical natures are linked to the same branch point [19]. Unique segregation properties are expected of these polymers [20, 21].

After tens of years of secondary importance, as reflected by an almost negligible number of publications, the first growth of interest in BCs was witnessed in the early 1970s for a series of pioneering works, see, for example, those by the Keller group [22, 23], that led to our physical understanding of BCs. Later on, only in the 1990s, such subjects became the focus of a great deal of research activity, the reasons being the larger availability of BCs with different architectures and chemi-

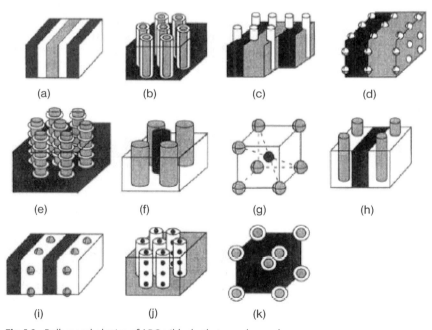

Fig. 1.3 Bulk morphologies of ABC triblocks that were known by 1995 (reproduced from [18], copyright (1995) with permission from The American Chemical Society).

cal compositions [24], and especially the closer collaboration between chemists on one side and materials scientists and physicists on the other. Apart from the large amount of work focused on advanced theoretical and computational methods, it is worth citing some milestone papers on the potential applications mainly based on their ability to form well-defined patterns, such as those on the preparation of membranes with tunable nanochannels [25–27], BC-based nanolithography [28, 29] or the last developments on the tailoring of hierarchically self-assembled hybrid materials (Fig. 1.4) [30].

However, most of the large-volume applications of the BCs that can be found nowadays in industrial developments do not take advantage of any particular nanostructures. For example, the use of BCs such as polystyrene-*block*-polybutadiene-*block*-polystyrene as thermoplastic elastomers (TPEs) is mainly due to their ability to prevent a macrophase separation process, retaining important features of their constituent homopolymers. Of course the material properties of such systems depend on the copolymer composition (hard to soft ratio), but no clear evidence on the impact of the periodic morphology has been evidenced. In addition, the marketplace for these TPEs is definitely reaching a plateau, even though it is still growing in volume, and future industrial developments are likely to be found in specialty polymers or polymer nanomaterials with high added value, where the material properties would be strongly related to the ordering, hierarchical or orientation properties of the nanodomains in a polymer [36, 37].

Indeed, there has been recent, increasing interest in block copolymers, at least on the applications front, mainly for the development of two approaches to the fabrication of nanomaterials, which take advantage of the spatial ordering and orientation of nanodomains: (1) via use of self-assembled copolymers as nanostructured materials, either "as they are" or through selective chemical isolation or processing of one or more components; (2) via template formation, including both the use of copolymer films directly as the template and the fabrication techniques, which rely on the preliminary formation of the template through an initial processing, followed by a nanoscale synthesis [38–41]. The most important and promising applications of BCs are presented and discussed in detail in the following chapters, along with practical information to resolve experimental difficulties and provide novice and experienced researchers with ideas as to where research efforts need to be invested.

The past decade has witnessed a rapid growth in block copolymer nanoresearch. In our view, sufficient foundations have now been built on which to base future applications. While BC nanomaterials, like inorganic or organic nanomaterials, may find their applications in areas such as biomedics, chemical separations, catalysis, nanoelectronics and other types of nanodevices, true breakthrough applications will arise only when the unique properties of BC soft nanomaterials, e. g., their size, the core-shell structure, shape diversity and solvent dispersibility in the case of block copolymer micelles, are taken advantage of. Some large, innovative companies, such as IBM, have already started to reap the benefits of their research results of the past decade for commercial product developments. Many more innovative companies are now pushing for their own product development, or collaborating

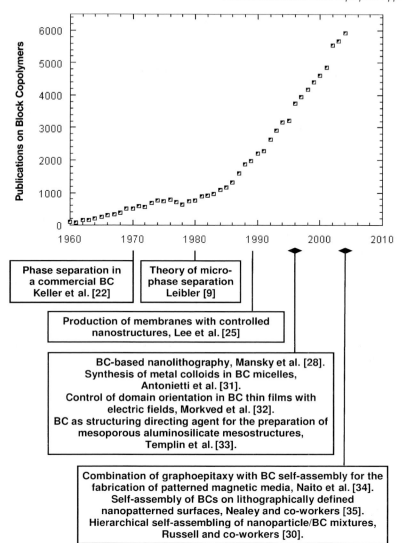

Fig. 1.4 Appearance of the number of scientific publications as a function of time (from Sci-Finder source, up to 2005).

on product development with academic researchers in this field. It is only a matter of years before we see the emergence of truly innovative BC nanoproducts infiltrating the world market. It is our hope that this book will serve as a prelude to the industrial practice of BC nanotechnology.

Acknowledgments

Funding for our investigations was provided by the French Ministry of Education and Research, the Centre National de la Recherche Scientifique CNRS and the Région Aquitaine (S. L.), the Ministerio de Ciencia y Tecnología (M. L.: MAT2005-07554-C02-01 and Strategic Action for Nanoscience and Nanotechnology, NAN2004-09195-C04-01) and NSERC of Canada (G. L.).

References

1 W. M. Tolles, *MRS Bull.* **2000**, *25*, 36.
2 M. Boncheva, G. M. Whitesides, *MRS Bull.* **2005**, *30*, 736.
3 M. Lazzari, C. Rodriguez-Abreu, J. Rivas, M. A. Lopez-Quintela, *J. Nanosci. Nanotechnol.* **2006**, *6*, 892.
4 C. T. Seto, G. M. Whitesides, *J. Am. Chem. Soc.* **1990**, *112*, 6409.
5 J. S. Manka, D. S. Lawrence, *J. Am. Chem. Soc.* **1990**, *112*, 2440.
6 U. Koert, M. M. Harding, J.-M. Lehn, *Nature (London)* **1990**, *346*, 339.
7 I. W. Hamley, *The Physics of Block Copolymers*, Oxford University Press, Oxford, **1998**.
8 I. W. Hamley, *Introduction to Soft Matter*, Wiley, Chichester, **2000**.
9 L. Leibler, *Macromolecules* **1980**, *13*, 1602.
10 F. S. Bates, *Science* **1991**, *251*, 898.
11 G. H. Fredrickson, F. S. Bates, *Annu. Rev. Mater. Sci.* **1996**, *26*, 501.
12 H.-A. Klok, S. Lecommandoux, *Adv. Mater.* **2001**, *13*, 1217.
13 A. K. Khandpur, S. Forster, F. S. Bates, I. W. Hamley, A. J. Ryan, W. Bras, K. Almdal, K. Mortensen, *Macromolecules* **1995**, *28*, 8796.
14 F. S. Bates, G. H. Fredrickson, *Phys. Today* **1999**, *52*, 32.
15 N. S. Cameron, M. K. Corbierre, A. Eisenberg, *Can. J. Chem.* **1999**, *77*, 1311.
16 J. F. Ding, G. J. Liu, M. L. Yang, *Polymer* **1997**, *38*, 5497.
17 J. Bang, S. M. Jain, Z. B. Li, T. P. Lodge, J. S. Pedersen, E. Kesselman, Y. Talmon, *Macromolecules* **2006**, *39*, 1199.
18 Z. Wei, Z. G. Wang, *Macromolecules* **1995**, *28*, 7215.
19 N. Hadjichristidis, *J. Polym. Sci. Part A: Polym. Chem.* **1999**, *37*, 857.
20 N. Hadjichristidis, H. Iatrou, M. Pitsikalis, S. Pispas, A. Avgeropoulos, *Progr. Polym. Sci.* **2005**, *30*, 725.
21 S. P. Gido, C. Lee, D. J. Pochan, S. Pispas, J. W. Mays, N. Hadjichristidis, *Macromolecules* **1996**, *29*, 7022.
22 A. Keller, E. Pedemonte, F. M. Willmouth, *Nature (London)* **1970**, *225*, 538.
23 M. J. Folkes, A. Keller, F. P. Scalisi, *Colloid. Polym. Sci.* **1973**, *251*, 1.
24 N. Hadjichristidis, S. Pispas, G. A. Floudas, *Block Copolymers: Synthetic Strategies*, Wiley-Interscience, New York, **2003**.
25 J. S. Lee, A. Hirao, S. Nakahama, *Macromolecules* **1989**, *22*, 2602.
26 G. J. Liu, J. F. Ding, A. Guo, M. Herfort, D. BazettJones, *Macromolecules* **1997**, *30*, 1851.
27 T. Hashimoto, K. Tsutsumi, Y. Funaki, *Langmuir* **1997**, *13*, 6869.
28 P. Mansky, P. Chaikin, E. L. Thomas, *J. Mater. Sci.* **1995**, *30*, 1987.
29 M. Park, C. Harrison, P. M. Chaikin, R. A. Register, D. H. Adamson, *Science* **1997**, *276*, 1401.
30 Y. Lin, A. Böker, J. He, K. Sill, H. Xiang, C. Abetz, X. Li, J. Wang, T. Emrick, S. Long, Q. Wang, A. Balazs, T. P. Russell, *Nature (London)* **2005**, *434*, 55.
31 M. Antonietti, E. Wenz, L. Bronstein, M. Seregina, *Adv. Mater.* **1995**, *7*, 1000.
32 T. L. Morkved, M. Lu, A. M. Urbas, E. E. Ehrichs, H. M. Jaeger, P. Mansky, T. P. Russell, *Science* **1996**, *273*, 931.
33 M. Templin, A. Franck, A. DuChesne, H. Leist, Y. M. Zhang, R. Ulrich, V. Schadler, U. Wiesner, *Science* **1997**, *278*, 1795.

34 K. Naito, H. Hieda, M. Sakurai, Y. Kamata, K. Asakawa, *IEEE Trans. Magn.* **2002**, *38*, 1949.
35 S. O. Kim, H. H. Solak, M. P. Stoykovich, N. J. Ferrier, J. J. de Pablo, P. F. Nealey, *Nature (London)* **2003**, *424*, 411.
36 T. P. Lodge, *Macromol. Chem. Phys.* **2003**, *204*, 265.
37 M. Muthukumar, C. K. Ober, E. L. Thomas, *Science* **1997**, *277*, 1225.
38 C. Park, J. Yoon, E. L. Thomas, *Polymer* **2003**, *44*, 6725.
39 M. Lazzari, M. A. Lopez-Quintela, *Adv. Mater.* **2003**, *15*, 1583.
40 I. W. Hamley, *Nanotechnology* **2003**, *14*, R39.
41 A.-V. Ruzette, L. Leibler, *Nat. Mater.* **2005**, *4*, 19.

2
Guidelines for Synthesizing Block Copolymers*

Daniel Taton and Yves Gnanou

2.1
Introduction

Macromolecular engineering of block copolymers is a vast multidisciplinary field which not only involves the concepts of molecular and macromolecular chemistry but also the physics and physico-chemistry of polymeric materials. In most cases the objective is to design and manipulate macromolecular objects that are capable of selective responses to an external stimulus such as pH, temperature, ionic strength, electric field, or any combination of these parameters [1–5].

Block copolymers occur in many applications as the result of their self-assembling properties, either in the solid state or in a selective solvent of one block, which provides a large variety of morphologies in the submicron region. The multifaceted role played by these materials make them useful as compatibilizers, viscosity modifiers, dispersants to stabilize colloidal suspensions, nanocarriers for the encapsulation and controlled release of drugs, templates for mineralization or, more generally, in nanoscience and nanotechnologies. Many of these aspects are covered in the subsequent chapters of this book.

Various areas of polymer science are thus concerned with block copolymers, from the molecular design of active species in macromolecular chemistry and macromolecular engineering, the physics of block copolymers in bulk or in solution to their use as specialty polymers in new applications. This also includes comprehensive theoretical and computational studies to predict the morphological behavior of block copolymers possessing incompatible blocks. For all these reasons, block copolymers have been investigated extensively in recent decades from both practical and theoretical aspects [1, 5].

The present chapter is devoted to the synthetic strategies for block copolymers, from traditional routes to more recent developments. In recent decades, advances in polymer chemistry through the development of "controlled/living" polymerization (CLP) techniques [6] have actually permitted block copolymers to be arranged

---

* A list of abbreviations is provided at the end of this chapter.

Block Copolymers in Nanoscience. Massimo Lazzari, Guojun Liu, Sébastien Lecommandoux
Copyright © 2006 WILEY-VCH Verlag GmbH & Co. KGaA, Weinheim
ISBN: 3-527-31309-5

in miscellaneous architectures, including not only linear (e. g., AB diblock, ABA and ABC triblock, ABABA pentablock, gradient copolymers, etc.) [1–7], but also branched systems (star-block, heteroarm, miktoarm, graft and other copolymers with an even more complex architecture) [4, 8]. Synthetic tools in macromolecular chemistry have reached such a maturity that they allow almost all types of block copolymers to be prepared, provided certain conditions are fulfilled so as to obtain chemically pure materials. Both the chemical nature and the overall composition of the blocks can be manipulated by using CLP techniques. In addition, control of the chemistry allows the introduction of specific functions within the blocks (H-bondings, charges, branching points, etc.), which are used to direct the self-assembly process and control the highly ordered mesostructures obtained [2]. These, in turn, allow the tailoring of block copolymers with complementary properties for the ultimate applications (hard/soft, crystallizable/amorphous, hydrophilic/hydrophobic, neutral hydrophilic/charged hydrophilic, zwitterionic, organic/inorganic, etc.).

The discussion is organized according to the type of CLP method. Owing to space limitations, it is not realistic to present an exhaustive list of all block copolymers that have been described so far in the literature. Even if restricted solely to the synthetic aspects, the number of publications on "block copolymers" is enormous (a few thousand: source *Science Finder*, September 2005). Many recent review articles, book chapters and highlights dealing with the general and specific aspects of block copolymers, or with specific classes of these materials are available in the literature and will serve as the main references of this chapter [1–9]. Our objective is to present the synthetic tool-box that is now available for preparing block copolymers and to provide the reader with some guidelines and golden rules to be followed with a view to obtaining structurally well-defined materials. The last section will briefly discuss the possibility of arranging block copolymers in more complex architectures through the combination of CLP methods with so-called branching reactions, which consists of introducing branching points throughout the architecture.

Among the linear block copolymers, AB diblock and ABA triblock copolymers consisting of only two components are the two most studied architectures, triblock systems being the most developed from a commercial point of view. More recently, attention has been paid to ABC triblock copolymers, due to their specific bulk morphologies, which differ from those observed with linear diblock copolymers through their higher complexity [7]. These examples most often consist of polymeric segments exhibiting a coil conformation (coil/coil systems). However, novel families of hybrid and innovating linear block copolymers have emerged. For instance, hybrid block copolymers comprising dendritic segments (entanglement-free) [10] or rod-like conformations based on synthetic polypeptides adopting an α-helix conformation [2, 11] or a rigid conducting block have been reported: dendron/coil, rod/dendron, rod/coil, rod/rod systems (see Fig. 2.1).

The first examples of "tailor-made" linear block copolymers by sequential "living" anionic polymerization date back to more than 60 years [12]. The traditional and most straightforward synthetic route to block copolymers is still the sequential

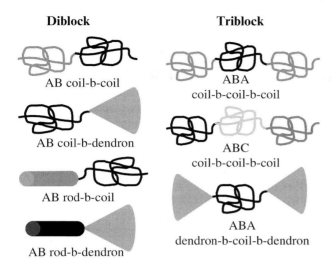

Fig. 2.1 The different types of linear block copolymers.

addition of two chain polymerizable monomers following a CLP mechanism, that is, when the active centers are essentially involved in propagation steps with very limited side events such as chain transfer and termination reactions [6]. After the first monomer has been consumed, active species (ionic, radical or organometallic) are still "alive" to initiate the polymerization of monomer B. Originally achieved with monomers such as styrenics and dienes by an anionic route [13], such a strategy has been further applied to other CLP methods, including cationic, radical and organometallic routes.

However, a distinction should be made between truly "living" and "controlled" polymerizations. According to the original definition by Swarcz [14], a "living" polymerization only implies two steps, which are initiation and propagation, with no chain breakings (an absence of irreversible chain termination and chain transfer steps). This, however, does not preclude slow initiation and/or slow exchanges between the different active species that can co-exist in the reaction mixture relative to the overall rate of propagation. In such a case, molar masses do not evolve linearly with monomer conversion, as in the "controlled" polymerizations case. Another penalty to pay is a broadening of the distribution of the molar masses. In other words, "living" polymerization does not necessarily lead to polymers of "controlled" dimensions (molar mass and PDI). Conversely, a "controlled" polymerization can tolerate a minimum degree of chain breakages, provided these do not significantly affect the overall molecular features of the final polymers. Thus, provided that initiation is complete and exchange between species of various reactivities is fast, the final dimensions of the polymer, i. e., its number average molar mass, is predicted by the initial concentration of the reagents and a molar mass distribution close to unity is achieved. A detailed comparison between living and controlled polymer-

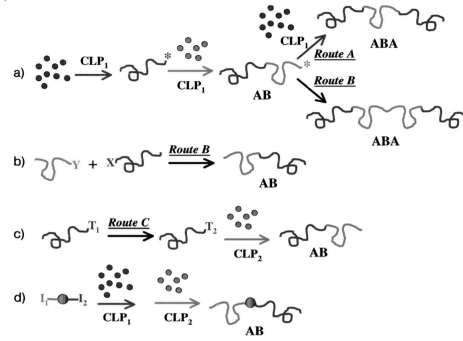

Fig. 2.2 Di- or triblock copolymer synthesis: (a) sequential "controlled/living" block copolymerization (sequential addition of monomers); (b) coupling of linear chains containing antagonist functions (X and Y); (c) switching from one polymerization method to another; and (d) use of a dual ("double-head") initiator consisting of two distinct initiating fragments (I_1 and I_2).

izations was the subject of a special issue of the *Journal of Polymer Science, Part A: Polymer Chemistry* in 2000 [15].

Figure 2.2 summarizes the possible routes to AB and ABA structures. In addition to the sequential addition of monomers (*route A*) either from monofunctional (both AB and ABA structures) or from difunctional initiators (ABA structure only), other strategies can be used. These include coupling of two preformed (co)polymer segments with antagonist functional end-groups (*route B*), combination of different modes of polymerization for the preparation of specific block copolymers that are not accessible from one polymerization mechanism only (*route C*), and one-pot initiation from dual bifunctional initiators for AB block copolymer synthesis (*route D*).

2.2
Free-radical Polymerization

The "uncontrolled" free-radical polymerization technique (RP) is generally believed to be useless for synthesizing block copolymers [16]. However, this can be achieved using a preformed end-functionalized polymer with an initiating fragment (azo-type or percarbonate). PEO with OH end-groups has been the most studied precursor for the preparation of block copolymers based on PS and PEO [17]. Interestingly, a PEO-azoinitiator could act as an initiator-surfactant (inisurf), the amphiphilic block copolymers formed *in situ* taking the role of a stabilizer for the emulsion surfactant-free polymerization of styrene [18]. "Double-head" radical initiators, which are compounds possessing two different types of initiating groups (see also Section 2.10), for instance initiators with a distinct rate of decomposition, could also be used for the preparation of block copolymers. One of the two radical initiating fragments first leads to a functionalized precursor polymer that can serve for subsequent emulsion polymerization of another monomer [19]. In both examples, it has to be mentioned that ill-defined materials are obtained (high PDIs and heterogeneous composition) because of the high probability of irreversible terminations between propagating chains, leading to mixtures of AB-type and ABA-type block copolymers. Thus, such procedures are not recommended when pure block copolymers are required.

2.3
Coupling Reactions of Homopolymers

The covalent coupling of two polymeric chains at their respective ends results in a diblock copolymer (Fig. 2.2, *route B*). This can be achieved by various chemical means and such a method can be expanded to the preparation of tri- and multiblock copolymers when α-ω-difunctional polymer precursors are used. This is a common synthetic route to multiblock copolymers from precursors derived by step growth polymerization [20, 21]. For example, α,ω-dihydroxypolyesters reacting with a α,ω-diisocyanatopolyurethane form block copolymers based on polyester and polyurethane blocks linked together via urethane groups. In such a case, molar masses of multiblock copolymers depend on stoichiometry between the preformed polymers. An even more complex composition can be obtained by starting from three or more polymeric precursors. Such a method implies that coupling reactions should be selective, fast and quantitative, conditions sometimes difficult to fulfill because of the incompatibility of the blocks, which decreases the reactivity between the antagonist functions. In other words, homopolymer precursors may contaminate the block copolymer materials.

A few examples of couplings of polymers derived by "controlled/living" chain polymerization have also been described. For example, anionic "living" PS chains were coupled with "living" cationic PEVE to form the corresponding PS-*b*-PEVE block copolymers [22]. Compounds consisting of PTHF and PS, PIB and PMMA were also prepared in the same way [23]. One can also combine this "coupling strategy" with that relying on the sequential "living" copolymerization (see Sections 2.4 to 2.8); in this case, "living" block copolymeric chains are linked onto difunctional coupling agents and produce ABA triblock copolymers. When an excess of living chains is employed, a fractionation step is required to separate ABA-type triblocks from AB-type diblocks. Typical examples are the synthesis of PS-b-PI-b-PS and PS-b-PtBA-b-PS triblock copolymers, where block copolymeric precursors are obtained by sequential anionic copolymerization, dichlorodimethylsilane, $(CH_3)_2SiCl_2$, and bis-(bromomethyl)benzene, $BrCH_2-\Phi-CH_2Br$ being used as coupling agents, respectively [24, 25].

To synthesize amphiphilic metallo-supramolecular block copolymers, Schubert and coworkers recently developed a coupling strategy that is not based on covalent linkages between preformed polymers, but on the formation of metal–ligand interactions of polymers that are end-functionalized by terpyridine units [26]. In this case, the prepolymers are coupled by adding ruthenium ions. The reversibility of the supramolecular bond allows the manipulation of nano-objects formed by self-assembly in aqueous solutions.

2.4
Sequential Anionic Polymerization

Anionic polymerization [27] has been the most popular approach to the formation of block copolymers until the emergence, in the mid-1990s, of "controlled/living" radical polymerization (C/LRP) techniques (see Section 2.8). The anionic active species, in most cases carbanions and oxanions, are highly reactive and special care is required to purify all chemicals from protic impurities to prevent chain breakages. For a long time just restricted to styrene and dienic monomers, anionic polymerization has been extended to other classes of monomers, among which some have been subjected to sequential block copolymerization (Fig. 2.3). Acidic monomers, such as acrylic and methacrylic acids, cannot be polymerized directly, and protected monomers are required in anionic polymerization. The details of anionic polymerization including the synthesis of block copolymers can be found in [26].

In some cases, it is essential to sequentially polymerize monomers in a certain order to access the targeted block copolymers. The "golden rule" is to follow the scale of reactivity, starting with the polymerization of the monomer that will form the higher reactive propagating center and then polymerize the other monomer. For example, polystyryl chains derived by anionic polymerization can initiate the polymerization of ethylene oxide, whereas oxanions generated from PEO chains are not capable of initiating the polymerization of styrene. In general, sequential anionic polymerization requires the following order of addition: dienes/styrene > vinylpyridines > (meth)acrylates > oxiranes > siloxanes [27]. However, "boosting" the nucleophilicity of a propagating chain end through its chemical modification, in order to enhance its reactivity, is an alternative route to move from a less to a more

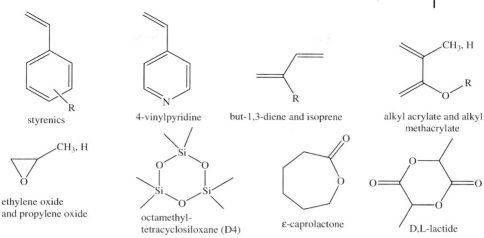

Fig. 2.3 Typical monomers suitable for anionic or coordinated anionic block copolymerization.

reactive species and thus reverse the order. An example of the latter strategy was provided by converting oxanions of PEO chains into silyl anions and subsequent polymerization of styrene or methacrylate for the successful synthesis of PEO-b-PS and PMMA-b-PEO block copolymers [28].

The sequential anionic polymerization of styrene and dienes is the prototypical example of the synthesis of well-defined block copolymers. The initiation step is typically achieved using alkyl lithium species and propagation steps proved to be "living" under suitable conditions [27]. As mentioned above, PS-b-PI-b-PS triblock copolymers referred to as thermoplastic elastomers are routinely produced by anionic polymerization, either by sequential addition of the two monomers or by chain coupling (Fig. 2.2, route A and route B). The choice of the experimental conditions (temperature, use of additives, solvent nature, etc.) has a dramatic impact on the regio- and stereoselectivity of the diene polymerization, which can be used to finely tune the properties of the resulting copolymers. For instance, the glass transition temperature of polybutadiene can be varied by the proportion of the 1,2-monomer units along the polymer backbone.

Synthesis of block copolymers composed of poly(meth)acrylates by anionic polymerization is more challenging because of the propensity of the anionic propagating centers to react with the carbonyl groups in termination reactions. However, well-defined block copolymers based on poly(meth)acrylate can be obtained by stabilizing the carbanionic active centers through specific μ-type, σ-type and μ/σ ligands used as additives (e. g., lithium-based additives, tetraalkylammonium or tetraalkylphosphonium salts, trialkylaluminum, etc.) [29, 30]. Among the amphiphilic block copolymers based on (meth)acrylic segments, the most common ones are based on a hydrophilic block of PMAA or PAA. Such materials can be prepared by sequential living anionic polymerization of *tert*-butyl methacrylate or *tert*-butyl acrylate, followed by a hydrolysis step of the *tert*-butyl groups [31]. As thoroughly investigated by Eisenberg's group, PS-*b*-PAA block copolymers synthesized by sequential anionic polymerization can form nanometer-sized polymeric micelles by self-assembly in water [32]. A further step was accomplished by Wooley's group who developed methods for cross-linking the core or the shell of polymeric micelles based on PAA. For instance, shell cross-linked micelles of PS-*b*-PAA were described, and hollow particles were created by ozonolysis of shell cross-linked micelles based on PI-*b*-PAA [33]. Recent advances in the design of polymeric architectures based on PAA or PMAA segments using CLP techniques have recently been reviewed by Mori and Muller [34].

Ethylene oxide (EO) is another anionically polymerizable monomer often employed to prepare amphiphilic block copolymers, e.g., PS-*b*-PEO, PB-*b*-PEO and PMMA-*b*-PEO. In this case, EO is generally polymerized after the vinylic monomer to comply with the scale of reactivity of the propagating species [27, 35]. Triblock copolymers, known as Pluronics®, based on PPO as the hydrophobic segregating block and PEO as the water soluble block represent a commercially important class of amphiphilic block copolymers used in various applications of aqueous systems (pharmaceutics, cosmetics, processing aids, additives in coatings, etc.) [36, 37]. These non-ionic polymeric surfactants are typically prepared by sequential anionic polymerization of ethylene oxide and propylene oxide [27].

2.5
Sequential Group Transfer Polymerization

Group transfer polymerization (GTP) is a CLP technique of choice for the synthesis of well-defined (meth)acrylate (co)polymers of narrow molar mass distribution. A silyl ketene acetal is generally employed as the initiator in the presence of either a nucleophilic or an electrophilic catalyst, e.g., tetrabutylammonium fluoride for methacrylayte or a Lewis acid based on aluminum for acrylates. The role of this catalyst is to selectively cleave the Si–O bond in order to generate an enolate-type species, which is the active form in GTP. Enolates are in fact in equilibrium with the dormant silyl ketenes and the fast and reversible exchange of $SiMe_3$ groups between dormant and active species ensures a "living/controlled" growth of the chains at room temperature in the case of alkyl acrylates and methacrylates [29, 38]. However, GTP cannot be applied to non-polar monomers such as styrene. Miscellaneous (meth)acrylic block copolymers have been synthesized by sequential GTP, as reviewed recently by Mori and Muller [34]. For instance, PMMA-*b*-PAA has been prepared by sequential GTP of methyl methacrylate and *tert*-butyl acrylate followed by hydrolysis of the *tert*-butyl groups [39]. Patrickios et al. reported the synthesis of diblock and triblock polyampholytes based on PDMAEMA, PMAA and PMMA by GTP [40]. Other examples of block copolymers, including amphiphilic and zwitterionic systems obtained via sequential GTP can be found in [34].

2.6
Sequential Cationic Polymerization

Development of "living" cationic polymerization was achieved in the 1980s, which further allowed the preparation of new types of block copolymers [41]. Monomers that can undergo "living" cationic block copolymerization are those bearing electron donating groups, including styrenics, vinyl ethers, isobutylene and oxazolines (Fig. 2.4). As with anionic polymerization, the solvent effect has a crucial role in the stability/reactivity of the cationic intermediates. In addition, relatively low temperatures are required to minimize the occurrence of termination and transfer reactions.

Numerous heterocyclic compounds can undergo cationic ring-opening polymerization but only a limited number of these monomers offer the possibility of CLP. As for vinyl monomers, it is the tendency of the corresponding cationic propagating species to undergo side reactions, which is the main limitation to the development of living systems. Oxazolines are, however, counter-examples as these cyclic monomers can form persistent oxazolinium salts under relatively non-demanding conditions [42]. Block copolymerization of two oxazolines possessing very distinct reactivities (e. g., monomers with R = fluoroalkyl groups are much less reactive than monomers containing non-fluorinated groups) in a one-shot feeding process can be achieved.

A simple synthetic approach to block copolymers is to sequentially polymerize monomers with similar reactivity, e. g., vinylic monomers of the same family but bearing different substituents. In this way, amphiphilic block copolymers can be prepared by polymerization of isobutyl vinyl ether with vinyl ether carrying a protected glucose derivative [43]. In contrast, block copolymerization of vinylic monomers possessing different reactivity (e. g., isobutylene and vinyl ethers) generally requires a specific initiating system for each monomer. Hence, crossover from one monomer to the other must be achieved by changing the reactivity of the species deriving from the first block. For instance, the Lewis acidity of the chain-ends can be manipulated by adding 1,1-diphenylethene, which cannot hompolymerize, onto "living" PIB chains prepared using $TiCl_4$, followed by the addition of $Ti(OEt)_4$. Such a modification of the initiating system allows the cross-over to the "living" cationic

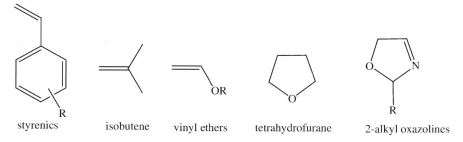

Fig. 2.4 Typical monomers suitable for cationic block copolymerization.

polymerization of vinyl ethers for the synthesis of PIB-*b*-PVE block copolymers, which would not be possible by adding the two monomers in this order, without this change in reactivity [44, 45].

2.7
Non-radical Metal-catalyzed Polymerization

Although polyolefins represent the most commercially important category of commodity polymers, control over the molecular characteristics (molar masses and PDIs) of poly(α-olefin) using single-site metallocene-based catalysts has only been developed recently [46, 47]. The "living" coordination polymerization of α-olefins by transition methods is still limited and application of metal-based catalysts to the preparation of block copolymers has not been fully explored.

Brookhart and coworkers described the preparation of PP-*b*-PHex and PP-*b*-POct di- and triblock copolymers employing a discrete nickel diimine catalyst [48]. The same group showed that ethylene could undergo "living" polymerization with palladium(II) diimine complexes (catalyst **1** in Fig. 2.5) for the production of branched, amorphous PEs [49]. Conditions for living polymerization of propylene, 1-hexene and 1-octadecene have also been demonstrated and PE-*b*-poly(octadecene) copolymers prepared.

Fujita et al. developed a new family of catalysts for olefin polymerization; these catalysts are based on non-symmetrical phenoxyimine chelate ligands combined with Group 4 transition metals [50]. For instance, titanium catalysts bearing fluorinated ligands (e. g., catalyst **2** in Fig. 2.5) not only induces the "living" polymerization of ethylene producing high molar mass PEs but also promotes the stereospecific "living" polymerization of propylene allowing access to highly syndiotactic monodisperse PP. These catalysts also served to prepare a number of polyolefin-based block copolymers (e. g., PE-*b*-(PE-*co*-PP), syndiotactic PP-*b*-(PE-*co*-PP), PE-*b*-PHex, etc.) [50, 51]. It is very likely that these catalytic systems will pave the way for synthetic developments of novel block copolymers based on polyolefins.

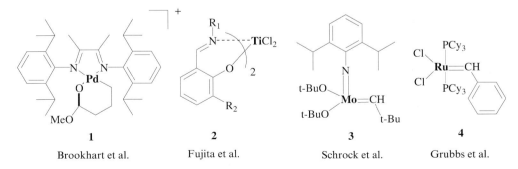

Fig. 2.5 Typical catalysts for "living" polymerization of olefins and for "living" ROMP.

Ring-opening metathesis polymerization (ROMP) is another metal-catalyzed polymerization method that can be controlled with specific catalysts such as **3** and **4** in Fig. 2.5, developed by Grubbs and Schrock (Nobel Prize 2005 with Chauvin), respectively, for certain cyclic olefins (mainly norbornene and cyclooctadiene derivatives) [52, 53]. ROMP was further applied for the synthesis of block copolymers, for instance water-soluble compounds exhibiting conducting properties [54], or block copolymers containing fluorescent moieties for use in light-emitting devices [55]. The latest advances in ROMP have concerned the development of catalysts for controlling the polymerization of functional group-containing monomers with no need of protection/deprotection chemistry [56] or for water-borne ROMP, allowing access to novel block copolymers for specific applications (e. g., as supports for chromatography or uses in light-emitting devices) [57].

2.8
Controlled Radical Polymerization

Radical processes are utilized extensively in industry with about 50 % of polymeric materials being produced by this chain polymerization [16]. Compared with ionic polymerization techniques, free radical processes offer the advantage of being applicable to a wider variety of monomers and are not as demanding as other chain addition mechanisms with respect to the purity of the reagents used. Radical processes can be implemented in an emulsion, suspension, solution or in bulk. In addition, radical growing species are highly tolerant of many functional groups, including acid, hydroxyl, amino, epoxide, etc., hence functional monomers can undergo radical polymerization without the help of protection chemistry. However, chain breakings occurring by irreversible terminations in "conventional" free radical polymerizations seriously limit their relevance in macromolecular synthesis. In recent years substantial research has been devoted to controlled/living radical polymerization (C/LRP) [58] methodologies that combine the advantages of truly "living" systems for the quality of the polymers formed with the ease inherent to radical processes. Among C/LRP systems, nitroxide-mediated polymerization (NMP) [59], atom transfer radical polymerization (ATRP) [60, 61] and reversible addition-fragmentation chain transfer (RAFT) [62, 63] have been investigated extensively. Each of these methods, however, involves drawbacks that still limit their industrial development. NMP often requires elevated temperatures and applies to a limited number of monomers. ATRP, although very powerful, is inoperative for monomers that can poison the catalyst (e. g., acrylic acid); moreover the removal of metallic ions from polymers is an unresolved issue at an industrial scale. As to the RAFT process, the preparation of the corresponding chain transfer agents suffers from drawbacks such as the use of carbon disulfide or iodine, two toxic reagents, and the final RAFT-derived (co)polymers are slightly colored and sometimes malodorous due to the presence of sulfur atoms at the chain ends. This shows the necessity to improve C/LRP methodologies before they can be implemented in industry.

At the laboratory scale, C/LRP methods have paved the way for their use in block copolymer synthesis and for macromolecular engineering in general. As in ionic procedures, monomers cannot be polymerized in an indiscriminate sequence. In C/LRP, block copolymerization should be performed in an order that privileges the fastest monomer. Also, blocking efficiency is sensitive to experimental conditions (temperature, monomer conversion, use of additives, etc.). For instance, it is recommended that the polymerization of the first monomer is discontinued before its total consumption so as to preserve its end functionality intact and thus prevent too high a concentration of dead blocks formed by termination reactions, the latter side events being unavoidable in any of the three C/LRP methods. It is because of this that C/LRP cannot be considered as a truly "living" system.

In order to avoid the premature termination of growing radicals, one has to lower the instantaneous concentration of the growing radicals by trapping them in a large reservoir of covalent dormant species and establish a fast equilibrium between the latter and a minute concentration of growth-active radical species. Should such a dynamic equilibrium be established, all chains would be given an equal opportunity to grow through constant inter-conversion between their active and dormant forms. This seminal idea was first conceptualized by Otsu and Yoshida in 1982 [64].

The possibility of preparing block copolymers by C/LRP has been demonstrated in many instances. A comprehensive description of block copolymers obtained by C/LRP can be found in recent review articles [14–18] or book chapters [65, 66]. Block copolymers, which were not accessible using existing techniques, could be prepared following most of the methods mentioned in Fig. 2.1 (sequential C/LRP, switching from one mechanism to another and use of double-head initiators), witnessing the extreme versatility of block copolymer synthesis by C/LRP. Worth noting are the possibilities now offered to incorporate hydrophilic blocks in amphiphilic copolymers without resorting to protection/deprotection chemistry *via* these C/LRP methodologies.

2.8.1
Atom Transfer Radical Polymerization (ATRP)

Discovered independently by Sawamoto and coworkers [67] and Wang and Matyjaszewski [68] in 1995, ATRP entails a reversible one-electron oxidation of a transition metal, abstracting a halogen from an organic halide (RX), thus generating a radical species (R$^\bullet$) that can attack a vinylic monomer and further propagate, followed by a one-electron reduction of the oxidized metal (Fig. 2.6). In the meantime, the metallic complex transfers its halogen to the growing chain thus affording a new dormant species. With this halide abstraction and subsequent transfer occurring repeatedly, polymer chains grow in-between each activation/deactivation cycle. Provided the latter process is rapid, all the chains can grow uniformly. The oxidized metal complexes X–Met^{n+1} generated by activation of the dormant alkyl halides behave as persistent radicals and therefore reduce the stationary concentration of the growing radicals, thereby minimizing the extent of the termination. Thus, as with NMP (see Section 2.8.2), ATRP is based on a

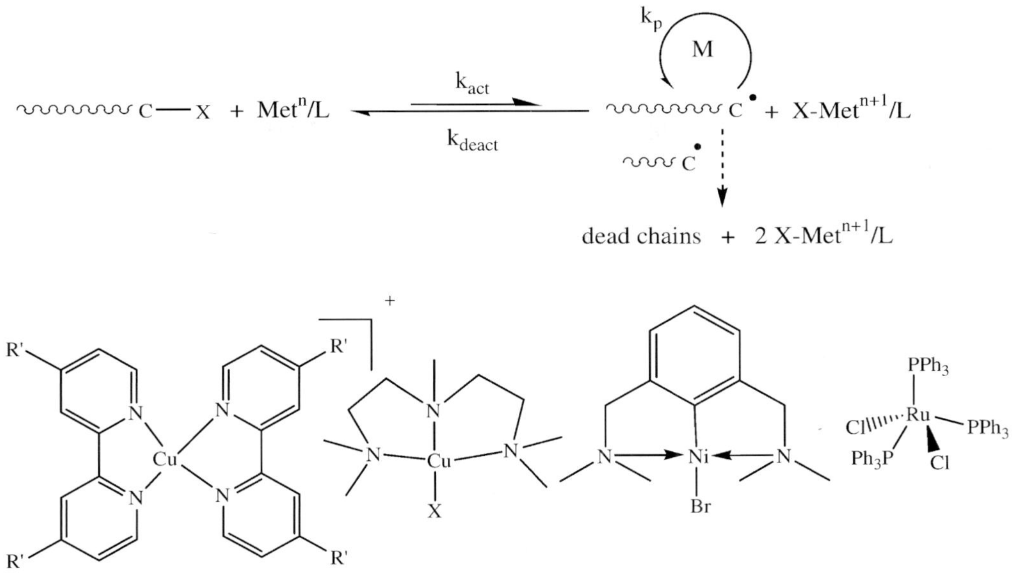

Fig. 2.6 Propagation step in ATRP and representative catalytic systems.

reversible termination process and is under the control of the "persistent radical effect" [69].

The initiating systems for ATRP consist of a transition metal/ligand complex and an initiator, typically an activated alkyl halide. The initiator RX determines the number of chains and provides the structure of the initiating moiety (R) and of the end (X) of the polymer chain, whereas the number of additions of monomer units, which determines the polydispersity of the polymer, depends on the activity of the catalyst. For the choice of the appropriate metal/ligand catalyst controlling polymerization, four criteria appear critical: (a) the metal center must have at least two accessible oxidation states separated by one electron, (b) it must exhibit affinity toward the halogen, (c) the coordination sphere around the metal must be able to accommodate the halogen upon oxidation and (d) the ligand must form a strong complex with the metal. As demonstrated by its extensive utilization in the synthesis of precision polymers, copper halides associated with a variety of nitrogen-based ligands proved the most versatile catalysts for macromolecular engineering, in particular for block copolymer synthesis.

However, ATRP often requires a tedious metal removal step, which still limits its industrial development, although solutions have been proposed to overcome this issue. These include the use of very powerful complexing agents, the development of a supported catalyst, biphasic catalysis, etc. [70].

In one of their first reports on ATRP in 1995 [68], Wang and Matyjaszewski described the synthesis of hard/soft block copolymers based on styrene and methyl acrylate units that were obtained by sequential polymerization. Other groups later showed how to take advantage of nickel(II) [71] or ruthenium(II) [72] catalytic systems to derive PMMA-*b*-PnBMA diblock copolymers in addition to PMMA-*b*-PnBMA-*b*-PMMA triblock copolymers. Many other examples of AB diblock copolymer synthesis have been reported in the literature by the sequential addition of two monomers or, alternatively, using ω-homopolymers as macroinitiators. ABA triblock copolymers have been synthesized using similar methods and difunctional ATRP initiators.

ABA-type triblocks aimed at being used as thermoplastic elastomers have been prepared using these methods with central soft blocks composed of butyl acrylate, methyl acrylate and 2-ethylhexyl acrylate units and hard outer blocks made of styrene, methyl methacrylate or acrylonitrile units [60, 61, 65, 66]. Also, ABC triblock as well as various tetra- and pentablock copolymers based on acrylates, methacrylates and styrenes have been prepared including amphiphilic structures with water-soluble polymers, such as those derived from 2-hydroxyethyl acrylate, acrylic acid, 2-dimethylaminoethyl methacrylate, 4-vinyl pyridine and others [7, 66].

The order of polymerization of monomers that belong to the same family is not an issue: in this case, one can access well-defined block copolymers with only a small proportion of the homopolymer contaminants. Interestingly, the same situation prevails for styrenics and alkyl acrylates that can be sequentially polymerized in either order by ATRP [73, 74]. The situation is obviously different for monomers exhibiting a distinct reactivity. For instance, the rate of initiation of a PMMA block by a chlorine-terminated PMA is slow as compared with the rate of propagation of methyl methacrylate, which results in a poorly defined block copolymer. One way to improve the blocking efficiency is to grow the PMMA block from a bromine terminated PMA precursor in the presence of CuCl so as to favor the initiation step over propagation ("halogen exchange") [75]. Thus, for the synthesis of block copolymers based on alkyl acrylate and alkyl methacrylate units, it is recommended to form the methacrylate-based block first and then cross over to the ATRP of the acrylate monomer, unless employing the "halogen exchange" technique. For the same reason, poly(methacrylate) or poly(acrylonitrile) blocks should not be grown from a polystyrene or a poly(alkyl acrylate) macroinitiator because of the poor blocking efficiency.

The "halogen exchange" technique was also applied by Jérome and coworkers for the synthesis of narrowly distributed PMMA-*b*-PnBA-*b*-PMMA triblock copolymers that were obtained from a difunctional PnBA macroinitiator [76]. The same triblock structures were also derived by these workers utilizing a four-step procedure based on the sequential ligated anionic polymerization of MMA, *t*BA and again MMA, followed by the transalcoholysis of the *t*BA units into the desired PnBA [76]. Obviously, ATRP was more straightforward than anionic polymerization with respect to the ease of the synthesis. Nevertheless, the mechanical and rheological properties as well as the morphology of the all-methacrylate based thermoplastic elastomers prepared by anionic polymerization were found to be better than those of the compounds prepared by ATRP. The poorer tensile properties of the ATRP-derived copolymers were probably due to the larger polydispersity of the PMMA outer blocks, although only minor differences could be detected by SEC or ^{1}H NMR analysis. Therefore, the level of structural and molecular control of materials derived by C/LRP techniques can approach that delivered by traditional anionic procedures but a thorough investigation of their physical behavior is required before their utilization in a particular application.

2.8.2
Nitroxide-mediated Polymerization (NMP)

Nitroxides are stable free radicals that are generally used in organic chemistry to trap transient radical species [69]. In NMP, nitroxides are used as persistent counter radicals to trap growing polymer chains in a reversible manner when the temperature is appropriately chosen. The structure of the nitroxide obviously plays a crucial role in the success of NMP. For instance, Gnanou, Tordo and coworkers developed an α-hydrogen bearing nitroxide, namely *N-tert*-butyl-*N*-[1-diethylphophono-(2,2-dimethylpropyl)] nitroxide (DEPN), of lower stability than TEMPO but which can accommodate alkoxyamines of much weaker C–O bonds than TEMPO [77]. Not only could styrene be polymerized at faster rates and under controlled conditions in the presence of DEPN and a radical initiator, but the polymerization of acrylates including functional monomers such as hydroxyethyl acrylate, acrylamides, dienes (butadiene and isoprene), and more recently methacrylates, was also found to be "living" in the presence of DEPN (Fig. 2.7) [77].

The approach of Hawker and coworkers was not that different; they developed preformed unimolecular alkoxyamines based on a similar α-hydrogen containing nitroxide, operating both as the initiator and controlling agent for NMP of the same above monomers and also for acrylonitrile, butadiene, etc. [78]. As conceptualized by Fischer, the number of dead chains eventually formed in NMP is exactly equal to the excess of nitroxide self-produced by the system to minimize termination reactions [69]. To minimize the number of dead chains to a level lower than that self-determined by the system, the nitroxide can be utilized in excess with respect to the initiator. In contrast, it is sometimes necessary to use additives

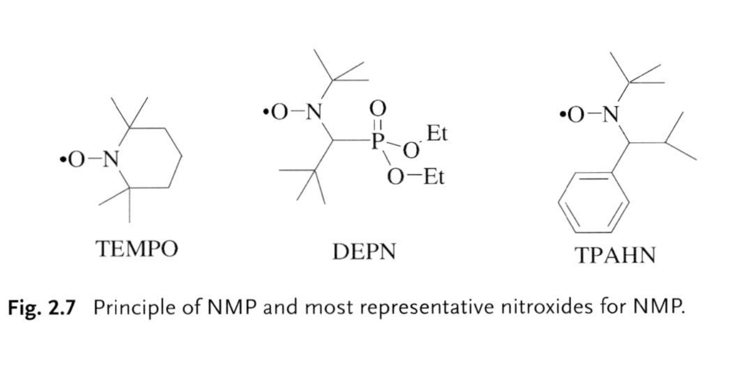

Fig. 2.7 Principle of NMP and most representative nitroxides for NMP.

(acids) as a means of reducing the amount of free nitroxide resulting from irreversible termination, and thus prevent the inhibition of the polymerization [79]. On these grounds, many examples of block copolymers derived by NMP (amphiphilic, hard/soft, etc.) have been provided in the literature. Hawker et al. [59] reviewed this field in 2001; other review articles are also available [65, 66]. Representative examples are discussed below.

Georges et al. reported the synthesis of PS-polydiene block copolymers by sequential NMP using TEMPO as the counter-radical [80]. The polymerization of isoprene had to be carried out at a fairly high temperature (145 °C) and only moderate conversion could be achieved: the PDIs were in the range of from 1.4 to 1.5, indicating that materials were ill-defined. Nitroxides based on imidazolidone developed by the group working with Rizzardo offer a few advantages over TEMPO and its derivatives in terms of molecular control when synthesizing random and block copolymers; for instance, soft/hard PnBA-b-PS [81]. As emphasized above, however, α-hydrogen containing alkoxyamines have considerably enlarged the range of block copolymers accessible by NMP, with respect to TEMPO-mediated NMP. For example, block copolymers such as PS-b-PI, PtBA-b-PDMA, etc. could be readily prepared by sequential NMP from this second generation of persistent nitroxides [78, 82, 83]. Again, the order of addition of the monomers in sequential NMP can be an issue when preparing specific block copolymers. For instance, crossing over from a PS precursor to an acrylate for styrene–acrylate block copolymers leads to contamination of the block structure by PS homopolymer chains. In contrast, the reverse pathway consisting of the polymerization of the acrylate monomer, followed by NMP of styrene allowed the preparation of well-defined block copolymers [59, 78].

Block copolymers containing a styrene-rich statistical copolymer as one of the blocks have also been derived by NMP. Provided the amount of styrene was sufficiently high in the initial monomer feed ratio, the statistical nitroxide-mediated copolymerizations of styrene and acrylonitrile, methyl acrylate, methyl methacrylate or vinyl carbazole could be achieved in a controlled fashion [84]. Other workers have also described the synthesis of hard/soft block copolymers based on styrene and alkyl (meth)acrylate blocks using TEMPO-type counter radicals; in either case the acrylic monomers could not be polymerized to complete conversion because of the lack of livingness of propagation of these monomers with this family of nitroxides [80, 85–87]. Rod-coil type copolymers as a particular category of hard/soft materials were, for instance, synthesized by sequential NMP of 4-acetoxystyrene and a liquid crystalline styrenic monomer in the presence of TEMPO [88]. Another example was provided by Hadziioannou and coworkers: an NMP-derived poly(styrene-co-chloromethylstyrene) block as the flexible coil was associated with a poly(p-phenylene vinylene) as the rigid-rod to form block copolymers with semiconducting properties [89].

The Armes group derived zwitterionic as well as acidic block copolymers by sequential NMP of charged hydrophilic monomers in aqueous solutions [90].

Finally, a wide variety of well-defined random and block copolymers comprising of styrenic (chloromethyl styrene, acetoxystyrene, styrene, etc.), isoprene and alkyl (meth)acrylate blocks have been synthesized using the α-hydrogen bearing nitroxides [59, 83].

2.8.3
Reversible Addition Fragmentation Chain Transfer (the RAFT Process)

An important development in the area of C/LRP came about in the late 1990s with the use of thiocarbonylthio compounds, of general structure Z–C(=S)SR, as reversible chain transfer agents (CTAs) in two independently discovered processes [62, 63, 91]. These have been termed RAFT, for Reversible Addition Fragmentation chain Transfer [92], and MADIX, for Macromolecular Design by Interchange of Xanthates [93]. From a mechanistic viewpoint (Fig. 2.8), both RAFT and MADIX processes are identical and only differ by the chemical nature of the CTA: RAFT prevails for CTAs in general, including dithioesters, dithiocarbamates, trithiocarbonates and xanthates, whereas MADIX refers to xanthates exclusively, where the Z group is an O-alkyl group (Z = OZ′).

In these polymerizations (Fig. 2.8), the CTA first reacts through its C=S double bond with propagating oligomers, leading to a transient radical. This radical intermediate undergoes a β-scission generating a new thiocarbonylthio compound and an expelled R• radical capable of re-initiating the polymerization. Increasing the stability of the radical intermediate may cause the polymerization to be retarded and/or inhibited, in particular when low molar mass polymers are targeted [62, 63, 91]. A fast equilibrium relative to the rate of propagation is highly desirable should polymers with narrow molar mass distributions be required. For this purpose, both the activating substituent Z and the R leaving group of the CTA, Z–C(=S)–S–R, should be chosen appropriately. Molar masses can be controlled

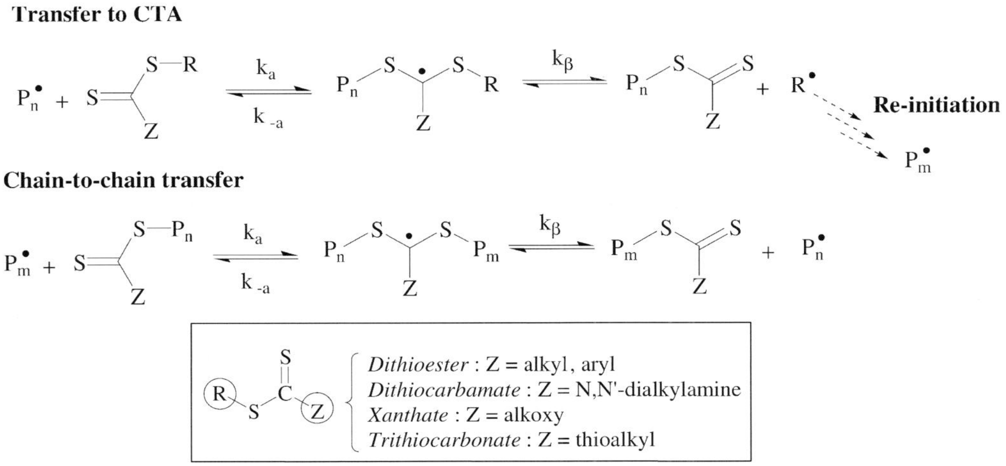

Fig. 2.8 Chain transfer events in RAFT polymerization and the main thiocarbonylthio compounds used in RAFT.

through the ratio of monomer to the RAFT/MADIX agent whereas the number of dead chains depends on the concentration of the radical source: this should be much lower than that of CTA.

Significant efforts have been directed at establishing the scope and limitations of RAFT and MADIX processes [62, 63, 91–93]. These include kinetic and mechanistic investigations, implementations with dispersed media or with homogeneous aqueous solutions and their use for macromolecular engineering. Among the advantages of RAFT/MADIX processes over other C/LRP is the possibility to polymerize a wide range of monomers, including hydrophilic ones such as acrylic acid or acrylamide directly in aqueous media [94, 95], or monomers with electron-donating substituents such as vinyl acetate or N-vinyl pyrrolidone. The MacCormick group developed specific RAFT agents to be used under aqueous solutions for the synthesis of a wide range of stimuli-responsive water-soluble (co)polymers [95].

Preparation of RAFT/MADIX agents, however, suffers from drawbacks: use of toxic reagents (CS_2, I_2, etc.). Likewise, not only the coloring of the compounds (pinkish and yellowish for dithioesters and xanthates, respectively) but also bad odors originate from thiocarbonylthio groups in the (co)polymers.

Route A (Fig. 2.2), based on the sequential addition of two monomers, was mostly used to access block copolymers by RAFT/MADIX [62, 63, 91–95]. However, efficient cross-over in RAFT-mediated sequential polymerizations requires that the rate of transfer to the terminal dithioester carried by a given precursor be higher than the rate of transfer to the dithioester generated at the end of the growing second block. In other words, it is essential that the first block grown provides the better leaving radical. For instance, to prepare well-defined methacrylate-*b*-acrylate diblock copolymers, one should polymerize the methacrylic monomer sequentially and then the acrylic one, this order of addition being imposed by the mechanism.

The synthesis of hard/soft block copolymers was successfully achieved in this way. In addition, ABA triblock copolymers, consisting of a soft central block and two external hard segments all based on methacrylates, could be derived by sequential polymerization from a bifunctional RAFT agent [96]. Alternatively, trithiocarbonate-containing compounds were also utilized to make PS-*b*-PnBA-*b*-PS triblock copolymers [97]. The two dithio functions concomitantly served as transferring sites in the sequential RAFT-mediated copolymerization of two corresponding monomers but, in this case, the dithio groups remained at the center of the final block copolymer. Alternatively, the chemical transformation of end group(s) of pre-formed polymers into thiocarbonylthio moieties allows further access to block copolymers by RAFT/MADIX (*route C*, Fig. 2.3). Recent overviews on RAFT/MADIX and their application to macromolecular engineering, including multiple examples of other block copolymer synthesis have been reported by Perrier and Takolpuckdee [63] and by Moad et al. [62].

2.9
Switching from One Polymerization Mechanism to Another

Using a pre-formed polymer that can be used as a macroinitiator for the growth of the second block by another mechanism is an alternative synthetic route to linear block copolymers (*route C*, Fig. 2.2). One resorts to this strategy when monomers that are to be paired in a targeted diblock structure do not polymerize by the same mechanism. With bifunctional precursors, ABA triblock copolymers can be obtained. This is referred to as a switch of mechanism polymerization: monomer A is polymerized by a CLP_1 mechanism via active center T_1, the latter being transformed into an active center T_2 from which monomer B can be polymerized by a CLP_2 mechanism. Switching of mechanism can either be achieved by an *in situ* conversion of the active center or after isolation of the first block followed by chemical modification of the active center. Examples of such a general strategy are numerous and it is not realistic to present all the possibilities. This field has formed parts of various review articles or book chapters that focus on a specific chain polymerization mechanism. For instance, Sawamoto discussed the possibility of combining "living" cationic polymerization (of vinyl ethers, isobutene or THF) with other polymerization mechanisms, including "living" anionic, etc. [41]. Other "switching" possibilities have been discussed in other review articles or book chapters [4, 5, 9, 27, 63, 65, 66]. A few representative examples are discussed below.

The switch from ionic processes to C/LRP is now common. For instance, phenylethyl chloride end-groups of PS chains obtained by "living" cationic polymerization of styrene can initiate the Cu-mediated ATRP of methyl acrylate, methyl methacrylate or styrene, affording the corresponding diblock copolymers [98]. Crossing-over from cationic ring opening polymerization of tetrahydrofuran to C/LRP of vinylic monomers has also been considered to derive block copolymers based on PTHF segments [99]. Synthesis of a hard/soft block copolymer through crossover from anionic polymerization to NMP or ATRP as an alternative route of the sequential anionic polymerization has also been contemplated [100].

Switching from C/LRP to ionic polymerization in this order is less common. Block copolymers of styrene and 1,3-dioxepane (DOP) or THF units were prepared by applying ATRP of styrene followed by the cationic ring-opening polymerization of DOP [101]. Starting from an α-hydroxy,ω-bromo PS precursor, Pan and coworkers cationically grew poly(DOP) blocks, after adding triflic acid to trigger polymerization.

Amphiphilic PNB-*b*-PVA block copolymers were synthesized by combining ROMP and aldol-group-transfer polymerization [102]. PNB-*b*-PS and PNB-*b*-PMA diblock copolymers as well as a polycyclopentadiene-*b*-PS diblock were also prepared by a switch-off mechanism involing ROMP [103]. To this end, *p*-bromomethylbenzaldehyde was used in a Wittig reaction in order to cap the polycycloolefin precursors with benzyl bromide at one chain end to trigger ATRP. Grubbs and coworkers [104] also contemplated the possibility of associating ATRP with ROMP to prepare block copolymers based on 1,4-PB: after carrying out the ROMP of 1,5-cyclooctadiene in the presence of an appropriate difunctional transfer agent, these workers took advantage of the halides present at the ends of the polybutadiene formed to grow either PS or PMMA blocks by ATRP.

Block copolymers made from PIB-b-PMMA were synthesized by combination of living cationic and living anionic polymerizations [105], PI-b-PLA by switching from living anionic polymerization to coordination-insertion polymerization [106], etc. Excellent thermoplastic elastomers were obtained upon associating polymers of high T_g (glass transition temperature) generated by ATRP with a rubbery cationically derived PIB in a triblock copolymer structure [107].

Combination of ring-opening polymerization (ROP) with other CLP procedures is also possible. For example, Schmidt and Hillmyer prepared diblock copolymers containing PI-b-PLA by switching from living anionic polymerization to aluminum alkoxide-initiated coordination-insertion ROP of lactide [108].

Condensation polymers have also been used as bifunctional ATRP macroinitiators for the synthesis of ABA triblock copolymers based on polysulfones [109], polyphenylenes [110], PDMS, etc. For instance, PS-b-PDMS-b-PS copolymers were obtained by polycondensation of α,ω-diamino-PDMS with a dichloro-azo compound to form an azo-containing PDMS macroinitiator first, followed by NMP in the presence of methoxy-TEMPO [111]. Dendrimers have been used as initiators for ATRP or NMP to yield hybrid linear/dendritic block copolymers [10].

NMP was also combined with RP for the preparation of a "library" of block copolymers based on styrene, tert-butylstyrene, 3,4-dimethoxystyrene, N-vinyl pyrrolidone and isopropylacrylamide units [112]. In order to synthesize block copolymers including vinyl acetate units, RP was combined with ATRP following different strategies [113]. For instance, vinyl acetate was polymerized from a bis-(halogenated)-azo initiator and PS blocks were grown by ATRP. Alternatively, ATRP of n-butylacrylate was performed first using the same azobis initiator, then PVAc blocks were grown under standard conditions.

Among the amphiphilic block copolymers, those based on PEO segments are the ones that have been studied most. Such materials can now be derived by combination of C/LRP techniques with anionic polymerization of ethylene oxide. For instance, PEO-based block copolymers were prepared by initially modifying an ω-hydroxy PEO by a functional alkoxyamine. The corresponding PEO-based macroinitiator was used to polymerize various monomers and give amphiphilic block copolymers [114]. As the polymerization proceeds at 130 °C, homoPS chains are formed by thermal self-initiation of styrene, which requires selective extraction with cyclohexane to get rid of these contaminants. Better defined diblock PEO-b-PS and triblock PS-b-PEO-b-PS structures were derived by ATRP of styrene from PEO macroinitiators with bromopropionate end functions, in the presence of a CuBr/dipyridyl system at 130 °C [115]. Following the same route, triblock PtBA-b-PEO-b-PtBA copolymers have been prepared [116]. The Armes group derived various double hydrophilic block copolymers (DHBC) possessing pH-responsive and/or temperature-responsive properties in water by ATRP of hydrophilic monomers (e. g., ω-methacrylate-PEO macromonomer, sodium methacrylate or sodium 4-vinyl benzoate) starting from a bromoester end-capped PEO macroinitiator [117]. RAFT methodology has also been applied to the preparation of PEO-*b*-PBzMA and PEO-*b*-PS amphiphilic block copolymers from ω-dithioester PEO precursors [96].

Many other amphiphilic block copolymers can be designed using the switch off mechanism. For instance, Wooley and coworkers described how to switch from anionic polymerization of ε-caprolactone to ATRP of *tert*-butylacrylate, which was followed by hydrolysis of the *tert*-butyl groups of the acrylate blocks thus yielding amphiphilic block copolymers PCL-*b*-PAA [118]. These compounds were found to self-organize into micelle-like structures, when water was present. The PAA-based shell part of these micelles was subsequently cross-linked by adding a diamine to form so-called "shell cross-linked knedel-like" nanoparticles containing, in this case, hydrolytically degradable crystalline core domains.

Last but not least, Chung [119] and Yanjarappa and Sivaram [120] discussed the possibility of switching from metallocene-mediated catalyzed polymerization of olefins to "controlled" radical polymerization of functional ethylenic monomers for the synthesis of polyolefin-based block copolymers. The corresponding methods include the introduction of either a reactive borane or a *para*-methylstyrene at the chain-ends of the olefin precursors.

2.10
Use of "Dual" Initiators in Concurrent Polymerization Mechanisms

"Dual" or "double-head" initiators contain two distinct sites for initiating the polymerization of monomers by different mechanisms, with no interference between the two mechanisms in the ideal case. Some examples of these dual compounds are provided in Fig. 2.9. This method of block copolymer synthesis requires that each initiating site in such dual initiators be stable and inert while the other one is active, that is, during the growth of the other block. The two monomers can even undergo a controlled polymerization process simultaneously, in a one-pot process without the need for a protection–deprotection step or functionalization of intermediates.

Sogah's group developed such multifunctional initiators, which were used in a one-pot simultaneous block polymerization of three monomers, i.e., styrene by NMP, phenyloxazoline (PhOx) by cationic ring-opening polymerization and ε-caprolactone by anionic ring-opening polymerization [121]. Hydrolysis of phenyl oxazoline units led to PS-*b*-polyethyleneimine amphiphilic block copolymers. Mecerreyes et al. used a "double-head" initiator that was capable of achieving the NMP of styrene and ring-opening polymerization of ε-caprolactone simultaneously in a one-pot procedure, for the formation of relatively well-defined PS-*b*-PCL block copolymers [122]. Interestingly, Grubbs and coworkers developed a dual ruthenium complex capable of initiating both ROMP and ATRP in a tandem polymerization for the production of block copolymers [123]. Other examples of the double-head initiators presented in Fig. 2.9 have been described [124, 125].

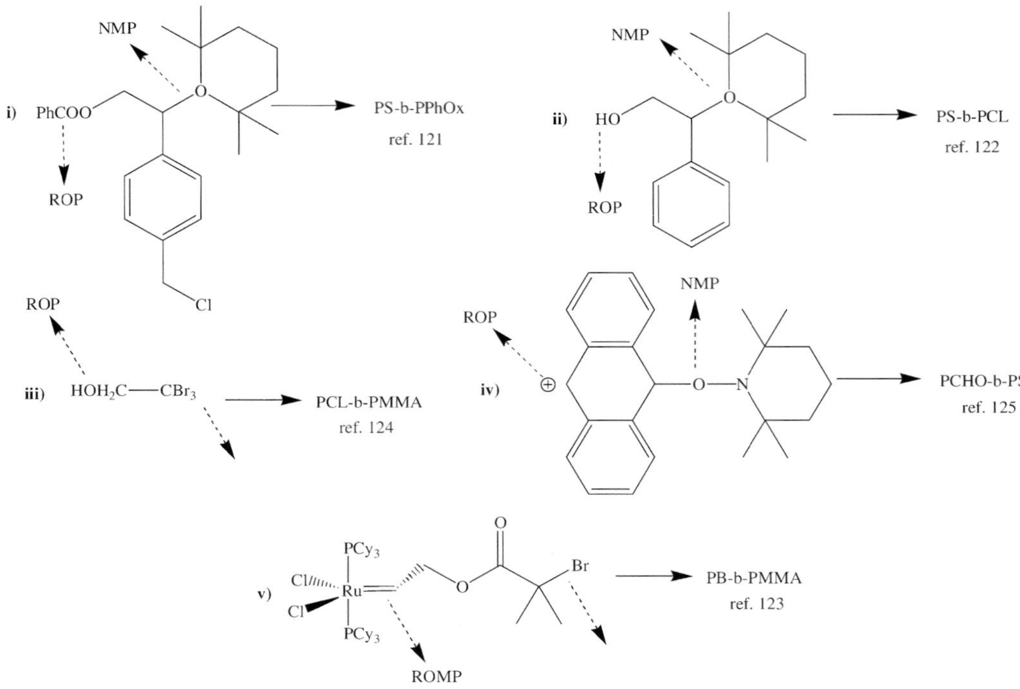

Fig. 2.9 Block copolymers from "dual initiators".

2.11
Chemical Modification of Pre-formed Block Copolymers

Pre-formed block copolymers obtained by any of the methods discussed above (see Fig. 2.2) can be chemically modified and form novel copolymeric materials. The molecular characteristics of the block copolymers obtained (chain length, composition and architecture) is fixed by the initial method used to synthesize the pre-formed block copolymer. The chemical modification is generally supposed to be sufficiently mild to avoid chain degradation, cross-linking. A representative example of such an approach is the catalytic hydrogenation of polydienes, to form PE-*alt*-PP if the PI block precursor contains a majority of 1,4 units [126]. Other examples including hydrolysis of *tert*-butyl groups of P*t*BA blocks, quaternization of PVP, hydrosilylation, epoxidation, fluorination, introduction of photoactive groups, mesogenic moieties, etc. can be found in [4].

2.12
Methods for the Synthesis of Block Copolymers with a Complex Architecture

Many of the methods discussed above have served as synthetic tools combined with "branching reactions" in order to design block copolymers with a branched architecture. It is beyond the scope of this chapter to review all the synthetic possibilities to access branched macromolecules consisting of block copolymers. Introduction of branching points throughout the structure is aimed at decreasing the entropy of the branched copolymers, which leads to the formation of shape-persistent nano-objects. In this respect, branched copolymeric materials are expected to possess distinct properties as compared with their linear counterparts.

Figure 2.10 shows the most common block copolymers with a complex architecture that can be obtained by combining CLP with selective methods of introduction of branching points. Many of these branched systems have been the subject of review articles or book chapters describing their synthetic methodologies and their relevant properties. For instance, Hadjichristidis et al. reviewed the field of (co)polymers with a complex architecture (including star-block copolymers, miktoarm stars, graft copolymers, etc.) obtained by living anionic polymerization [8]. An earlier overview by the same group on miktoarm stars is also available [127]. Charleux and Faust reviewed the field of block copolymers with a star-like architecture derived by living cationic polymerization [128]. General synthetic methods of dendrimers consisting of true polymeric chains between the branching point, including dendrigrafts and dendrimer-like polymers, was discussed by Teestra and Gauthier [129]. Sections devoted to the synthesis of block copolymers with a branched architecture involving C/LRP methodologies are also available [7, 59–63, 65, 66].

A typical branching reaction, for instance, consists of using multifunctional initiators; when combined with sequential CLP, this leads to star-block copolymers following a divergent ("core-first") approach. In this case, the branching point corresponds to the core of the star. For this reason, star-shaped polymers are often viewed as models of branched polymer structures, structurally well-defined stars being useful in providing acute insight into how branching affects the overall properties of polymers in solution or in the melt. Convergent ("arm-first") approaches are also possible to access star-like copolymers. For instance, coupling "living" linear copolymeric chains onto a multifunctional comonomer (a divinylic comonomer is commonly employed) leads to star-block copolymers composed of a microgel-like core. Alternatively, one can use a precursor bearing multiple complementary functions to that of the living chains so as to obtain arm-first star polymers. If block copolymers are deactivated, star-block copolymers are obtained.

Successive deactivation of different homopolymers onto the multifunctional precursor is a method of choice for miktoarm star synthesis [127]. In this case, chlorosilane derivatives are often employed.

For graft copolymer synthesis, one can apply different methods among which are the "grafting onto", the "grafting from" and the "macromonomer" methods. In the first case, "living" linear chains obtained by CLP are deactivated onto a pre-formed

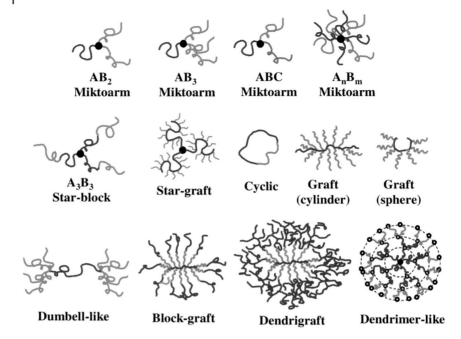

Fig. 2.10 Common representative block copolymers with a branched architecture.

macromolecular backbone possessing functional side groups. It is worth mentioning that reiteration of the same process, that is, "grafting the grafts", permits dendrigrafts to be synthesized (Fig. 2.10) [129]. In the "grafting from" approach, a macromolecular precursor serves as a multifunctional initiator for CLP, from which the side chains are grown. The "macromonomer", sometimes called "grafting through", requires the use of a preformed polymer containing a group at one end of the chain that can be polymerized. Grafting occurs through homopolymerization of this macromonomer or by copolymerizing it with a "regular" monomer. Depending on the chain length of the backbone, graft copolymers can adopt either a star-like or a cylinder-like conformation in solution (Fig. 2.10).

As for dendrimer-like (co)polymers, they are synthesized in most cases following a divergent approach by repeating two elementary steps, i. e., CLP from multifunctional precursors and chain-end functionalization, in order to create at least two initiating (geminal) sites per arm from which upper generations can be grown [129].

Other complex architectures can be generated if one combines several branching reactions and CLP. For instance, "star-graft copolymers" (Fig. 2.10) are obtained by first using a multifunctional initiator in a CLP, each branch of the star polymer being a further graft copolymer when growing additional chains by the "grafting from" approach.

Table 3.1: (continued)

Hydrophobic block (B)	Hydrophilic block (A)	Vesicle-forming conditions
(phenyl-diethyl-silane polymer)	PEO	BABAB; water/THF; $f_A = 0.5$ [19]
polystyrene	poly(4-vinylpyridine)	AB, ABA; dioxane/water; $f_A = 0.2$ [20]
polystyrene	PEO	AB; DMF/water; $f_A = 0.03$ [21]
poly(methyl-methacrylate-co-)	PEO	AB; water; $f_A = 0.3$ [22]
poly(ε-caprolactone)	PEO	AB; water-blended with PBD-PEO; $f_A = 0.4$ [22]
poly(butylene oxide)	PEO	AB ABA; water; $f_A = 0.2$–0.4 [22, 24]
polybutadiene	poly(glutamic acid)	AB; water/NaOH; $f_A = 0.9$ [25]; $f_A = 0.3$–0.8 [26]

Table 3.2: Chemistry of block copolymers able to assemble into vesicular structures in non-aqueous media. The nature of the A and B blocks is shown with information on the architecture and the experimental conditions.

Block (A)	Block (B)	Vesicle-forming conditions
(4-phenylquinoline polymer)	(polystyrene)	AB; 1:1–1:4 (TFA:DCM, 25 °C) Giant [27]
(functionalized polystyrene with pyridine/amide groups, n=10)	(functionalized polystyrene with thymine groups, n=10)	*Complementary homopolymers A + B; CHCl$_3$ 1:1 mixture Giant [28]
(norbornene copolymer with CO$_2$CH$_3$ and diamidopyridine side groups)	(norbornene copolymer with CO$_2$CH$_3$ and thymine/amide side groups)	*Complementary homopolymers A + B; CHCl$_3$ 1:1 mixture 5×10^{-3} mol L^{-1} [29]
(polystyrene)	(poly(cinnamoyl ethyl acrylate))	AB; THF/Cyclopentane (cyclopentane > 85 %) [30]
(polyisoprene/polybutadiene)	(poly(cinnamoyl ethyl acrylate))	AB; Hexane:THF 9:1; [30]

Table 3.2: (continued)

Block (A)	Block (B)	Vesicle-forming conditions
[structure]	[structure]	AB, cyclic AB; phthalates, n-alkanes [31, 32]

The chemistry and architecture of copolymers play a fundamental role in the self-assembly pathways, stability and functionality of polymer vesicles. The coexistence of different aggregates and their morphology depend strongly on the molar mass distribution of the block copolymers. Copolymer asymmetry can be used to direct thermodynamically driven assemblies in polymeric vesicles, such as with ABC triblock copolymers [33] or with ABCA tetrablocks [34, 35]. At the same time, owing to the preferential disposition of polymer chains in the inner or outer membrane layers, high polydispersity may contribute to the thermodynamic stabilization of polymersomes [36, 37].

Addition of recognition sites or fluorescent labels on the surface of polymersomes can be facilitated by the introduction of reactive end-groups and subsequent post-polymerization reactions. As the assembly behavior of the copolymers could be influenced by the change in f following functionalization, it is convenient to blend the functionalized and the native copolymers at ratios that do not affect vesicle formation.

The chemistry of hydrophilic blocks will dictate the properties of the vesicles in water, such as adhesiveness and, in biological environments, the ability to resist protein adsorption, which is a key property for "stealth" drug delivery systems.

As mentioned previously, amorphous and rubbery hydrophobic blocks will confer to the membranes the liquid-like character typical of lipid bilayers. Moreover, the apolar region of the membrane will dictate the permeability of solutes and water. Lipid membranes have characteristic permeabilities to different solutes and the presence or absence of unsaturation in the hydrocarbon chains or of cholesterol in the bilayer can contribute to small changes in permeability. In copolymers the presence of heteroatoms in the hydrophobic block, such as oxygen [in poly(butylenes) and poly(propylene oxide)], silicon (in silicones) or sulfur [in poly(propylene sulfide)] can produce similar effects. In hydrocarbon block copolymer vesicles, permeability dramatically drops when compared with low MW analogue lipid membranes due to the increased membrane thickness, but the presence of heteroatoms could introduce selectivity for certain molecules, such as glucose [38].

3.3
Block Copolymer Vesicle Formation in Water

As introduced above, amorphous hydrophobic blocks with a low glass transition temperature possess the necessary mobility to allow the system to rearrange as the selective solvent for the hydrophilic block (water) is added. The degree of crystallinity of the hydrophilic block can obviously also play a role in the kinetics of vesicle formation. Several studies have reported the effect of kinetics on the morphology and size of aggregates. Thermodynamic considerations are not always sufficient to predict phase behavior of amphiphilic block copolymers in a binary system, and the higher the molecular weights the stronger the kinetic trapping of the structures will be.

Polymer vesicles can be prepared according to the following basic methods, which were developed for liposomes:

- Film hydration: A thin film of the copolymer, deposited in a round-bottomed flask by solvent evaporation, is exposed to excess water and allowed to swell at controlled temperature and agitation speed.
- Co-solvent method: Concentrated copolymer solution in a water-miscible solvent is added dropwise to water under vigorous stirring. Removal of organic solvent can be performed by dialysis or freeze-drying.
- Detergent method: The polymer is dispersed in water with the help of a surfactant with high-CMC. Under stirring, the detergent is removed by dialysis or by using appropriate resins.

Among the hydrophobes that have sufficiently low T_g to form vesicular aggregates by the film rehydration method, one can cite: poly(butadiene) and its hydrogenation product poly(ethyl ethylene), poly(dimethyl siloxane) (PDMS), poly(propylene sulfide) (PPS) (Fig. 3.2), poly(butylene oxide) (PBO) and poly(propylene oxide) (PPO). As a general rule, the higher the MW of the macroamphiphiles the more difficult is the formation with this method at room temperature and hydration has to be coupled to additional treatments such as freeze-thawing, heating and sonication. When block copolymers have low MWs, spontaneous formation of vesicles can occur in a short time and without any additional energetic input. The process can be imaged by optical or confocal fluorescence microscopy (Fig. 3.3) and kinetics information derived from the evolution of sprouting vesicles [39].

The size control of vesicles formed in water can be easily achieved by extrusion through filters with pores of a given size. Polycarbonate filters are often used due to the cylindrical shape of the pores that give the best results in terms of polydispersity.

Although the T_g of the hydrophobic block is vital to the spontaneous vesicle formation in water, the nature and MW of the hydrophile plays an important role in determining the morphology of aggregates. Non-ergodic behavior has been observed for PB-PEO copolymers at sufficiently high MWs [40] reflecting the slower chain dynamics and thus the kinetic trapping of microphases.

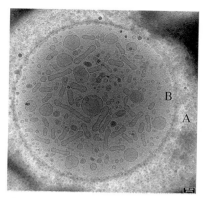

Fig. 3.2 Cryo-TEM micrograph of a 1%-wt. PEG_{16}-PPS_{50}-PEG_{16} dispersion in water after extrusion through a 100-nm polycarbonate membrane. The holey carbon film (marked with A) is where the vitrified solution is embedded (marked with B). Many tubular and spherical vesicles can be identified (Müller and Napoli, unpublished results).

More complex architectures such as ABA [15], ABC [33], ABCA [34] and ABABA [19] were shown to form vesicles in aqueous media.

Preparation of polymer vesicles from glassy or crystalline block copolymers requires the presence of a cosolvent to lower the T_g below room temperature. Pioneering work with PS-PAA and PS-PEO [13, 21] copolymers showed vesicular structures of colloids formed by adding water to a DMF or dioxane polymer solution.

Nolte and coworkers reported vesicle formation from poly(methylphenylsilane)-PEO pentablock copolymers in THF by dropwise addition of water [19]. The same group reported the effect of solvent composition [41] on the copolymer aggregation behavior and thus on the conformation of the PMPS backbone resulting in tunable optoelectronic properties. Amphiphilic diblock copolymers derived from isocyanopetides and styrene generated giant vesicles up to 100 µm in diameter by the electroformation method [42]. In this study polystyrene$_{40}$-b-poly(L-isocyano-alanine(2-thiophen-3-ethyl)amid) (PS-PIAT) was used as the diblock copolymer and THF as the cosolvent.

In other reports solvents immiscible with water were used to form polymeric vesicles. Feijen and coworkers reported vesicle formation for diblock copolymers of PEO and polyesters or poly(carbonates) with both water-miscible and immiscible solvents [43]. Organic solvent removal remains vital for further characterization of and experiments with vesicles formed in water-immiscible solvents and in some instances this is very difficult to achieve.

The co-solvent method, also known as the solvent injection method, not only permits vesicle formation for glassy or crystalline block copolymers but also speeds up the process and opens the way to the large-scale production of polymer vesicles. Förster and coworkers reported the use of ink-jet printing to produce polymer vesicles with narrow size distributions. The average size of vesicles produced depends

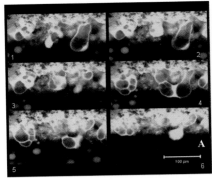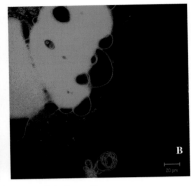

Fig. 3.3 Confocal laser scanning micrograph taken at an interface of water and (A) a film of $EO_{16}BO_{22}$ (pre-stained with Rhodamine B octadecyl ester perchlorate) [24], (B) a film of PEG_{16}-PPS_{25} (pre-stained with hydrophobic BODIPY) showing the evolution from swollen lamellar structure to vesicles. (A: with permission from ACS Publications Division; B: Napoli and Rigler, unpublished results).

on the initial polymer concentration in the organic phase and the type of nozzle used [44].

3.4
Block Copolymer Vesicle Formation in Organic Solvents

Block copolymers can also assemble into vesicular objects when dispersed in an organic solvent that is selective for one of the blocks. Such properties, now universally accepted, were observed more than 30 years ago in blends of a diblock copolymer (A_x-B_y) with its corresponding homopolymer (A_n or B_n).

The compartmentalization of vesicles in organic solvents can be exploited for non-biological applications and expands the self-assembly principles of block copolymers. In Table 3.2 examples of block copolymers that form vesicles in non-aqueous media are shown.

One of the first examples of polymeric vesicles formed in a non-aqueous solvent was reported by Jenekhe and Chen [27]. Poly(phenylquinoline)-block-polystyrene rod-coil diblock copolymers were observed to aggregate into spherical large particles when dispersed in a mixture of a selective solvent for PPQ (trifluoroacetic acid) and various amounts of dichloromethane or toluene. Optical and fluorescence microscopy suggested the possible vesicular morphology and the ability to encapsulate fullerenes confirmed the presence of a cavity.

Rotello and coworkers introduced a novel route to vesicle formation in dichloromethane by mixing two random copolymers bearing a host and a guest group able to complex via hydrogen bonding, respectively (Fig. 3.4). Complementarity between polymers was achieved using a diaminopyridine-thymine three-point hydrogen-bonding interaction. Vesicular nature of the aggregates was proven by con-

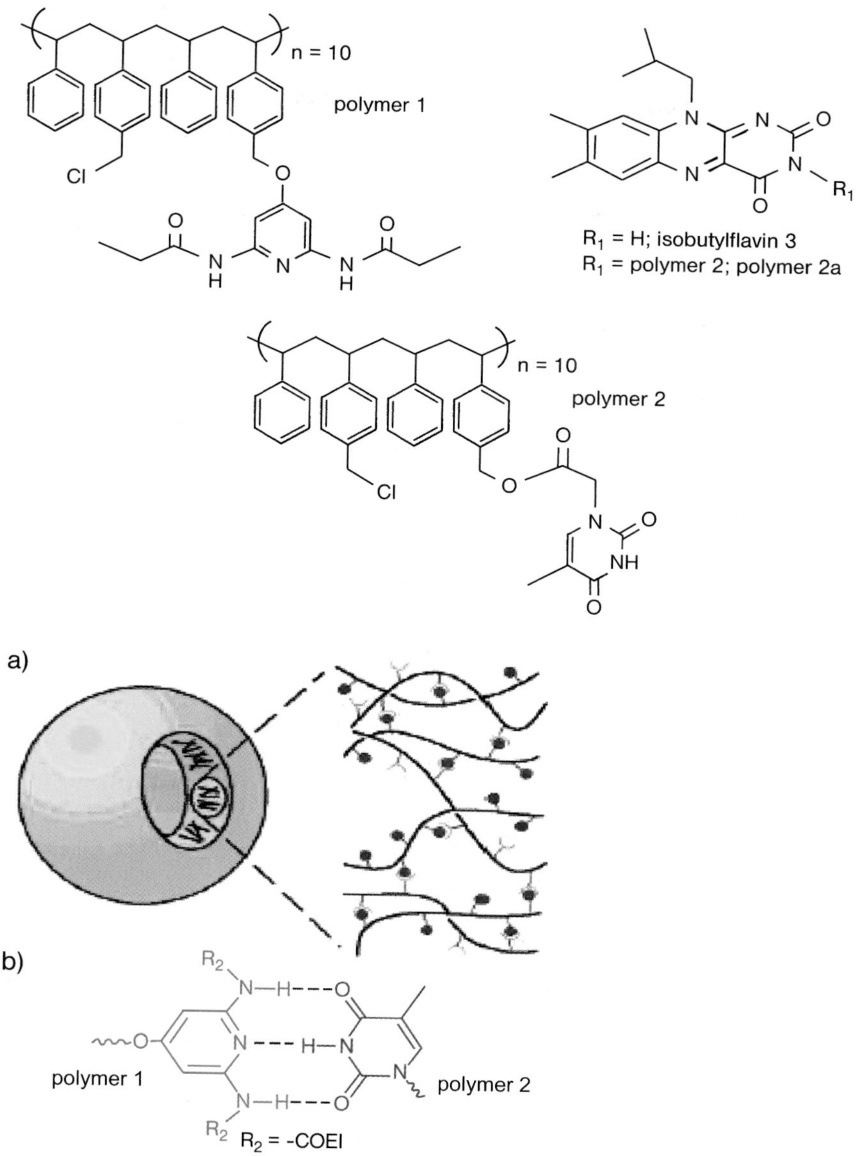

3.4 Block Copolymer Vesicle Formation in Organic Solvents

R_1 = H; isobutylflavin 3
R_1 = polymer 2; polymer 2a

Fig. 3.4 Schematic representation of diaminopyridine and thymine based polymers and their assembly into vesicular objects (a), and (b) the corresponding recognition dyad [28] (with permission from ACS Publications Division).

focal scanning laser microscopy (CSLM) using a fluorescent flavin-functionalized polymer. To extend the recognition induced polymersome (RIP) assembly to different backbones, the same group reported [29] formation of vesicles in $CHCl_3$ from polynorbornene-based copolymers featuring complementary side 2,6-diacyl-diaminopyridine and uracil groups. Although a thorough characterization of membrane properties and morphology has not been reported, this approach to colloidal objects in non-polar solvents deserves attention for potential application in confined reaction systems.

Schlaad and coworkers elaborated a vesicle-assembling strategy from copolymers having blocks with opposite charges [45]. Quaternized poly(4-vinyl pyridine) (P4VP)-PS (positively charged) and PBD-poly(t-Bu-methacrylate) (negatively charged) assembled into "polyion complex" vesicles in THF, a solvent for PS and PBD but not for the charged blocks.

Another interesting report of self-assembly mediated by complexation of copolymers in a non-polar solvent has been reported by Jiang and coworkers [46]. A diblock copolymer of PS and P4VP was reported to self-assemble into vesicular structures when, prior to dispersion in chloroform (a solvent for PS and P4VP), P4VP is allowed to be complexed by formic acid, which alters the block solubility in $CHCl_3$. The vesicular structure of the objects seems, however, to be questionable as the same group later reported the formation of micellar aggregates using the same principle and block copolymer [47].

Self-assembly of di- and tri-block copolymers of rigid poly(3-hexylthiophene) and of poly(butyl acrylate) was investigated by de Cuendias et al. [48] in a selective solvent for the flexible blocks. The opto-electronic properties conferred by the packing of the thiophene blocks were studied by fluorescence spectroscopy.

Polyisoprene (PI)-block-poly(2-cinnamoylethyl methacrylate) (PCEMA) and polystyrene-block-poly(2-cinnamoylethyl methacrylate) can form vesicular aggregates when exposed to a mixture of hexane (good solvent for PI) and THF or cyclopentane and THF, respectively [30]. The phase behavior of the PCEMA-based copolymers is determined by the relative length of the blocks and by the volume fraction of the non-solvent for PCEMA, as occurs for tertiary systems of low MW surfactants.

Putaux et al. reported the vesicle-forming properties of linear and cyclic PS-PI diblock copolymers in n-heptane or n-decane, which are selective solvents for PI [32]. Interestingly, the concentration-dependent transition from micelles to vesicles could be observed with cryo-TEM as it occurred at relatively high concentrations. The cyclization increased the heterogeneity of the aggregates but did not impair the ability to form vesicles. In another recent paper, Kesselman et al. reported on vesicles from PS-PI copolymer in phthalate solvents [31]. In this case, "expert" cryo-fixation of the solution allowed direct visualization of the vesicular morphology and the effect of solvent quality for PS on the morphology of the aggregates.

3.5
Properties of Polymer Vesicles

3.5.1
Morphology and Size of Polymer Vesicles

The shape, lamellarity and size of polymer vesicles are key to defining their application field.

While optical microscopy is suitable to study properties of giant vesicles, submicron sized vesicles require electron microscopy to fully appreciate their morphological complexity.

Yu and Eisenberg reported a transmission electron microscopy (TEM) investigation of different PS-PEO copolymer bilayer morphologies (Fig. 3.5) induced by a change in ionic strength, polymer concentration or by the preparation conditions (DMF–water content) [21]. First the copolymer was diluted in DMF or water–DMF mixtures of varying water content and then more water was added as a precipitant for the hydrophobic block. Self-assembly took place at some critical water content, which depended on the relative and absolute block lengths and the nature of the polymer. The addition of water was continued until a predetermined point in the phase diagram had been reached, depending on the desired morphology. The system was then quenched by the addition of a large excess of water to kinetically

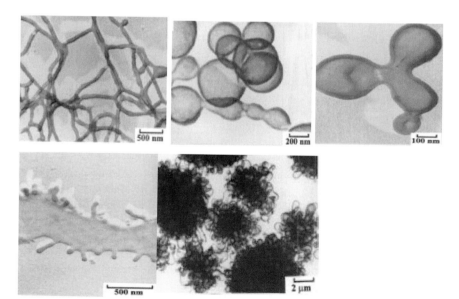

Fig. 3.5 Different morphologies from PS-PAA block copolymers at various THF/water ratios. From the upper left corner clockwise: worms, vesicles, "starfish" vesicles, large compound vesicles (LCV) and lamellae (with permission from ACS Publications Division) [49].

freeze the morphology, and the colloidal solution was dialyzed against water for the removal of the organic solvent.

Although the complete removal of the organic solvent is reported to occur via dialysis, non-detectable traces might remain in the hydrophobic part of the vesicle wall and maintain membrane fluidity. Shapes of vesicular aggregates ranged from tubular to spherical and from more exotic large compound (LCV) and starfish vesicles to "simpler" extended lamellae.

They also showed that not only the water content regulates the morphology but also the addition of ions (µM concentrations of $CaCl_2$ or HCl and mM concentrations of NaCl) induced a change of morphology [50]. They suggested that these changes were induced by decreased steric and electrostatic repulsion. The effect of organic solvent on the morphology has been studied and exploited to achieve [49b] size-control of crew-cut aggregates. A change in the volume ratio THF/dioxane/water in which PS-PAA copolymers are dispersed dramatically affects their size for dioxane-rich mixtures and less so for THF-rich solutions.

Won et al. [49c] gave a good example of the power of Cryo-TEM in high-resolution imaging of colloidal polymer aggregates. With this technique it is possible to rapidly freeze and image the aqueous suspension allowing direct visualization of the aggregates without the risk of artifacts that could occur in conventional TEM from drying and/or staining. Vesicular and micellar aggregates in dispersions of PEO-PBD, PEO-PEE and PEO-PEE-PEO copolymers in water were easily recognized. Coexistence of worm-like micelles with vesicles and spherical micelles has been observed on opposite sides of the phase diagram. Such behavior cannot be explained just by the intrinsic polydispersity of synthetic macromolecules but kinetic trapping of structures occurring during polymer hydration also has to be considered.

Cryo-TEM has been employed to give insight into the oxidative vesicle-to-micelle transition in PEG-PPS copolymer suspensions [16]. The hypothesis of the oxidation-induced increase in the curvature of the aggregates was confirmed by imaging the coexistence of worms, spheres and vesicles in a cryo-fixed reaction solution.

The local membrane shape of PEO_{20}-PB_{32} giant vesicles has been investigated by Haluska et al. [51] revealing a high-genus morphology that was persistent with time. In this case the spontaneous formation of large pores in the membrane could be explained in terms of bending elastic energy. These workers actually explain the surprisingly high genus of the polymeric membranes as being due to the different conditions (as a result of sample preparation) the PEG chains experience on the two sides of the bilayer.

A rapid temperature drop induced beading in tubular polymersomes from the same copolymer, as observed and quantitatively analyzed using phase-contrast microscopy [52]. These experimental results were later confirmed by numerical calculations based on the elastic energy model [53].

Although there is sufficient experimental data to elucidate the morphology and behavior of polymer vesicles, there are an increasing number of theoretical studies addressing block copolymer vesicle formation using self-consistent field theory [54–56] or coarse-grain molecular dynamics [57–59].

Light scattering has been extensively used with polymer vesicle suspensions to determine vesicle size distribution, their disruption by dilution [15] or detergent exposure [60] and morphology evolution with thermal treatment [37]. The average hydrodynamic radius R_h determined by dynamic measurements (DLS) gives an indication of the size of colloidal objects. When this is coupled with static light scattering (SLS) measurements, which give information on the radius of gyration R_g of aggregates, the morphology of vesicles can be extrapolated. In fact, the ratio R_g/R_h approaches unity for spherical vesicles with molecularly thin membranes [15].

Nuclear magnetic resonance techniques can also be used successfully to determine membrane properties such as permeability and hydration limit (water content along the polymer chains forming the membrane) in addition to diffusion coefficients of the vesicles [61–63].

3.5.2
Membrane Properties

Membrane properties of polymersomes reflect the chemical composition, the block length ratio and the MW of block copolymers from which they are formed. The extent of segregation between different blocks and the interaction between single blocks and liquids can be estimated by the Flory effective interaction parameter, χ. It has to be emphasized that traces of solvent in the inner region of the membrane can dramatically affect the permeability and mechanical properties of the vesicles. Conformation of the solvated membrane corona can have effects on vesicle adhesion and fusion. In this section we will discuss the properties of various copolymer membranes and the techniques employed to capture their key features.

3.5.2.1 Polymer Membrane Thickness

Cryo-TEM has been used to obtain high-resolution images of membranes and thus to measure their thickness. Increased membrane thicknesses (d), up to $\approx$20 nm, were measured with increasing copolymer molecular weight. A scaling factor of $\approx$0.5 was determined by fitting the measured values versus the molar mass of the hydrophobic blocks (M_h) [64]. It is interesting to note that, in general, block copolymers are expected to be in the strong segregation limit (SSL). In such a configuration, determined by a balance of interfacial tension and chain entropy, the scaling of $d \approx (M_h)^\alpha$ yields $\alpha = 2/3$. This scaling factor has been discussed in more recent work by Battaglia and Ryan [24] for block copolymers of butylene oxide and ethylene oxide in which different copolymer architectures are able to form vesicles in water. TEM and small-angle X-ray scattering were combined to measure the membrane thickness of polymersomes, and a scaling factor $\alpha = 2/3$ with M_h or N_h (degree of polymerization of the hydrophobic block) was found, in agreement with the SSL regime observed previously for bulk microphases (Fig. 3.6).

The degree of polymerization (DP) affects membrane thickness and at the same time has a large influence on lateral mobility of polymer chains within the mem-

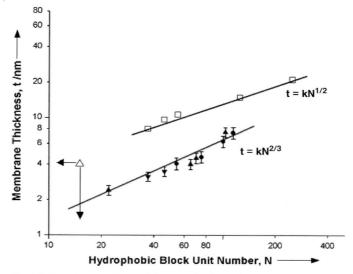

Fig. 3.6 Experimental data and fitting of the dependence of vesicle membrane thickness (t in nm) from the degree of polymerization (expressed as block unit number, N) of the hydrophobic block (with permission from ACS Publications Division) [24].

brane. Entanglements are responsible for reduced mobility, as investigated by Lee et al. [65] using fluorescence recovery after photobleaching (FRAP), for copolymers with sufficiently high MW. In such a situation the mobility of chains can no longer be described by the Rouse model (diffusion-like), instead reptation-like motion is the predominant mode of chain lateral rearrangement.

Similar considerations explain the resistance of polymer vesicles to detergents. Liposomes would normally be destroyed when exposed to a sodium dodecyl sulfate (SDS), or other strong detergent, via formation of mixed micelles that require flip-flop lipid transport. In polymersomes with thicker membranes this process is hindered by the limited mobility of high MW polymer chains, resulting in better resistance to dissolution. Combination of experimental data and theoretical modeling suggests that polymer vesicle dissolution is mediated by the diffusion of detergent molecules across the membrane [60].

3.5.2.2 Mechanical Properties of Polymer Vesicles

As introduced at the beginning of the chapter, polymer vesicle design and characterization profit from the large amount of information available for lipid systems and natural membranes. The first attempt to directly compare the elastic properties of copolymer and lipid membranes by the micropipette aspiration method (Fig. 3.7) was reported by Discher et al. [5]. In this paper a PEO-PEE diblock copolymer (M_n = 3900 g mol^{-1}) membrane appeared, under micromanipulation, to be qualitatively similar to a fluid lipid bilayer. However, the bending modulus (K_b)

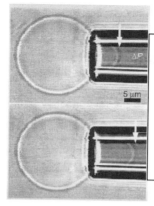

$$\tau = \frac{\Delta P \cdot R_p}{2(1 - R_p/R_s)}$$

Rp, Rs are the respective radii of the micropipette and the outer spherical contour

$$K_b \approx k_B T \ln(\tau)/(8\pi\alpha) + \text{constant}$$

Fig. 3.7 A polymersome aspired into the micropipette maintains its spherical shape and a protruded portion enters the pipette tip (with permission from The American Association for the Advancement of Science) [5].

of the polymersome membrane, determined by plotting the tension (τ) versus the change in vesicle-projected area ($\Delta A/A_0 = \alpha$), was found to be almost an order of magnitude higher. Moreover, the giant polymersomes could sustain a greater strain before rupture.

Dimova et al. [66] also investigated the viscosity properties of PEO-PB polymersomes using the falling-ball method and optical-trap dynamometry. With the first method the surface viscosity is measured by monitoring a latex bead ≈ 5 μm) incorporated into the membrane of the giant vesicle ≈ 50 μm) falling along the membrane by gravity. The latter makes use of the radiation pressure force of an optical trap to attract the latex particle from a defined start position ≈ 1 particle radius away). When compared with lipid (SOPC and DMPC) vesicles, the interlayer viscous drag of polymersomes was found to be about 500 times larger.

Bermudez et al. [67] studied the effect of copolymer MW (membrane thickness d) on the bending rigidity (K_c) of polymersome membranes and found that the scaling between K_c and d was nearly quadratic, in agreement with established theories for bilayers. Surprisingly, vesicles from PEO_{26}-PBD_{46} gave lower K_c values when compared with PEO_{40}-PEE_{37} although having a thicker hydrophobic membrane core. These workers suggested that the PEO chains contribute to the overall membrane thickness. Under the SSL regime, PEO chains would probably adopt a brush conformation.

Polarity-sensitive hydrophobic fluorescent probes, such as Laurdan, can be used to measure the bending contribution to hydration in polymeric membranes. The dye localizes in the membrane core, and when water molecules access the membrane, thereby increasing the local polarity, the emission band of Laurdan shifts to the red. The technique was employed [68] to evaluate the effect of temperature on water permeability of DMPC and SOPC membranes, which showed a marked shift at their liquid crystalline transition temperatures (T_{LC}, respectively, 10 and 24 °C)

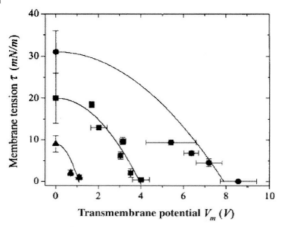

Fig. 3.8 Membrane tension τ versus transmembrane potential V_m. Rupture results are plotted for SOPC (triangles), EO_{40}-EE_{37} (squares) and EO_{51}-BD_{126} (circles) vesicles (with permission from APS Associate Publisher) [69].

compared with PEO-PEE polymersomes that had a minor red shift of Laurdan between 5 and 60 °C, well above their T_g. An increase in temperature results in thermal expansion of membranes and probe hydration increases concomitantly, a phenomenon observable across membranes with different permeabilities.

Resistance of polymeric membranes to electromechanical stimuli has been investigated by Aranda-Espinoza et al. [69] by coupling the micropipette aspiration method with electroporation. The molecular weight dependence of membrane thickness is also reflected by the critical potential V_c required for vesicle rupture at each applied tension (Fig. 3.8).

In similar experiments the evolution of electroformed pores in polymersomes with increasing membrane thickness was measured by phase contrast microscopy (Fig. 3.9) [70]. Vesicles made from low MW PEO-PEE and PEO-PBD showed unstable pores that kept growing and eventually disrupted the vesicle. Surprisingly, the membrane fragments reassembled into smaller vesicles, a process that could not be observed in the timescale of the experiment in higher MW copolymer vesicles. Owing to chain entanglements and thus increased membrane viscosity, the highest MW PEO-PBD (M_n = 20 kDa) vesicles are characterized by the formation of smaller pores that do not evolve to vesicle rupture.

Evaluation of the resistance to an electric potential can be measured in black lipid membranes using an electrochemical cell separated by a septum. Nardin et al. used this setup to study the electroporation mechanism in a free-standing PMOXA-PDMS-PMOXA polymer membrane and to compare its rupture potential before and after cross-linking by terminal methacrylate groups [71].

An insight into the various contributions to mechanical response of copolymer (PB_{407}-PEO_{286}) vesicles by the hydrophilic corona and the hydrophobic region has been given by atomic force microscopy investigations [72]. When approaching vesi-

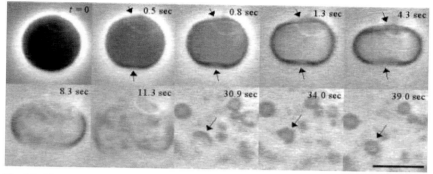

Fig. 3.9 Time sequences of an OE7 ($d = 8$ nm) vesicle following electroporation. Arrows in the top panels emphasize growing pores while those in the bottom panels point at a resealing membrane fragment. The scale bar is 10 μm (with permission from EDP Sciences) [70].

cles in non-contact mode, the tip experienced a first repulsion force given by the PEO corona at ≈60 nm from the substrate and, at an applied force of ≈2 nN and a distance of ≈20 nm from the substrate a "jump-in" event occurred. The latter is probably related to the tip insertion into the polymer bilayer and the subsequent linear spring-like response suggests the compression of the hydrophobic core.

Dependence of mechanical properties of polymersomes with the molar mass of the constituent copolymers has been addressed by Bermudez et al. [64] who found that the area elastic modulus K_a is not dependent on the MW but is only determined by the chemical composition of the interface. At the same time the maximal areal strain at rupture depends on the M_h and scales with a power-law exponent of 0.6 ($\Delta A/A_0 \approx M_h^{0.6}$).

3.5.2.3 Adhesion of Polymer Vesicles

Adhesion of particles to biologically relevant surfaces, such as the cell surface, is a fundamental process that mediates internalization or fusion with the membrane. Some potential applications of polymersomes, for example drug delivery, will benefit from the understanding of their adhesion behavior. On the other hand, the structure of the brushy corona around polymer vesicles contributes with steric repulsive forces to minimize contact and prevent adhesion between polymer membranes. Therefore, to investigate polymersome response to adhesion, copolymers modified with chemical moieties complementary to specific receptors have been used in fundamental studies.

Polymer vesicles have been used to measure adhesion forces driven by receptor–ligand interactions, such as biotin–streptavidin. The group of Hammer addressed [73] this interaction by using micropipette manipulation, where biotin–lysin coated polymersomes — obtained by mixing vesicle-forming PEO-PBD native and biotynilated copolymers — were exposed to avidin-coated beads. The effect of the functionalized polymer length relative to the native copolymer and of the biotin density was evaluated by membrane tension measurements and imaged by op-

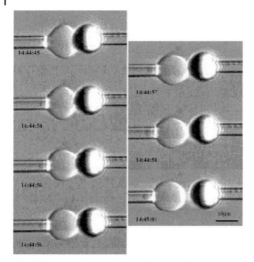

Fig. 3.10 Micropipette manipulation of a polymersome (left) and a latex bead, respectively, functionalized with biotin and avidin (with permission from ACS Publications Division) [73].

tical microscopy. After contact and adhesion, a pressure ΔP was applied to the micropipette holding the vesicle and the critical tension to peel off was measured upon contact loss (Fig. 3.10).

The results showed that adhesion strength increases when the functionalized copolymer is longer than the surrounding native copolymer, meaning that the ligand can fully interact with the receptor. At the same time, maximal adhesion is reached at a plateau of biotin surface density, 55 mol-% for the vesicles with longer functionalized chains and 10 mol-% when the native and biotinylated chains have the same MW.

Further investigations from the same group [74] tried to better elucidate the mechanism and proposed a model to help design functionalized polymersomes with predetermined ligand-induced adhesiveness. The strength of the receptor–ligand bond represents only one side of the problem, and engineering of ligands display might yield the optimal properties for the desired application.

3.5.2.4 Fusion and Fission of Polymer Vesicles

In living systems, fusion and fission of lipid vesicles are processes fundamental to many important functions of the cell, from neurotransmission to active defense from pathogens. Natural lipids are designed to permit, under certain conditions (e. g., pH, ionic strength), the fusion or the fission of vesicles formed by their assembly. On the other hand, polymer vesicles have been designed to be tougher than liposomes and their fusion/fission behavior has never been included in the design stage [75].

Extending the concepts developed for liposome fusion, as a first step to fusion we can imagine the intimate contact between the two fusing vesicles. In general, for polymer vesicles there is a strong steric hindrance to contact and fusion due to the solvated polymer brushes on the outer wall. Despite this barrier, fusion of polymersomes has been reported to occur given the right chemistry and conditions. Luo and Eisenberg proposed a fusion/fission mechanism for PS-PAA diblocks in different dioxane–water mixtures to explain the appearance — observed through TEM — of intermediate morphologies where two vesicles closely interact or share a membrane wall [36]. The same group [76] reported the effect of solvent composition on the fusion rate of vesicles from the same copolymers as measured by turbidimetry.

Real-time visualization of polymer vesicles fusion and fission has been reported for hyperbranched poly(3-ethyl-3-oxetanemethanol) core and many oligo(ethylene oxide) arms (HBPO-star-PEO) [77–79]. Although it is difficult to imagine a bilayer-like packing of such dendritic polymers, these workers support the fusion process using the proximity of the liposomes model [80].

3.6
Functional Polymer Vesicles

The properties of the polymer vesicles can be altered by designing degradable, cross-linkable amphiphilic copolymers or by blending inert with reactive copolymers. With a bottom-up approach one can tailor properties of the polymersomes at the molecular engineering stage.

Nardin and coworkers used methacrylate end-functionalized PMOXA-PDMS-PMOXA copolymers to stabilize membranes and vesicles formed in water [15, 71]. Cross-linked vesicles are able to withstand dehydration–rehydration cycles and exposure to organic solvents.

Discher et al. used the unsaturation in the hydrophobic poly(butadiene) block to cross-link PEO-PBD polymersomes and observed extreme resistance to drying, organic solvent exposure and osmotic stress [81]. The use of micropipette manipulation gave an insight into the dramatically improved mechanical properties and resistance to rupture when compared with non-cross-linked vesicles.

Degradability of block copolymers was introduced to destabilize vesicles in response to a chemical or physical stimulus. Controlled release via vesicle disruption is very attractive in biomedical applications where stability of the carrier has to be coupled to prompt content release in the target tissue or to sustained release with time. For this reason many research groups started to investigate vesicle-forming properties of block copolymers that can be degraded upon a change in pH by hydrolysis [22, 43] or by exposure to oxidants [16, 38].

By blending PEO-PBD with PEO-poly(lactic acid) (PLA), Ahmed et al. [82] investigated stabilized polymersomes that can become porous by selective degradation of the PLA block. These workers studied the membrane cross-linking at the air/water interface in a Langmuir trough setup and deposited copolymer monolayers by AFM.

The oxidative degradation of PEG-poly(propylene sulfide) block copolymer vesicles results from the conversion of the thioethers in the hydrophobic block into sulfoxide and sulfone groups. Increased hydrophilicity of the PPS block drives the vesicle conversion into aggregates with higher curvature (micelles) before all the block copolymer chains eventually dissolve in water [16]. A more recent report addressed the vesicle-to-micelle transition at the molecular level by monitoring the oxidation of polymer monolayers at the air/water interface in a Langmuir balance [83]. Combination of the monolayer experiments with the vesicle degradation kinetics studies yielded a common dissolution mechanism for block copolymers in the two geometries.

Thermoresponsive cross-linked polymer vesicles were reported to form from PNIPAAM-poly(D,L lactic acid) in water [84]. In this case, the temperature responsiveness of PNIPAAM is coupled to the hydrolyzability of the hydrophobic polylactide block to give vesicles multiple functionality.

3.7
Biohybrid Polymer Vesicles

Although block copolymer membranes can be two- to ten-fold thicker than conventional lipid bilayers, they can be regarded as mimetic of biological membranes and can be used as a matrix for membrane proteins. Moreover, incorporation of protein functionality into polymer vesicles can be achieved by rational design of amphiphilic polypeptide-based copolymers. Secondary structure features such as helicity are introduced into the peptide block by polymerization of appropriate amino acid derivatives.

3.7.1
Polypeptide-based Copolymer Vesicles

One of the first reports on vesicular self-assembly of a peptide-polymer conjugate is the one from Kimura et al. [85] where a naturally occurring hydrophobic peptide, gramicidin A, was coupled to a PEG chain. The 15-mer peptide adopts an antiparallel double-helix conformation and has antibiotic activity due to its ability to form channels in lipid membranes. Resistance to solubilization of the "peptosomes" by Triton X-100 was measured to be $\approx$10 times higher than DMPC liposomes. The same group reported [86] the formation of vesicles in water from a short glycopeptide composed of lactose as the hydrophilic block and a sequence of (Ala-2-aminoisobutyric acid)$_n$ with $n = 4$. Although such materials cannot be strictly considered polymers, they represent a first approach to naturally derived macroamphiphiles able to assemble into vesicles and keep their secondary structure.

Controlled polymerization of α-amino acid N-carboxy anhydrides [87] opened the way to a wide range of polypetide-based block copolymer architectures.

Poly(butadiene)-*b*-poly(L-glutamate) (PBD-PG) of different weight ratios in PG have been investigated for vesicle formation in aqueous media at different pH values. Interestingly, two reports appeared almost at the same time discussing the effect of the pH-induced helix-to-coil transition (4.5 < pH < 11.5) of the PG block onto the vesicle morphology and size. Although the architectures analyzed differ in the ratio between PB and PG, it is interesting to note that for the largest PG content (DP = 100) the transition is reflected in an increase in hydrodynamic radius [25]. No effect was measured by Kukula et al. [26] in copolymers with shorter PG blocks.

In a more recent report [88], addressing the detailed structural characterization of PBD-PG block copolymers dispersions, increase in size with pH of both micelles and vesicles has been measured by light and neutron scattering.

Diethylene glycol modified poly(L-lysine) (PLK) and poly(L-leucine) (PLL) retain their α-helix conformation when coupled into a linear amphiphilic copolymer (Fig. 3.11). At a given weight fraction of the hydrophobic L-leucine block, the fully peptidic copolymers could assemble into large vesicles, as demonstrated by CSLM [89].

When L-lysine (K) monomers were introduced in the hydrophobic block, vesicles could be destabilized under acidic conditions due to the increased hydrophilicity of protonated lysines.

The same group reported vesicle formation from diblock copolymers of charged L-K and L-L [90]. The effect of the relative chain length of the blocks on aggregate morphology as well as the minimum block length of L-L for α-helical conformation was investigated. $K_{60}L_{20}$ resulted as the ideal composition for vesicular structures (Fig. 3.12) and the partial effect of the stereochemistry on self-assembly was re-

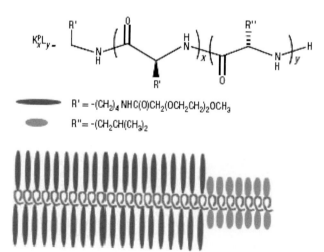

Fig. 3.11 Structure of the helical diblock polypeptide based on hydrophobic poly(L-leucine) (right end of the copolymer) and the hydrophilic diethylene glycol modified poly(L-lysine) (left end of the copolymer) (with permission from Nature Publishing House) [89].

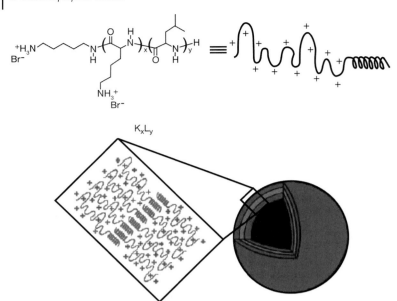

Fig. 3.12 Scheme of poly(L-lysine)-poly(L-leucine) assembly into vesicles (with permission from ACS Publications Division) [90].

ported. Interestingly, dispersion of a solute prior to suspension extrusion leads to its efficient entrapment.

3.7.2
Protein Incorporation into Polymer Vesicles

From pioneering work in our laboratory, it is now well established [91] that membrane proteins can be efficiently incorporated into polymeric membranes and remain functional. Nardin et al. reported the incorporation of OmpF [92], a size-exclusive channel trans-membrane protein, into PMOXA-PDMS-PMOXA copolymer membranes. To demonstrate the OmpF functionality, an enzyme (β-lactamase), for β-lactam antibiotics such as ampicillin, was encapsulated into the vesicles. When exposed to ampicillin, which is small enough to diffuse through OmpF, production of ampicillinoic acid was detected with the iodine test. Vesicles without OmpF showed no substrate conversion.

LamB, another channel forming protein, which also serves as a receptor for λ phage to trigger the ejection of λ phage DNA, was incorporated into the PMOXA-PDMS-PMOXA membrane to demonstrate that DNA translocation is possible across a completely synthetic copolymer membrane (Fig. 3.13). The incorporation and release of the DNA could be followed by fluorescence measurements using YO-PRO-1, an intercalating dye forming a fluorescent complex with double-stranded DNA.

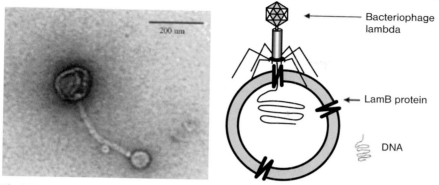

Fig. 3.13 TEM image of negatively stained vesicle bearing LamB with a docked λ phage (left). A schematic representation of the docking process (right) (with permission from The National Academy of Sciences, USA) [93].

Membrane proteins usually have an extracellular and a cytoplasmic side and their function depends on their correct orientation across the membrane. In a synthetic environment, such as in a block copolymer membrane of a vesicle encapsulating an enzyme, orientation is also key to functionality. For this reason Stoenescu and Meier explored the possibility of driving the correct protein insertion via the introduction of asymmetry into polymeric membranes. PEG-PDMS-PMOXA ABC triblock copolymers were used to form vesicles where the different chain length and the incompatibility between the A and C blocks drive the segregation of the A block in the inner side of the membrane [33]. This results in a higher percentage of correct aquaporin 0 insertion, as measured by response to an osmotic stress, when compared with symmetric ABA membranes [94].

Ionophores, such as alamethicin, were used to make calcium ions permeate through the membrane of phosphate encapsulated PMOXA-PDMS-PMOXA giant vesicles. Calcium phosphate crystallization occurred intravesicularly, as shown by TEM [95].

Although experimental evidence proved the incorporation of membrane proteins into copolymer vesicles, the paradox of a large thickness mismatch between protein transmembrane domains (designed for lipids) and copolymer chain length was only recently addressed by theoretical work [96]. Using a mean-field model, it has been possible to determine to what extent the protein perturbs the polymer chains arrangement in the membrane and, most importantly, that this insertion is thermodynamically permitted at the expense of gradual chain conformational rearrangement (Fig. 3.14).

Encapsulation of active proteins, such as enzymes, is another route towards inserting protein functionality into polymer vesicles. Hubbell and coworkers [38] reported the encapsulation of glucose oxidase into PEG-PPS vesicles and its activity when exposed to extravesicular β-D-glucose. In addition to providing insight into initial glucose permeability across the polymer membrane, the system proved

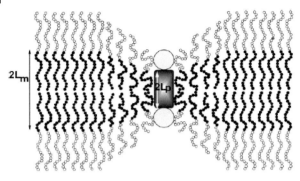

Fig. 3.14 Model of polymer chains relaxation to incorporate a membrane protein across a polymer bilayer (with permission from the Biophysical Society) [96].

the oxidation sensitivity of the PPS membrane by biological means. Hydrogen peroxide produced in the reaction attacked the thioethers in the hydrophobic domain thus disrupting the vesicles.

3.8
Potential Applications of Polymer Vesicles

The hollow morphology offers a double approach to molecule encapsulation, i.e., the inner solvent volume and the membrane core. These two domains are characterized by different physico-chemical environments and can serve to trap molecules with different polarities and solubility in the medium where the vesicles are dispersed.

This ability to release encapsulated molecules in response to a stimulus opens the way to diverse applications such as medical (drug delivery, diagnostics), electronics (chip nanopatterning) and catalysis (enzyme entrapment).

The interest in the use of polymersomes in biomedicine stems mainly from their similarity with stealth liposomes [97] and the unique properties discussed in the previous sections that confer on them some advantages over lipid vesicles. Toughness, limited membrane permeability and denser hydrophilic polymer coatings are sought after properties in sustained drug delivery applications, where it is very important to extend the circulation time of the drug carrier.

In vitro resistance of polymersomes to protein adsorption [98] gave the first indication of the potential to exploit their properties in long-circulating drug delivery systems. Lack of cytotoxicity reflects the extremely low critical micellar concentration (CMC) of vesicle-forming copolymers.

The same group evaluated the *in vivo* performance of polymersomes with varying PEG chain lengths [99]. Interestingly, the previous limit in blood circulation times, set by PEGylated liposomes, was almost doubled. The denser PEG layer in

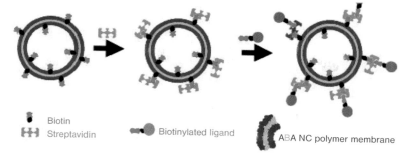

Fig. 3.15 Schematic of vesicle functionalization using the biotin-streptavidin couple.

block copolymer membranes is believed to be responsible for the improved *in vivo* performance of polymer vesicles when compared with stealth liposomes.

Cell-targeting functionality has been tested *in vitro* by Broz et al. [100] using PMOXA-PDMS-PMOXA polymersomes bearing biotin groups. The use of streptavidin permitted the coupling of a ligand for the scavenger receptor A1(SRA1) (Fig. 3.15) from macrophages, and specific binding of fluorescently labeled nanocontainers on COS-7 and THP-1 cell lines could be monitored by fluorescence microscopy. Compared with control experiments that supported the stealth character of native vesicles, functionalized polymer vesicles could target SRA1 expressing macrophages and be internalized.

Polymer vesicles are also suitable for transporting and targeting imaging agents such as fluorescent dyes and paramagnetic contrast agents. Hammer and collaborators [101] reported the encapsulation of porphyrin-based near-infrared (NIR) fluorophores into PBD-PEO block copolymer vesicles. High-loading capacity and direct injection of NIR-emissive polymersomes permitted the *in vivo* visualization of tumors in rats. Coupling of targeting moieties to NIR-emissive polymersomes might allow early detection of tumors using non-invasive techniques.

Magnetic colloids were obtained from aqueous dispersions of hydrophobic iron oxide nanoparticles and block copolymers of PGA and PBD. Lecommandoux and coworkers [102] reported the formation of magnetic micelles and vesicles and characterized their response to a magnetic field as low as 290 G.

The similarity of polymersomes with cells can be exploited to design stable minimal cells that not only embed membrane proteins but also carry functional proteins for medical applications. Arifin and Palmer reported [103] recently on the encapsulation of hemoglobin into polymersomes. The oxygen binding properties of such constructs suggest a novel approach to oxygen therapeutics.

Encapsulation of enzymes into polymer vesicles and controlled substrate permeability across polymeric membrane are being investigated to exploit applications in biocatalysis. Membrane protein incorporation into polymer nanocontainers offers an elegant way to control substrate and product diffusion. Although the field is young, the research efforts to bring polymersomes to real life applications are growing almost exponentially.

3.9
Concluding Remarks

Polymer vesicles are a classical example of a bottom-up approach to nanosciences and represent an exciting area of multidisciplinary research. We have tried to give an overview of the field that does not pretend to be complete but aims at raising the interest of the reader to go on to further readings. Several reviews have appeared in the last few years, which complement our perspective [75, 104–112].

Acknowledgements

The authors wish to thank all the collaborators and colleagues who have contributed to the work presented here. Special thanks are due to Dr. Tom Weeks and Dr. Almut Mecke for proofreading the manuscript.

References

1 TANFORD, C. Hydrophobic effect and organization of living matter. *Science* 200, 1012–1018 (1978).
2 ISRAELACHVILI, J. N., MITCHELL, D. J., NINHAM, B. W. Theory of self-assembly of hydrocarbon amphiphiles into micelles and bilayers. *J. Chem. Soc., Faraday Trans. 2* 72, 1525–1568 (1976).
3 HAJDUK, D. A., KOSSUTH, M. B., HILLMYER, M. A., BATES, F. S. Complex phase behavior in aqueous solutions of poly(ethylene oxide)-poly(ethylethylene) block copolymers. *J. Phys. Chem. B* 102, 4269–4276 (1998).
4 WON, Y. Y., DAVIS, H. T., BATES, F. S. Giant wormlike rubber micelles. *Science* 283, 960–963 (1999).
5 DISCHER, B. M., WON, Y. Y., EGE, D. S., LEE, J. C. M., BATES, F. S., DISCHER, D. E., HAMMER, D. A. Polymersomes: Tough vesicles made from diblock copolymers. *Science* 284, 1143–1146 (1999).
6 MITCHELL, D. J., TIDDY, G. J. T., WARING, L., BOSTOCK, T., MCDONALD, M. P. Phase-behavior of polyoxyethylene surfactants with water — Mesophase structures and partial miscibility (cloud points). *J. Chem. Soc., Faraday Trans. 1* 79, 975–1000 (1983).
7 YANG, J., WEGNER, G., KONINGSVELD, R. Phase-behavior of ethylene oxide-dimethylsiloxane Peo-Pdms-Peo triblock copolymers with water. *Colloid Polym. Sci.* 270, 1080–1084 (1992).
8 YANG, J. L., WEGNER, G. Polymerization in lyotropic liquid-crystals. 1. Phase-behavior of photo-cross-linkable ethylene-oxide (Peo)-dDimethylsiloxane (Pdms) triblock copolymers Peo-Pdms-Peo in aqueous-solution. *Macromolecules* 25, 1786–1790 (1992).
9 YANG, J. L., WEGNER, G. Polymerization in lyotropic liquid-crystals. 2. Synthesis of photo-cross-linkable Peo-Pdms-Peo Triblock copolymers and permanent stabilization of their supermolecular structures in lyotropic mesophases. *Macromolecules* 25, 1791–1795 (1992).
10 MAASSEN, H. P., YANG, J. L., WEGNER, G. The structure of poly(ethylene oxide)-poly(dimethylsiloxane) triblock copolymers in solution. *Makromol. Chem. Macromol. Symp.* 39, 215–228 (1990).
11 WANKA, G., HOFFMANN, H., ULBRICHT, W. Phase-diagrams and aggregation behavior of poly(oxyethylene)-poly(oxypropylene)-poly(oxyethylene) triblock copolymers in aqueous-solutions. *Macromolecules* 27, 4145–4159 (1994).

12 VAN HEST, J. C. M., DELNOYE, D. A. P., BAARS, M. W. P. L., VAN GENDEREN, M. H. P., MEIJER, E. W. Polystyrene-dendrimer amphiphilic block-copolymers with a generation-dependent aggregation. *Science* 268, 1592–1595 (1995).

13 ZHANG, L. F., EISENBERG, A. Multiple morphologies of crew-cut aggregates of polystyrene-B-poly(acrylic acid) block-copolymers. *Science* 268, 1728–1731 (1995).

14 RIESS, G. Micellization of block copolymers. *Progr. Polym. Sci.* 28, 1107–1170 (2003).

15 NARDIN, C., HIRT, T., LEUKEL, J., MEIER, W. Polymerized ABA triblock copolymer vesicles. *Langmuir* 16, 1035–1041 (2000).

16 NAPOLI, A., VALENTINI, M., TIRELLI, N., MULLER, M., HUBBELL, J. A. Oxidation-responsive polymeric vesicles. *Nat. Mater.* 3, 183–189 (2004).

17 ZIPFEL, J., LINDNER, P., TSIANOU, M., ALEXANDRIDIS, P., RICHTERING, W. Shear-induced formation of multilamellar vesicles ("onions") in block copolymers. *Langmuir* 15, 2599–2602 (1999).

18 SCHILLEN, K., BRYSKHE, K., MEL'NIKOVA, Y. S. Vesicles formed from a poly(ethylene oxide)-poly(propylene oxide)-poly(ethylene oxide) triblock copolymer in dilute aqueous solution. *Macromolecules* 32, 6885–6888 (1999).

19 HOLDER, S. J., HIORNS, R. C., SOMMERDIJK, N., WILLIAMS, S. J., JONES, R. G., NOLTE, R. The first example of a poly(ethylene oxide) poly(methylphenylsilane) amphiphilic block copolymer: vesicle formation in water. *Chem. Commun.* 1445–1446 (1998).

20 ZHU, J. T., JIANG, Y., LIANG, H. J., JIANG, W. Self-assembly of ABA amphiphilic triblock copolymers into vesicles in dilute solution. *J. Physical Chemistry B* 109, 8619–8625 (2005).

21 YU, K., EISENBERG, A. Bilayer morphologies of self-assembled crew-cut aggregates of amphiphilic PS-b-PEO diblock copolymers in solution. *Macromolecules* 31, 3509–3518 (1998).

22 AHMED, F., DISCHER, D. E. Self-porating polymersomes of PEG-PLA and PEG-PCL: hydrolysis-triggered controlled release vesicles. *J. Control. Release* 96, 37–53 (2004).

23 HARRIS, J. K., ROSE, G. D., BRUENING, M. L. Spontaneous generation of multilamellar vesicles from ethylene oxide/butylene oxide diblock copolymers. *Langmuir* 18, 5337–5342 (2002).

24 BATTAGLIA, G., RYAN, A. J. Bilayers and interdigitation in block copolymer vesicles. *J. Am. Chem. Soc.* 127, 8757–8764 (2005).

25 CHECOT, F., LECOMMANDOUX, S., GNANOU, Y., KLOK, H. A. Water-soluble stimuli-responsive vesicles from peptide-based diblock copolymers. *Angew. Chem., Int. Ed. Engl.* 41, 1339–1343 (2002).

26 KUKULA, H., SCHLAAD, H., ANTONIETTI, M., FORSTER, S. The formation of polymer vesicles or "peptosomes" by polybutadiene-block-poly(L-glutamate)s in dilute aqueous solution. *J. Am. Chem. Soc.* 124, 1658–1663 (2002).

27 JENEKHE, S. A., CHEN, X. L. Self-assembled aggregates of rod-coil block copolymers and their solubilization and encapsulation of fullerenes. *Science* 279, 1903–1907 (1998).

28 ILHAN, F., GALOW, T. H., GRAY, M., CLAVIER, G., ROTELLO, V. M. Giant vesicle formation through self-assembly of complementary random copolymers. *J. Am. Chem. Soc.* 122, 5895–5896 (2000).

29 DRECHSLER, U., THIBAULT, R. J., ROTELLO, V. M. Formation of recognition-induced polymersomes using complementary rigid random copolymers. *Macromolecules* 35, 9621-9623 (2002).

30 DING, J. F., LIU, G. J., YANG, M. L. Multiple morphologies of polyisoprene-block-poly(2-cinnamoylethyl methacrylate) and polystyrene-block-poly(2-cinnamoylethyl methacrylate) micelles in organic solvents. *Polymer* 38, 5497–5501 (1997).

31 KESSELMAN, E., TALMON, Y., BANG, J., ABBAS, S., LI, Z., LODGE, T. P. Cryogenic transmission electron microscopy imaging of vesicles formed by a polystyrene-polyisoprene diblock copolymer. *Macromolecules* 38, 6779–6781 (2005).

32 Putaux, J. L., Minatti, E., Lefebvre, C., Borsali, R., Schappacher, M., Deffieux, A. Vesicles made of PS-PI cyclic diblock copolymers: In situ freeze-drying cryo-TEM and dynamic light scattering experiments. *Faraday Discuss. R. Soc. Chem.* 128, 163–178 (2005).

33 Stoenescu, R., Meier, W. Vesicles with asymmetric membranes from amphiphilic ABC triblock copolymers. *Chem. Commun.* 3016–3017 (2002).

34 Brannan, A. K., Bates, F. S. ABCA tetrablock copolymer vesicles. *Macromolecules* 37, 8816–8819 (2004).

35 Gomez, E. D., Rappl, T. J., Agarwal, V., Bose, A., Schmutz, M., Marques, C. M., Balsara, N. P. Platelet self-assembly of an amphiphilic A-B-C-A tetrablock copolymer in pure water. *Macromolecules* 38, 3567–3570 (2005).

36 Luo, L. B., Eisenberg, A. Thermodynamic size control of block copolymer vesicles in solution. *Langmuir* 17, 6804–6811 (2001).

37 Bryskhe, K., Jansson, J., Topgaard, D., Schillen, K., Olsson, U. Spontaneous vesicle formation in a block copolymer system. *J. Phys. Chem. B* 108, 9710–9719 (2004).

38 Napoli, A., Boerakker, M. J., Tirelli, N., Nolte, R., Sommerdijk, N., Hubbell, J. A. Glucose-oxidase based self-destructing polymeric vesicles. *Langmuir* 20, 3487–3491 (2004).

39 Battaglia, G., Ryan, A. J. The evolution of vesicles from bulk lamellar gels. *Nat. Mater.* 4, 869–876 (2005).

40 Jain, S., Bates, F. S. On the origins of morphological complexity in block copolymer surfactants. *Science* 300, 460–464 (2003).

41 Sommerdijk, N. A. J. M., Holder, S. J., Hiorns, R. C., Jones, R. G., Nolte, R. J. M. Self-assembled structures from an amphiphilic multiblock copolymer containing rigid semiconductor segments. *Macromolecules* 33, 8289–8294 (2000).

42 Vriezema, D. M., Kros, A., de Gelder, R., Cornelissen, J. J. L. M., Rowan, A. E., Nolte, R. Electroformed giant vesicles from thiophene-containing rod-coil diblock copolymers. *Macromolecules* 37, 4736–4739 (2004).

43 Meng, F. H., Hiemstra, C., Engbers, G. H. M., Feijen, J. Biodegradable polymersomes. *Macromolecules* 36, 3004–3006 (2003).

44 Förster, S. in Polymer Vesicles Meeting (Schloss Beuggen (Germany), 2005).

45 Schrage, S., Sigel, R., Schlaad, H. Formation of amphiphilic polyion complex vesicles from mixtures of oppositely charged block ionomers. *Macromolecules* 36, 1417–1420 (2003).

46 Peng, H. S., Chen, D. Y., Jiang, M. Self-assembly of formic acid/polystyrene-block-poly(4-vinylpyridine) complexes into vesicles in a low-polar organic solvent chloroform. *Langmuir* 19, 10989–10992 (2003).

47 Yao, X. M., Chen, D. Y., Jiang, M. Formation of PS-b-P4VP/formic acid core-shell micelles in chloroform with different core densities. *J. Phys. Chem. B* 108, 5225–5229 (2004).

48 de Cuendias, A., Le Hellaye, M., Lecommandoux, B., Cloutet, E., Cramail, H. Synthesis and self-assembly of polythiophene-based rod-coil and coil-rod-coil block copolymers. *J. Mater. Chem.* 15, 3264-3267 (2005).

49 a) Yu, Y. S., Zhang, L. F., Eisenberg, A. Morphogenic effect of solvent on crew-cut aggregates of apmphiphilic diblock copolymers. *Macromolecules* 31, 1144–1154 (1998).
b) Choucair, A., Lavigueur, A., Eisenberg, A. Polystyrene-b-poly(acrylic acid) vesicle size control using solution properties and hydrophilic block length. *Langmuir* 20, 3894–3900 (2004).
c) Won, Y. Y., Brannan, A. K., Davis, H. T., Bates, F. S. Cryogenic transmission electron microscopy (Cryo-TEM) of micelles and vesicles formed in water by poly(ethylene oxide)-based block copolymers. *J. Phys. Chem. B* 106, 3354–3364 (2002).

50 Zhang, L. F., Yu, K., Eisenberg, A. Ion-induced morphological changes in "crew-cut" aggregates of amphiphilic block copolymers. *Science* 272, 1777–1779 (1996).

51 Haluska, C. K., Gozdz, W. T., Dobereiner, H. G., Forster, S., Gompper, G. Giant hexagonal superstructures in diblock-copolymer membranes. *Phys. Rev. Lett.* 89, (2002).

52 Reinecke, A. A., Dobereiner, H. G. Slow relaxation dynamics of tubular polymersomes after thermal quench. *Langmuir* 19, 605–608 (2003).

53 Gozdz, W. T. Spontaneous curvature induced shape transformations of tubular polymersomes. *Langmuir* 20, 7385–7391 (2004).

54 Sevink, G. J. A., Zvelindovsky, A. V. Self-assembly of complex vesicles. *Macromolecules* 38, 7502–7513 (2005).

55 Fraaije, J. G. E. M., van Sluis, C. A., Kros, A., Zvelindovsky, A. V., Sevink, G. J. A. Design of chimaeric polymersomes. *Faraday Discuss. R. Soc. Chem.* 128, 355–361 (2005).

56 Fraaije, J. G. E. M., Zvelindovsky, A. V., Sevink, G. J. A. Computational soft nanotechnology with mesodyn. *Mol. Simulat.* 30, 225–238 (2004).

57 Srinivas, G., Discher, D. E., Klein, M. L. Self-assembly and properties of diblock copolymers by coarse-grain molecular dynamics. *Nat. Mater.* 3, 638–644 (2004).

58 Nielsen, S. O., Lopez, C. F., Srinivas, G., Klein, M. L. Coarse grain models and the computer simulation of soft materials. *J. Phys.-Condensed Matt.* 16, R481–R512 (2004).

59 Srinivas, G., Klein, M. L. Coarse-grain molecular dynamics simulations of diblock copolymer surfactants interacting with a lipid bilayer. *Mol. Phys.* 102, 883–889 (2004).

60 Pata, V., Ahmed, F., Discher, D. E., Dan, N. Membrane solubilization by detergent: Resistance conferred by thickness. *Langmuir* 20, 3888–3893 (2004).

61 Valentini, M., Napoli, A., Tirelli, N., Hubbell, J. A. Precise determination of the hydrophobic/hydrophilic junction in polymeric vesicles. *Langmuir* 19, 4852–4855 (2003).

62 Valentini, M., Vaccaro, A., Rehor, A., Napoli, A., Hubbell, J. A., Tirelli, N. Diffusion NMR spectroscopy for the characterization of the size and interactions of colloidal matter: The case of vesicles and nanoparticles. *J. Am. Chem. Soc.* 126, 2142–2147 (2004).

63 Rumplecker, A., Forster, A., Zahres, M., Mayer, C. Molecular exchange through vesicle membranes: A pulsed field gradient nuclear magnetic resonance study. *J. Chem. Phys.* 120, 8740–8747 (2004).

64 Bermudez, H., Brannan, A. K., Hammer, D. A., Bates, F. S., Discher, D. E. Molecular weight dependence of polymersome membrane structure, elasticity, and stability. *Macromolecules* 35, 8203–8208 (2002).

65 Lee, J. C. M., Santore, M., Bates, F. S., Discher, D. E. From membranes to melts, rouse to reptation: Diffusion in polymersome versus lipid bilayers. *Macromolecules* 35, 323–326 (2002).

66 Dimova, R., Seifert, U., Pouligny, B., Forster, S., Dobereiner, H. G. Hyperviscous diblock copolymer vesicles. *Eur. Phys. J. E* 7, 241–250 (2002).

67 Bermudez, H., Hammer, D. A., Discher, D. E. Effect of bilayer thickness on membrane bending rigidity. *Langmuir* 20, 540–543 (2004).

68 Lee, J. C. M., Law, R. J., Discher, D. E. Bending contributions to hydration of phospholipid and block copolymer membranes: Unifying correlations between probe fluorescence and vesicle thermoelasticity. *Langmuir* 17, 3592–3597 (2001).

69 Aranda-Espinoza, H., Bermudez, H., Bates, F. S., Discher, D. E. Electromechanical limits of polymersomes. *Phys. Rev. Lett.* 8720, art. No-208301 (2001).

70 Bermudez, H., Aranda-Espinoza, H., Hammer, D. A., Discher, D. E. Pore stability and dynamics in polymer membranes. *Europhys. Lett.* 64, 550–556 (2003).

71 Nardin, C., Winterhalter, M., Meier, W. Giant free-standing ABA triblock copolymer membranes. *Langmuir* 16, 7708–7712 (2000).

72 Li, S. L., Palmer, A. F. Structure and mechanical response of self-assembled poly(butadiene)-b-poly(ethylene oxide) colloids probed by atomic force microscopy. *Macromolecules* 38, 5686–5698 (2005).

73 Lin, J. J., Silas, J. A., Bermudez, H., Milam, T. M., Bates, F. S., Hammer, D. A. The effect of polymer chain length and surface density on the adhesiveness of functionalized polymersomes. *Langmuir* 20, 5493–5500 (2004).

74 Lin, J. J., Bates, F. S., Hammer, D. A., Silas, J. A. Adhesion of polymer vesicles. *Phys. Rev. Lett.* 95, (2005).

75 Discher, D. E., Eisenberg, A. Polymer vesicles. *Science* 297, 967–973 (2002).

76 Choucair, A. A., Kycia, A. H., Eisenberg, A. Kinetics of fusion of polystyrene-b-poly(acrylic acid) vesicles in solution. *Langmuir* 19, 1001–1008 (2003).

77 Zhou, Y. F., Yan, D. Y. Real-time membrane fusion of giant polymer vesicles. *J. Am. Chem. Soc.* 127, 10468–10469 (2005).

78 Zhou, Y. F., Yan, D. Y. Real-time membrane fission of giant polymer vesicles. *Angew. Chem., Int. Ed. Engl.* 44, 3223–3226 (2005).

79 Zhou, Y. F., Yan, D. Y. Supramolecular self-assembly of giant polymer vesicles with controlled sizes. *Angew. Chem., Int. Ed. Engl.* 43, 4896–4899 (2004).

80 Mayer, A. What drives membrane fusion in eukaryotes? *Trends Biochem. Sci.* 26, 717–723 (2001).

81 Discher, B. M. Bermudez, H., Hammer, D. A., Discher, D. E., Won, Y. Y., Bates, F. S. Cross-linked polymersome membranes: Vesicles with broadly adjustable properties. *J. Phys. Chem. B* 106, 2848–2854 (2002).

82 Ahmed, F., Hategan, A., Discher, D. E., Discher, B. M. Block copolymer assemblies with cross-link stabilization: From single-component monolayers to bilayer blends with PEO-PLA. *Langmuir* 19, 6505–6511 (2003).

83 Napoli, A., Bermudez, H., Hubbell, J. A. Interfacial reactivity of block copolymers: understanding the amphiphile-to-hydrophile transition. *Langmuir* 21, 9149–9153 (2005).

84 Hales, M., Barner-Kowollik, C., Davis, T. P., Stenzel, M. H. Shell-cross-linked vesicles synthesized from block copolymers of poly(D,L-lactide) and poly(N-isopropyl acrylamide) as thermoresponsive nanocontainers. *Langmuir* 20, 10809–10817 (2004).

85 Kimura, S., Kim, D. H., Sugiyama, J., Imanishi, Y. Vesicular self-assembly of a helical peptide in water. *Langmuir* 15, 4461–4463 (1999).

86 Kimura, S., Muraji, Y., Sugiyama, J., Fujita, K., Imanishi, Y. Spontaneous vesicle formation by helical glycopeptides in water. *J. Coll. Interf. Sci.* 222, 265–267 (2000).

87 Deming, T. J. Facile synthesis of block copolypeptides of defined architecture. *Nature (London)* 390, 386–389 (1997).

88 Checot, F., Brulet, A., Oberdisse, A., Gnanou, Y., Mondain-Monval, O., Lecommandoux, S. Structure of polypeptide-based diblock copolymers in solution: Stimuli-responsive vesicles and micelles. *Langmuir* 21, 4308–4315 (2005).

89 Bellomo, E. G., Wyrsta, M. D., Pakstis, L., Pochan, D. J., Deming, T. J. Stimuli-responsive polypeptide vesicles by conformation-specific assembly. *Nat. Mater.* 3, 244–248 (2004).

90 Holowka, E. P., Pochan, D. J., Deming, T. J. Charged polypeptide vesicles with controllable diameter. *J. Am. Chem. Soc.* 127, 12423–12428 (2005).

91 Meier, W., Nardin, C., Winterhalter, M. Reconstitution of channel proteins in (polymerized) ABA triblock copolymer membranes. *Angew. Chem., Int. Ed. Engl.* 39, 4599, (2000).

92 Nardin, C., Widmer, J., Winterhalter, M., Meier, W. Amphiphilic block copolymer nanocontainers as bioreactors. *Eur. Phys. J. E* 4, 403–410 (2001).

93 Graff, A., Sauer, M., Van Gelder, P., Meier, W. Virus-assisted loading of polymer nanocontainer. *Proc. Natl. Acad. Sci. USA* 99, 5064–5068 (2002).

94 Stoenescu, R., Graff, A., Meier, W. Asymmetric ABC-triblock copolymer membranes induce a directed insertion of membrane proteins. *Macromol. Biosci.* 4, 930–935 (2004).

95 Sauer, M., Haefele, T., Graff, A., Nardin, C., Meier, W. Ion-carrier controlled precipitation of calcium phosphate in giant ABA triblock

copolymer vesicles. *Chem. Commun.* 2452–2453 (2001).
96 Pata, V., Dan, N. The effect of chain length on protein solubilization in polymer-based vesicles (polymersomes). *Biophys. J.* 85, 2111–2118 (2003).
97 Lasic, D. D., Martin, F. J. (eds.) Stealth liposomes, CRC Press, Boca Raton, FL, 1995.
98 Lee, J. C. M., Bermudez, H., Discher, B. M., Sheehan, M. A., Won, Y. Y., Bates, F. S., Discher, D. E. Preparation, stability, and in vitro performance of vesicles made with diblock copolymers. *Biotechnol. Bioeng.* 73, 135–145 (2001).
99 Photos, P. J., Bacakova, L., Discher, B., Bates, F. S., Discher, D. E. Polymer vesicles in vivo: correlations with PEG molecular weight. *J. Control. Release* 90, 323–334 (2003).
100 Broz, P., Benito, S. M., Saw, C., Burger, P., Heider, H., Pfisterer, M., Marsch, S., Meier, W., Hunziker, P. Cell targeting by a generic receptor-targeted polymer nanocontainer platform. *J. Control. Release* 102, 475–488 (2005).
101 Ghoroghchian, P. P., Frail, P. R., Susumu, K., Blessington, D., Brannan, A. K., Bates, F. S., Chance, B., Hammer, D. A., Therien, M. J. Near-infrared-emissive polymersomes: Self-assembled soft matter for in vivo optical imaging. *Proc. Natl. Acad. Sci. USA* 102, 2922–2927 (2005).
102 Lecommandoux, S. B., Sandre, O., Checot, F., Rodriguez-Hernandez, J., Perzynski, R. Magnetic nanocomposite micelles and vesicles. *Adv. Mater.* 17, 712, (2005).
103 Arifin, D. R., Palmer, A. F. Polymersome encapsulated hemoglobin: a novel type of oxygen carrier. *Biomacromolecules* 6, 2172–81 (2005).
104 Kita-Tokarczyk, K., Grumelard, J., Haefele, T., Meier, W. Block copolymer vesicles — using concepts from polymer chemistry to mimic biomembranes. *Polymer* 46, 3540–3563 (2005).
105 Vriezema, D. M., Aragones, M. C., Elemans, J. A. A. W., Cornelissen, J. J. L. M., Rowan, A. E., Nolte, R. J. M. Self-assembled nanoreactors. *Chem. Rev.* 105, 1445–1489 (2005).
106 Taubert, A., Napoli, A., Meier, W. Self-assembly of reactive amphiphilic block copolymers as mimetics for biological membranes. *Curr. Opin. Chem. Biol.* 8, 598–603 (2004).
107 Opsteen, J. A., Cornelissen, J. J. L. M., van Hest, J. C. M. Block copolymer vesicles. *Pure Appl. Chem.* 76, 1309–1319 (2004).
108 Forster, S., Konrad, M. From self-organizing polymers to nano- and biomaterials. *J. Mater. Chem.* 13, 2671–2688 (2003).
109 Dalhaimer, P., Bates, F. S., Aranda-Espinoza, H., Discher, D. Synthetic cell elements from block copolymers — hydrodynamic aspects. *C. R. Phys.* 4, 251–258 (2003).
110 Antonietti, M., Forster, S. Vesicles and liposomes: A self-assembly principle beyond lipids. *Adv. Mater.* 15, 1323–1333 (2003).
111 Hammer, D. A., Discher, D. E. Synthetic cells-self-assembling polymer membranes and bioadhesive colloids. *Annu. Rev. Mater. Res.* 31, 387–404 (2001).
112 Discher, B. M., Hammer, D. A., Bates, F. S., Discher, D. E. Polymer vesicles in various media. *Curr. Opin. Coll. Interface Sci.* 5, 125–131 (2000).

4
Block Copolymer Micelles for Drug Delivery in Nanoscience

Younsoo Bae, Horacio Cabral, and Kazunori Kataoka

Polymers are flexible chain-like chemical substances composed of a large number of repeating units known as monomers. These monomers are linked to each other through a covalent bond with a strong binding energy (ca. 100 kcal mol^{-1}), and this is the reason why polymers are generally characterized by their high molecular weights and stable backbones. Between the polymer chains, however, van der Waals forces and hydrogen bonding with a slightly weaker binding energy (tens of kcal mol^{-1}) are predominant. Accordingly, molecular interactions and the physico-chemical properties of polymers can change due to the chemical structures and sequences of the monomers. Interestingly, natural polymers, such as proteins, starches and cellulose, take advantage of this. All peptides and proteins are built up through a combination of amino acids. With different sequences of amino acids, they have distinct conformations with which to store the biological information. For these reasons, synthetic polymers have been designed and prepared to mimic or even to surpass natural polymers by functionalizing their constituent monomers. This ease in determining the sequences and types of various monomers is one of the significant advantages of synthetic polymers, and indeed a large number of studies have been carried out over the past few decades to provide new properties and functionalities to synthetic polymers based on these rationales.

Thus, when preparing polymers, two or more different types of monomers can be lined up randomly, or form blocks and grafting branches (Fig. 4.1). These polymers are referred to as random-, block- and graft-copolymers according to their sequences of monomers; different monomer sequences induce characteristic physico-chemical properties that affect the biological activities, even though the obtained polymers have equivalent molecular weights. In general, each block of the block copolymers maintains its specific properties while connected in a single polymer chain, and immiscibility between blocks often induces a change in the higher-ordered structures. For instance, it is well known that amphiphilic block copolymers comprising hydrophilic and hydrophobic blocks can form a nanostructure in an aqueous solution when the miscibility between the hydrophilic and hydrophobic blocks is sufficiently poor to separate the phases which eventually self-assemble into supramolecular assemblies [1]. This gives rise to a spherical nanostructure characterized by its core-shell structure, termed a polymeric micelle

---

Block Copolymers in Nanoscience. Massimo Lazzari, Guojun Liu, Sébastien Lecommandoux
Copyright © 2006 WILEY-VCH Verlag GmbH & Co. KGaA, Weinheim
ISBN: 3-527-31309-5

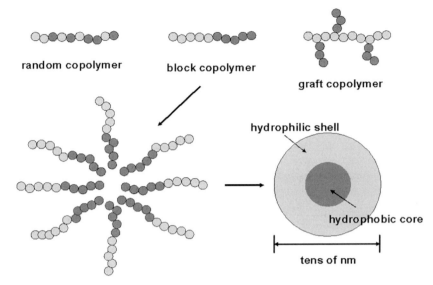

amphiphilic block copolymers self-assemble into a spherical supramolecular nanostructure called a polymeric micelle in water

Fig. 4.1 Block copolymer micelles.

(Fig. 4.1). The size of the polymeric micelles usually ranges from 20 to 100 nm, which is similar to that of viruses and lipoproteins. In fact, this size range is critical for extravasation of polymeric micelles from the blood compartment into the target tissues, most typically solid tumors, as described in detail below. One of the advantages of polymeric micelles is that the micelles can be used as nanocarriers for small compounds, and many examples of successful applications can also be found in their usage for drug delivery [2, 3].

Hence, in this chapter, we will focus on the design of block copolymers to prepare polymeric micelles by which drugs are delivered to a specific site in the body, reducing any side effects while increasing the therapeutic efficacy. We will also briefly review our recent approaches in the field of nanobio-science for drug delivery to help in the understanding of the concept.

Adriamycin (ADR), or doxorubicin, is an anticancer drug widely used in the treatment of leukemia, breast carcinoma and other solid tumors. However, it is frequently accompanied by toxic side effects, such as cardiotoxicity, myolosuppression and nephrotoxicity [4–6]. As with other general types of anticancer drugs, ADR is a low molecular weight compound that easily penetrates blood vessels to access normal tissue as well as malignant tissues, so that it undergoes non-specific drug distribution in the body. For these reasons, tumor-specific drug delivery of ADR is strongly required. However, a polymeric drug delivery system should be based on pathological rationales with respect to the tumor tissues. It has been reported that compounds with high molecular weight can selectively accumulate in solid tumors, owing to the so-called enhanced permeability and retention (EPR) effect, which explains why abnormal vasculatures with increased vessel permeability and poor lymphatic drainage lead to the tumor tissues taking up large molecules more efficiently than normal tissues [7, 8]. As a result, high concentrations of polymers build up in these tissues and reach levels around 10–1000-times higher than would be normally seen after intravenous administration of small molecules [9]. The EPR effect, however, cannot be achieved without overcoming non-specific uptake of the compounds by the reticuloendothelial system (RES) in the body. The RES mainly involves macrophages and endothelial cells which eliminate immune complexes. In order to avoid uptake by the RES, polymers should be highly hydrophilic and biocompatible to prevent interaction with the proteins that make them easily recognizable by the RES. Consequently, many scientists have attempted to conjugate hydrophobic ADR to hydrophilic polymers that increase water solubility and molecular weight [10–12]. These polymer–drug conjugates, however, often become hydrophobic and precipitate in water as the drug-loading contents increase.

In order to clarify these problems, our research group designed various types of self-assembling block copolymers composed of segments for both binding hydrophobic drugs and increasing hydrophilicity, so that the block copolymers form a drug-loaded polymeric micelle, which facilitates tumor-specific accumulation through the EPR effects while avoiding the RES uptake [13, 14].

In Fig. 4.2, an illustration of an ADR-loaded polymeric micelle and the chemical structure of its constituent block copolymers are shown. For the synthesis of these block copolymers, α-methoxy ω-amino poly(ethylene glycol) (PEG) was used as

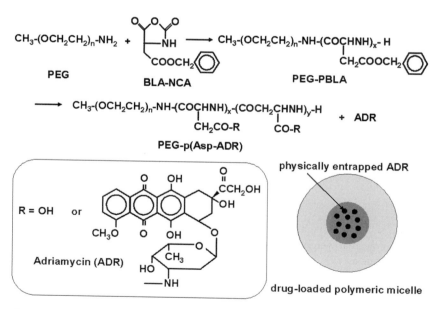

Fig. 4.2 Preparation of the polymeric micelles for the delivery of ADR.

a macroinitiator to polymerize β-benzyl L-aspartate N-carboxy anhydride (BLA-NCA). The synthesized poly(ethylene glycol)–poly(β-benzyl L-aspartate) (PEG-PBLA) block copolymers showed a narrow distribution of molecular weights, and light scattering measurements revealed that PEG-PBLA block copolymers self-assembled into micelles in water [14]. The micelles can be considered as a type of PEG-shielded (or pegylated) nanoparticle because it is generally accepted that PBLA blocks form a core and a PEG shell surrounds it. Subsequently, the core-forming PBLA blocks were modified further by conjugating ADR molecules to enhance π–π interactions between the drug molecules that are permanently conjugated to the polymers and added in addition during the micelle formation. As shown, benzyl groups of the PBLA block were deprotected, and ADR was conjugated to the poly(aspartic acid) blocks at its sugar amino group. The obtained poly(ethylene glycol)–poly(aspartic acid adriamycin) [PEG-p(Asp-ADR)] block copolymers were mixed with free ADR through physical entrapment. The size of the micelles was approximately 50 nm, with a narrow distribution. It is worth noting that the micelles were very stable in water [15]. The critical micelle concentration (CMC) of the polymeric micelles was 10^{-6}–10^{-7} M, which means that the polymeric micelles are 1000-fold more stable than the micelles from low molecular weight surfactants and that they can remain stable even under the dilute conditions in the blood after intravenous (i. v.) injection [16].

Subsequent studies over the past few decades have revealed that this approach provided a large number of advantageous pharmaceutical properties, such as high water solubility, high drug-carrying capacity, simple sterilization by microfiltration and prolonged storage in a freeze-dried state [14]. Moreover, animal experiments using tumor-bearing mice clearly showed that ADR-loaded micelles circulated for a long period of time after injection and accumulated effectively in cancer tissues by suppressing protein adsorption in the blood through pegylation [17]. Note that without free drugs these micelles were non-toxic even when administered repeatedly, because they were prepared from the PEG-p(Asp-ADR) block copolymers that have a molecular weight small enough to be removed from the body, mainly through the bile route. Therefore, these studies clearly demonstrated that block copolymers are useful tools, particularly in preparing a supramolecular nanodevice, such as a polymeric micelle that provides a unique and effective formulation for cancer treatment by reducing toxicity while maintaining the biological activity of existing drugs [18].

It is clear that hydrophobic ADR can be entrapped stably in the core of the polymeric micelles from PEG-p(Asp-ADR) block copolymers, as described above. However, it should be noted that drug release from the micelles still remains controlled. In actuality, drug release control is very important in terms of pharmacokinetics, which is concerned with absorption, distribution, metabolism and excretion of a drug in the body. Pharmacokinetics, therefore, determines injection timing, amount and dose of a drug so that drug concentration in the body can be maintained according to the disease state. Consequently, it is expected that, if we control drug release from the micelles, we might be able to successfully achieve a pinpoint attack at the limited tumor sites. In addition, it is expected that possible drug leakage from the micelles during circulation in the blood would be completely eliminated.

Based on this background, our research group has recently developed a pH-sensitive polymeric micelle whose drug release is triggered by intracellular proton concentration [19]. In order to prepare the pH-sensitive micelles, the drug-binding linkage was newly designed, while the polymer backbone of block copolymers was the same as the micelles from PEG-p(Asp-ADR) block copolymers. However, this time no free ADR was added additionally, but the drugs conjugated to the pH-sensitive linkers were designed to be released in response to a change in pH. It must be emphasized that the drugs become inactive temporarily when they are conjugated to the polymers [20]. Accordingly, drug release control leads to the selective biological effects of the micelles being achieved. As shown in Fig. 4.3, a hydrazide group was introduced to the core-forming segment of the block copolymers as a drug-binding linker, and ADR was conjugated to the hydrazide groups at its C-13 carbonyl through a Schiff-base bond. A Schiff-base bond is known to be stable under physiological conditions (pH 7.4), but cleavable under acidic conditions (pH 4–5). Such acidic conditions are found in the intracellular vesicles such as endosomes and lysosomes [21–23]. Consequently, the pH-sensitive micelles can entrap the loaded ADR stably in the blood but release the drugs selectively after they enter the targeted cells. In vitro and in vivo experiments revealed that the bioavailability of ADR was significantly enhanced by this system [24].

Figure 4.4 clearly shows that the pH-sensitive micelles suppressed tumor growth effectively in the tumor-bearing mice. It should be noted that the anti-tumor activity of the micelles in Fig. 4.4 is compared with that of free ADR in terms of

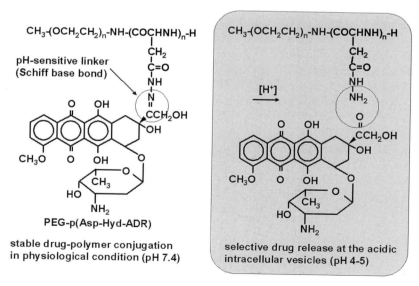

Fig. 4.3 Preparation of the pH-sensitive micelles for the intracellular delivery of ADR.

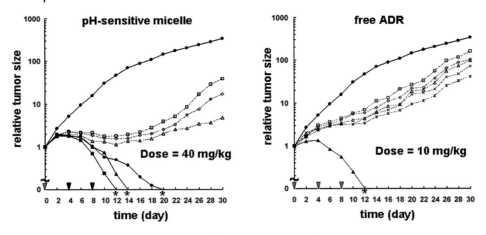

Fig. 4.4 In vivo anti-tumor activity of the pH-sensitive micelles.

maximum tolerated dose (MTD). In contrast to free ADR that led the mice to die with a 15 mg kg^{-1} dose, the micelles were safely injectable up to a 40 mg kg^{-1} dose, and the tumors had completely disappeared in three of six mice. Specifically, MTD was increased about 4-fold or more and the effective dose (ED) was shown to be between 20 and 40 mg kg^{-1}. Surprisingly, the mice treated with the pH-sensitive micelles did not show any critical body weight loss, which would have indicated toxic side effects of the drugs. This result is worth noting because it is totally distinct from that of the mice administered with free ADR, which showed severe body weight loss even though the tumors regressed. These results indicate that the bioavailability of the loaded drug ADR, is significantly improved by intracellular environment-sensitive drug release control. In particular, such an effective and low toxic drug delivery system was realized with the least amount of polymers, increasing the delivery efficiency. Such enhanced safety and effectiveness of the micelles are probably due to the fact that their systemic drug leakage was successfully suppressed, as it was designed to be.

So far we have seen how the polymeric micelles from the block copolymers whose core-forming blocks are functionalized affect the bioavailability of the loaded drugs in terms of toxicity and anti-tumor activity. Next, we will take a look at how the modification of the shell-forming blocks of the block copolymers broadens the possibility of the polymeric micelles and provides a new functionality, so-called active drug-targeting. At the beginning of the chapter, we described how the PEG shell of the micelles plays an important role in achieving prolonged circulation in the blood [25]. This implies that if we conjugate certain molecules that have charges or an affinity relative to other molecules in the body on the surface of the micelles, we can expect the micelles to behave as a molecular sensor in the body. One of the most successful applications based on this hypothesis would be to actively guide the micelles to the tumor cells through ligand–receptor interaction. For example,

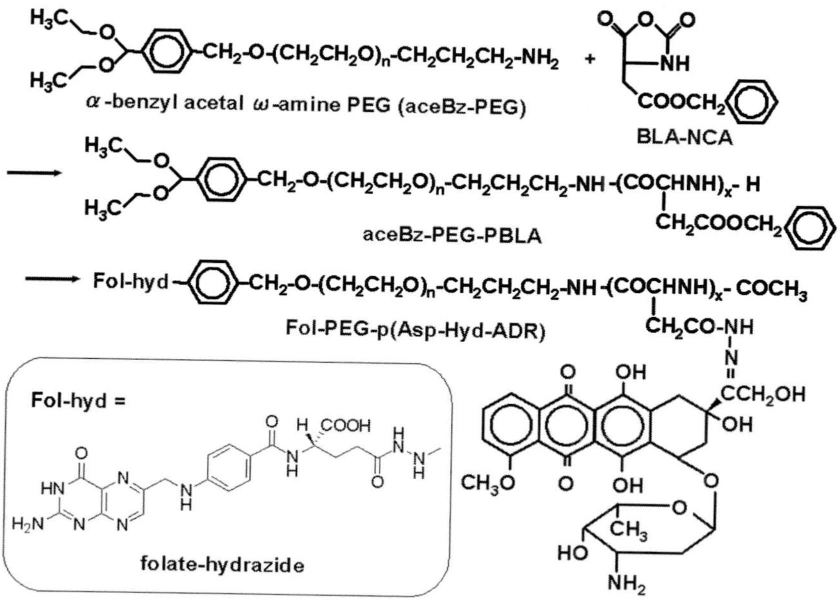

Fig. 4.5 Preparation of the multifunctional polymeric micelles for active drug delivery.

folic acid (Fol), a B-complex vitamin, has recently drawn attention because its receptors are over-expressed selectively on the surface of cancer cells; the compounds conjugated with folate show a higher affinity for the cancer cells [26–28]. Our recent research showed that the pH-sensitive micelles become more effective in terms of the 50% cell-growth inhibitory concentrations (IC50) of drugs by conjugating folate on their surfaces [29]. To prepare these micelles, hetero-functional α-(benzyl acetal)-ω-amino-PEG (aceBz-PEG) and folic acids with the γ-carboxyl group functionalized with hydrazide groups (Fol-hyd) were synthesized (Fig. 4.5). Similar to the synthetic method for PEG-PBLA block copolymers, aceBz-PEG was used as a macroinitiator to prepare aceBz-PEG-PBLA, and Fol-hyd was conjugated at the end of the shell-forming PEG, followed by drug conjugation to the core-forming segments through pH-sensitive linkers. Folate-conjugated block copolymers self-assembled into 60-nm micelles.

Figure 4.6 demonstrates that the IC50 values of folate-conjugated pH-sensitive micelles (FM) against human pharyngeal cancer KB cells are significantly reduced and are almost the same as that of the free drugs. KB cells are well known because of their over-expression of folate receptors (FR), and therefore, a significant increase in the cytotoxic activity of the micelles would be due to the augmented cellular uptake through folate–FR interaction.

The folate–FR interaction of the micelles was supported by results from surface plasmon resonance (SPR) and flow cytometric (FCM) measurements. SPR showed that FM interacts strongly and selectively with folate-binding proteins (FBP) on the

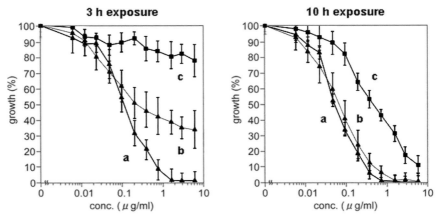

Fig. 4.6 Enhanced cytotoxicity of the micelles against cancer cells through the conjugating folate on their surface.

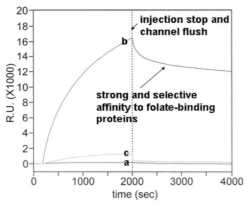

Fig. 4.7 Surface plasmon resonance (SPR) measurements to evaluate binding properties of folate-conjugated micelle.

surface of a gold chip (Fig. 4.7). Once FM binds with FBP, it is very difficult to detach them by simply washing with buffer solution, while the micelles without folate-conjugation show no interaction with FBP.

Such characteristic properties were also observed in the living cells, which were confirmed by FCM experiments. The folate-conjugated micelles were effectively taken up by KB cells even with a short exposure time (3 h), and the cellular uptake of the micelles increased as a dependence on time over 24 h (Fig. 4.8).

These results suggest that the high affinity of FM for FBP increased its cellular uptake and that this resulted in high cytotoxicity of the micelles. Accordingly, we can appreciate that the surface modification of the micelles can be achieved by functionalizing the shell-forming segment of the block copolymers, and such

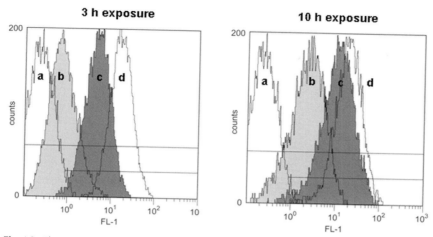

Fig. 4.8 Flow cytometric (FCM) analysis of the cellular uptake of the folate-conjugated micelles by KB cancer cells.

modification changes the behavior of the micelles, as does an intelligent nanodevice that selectively interacts with living cells. This system should be useful for the treatment of various types of cancers that over-express folate receptors. Moreover, this approach is very promising because many cancer cells are known to over-express folate receptors on their surfaces. We can deliver a variety of drugs other than ADR through covalent linkage or by simple physical and electrostatic entrapment.

However, the driving force of micellization is a combination of intermolecular forces, such as hydrophobic interaction, electrostatic interaction, metal complexation and hydrogen bonding of the constituent block copolymers. Of these, our studies have shown that metal complexation is useful to prepare the polymeric micelles ready for the delivery of platinum drugs. The following examples are some of the recent studies related to this topic [30, 31].

The platinum drugs represent a unique and important class of anti-tumor compounds (Fig. 4.9). Alone or in combination with other chemotherapeutic drugs, *cis*-diaminedichloroplatinum(II) (cisplatin, CDDP) and its analogues have made a significant impact on the treatment of a variety of solid tumors. Although early clinical trials demonstrated significant activity against several tumor types, particularly testicular tumors, the severe renal and gastrointestinal toxicity caused by the drug almost led to its abandonment [32]. The unique activity and toxicity profile observed with cisplatin has stimulated the development of new platinum analogues that are less toxic and more effective against a variety of tumor types, including those that have developed resistance to cisplatin. Nevertheless, even though enormous efforts have been devoted to developing improved cisplatin analogues, only a limited number of these analogues reached their final approval. Among these compounds, dichloro(1,2-diaminocyclohexane) platinum(II) (DACH-Pt) showed a

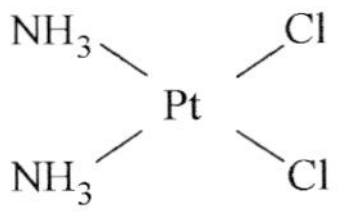

cis(diaminedichloro)platinum(II)

CDDP

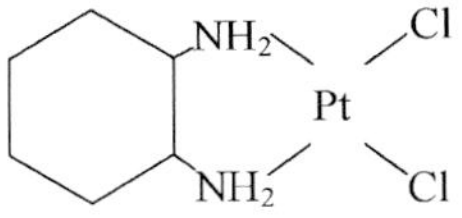

Dichloro(1,2-diaminocyclohexane)platinum(II)

DACHPt

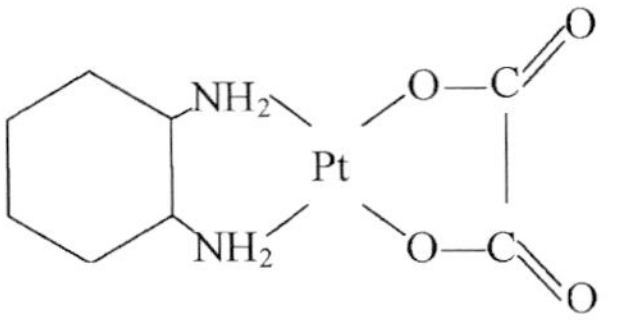

trans-l DACH oxalato platinum(II)

Oxaliplatin

Fig. 4.9 Chemical structures of CDDP, DACH-Pt and oxaliplatin.

broad and distinctly different spectrum of activity from that of cisplatin, such as low toxicity, while it maintained anti-tumor activity even in many cisplatin-resistant cancers [33, 34].

On the other hand, the solubility of DACH-Pt in water is much lower than CDDP (0.25 mg mL^{-1} for DACH-Pt and 1.2 mg mL^{-1} for CDDP) [35]. Thus, oxalate 1,2-diaminocyclohexane platinum(II) (oxaliplatin) was developed in 1977 to enhance the water solubility of DACH-Pt [36]. Oxaliplatin possesses an oxalate group, which is displaced by water and nucleophiles (such as Cl$^-$ and HCO$_3^-$ ions) in biological media to activate the drug, and it also has a non-hydrolyzable diaminocyclohexane (DACH) ligand, which is maintained in the final active cytotoxic metabolites of the drug [36]. As with ADR and other low molecular mass compounds, platinum drugs are known to be rapidly distributed in the whole body and undergo fast renal clearance [37, 38].

To clarify these problems, we incorporated CDDP in the micelles as shown for the case of ADR. However, in order to load CDDP in the micelles, a new drug-loading formulation needs to be employed not only to conjugate CDDP to the core-forming segment of the block copolymers, but also to liberate the drugs from the micelles. Recent studies have shown that formulations employing amine, thiol and carboxylic groups can control the conjugation and release of the drugs due to the reversibility of the coordination bonds with Pt(II), which are extremely stable but cleavable in the presence of chloride ions or other ligands [39]. The micelles incorporating platinum drugs were prepared from poly(ethylene glycol)-poly(glutamic acid) (PEG-pGlu) block copolymers (Fig. 4.10). CDDP-loaded micelles were 20 nm in size with a narrow distribution, while DACH-Pt-loaded micelles had a 40-nm

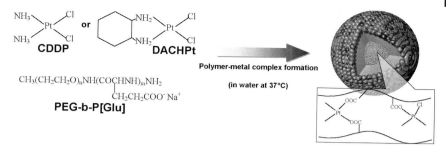

Fig. 4.10 Preparation of the polymeric micelles for the delivery of platinum drugs.

size. Both CDDP- and DACH-Pt-loaded micelles were stable in distilled water at extremely dilute concentrations, showing no dissociation and precipitation.

Such exceptionally stable micelles, however, release the drugs and become dissociated when they are incubated in media containing chloride ions. The time-dependent drug release profile of the micelles in 10 mM PBS with 150 mM NaCl at 37 °C is shown in Fig. 4.11. Interestingly, the cisplatin-loaded micelles released the platinum complexes in a sustained manner for more than 100 h [40]. In contrast, the release profile presented for DACH-Pt-loaded micelles showed some interesting differences. During the first 15 h of exposure, the DACH-Pt-loaded micelles presented an induction period with a very low amount of released platinum complexes [41]. This phase may be related to the more hydrophobic nature of the core of DACH-Pt-loaded micelles, which may hinder the diffusion of water, ions and platinum complexes. However, after this induction period, DACH-Pt-loaded micelles released the drug in a sustained way, and this release rate was found to be comparable to that of cisplatin-loaded micelles, probably because a sufficient amount of the DACH-Pt-p(Glu) complexes had been cleaved by the chloride ions to increase the hydrophilicity of the micelle core and to ease the diffusion of the platinum complexes.

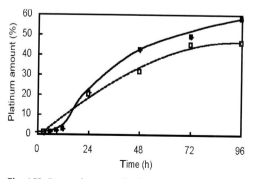

Fig. 4.11 Drug release profile for DACH-Pt-loaded micelles (closed symbols) and CDDP-loaded micelles (open symbols) in 10 mM PBS plus 150 mM NaCl (pH 7.4, 37 °C).

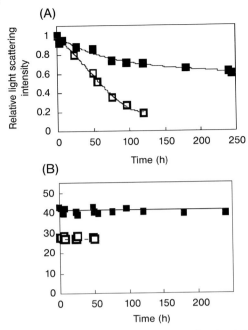

Fig. 4.12 Kinetic stability of the platinum drug-loaded micelles in 10 mM PBS plus 150 mM NaCl (pH 7.4, 37 °C): (A) variation of the relative light scattering intensity; (B) micelle average hydrodynamic diameter obtained by dynamic light scattering. Closed symbols: DACH-Pt-loaded micelle; open symbols: CDDP-loaded micelle.

When designing a drug carrier, the premature drug release from the system is an undesirable event that can lead to toxic effects due to the presence of the free drug in the bloodstream. On the other hand, the unique profile of drug release shown by the DACH-Pt-loaded micelle, which may still be rare among the other drug delivery systems for oxaliplatin active complexes, might be advantageous for *in vivo* use, as the drug release is expected to be hastened after the micelles reach tumor sites. Because the polymer-metal complex formation is the driving force for the formation of CDDP-micelles and DACH-Pt-micelles, the micelle structure may become unstable due to the release of platinum complexes. According to the light scattering measurements (Fig. 4.12), CDDP-micelles release the drugs maintaining a 25 nm size up to 50 h but dissociate within 100 h. On the other hand, the stability of the DACH-Pt-micelles was significantly elevated, and the size was maintained for 240 h. This higher stability of DACH-Pt-micelles might be due to the bulky and hydrophobic nature of the DACH group in DACH-Pt. Thus, once the micelle is formed, the higher hydrophobicity of the micellar core could lead to increased kinetic stability, maintaining the micelles for a prolonged time period. The high kinetic stability of both polymeric micelles seems to be appropriate for systemic drug delivery, in addition, the drug release from the micelles was not compromised.

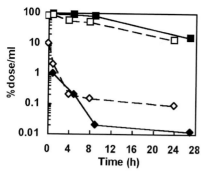

Fig. 4.13 Fig. 4.13 Time profile of platinum concentration in plasma for free CDDP (◇), free oxaliplatin (◆), CDDP-loaded micelles (□) and DACH-Pt-loaded micelles (■).

In general, an elevated kinetic stability in biological media during circulation is a desired property of drug carriers because it is correlated with a prolonged lifetime in the bloodstream and a higher bioavailability. This is also shown in the case of platinum drug-loaded micelles. Cisplatin and oxaliplatin are cleared very fast from the bloodstream, and this is mostly as a result of renal clearance. For cisplatin, 15–25 % of the dose was rapidly excreted in the first 2–4 h and 20–75 % within 24 h after injection [37]. In the case of oxaliplatin, at the end of a 2 h infusion of oxaliplatin, only 15 % of the administered platinum was present in the systemic circulation [38]. By micellization of platinum drugs, the bloodstream circulation will be extended while the drugs are protected from inactivation, which is due to irreversible binding between the plasma proteins and platinum drugs. The platinum level in plasma after administration of free platinum drugs (cisplatin and oxaliplatin) or platinum drug-loaded micelles is shown in Fig. 4.13. Cisplatin-loaded micelles upheld approximately 60 % of the injected dose in the plasma up to 8 h and 13 % of the Pt level in plasma 24 h after injection. The amount of platinum in the DACHPt-loaded micelles at 8 h was determined to be over 80% of the injected dose and more than 16 % even at 24 h after injection. The increased residence time of platinum drug-loaded micelles in the bloodstream seems to be closely connected with the high kinetic stability of the micelles shown in Fig. 4.12. Moreover, high and selective accumulation of drug carriers at the tumor site is essential for drug-targeting success. Because polymeric micelles accumulate preferentially in solid tumors due to the EPR effect, platinum drug-loaded micelles are also expected to augment the platinum drug concentration in tumor tissues.

Accumulation of free drugs and platinum drug-loaded micelles in normal tissues (kidney, liver, spleen and muscle) and in solid tumors (Lewis lung carcinoma, LLC, for the cisplatin system and murine colon adenocarcinoma, C-26, for oxaliplatin system) is shown in Fig. 4.14. Free drugs were rapidly distributed to each organ in accordance with their rapid plasma clearance. The greatest part of free cisplatin appeared to be excreted through glomerular filtration because the highest Pt level was detected in the kidney at 3 min after injection and corresponds to

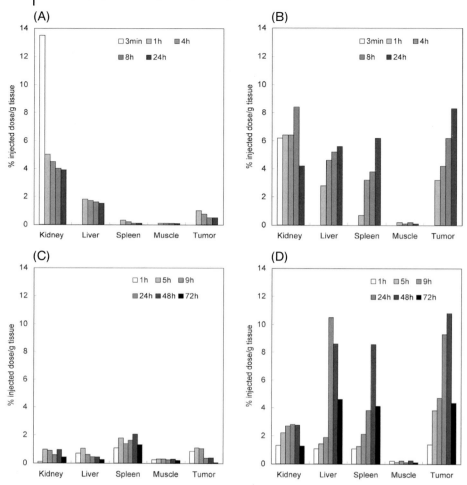

Fig. 4.14 Biodistribution of platinum drug-loaded micelles:
(A) free CDDP; (B) CDDP-loaded micelle; (C) free oxaliplatin;
(D) DACH-Pt-loaded micelle.

previous results. Moreover, it has been determined that the maximum peak of urinary cisplatin concentration is associated with the drug nephrotoxicity much more accurately than the total renal platinum concentration. Therefore, such a rapid accumulation and clearance of free cisplatin in the kidney may be associated with its renal toxicity.

In contrast, the cisplatin-loaded micelles did not show such a high maximum concentration in the kidney, suggesting that renal toxicity might be restricted for the micelles compared with that for free cisplatin. Furthermore, the remarkably prolonged blood circulation of cisplatin-loaded micelles enhanced the cumulative accumulation in each organ and solid tumor. Thus, the Pt levels in the liver, spleen and tumor continued to increase up to at least 24 h after administration. Consequently, the cisplatin-loaded micelles presented 4-, 39- and 20-fold higher accumulation in the liver, spleen and tumor, respectively, than free cisplatin 24 h after injection. As a result, the cisplatin-loaded micelles presented a higher accumulated dose in tumor tissue than in normal organs. The area under the curve (AUC) ratios between the tumor and the organs was determined to be greater than 1, showing the preferential accumulation of the micelles at the tumor site.

For DACHPt-loaded micelles, the accumulation at the tumor at 24 h was similar to that of cisplatin-loaded micelles. However, the former maintained a higher platinum concentration in the plasma than the latter over a similar time range probably due to the enhanced kinetic stability of the DACHPt-loaded micelles. Eventually, the maximum accumulation of DACHPt-loaded micelles was observed at 48 h with 11% of the dose per g of tissue.

Another remarkable fact is the high accumulation in the liver of DACHPt-loaded micelles compared with cisplatin-loaded micelles. This behavior is probably due to the higher hydrophobic formulation of DACHPt-loaded micelles, which apparently leads to an augmented entrapment of the carrier in the sinusoidal tissue of the liver [42, 43]. However, the ratio of the accumulated dose of DACHPt-loaded micelles in tumor tissue to the accumulated dose in the liver was determined to be higher than 1 (1.5-times). In addition, similar results were observed for other tumor-to-organ ratios (2.5-times tumor/spleen and 3.2-times tumor to kidney). Thus, it can be concluded that the micelles accumulated preferentially at the tumor site leading to ideal passive targeting.

As we have reviewed in this chapter, polymeric micelles have great potential as a means of functional supramolecular nano-assembly for the delivery of drugs. Some of the polymeric micellar formulations have already been in clinical trial, for example, the ADR-loaded and paclitaxel-incorporated micelle systems at the National Cancer Center Hospital in Japan.

The micelles have specific advantages over other types of carrier systems as follows: (a) longevity in blood circulation due to the effective stealth property of the dense PEG layer as a shell for the nanostructure; (b) wide availability in encapsulating drugs with different properties through the modulation of hydrophobic interaction, electrostatic interaction and hydrogen bonding; (c) low accumulating toxicity due to the self-dissociating property to give monomer molecules with a molecular weight less than the threshold of kidney excretion; and (d) ease in the control of size by modulating the chemical composition of the block copolymers.

The final goal of the polymeric micelles would be to achieve a safer and more effective medical treatment considering the quality of life (QOL). For these reasons, a tumor-targeting nanodevice that selectively functions at a suitable location will also be required. The polymeric micelles, therefore, will be implemented with versatile properties not only to deliver biologically active materials but also to interact with the cells by detecting their abnormalities (biosensors), diagnosing types of disease (nanoprocessors) and taking suitable actions (nanoeffectors) in the future, and block copolymers are inevitable for the materialization of such multifunctional nanodevices.

References

1 J. M. Lehn (2002) *Science* 295 2400–2403.
2 M. L. Adams, A. Lavasanifar, G. Kwon (2003) *J. Pharm. Sci.* 92, 1343–1355.
3 K. Kataoka, A. Harada, Y. Nagasaki (2001) *Adv. Drug. Deliver. Rev.* 47, 113–131.
4 T. I. Ghose, A. H. Blair, P. N. Kulkarni (1983) *Methods Enzymol.* 93, 280–333.
5 A. Goldin, A. Rahman, A. M. Casazza, F. Guiliani, A. DiMarco, N. O. Kaplan, P. S. Schein (1983) in *Anthracyclines: Current Status and Future Developments*, eds. G. Mathe, R. Maral, R. DeJager, Masson, New York, pp. 65–71.
6 A. M. Casazza (1975) Immunodepressive Activity of Daunomycin and Adriamycin, in *Adriamycin Review*, European Press, Ghent, pp. 123–131.
7 H. Maeda, J. Wu, T. Sawa, Y. Matsumura, K. Hori (2000) *J. Control. Release* 65, 271–284.
8 Y. Matsumura, H. Maeda (1986) *Cancer Res.* 46, 6387–6392.
9 H. Maeda (2001) *Adv. Drug. Deliv. Rev.* 46, 169–185.
10 A. J. M. D'souza, E. M. Topp (2004) *J. Pharm. Sci.* 93, 1962–1979.
11 R. Duncan (2003) *Nat. Rev. Drug. Discov.* 2, 347–360.
12 J. Kopecek, P. Kopeckova, T. Minko, Z. R. Lu, C. M. Peterson (2001) *J. Control. Release* 74, 147–158.
13 M. Yokoyama, T. Okano, Y. Sakurai, S. Fukushima, K. Okamoto, K. Kataoka (1999) *J. Drug Target.* 7, 171–186.
14 K. Kataoka, G. Kwon, M. Yokoyama, T. Okano, Y. Sakurai (1993) *J. Control. Release* 24, 119–132.
15 M. Yokoyama, M. Miyauchi, N. Yamada, T. Okano, Y. Sakurai, K. Kataoka, S. Inoue (1990) *J. Control. Release* 11, 269–278.
16 M. Yokoyama, T. Okano, Y. Sakurai, H. Ekimoto, C. Shibazaki, K. Kataoka (1991) *Cancer Res.* 51, 3229–3236.
17 G. S. Kwon, S. Suwa, M. Yokoyama, T. Okano, Y. Sakurai, K. Kataoka (1994) *J. Control. Release* 29, 17–23.
18 T. Nakanishi, S. Fukushima, K. Okamoto, M. Suzuki, Y. Matsumura, M. Yokoyama, T. Okano, Y. Sakurai, K. Kataoka (2001) *J. Control. Release* 74, 295–302.
19 Y. Bae, S. Fukushima, A. Harada, K. Kataoka (2003) *Angew. Chem., Int. Ed. Engl.* 42, 4640–4643.
20 H. Ringsdorf (1975) *J. Polym. Sci. Polym. Symp.* 51, 135–153.
21 K. D. Jensen, A. Nori, M. Tijerina, P. Kopeckova, J. Kopecek (2003) *J. Control. Release* 87, 89–105.
22 D. Willner, P. A. Trail, S. J. Hofstead, H. D. King, S. J. Lasch, G. R. Braslawsky, R. S. Greenfield, T. Kaneko, R. A. Firestone (1993) *Bioconjugate Chem.* 4, 521–527.
23 T. Kaneko, D. Willner, I. Monkovic, J. O. Knipe, G. R. Braslawsky, R. S. Greenfield, D. M. Vyas (1991) *Bioconjugate Chem.* 2, 133–141.
24 Y. Bae, N. Nishiyama, S. Fukushima, H. Koyama, Y. Matsumura, K. Kataoka (2005) *Bioconjugate Chem.* 16(1), 122–130.
25 Y. Yamamoto, Y. Nagasaki, Y. Kato, Y. Sugiyama, K. Kataoka (2001) *J. Control. Release* 77, 27–38.
26 S. D. Weitman, R. H. Lark, L. R. Coney, D. W. Fort, V. Frasca, V. R. Zurawski, B. A. Kamen (1992) *Cancer Res.* 52, 3396–3401.
27 B. Stella, S. Arpicco, M. T. Peracchia, D. Desmaele, J. Hoebeke, M. Renoir, J. D'angelo, L. Cattel, P. Couvreur (2000) *J. Pharm. Sci.* 89, 1452–1464.
28 R. J. Lee, P. S. Low (1994) *J. Biol. Chem.* 269, 3198–3204.
29 Y. Bae, W. D. Jang, N. Nishiyama, S. Fukushima, K. Kataoka (2005) *Mol. BioSyst.* 1, 242–250
30 N. Nishiyama, S. Okazaki, H. Cabral, M. Miyamoto, Y. Kato, Y. Sugiyama, K. Nishio, Y. Matsumura, K. Kataoka (2003) *Cancer Res.* 63, 8977–8983.
31 N. Nishiyama, Y. Kato, Y. Sugiyama, K. Kataoka (2001) *Pharmaceut. Res.* 18, 1035–1041.
32 B. A. Chabner, C. E. Myers (1989) in *Cancer: Principles and Practice of Oncology*, eds. V. T. DeVita, S. Hellman, S. A. Rosenberg, 3rd edn, JB Lippincott, Philadelphia.

33 O. Rixe, W. Ortuzar, M. Alvarez, R. Parker, E. Reed, K. Paull, T. Fojo (1996) *Biochem. Pharmacol.* 52, 1855–1865.

34 E. Raymond, S. Faivre, S. Chaney, J. Woynarowski, E. Cvitkovic (2002) *Mol. Cancer Ther.* 1, 227–235.

35 Y. Kidani, M. Iigo, K. Inagaki, A. Hoshi, K. Kuretani (1978) *J. Med. Chem.* 21, 1315–1318.

36 Y. Kidani, K. Inagaki, R. Saito, S. Tsukagoshi (1977) *J. Clin. Hematol. Oncol.* 7, 197–202.

37 M. A. Graham, G. F. Lockwood, D. Greensdale, S. Brieza, M. Bayssas, E. Gamelin (2000) *Clin. Cancer Res.* 6, 1205–1218.

38 Z. H. Siddik, D. R. Newell, F. E. Boxall, K. R. Harrap (1987) *Biochem. Pharmacol.* 36, 1925–1932.

39 M. E. Howe-Grant, S. J. Lippard (1980) in *Metal Ions in Biological Systems*, ed. H. Sigel, 3rd edn, Marcel Dekker, New York.

40 N. Nishiyama, S. Okazaki, H. Cabral, M. Miyamoto, Y. Kato. Y. Sugiyama, K. Nishio, Y. Matsumura, K. Kataoka (2003) *Cancer Res.* 63, 8977–8983.

41 H. Cabral, N. Nishiyama, S. Okazaki, H. Koyama, K. Kataoka (2005) *J. Control Release* 101, 223–232.

42 J. E. van Montfoor, B. Hagenbuch, G. M. M. Groothuis, H. Koepsell, P. J. Meier, D. K. F. Meijer (2003) *Curr. Drug Metab.* 4, 185–211.

43 H. Ayhan, A. Tuncel, N. Bor, E. Piskin (1995) *J. Biomater. Sci.-Polym. E.* 7, 329–342.

5
Stimuli-responsive Block Copolymer Assemblies*

Jean-François Gohy

5.1
Introduction

Block copolymers have been the focus of much interest during the last 30 years because their constituent blocks are generally immiscible, leading to phase separated structures. As the different blocks are linked together by covalent bonds, the phase separation process is spatially limited and results in self-assembled structures whose characteristic sizes are of macromolecular dimension and thus range from ca. 10 to 100 nm [1]. This phase separation has motivated the used of block copolymers in nanotechnology, as discussed thoroughly in other chapters of this book. Depending on the processing conditions, different types of nanomaterials have been obtained. In bulk, the microdomains can form more or less long-range ordered structures, such as cubic arrays of spheres or hexagonally-packed cylinders. Such regular structures can also be considered in thin films, provided that the thickness of the film is greater than the characteristic period of the ordered structure. Self-assembly of block copolymers can also be considered in a solvent that is selective for one of the blocks. In this instance block copolymer micelles are obtained in which the insoluble blocks form the micellar core and the soluble ones constitute the corona.

Stimuli-responsive polymers are polymers that undergo relatively large and abrupt, physical or chemical changes in response to small external changes in the environmental conditions. Such polymers are widely found in living systems. Indeed, proteins, polysaccharides and nucleic acids are typically stimuli-responsive polymers. The basic operating principle of a stimuli-responsive polymer consists in the recognition of a signal, the evaluation of the magnitude of this signal and then in changing its chain conformation in response to the signal [2]. Stimuli can be classified as either physical or chemical. Typical chemical stimuli are pH changes, ionic strength variation and addition of chemical agents. At the molecular level, these stimuli will modulate the interactions between polymer

---

* A list of abbreviations is provided at the end of this chapter.

Block Copolymers in Nanoscience. Massimo Lazzari, Guojun Liu, Sébastien Lecommandoux
Copyright © 2006 WILEY-VCH Verlag GmbH & Co. KGaA, Weinheim
ISBN: 3-527-31309-5

chains or between the polymer chains and the solvents. Physical stimuli are temperature changes, electric or magnetic field variations and mechanical stress. Such stimuli will alter molecular interactions at critical onset points. Some polymers can combine two or more stimuli-responsive properties, while two or more signals can be simultaneously applied in order to induce a response in the so-called dual-responsive polymer systems [3]. Recently, biochemical stimuli have been considered as another category, which involves the responses to antigens, enzymes, ligands or other biochemical agents [4].

Whenever a stimuli-responsive polymer is introduced into a block copolymer architecture, a stimuli-responsive block copolymer is obtained. Considering the numerous possibilities in designing block copolymer architectures and as stimuli-responsive copolymers can be processed in various forms, an infinite variety of stimuli-responsive materials can be predicted. However, few possibilities have been exploited so far. In the following, we will concentrate on the main systems investigated to date, which are often based on simple linear AB block copolymers containing at least one stimuli-responsive block. As the usual stimuli (temperature, pH, ionic strength) affect the solubility of water-soluble systems, aqueous block copolymer micelles will essentially be discussed. Some specific examples of stimuli applied to bulk materials or thin films from block copolymers will also be briefly discussed. The aim of this chapter is not to present an exhaustive review of the field of stimuli-responsive copolymers but rather to illustrate the basic principle of operation of such systems with a few selected examples. For more complete information on the general field of stimuli-responsive polymers, the reader is directed to review articles, such as the recent ones from Gil and Hudson [5] and Lecommandoux and coworkers [6].

In the following, block copolymers will be designated by the acronym A-B, where A and B are the abbreviations associated with the various constituent blocks. For example, a poly(styrene)-block-poly(4-vinylpyridine) diblock copolymer will be referred to PS-P4VP.

5.2
Stimuli-sensitive Micellization

Stimuli-sensitive micellization is one of the early examples of stimuli-responsive micellization and has been typically implemented for water-based systems. In this respect, double-hydrophilic, i.e., AB block copolymers containing two water-soluble blocks have been considered. Under normal conditions, double hydrophilic block copolymer chains have no tendency to aggregate in aqueous media and are thus observed as unimers. However, once an adequate stimulus is applied, one of the hydrophilic blocks becomes hydrophobic. The initial double hydrophilic copolymer is then transformed into an amphiphilic copolymer and micelle formation is observed (see Fig. 5.1). The field of double hydrophilic block copolymers has recently been reviewed by Cölfen [7]. This transition is generally observed to be reversible and results in a sharp modification of the macroscopic

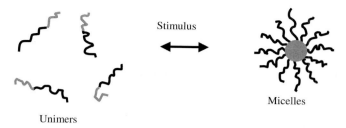

Fig. 5.1 Schematic principle of operation for a stimuli-sensitive micellization process.

characteristic features of the aqueous medium. For example, one can expect a sharp modification to the viscosity of the solution once the system is shifted from unimers to micelles.

5.2.1
Temperature-sensitive Micellization

Temperature-responsive polymer blocks exhibiting either an LCST or UCST behavior have been combined with other hydrophilic blocks. Interestingly, the switchable property has been utilized as a basic concept to manufacture an attractive stimulus-sensitive system that might loose or gain its amphiphilic property according to temperature changes, resulting in temperature responsive micelle formation.

One of the first examples of stimuli-responsive micellization is found in PEO-PPO and PEO-PPO-PEO block copolymers, also known as Pluronics™. In these copolymers, micellization is driven by an increase in temperature. This arises from the fact that dehydration of the PPO block is taking place with increasing temperature. The initial double hydrophilic copolymer is then transformed into an amphiphilic one above a critical temperature, which has been referred to as the critical micelle temperature or CMT, and is a very useful value for PEO-PPO-PEO copolymers. The value of the CMT ranges from 20 to 50 °C in commercially available PEO-PPO-PEO copolymers. The CMT increases whenever the copolymer concentration is increased [8]. At higher temperatures, dehydration of the PEO blocks is then observed, resulting in a completely hydrophobic system, which precipitates out from the solution.

Pluronic copolymers have been the topic of many investigations and are used in several commercial applications. Their micellization behavior has been studied quite extensively and summarized in the review articles by Chu and Zhou [8], Almgren et al. [9], Hamley [1], Booth and Atwood [10] and Wanka et al. [11]. The PPO block can be replaced by other hydrophobic blocks. As an example, PBO was introduced as a block that is more hydrophobic than PPO. The CMC of the related PBO-PEO copolymers was found to be very low [12]. Many other thermo-sensitive double-hydrophilic block copolymers have been considered in the literature. Most are based on polymer blocks exhibiting an LCST in water, such as PMVE, PDMAEMA or PNIPAM. These blocks have generally been associated with

PEOs. At temperatures below the LCST, both blocks are soluble and no aggregation is observed. At temperatures above the LCST, the thermo-sensitive blocks turn hydrophobic and micellization takes place. A typical example of such a copolymer is PEO-PNIPAM [13, 14].

It should be noted that the relative lengths of the PEO and PNIPAM blocks have an influence on the LCST of PNIPAM. Indeed, the LCST of PNIPAM was shifted to higher temperatures for sufficiently long PEO blocks (MW of 1900 g mol^{-1}), while it was not affected for short blocks (MW of 550 g mol^{-1}) [15]. Above the LCST, the micellar characteristic features were affected by the length of the PNIPAM block, the PNIPAM/PEO molar ratio and the copolymer concentration [14]. Another method to control the LCST of PNIPAM consists in using a PNIPAM-based random copolymer instead of pure PNIPAM. This strategy is illustrated for poly(NIPAM-co-HEMA-lactate) blocks associated with PEO [15]. These copolymers showed an increase hydrophilicity after the hydrolysis of the hydrophobic lactate groups in the poly(NIPAM-co-HEMA-lactate) blocks, which increased the LCST to above body temperature (Fig. 5.2) [16]. Hydrolytically sensitive micelle formation was therefore designed, where the CMT was controlled by the hydrolysis of the lactate ester group. Therefore, this block copolymer formed micellar structures at physiological temperatures, which could be disassembled after hydrolysis of the lactate groups.

Other thermo-responsive micellization examples have been reported for systems based on PEOVE-PMOVE [17], PMVE-PVA [18] and PMVE-PMTEGVE [19] copolymers. Thermo-sensitive micellization was also considered for more complex ABC triblock copolymer architectures. In this context, the temperature-dependent

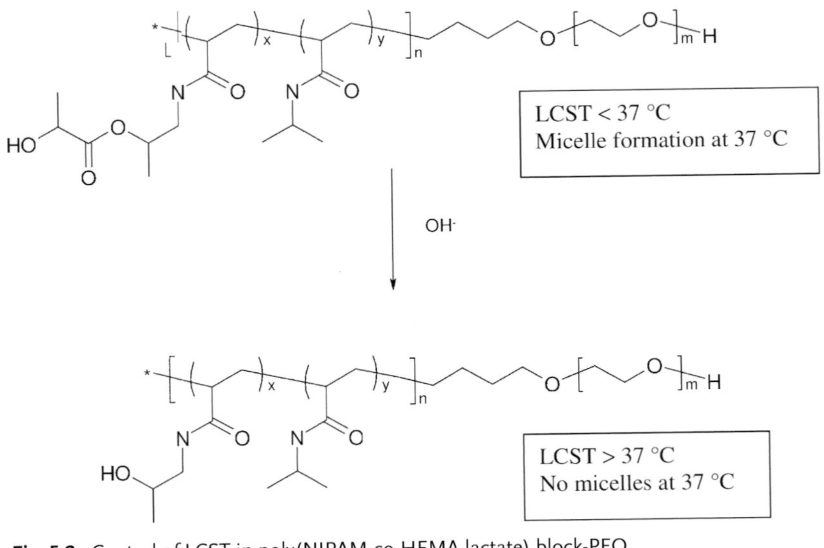

Fig. 5.2 Control of LCST in poly(NIPAM-co-HEMA-lactate) block-PEO copolymers by hydrolysis of the HEMA-lactate into HEMA.

association of PEOVE-PMOVE-PEOEOVE triblock copolymers was recently investigated. These copolymers form micelles in water when the temperature is increased above the LCST of the PEOVE block. The micelles then associate to form a physical gel above the LCST of the PEOEOVE block and finally precipitate when the temperature is higher than the LCST of the PMOVE block [20].

This research field is extremely active due to the potential biomedical applications of thermosensitive micelles for controlled drug delivery.

5.2.2
pH-sensitive Micellization

pH has been widely used as a stimulus for reversible micellization in aqueous media. Double hydrophilic block copolymers containing weak acidic or basic blocks are ideal candidates provided that the neutral form of these blocks is hydrophobic while the ionized state is water soluble. A typical example of such a pH-responsive micellization has been reported for P2VP-PEO copolymers that exist as unimers at pH < 5 and form micelles at higher pH values [21]. This behavior is directly linked to the deprotonation of the P2VP blocks as pH is increased. In this respect, P2VP blocks are protonated and positively charged at low pH and thus water soluble. Deprotonation results in hydrophobic P2VP blocks, which then aggregate into micellar cores. Although the deprotonation process is continuous, it is more significant around the pK_a of the P2VP blocks, which explains why micellization is observed around pH = 5.

The hydrophobization of the P2VP blocks results in micelles rather than in precipitated particles because the P2VP blocks are linked to water-soluble PEO segments. The beneficial role of the copolymer architecture is thus clearly evidenced in this case. It should be noted that this pH-sensitive process is completely reversible, the system can, in principle, be shifted infinitely between the two states. A critical factor that could affect the process is the purity of the starting copolymer. Indeed, the presence of P2VP homopolymer leads to the so-called anomalous micellization process [22], which is a combination of precipitation and micellization. The formation of inorganic salts, as the pH is changed many times in the same vessel by continuous addition of base and acid, could also affect the reversibility of the pH-sensitive process, because the solubility of both the protonated P2VP and PEO blocks is affected by the ionic strength.

The careful control in the progress of the P2VP block deprotonation can be used as a tool to control the morphology of the micelles formed accordingly. This concept is illustrated for asymmetric P2VP-PEO copolymers containing a major P2VP block. Although these copolymers precipitated out from the solution when they were totally deprotonated, micellar structures were observed in a limited pH range located near the pK_a of P2VP. Moreover, a transition from spherical to rodlike micelles and finally vesicles was observed for these samples as the degree of deprotonation was increasing, in agreement with the resulting changes in the hydrophilic/hydrophobic balance of the copolymer [23]. Giant vesicles with diameters of 5–10 μm were also recently obtained by Förster et al. from PEO-P2VP copoly-

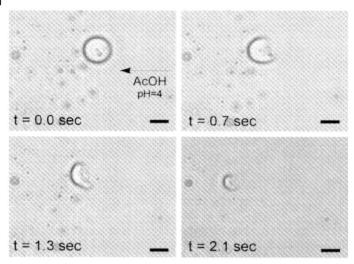

Fig. 5.3 Monitoring the disintegration process of vesicles from a PEO-P2VP copolymer by addition of acetic acid (reprinted with permission from [23]).

mers [24]. These vesicles were stable down to pH 4.5, below which the P2VP-block is protonated and the vesicles dissolved (see Fig. 5.3).

Micellization for P2VP-PEO and the parent P4VP-PEO copolymers has been induced at low pH by the addition of noble metal salts [25, 26]. The driving force of such a micellization is the coordination of vinylpyridine (VP) units with metal ions. The reversibility of such a micellization process has, however, not yet been addressed. One can imagine that the addition of a stronger complexant towards the noble metal ions would be required in order to reverse the process. However, reduction of the metal ions embedded in the P2VP-PEO and P4VP-PEO micelles finally resulted in the formation of well-defined metal nanoparticles. Templating is thus the favored application in this case.

Another example of pH-driven micellization is illustrated in PEO-P(M)AA double hydrophilic block copolymers. At low pH, hydrogen bonds form between P(M)AA and PEO blocks, which are further disrupted by an increase in pH due to the ionization of the (M)AA units [27]. Micellization was observed for an asymmetric PMAA-PEO diblock copolymer containing a major PEO block [28].

In this case, intra- and/or intermolecular hydrogen bonding between the PMAA blocks and PEO segments was thought to result in micellar cores, which were stabilized by a corona formed by the excess of PEO segments. The process of aggregation is thus proceeding differently than for the previous P2VP-PEO system and is illustrated in Fig. 5.4. These micelles disintegrated whenever the pH was raised above 5 as a result of the ionization of the PMAA blocks. Furthermore, these micelles showed a thermo-responsive behavior related to the limited thermal stability of the PMAA/PEO hydrogen-bonded complexes. In this respect, an increase in temperature resulted in the breaking of the initial hydrogen-bonded

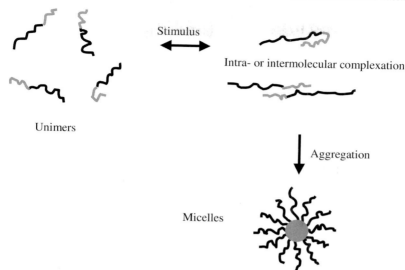

Fig. 5.4 Schematic representation of a stimuli-responsive system where the stimulus triggers the formation of intra- or intermolecular insoluble complexes between the constituent blocks of the copolymer. These complexes self-associate into micellar structures in a subsequent step.

complexes. At the same time, the released PMAA blocks reached their LCSTs in water and a new type of micelle is formed, containing an insoluble PMAA core and a PEO corona. The association behavior of grafted copolymers based on P(M)AA and PEO have also been studied recently. Although hydrogen-bonded structures were detected in these systems, pH sensitive ill-defined aggregates were observed rather than micelles [29, 30].

pH-sensitive micellization can be observed in double hydrophilic diblock copolymers containing two ionizable blocks. This situation is illustrated by PSS-PSCNa copolymers, which form micelles at low pH whenever the PSCNa block is protonated, and are therefore less soluble than the ionized form [31].

Double hydrophilic block copolymers containing two basic blocks have also been studied, as reported by Armes and coworkers for PDMAEMA-PDEAEMA copolymers [32] and Gohy et al. for P2VP-PDMAEMA diblocks [33]. pH-induced micellization has been observed in the two cases. However, the relatively large difference between the pK_as of the P2VP and PDMAEMA blocks allowed the formation of three different association states, depending on pH: at low pH, both the P2VP and PDMAEMA blocks were protonated and water soluble; at intermediate pH, polyelectrolyte block copolymer micelles were formed with an insoluble P2VP core and a positively charged PDMAEMA corona; at high pH, micelles were formed with the same insoluble P2VP core and thermo-responsive uncharged PDMAEMA coronal chains. Recent progress in the RAFT polymerization process has allowed the

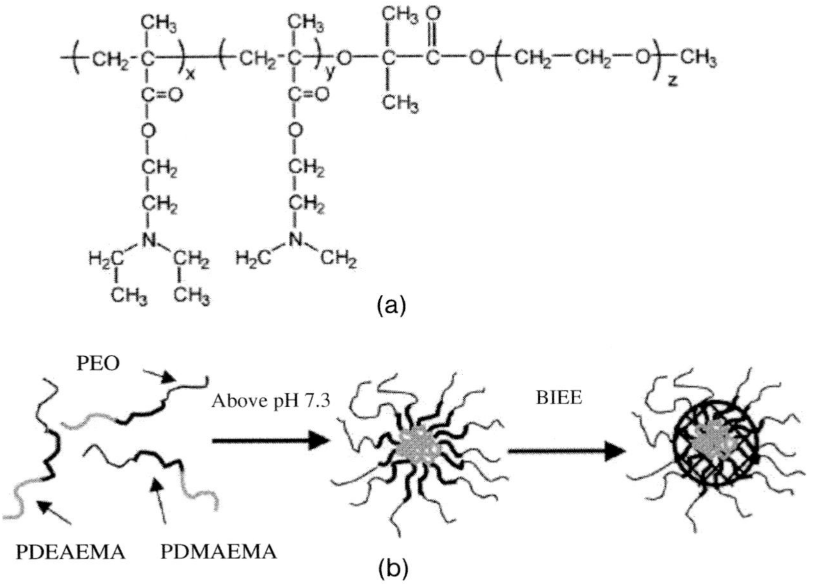

Fig. 5.5 pH-sensitive core-shell corona micelles from a PDEAEMA-PDMAEMA-PEO triblock terpolymer. The shell of these micelles can be selectively cross-linked leading to very interesting nano-objects.

easy synthesis of double amphiphilic block copolymers, characterized by similar pH-responsive behaviors to those described above [34–36].

Triblock terpolymers have also been considered for pH-sensitive micellization. This concept is illustrated for monomethoxy-capped PEO-PDMAEMA-PDEAEMA copolymers, which form micelles above pH 7.3 (Fig. 5.5) [37]. These micelles can be further stabilized by shell cross-linking. This can be achieved by using 1,2-bis-(2-iodoethoxy)ethane (BIEE) as the cross-linker. The cross-linking process locks the micellar structure and therefore impedes unimers from being obtained. However, it results in a very interesting three-layer nano-object in which the central core can still be changed from hydrophobic to hydrophilic. This peculiar stimuli-responsive behavior was used to advantage to encapsulate a drug and control its release. The drug was loaded into the micelles by increasing the pH of a copolymer–drug solution from 2 to 8. The drug release can either be slow when the micelles are stored at pH 8, or rapid when the micelles are dissociated by bringing the pH back to 3.

5.2.3
Ionic Strength Sensitive Micellization

Double hydrophilic copolymers containing positively and negatively charged blocks have been referred to as polyampholytic systems. Interpolyelectrolyte complexes (IPEC) between the oppositely charged blocks are observed in these systems, which

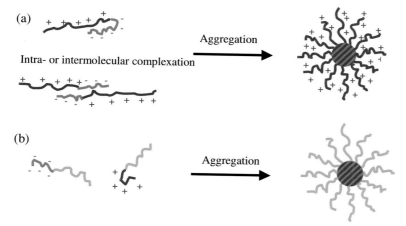

Fig. 5.6 Schematic representation of ionic strength sensitive micellization. (a) An asymmetric polyampholytic diblock is used. IPEC formation is observed. The insoluble complexes further self-assemble into micelles that are stabilized by the excess of uncomplexed charged blocks. (b) Two diblocks bearing opposite charges are mixed, resulting in the formation of IPEC, which further self-assemble into micelles. The charged blocks are linked to uncharged soluble blocks (in yellow), which allow the stabilization of the micelles.

can lead to insoluble material in aqueous solution. Solubility can, however, be restored whenever these electrostatic interactions are screened by the addition of salt. Ionic strength sensitive systems can therefore be designed from IPEC-containing copolymers. The operating principle of such ionic strength sensitive systems is comparable to the schematic representation depicted in Fig. 5.6a. Micellar systems have been observed for asymmetric PDMAEMA-PMAA copolymers containing an excess of negative or positive charges. These micelles contain cores formed by the insoluble interpolyelectrolyte complexes, which are surrounded by a corona formed by the uncomplexed segments [39–41]. As the degree of ionization of both the PDMAEMA and PMAA blocks depends on pH, the polyampholytic character is only observed in a restricted pH region in which both the constituent blocks are partially or totally ionized. The PDMAEMA block is characterized by an LCST in water, hence temperature-responsive systems can also be built from such copolymers [41]. Very recently, more sophisticated PDMAEMA-PMAA copolymers bearing a fullerene end-group have been prepared, which introduce multi-responsive systems that combine electrostatic, hydrophobic and charge transfer interactions [42].

Ionic strength responsive IPEC can also form in mixtures of two mutually interacting (co)polymers (see Fig. 5.6b). This strategy is exemplified for P2VP-PEO/PSS mixtures at low pH. In this case, IPECs are formed between the positively charged P2VP blocks and the negatively charged PSS ones. These complexes self-assemble into micelles stabilized by PEO coronal chains. Addition of a sufficient amount of salt results in the screening of the electrostatic interactions and in the breaking of the micelles [43].

Changes in ionic strength can not only screen electrostatic interactions in IPEC but also affect the solubility of non-ionic water-soluble polymers. For example, the water solubility of polyethers, such as PEO, critically depends on the ionic strength. This effect has been explained on the basis of a competition between the metallic cations of the added salt that can be complexed to PEO and the initial hydrogen bonds between the water molecules and PEO. Addition of a sufficient amount of salt then results in the insolubilization of the polymer. Such an effect is demonstrated in the work of Patrickios et al. who used ABC triblock copolymers containing an insoluble PEVE block, a thermo-responsive PMVE mid-block and a water-soluble PMTGVE outer block. While in aqueous solutions only unimers were found, addition of salt led to aggregates.

5.3
Stimuli-responsive Micelles

In the former section, the effect of external stimuli on the unimer–micelle equilibrium was essentially discussed. Stimuli can, however, affect other characteristic features of block copolymer micelles, such as the size of one of their specific compartments (core or corona) or their morphology, as depicted in Fig. 5.7. These different features will be discussed in the present section.

Stimuli-sensitive micellar systems can be designed from classical amphiphilic diblock copolymers containing a hydrophobic block linked to a stimuli-responsive water-soluble block. This type of system forms micelles containing hydrophilic stimuli-responsive coronal chains that can become hydrophobic. The application of the stimulus results, in this case, in the complete hydrophobization of the copolymer, and precipitation or flocculation of the micelles is generally observed through intermicellar aggregation. Typical examples of such systems that have been reported are copolymers containing a thermo-sensitive water-soluble PNI-

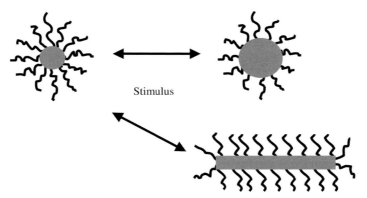

Fig. 5.7 Micelles showing a change in the size of one of their compartments or a morphological transition upon application of a stimulus.

PAAM block linked to hydrophobic blocks [45, 46]. Another related example has been reported for PIBVE-PMVE block copolymers in water. At room temperature, the PIBVE blocks are hydrophobic while the PMVE are hydrophilic. Micelles containing a PIBVE core surrounded by a PMVE corona are thus observed. Once the temperature is raised above the LCST of the PMVE blocks, intermicellar aggregation is observed. This aggregation process was demonstrated to be reversible and the initial micelles were again observed as the solution was cooled down to room temperature [47].

Typical examples in which a stimulus triggers a change in the micellar dimension (Fig. 5.7) are reported for ABC terpolymer-based micelles. PS-P2VP-PEO micelles consisting of a PS core, a pH-sensitive P2VP shell and a PEO corona are an illustration of this concept [48, 49]. At pH values above 5, the P2VP blocks are hydrophobic and collapsed onto the PS core, while at pH values below 5, they are protonated and adopt a stretched conformation because of the mutual electrostatic repulsions. These two micellar states have been characterized by DLS and visualized by TEM and AFM, as shown in Fig. 5.8 for spherical PS-P2VP-PEO micelles. Moreover, the transition between the two types of micelles was entirely reversible and could be repeated many times.

This concept was recently extended to the so-called metallo-supramolecular PS-P2VP-[Ru]-PEO triblock terpolymer, where -[Ru]- represents a ruthenium(II)-terpyridine bis-complex located at the junction between the P2VP and PEO blocks [50]. In addition to the pH-sensitive P2VP block, these copolymers contain a stimuli-sensitive Ru complex. Indeed, the complex can be opened by adding a large excess of a strong competing ligand. This releases the PEO block and yields core-shell micelles bearing free terpyridine moieties on the surface of the shell. The micelles could then be functionalized by re-forming metal complexes between the terpyridines of the shell and the desired molecules bearing a terpyridine group [51].

The metallo-supramolecular approach can thus be seen as another method of introducing stimuli-responsive properties into the micellar system. Other pH-responsive micelles have been reported for ABC terpolymer-based micelles, as illustrated by the work of Giebeler and Stadler on PS-P2(4)VP-PMAA micelles. This example is characterized by a rather complex behavior as IPEC formation is expected to occur between the P2(4)VP and PMAA blocks near the isoelectric point of the copolymer.

Morphological changes can be also triggered by stimuli in block copolymer micelles (Fig. 5.7). The simplest approach to observing morphology changes in block copolymer micelles consists in changing one variable of the phase diagram of such a system. This implies working with block copolymer micelles at the thermodynamic equilibrium. Such morphological changes have been reported for PPO-PEO and PBO-PEO copolymers dissolved in water at fairly high concentration, i.e., above 20 wt-% [53–55]. The kinetics for equilibration between different morphologies can, however, be rather slow. In this respect, weeks or months are sometimes needed to see whether the micelles reached equilibrium. In a PPO-PEO/water system, the phase sequence can go through L_1 (disordered phase) to I_1 (cubic) to H_1

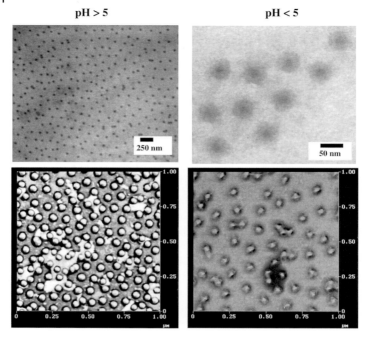

Fig. 5.8 pH-sensitive shell in core-shell-corona micelles from PS-P2VP-PEO. At pH > 5, the P2VP shell is hydrophobic and collapsed onto the PS core, its thickness is about 7.5 nm. At pH < 5, the P2VP shell is protonated and hydrophilic, it extends into the solution (thickness about 15 nm). The micelles can be reversibly shifted from one of these two states by repeated pH changes. Top images are TEM pictures in which a selective staining agent for both the PS and P2VP blocks has been added (RuO_4). Bottom images are AFM phase images (scale: white 45°, black 0°). Right images are recorded at pH > 5 while the left images correspond to pH < 5.

(hexagonal) and then to L_α (lamellar) with increasing polymer concentration or increasing temperature. In a PPO-PEO/water/oil ternary system, along the water to copolymer and then copolymer to oil sides, the phase can change from L_1 to I_1 to H_1 to V_1 to L_α and then to V_2 to H_2 to I_2 to L_2. These phase transitions can simply be ascribed to the changes in the volume fraction of the oil or the water phase.

Morphological changes induced by the addition of water to a PS-PAA copolymer dissolved in dioxane/water mixtures have been studied by Shen and Eisenberg [56]. Such copolymers are characterized by a large hydrophobic PS block and are reported for the formation of the so-called "crew-cut" micelles [57]. Compared with the PEO-based systems described above, morphological transitions occur at low polymer concentrations (0.1–10 wt-% polymer) and relatively low water contents (0–45 wt-% water). The morphologies encountered are in the order of spheres, rods and vesicles. On the basis of the reversibility of both the morphology and size of aggregates as a function of the water content, it was suggested that the various observed morphologies, including vesicles, are true equilibrated thermodynamic

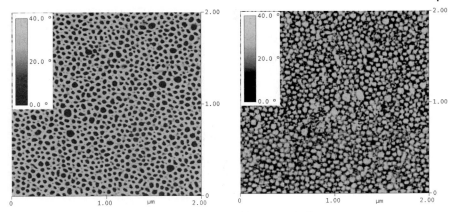

Fig. 5.15 AFM phase images of a PS$_{375}$-[Ru]-PEO$_{225}$ film (74 nm thickness) obtained by spin coating on silicon, showing the PEO cylindrical microdomains oriented normal to the film surface (left), and of the same film after creation of the pores by treatment with a Ce(SO$_4$)$_2$ solution (right). The phase scale is shown in the inset (reprinted with permission from [81]).

of metallo-supramolecular copolymers and is tentatively attributed to an ordering effect resulting from the charged ruthenium(II)–terpyridine bis-complexes.

5.6
Stimuli-responsive Block Copolymers in the Bulk

Stimuli-responsive properties have recently been imparted to block copolymer bulk materials. Temperature is an appropriate stimulus to address such materials because it does not require the use of a solvent to bring the material in contact with the stimulus. In this respect, the common pH or ionic strength stimuli are not appropriate to address the bulk materials. The effect of temperature on the general properties of bulk block copolymers is an obvious feature, which has been documented for some time. Indeed, the mechanical properties, e.g., loss or storage moduli, will be modified once the glass transition temperature of the constituent blocks is crossed. Moreover, the thermal energy will overcome at some point the Gibbs free energy associated with the phase separation process. As a result, the different constituent blocks of the copolymer will become miscible and the phase-separated structure will no longer be observed. Such a behavior is reversible, very well documented and referred to as the order–disorder transition. This issue will not be considered further, although it can be seen in principle as a stimulus-responsive behavior. In the following, we would like to highlight two examples of temperature-responsive behavior occurring in block copolymers before reaching the order–disorder transition.

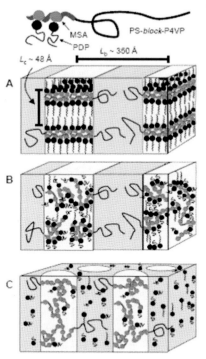

Fig. 5.16 Schematic representations of the different hierarchical organizations found in PS-P4VP-MSA-PDP complexes at temperatures below 100 °C (A), between 100 and 150 °C (B), above 150 °C (C) (reprinted with permission from [82]).

The first example is related to the use of block copolymer–surfactant complexes, which have already been considered within the framework of block copolymer thin films. This system is based on PS-P4VP copolymers whose P4VP blocks have been protonized by MSA. PDP has been further complexed via hydrogen-bonding interaction to the P4VP/MSA pairs (see Fig. 5.16) [82]. The present system thus differs from the previously described one (see Section 5.5) by the presence of the MSA molecule. The idea behind the use of MSA is to create charged species on the P4VP backbone in order to introduce ionic conductivity within the P4VP-MSA-PDP domains. At room temperature, a hierarchical layer-into-layer structure is observed (Fig. 5.16), according to the relative volume fractions of the PS and P4VP-MSA-PDP blocks. This lamellar morphology was observed to persist up to 100 °C. Between 100 and 150 °C, the lamellar morphology related to the block copolymer structure was still observed but the second layered structure had disappeared (Fig. 5.16). Therefore, an order–disorder transition within the polymer–amphiphile complex layers took place at 100 °C. Above 150 °C, the disordered P4VP-MSA-PDP domains formed hexagonal cylindrical domains embedded in a PS-PDP matrix. This transition is due to the fact that PDP became miscible with PS above about 130 °C and that the hydrogen bonding between PDP and P4VP-MSA diminished strongly at increased temperatures. This result implies that PDP will decouple from the P4VP-MSA backbone and gradually diffuse into the PS phase, thus effectively increasing the volume fraction of the PS-containing domains at the expense of the P4VP-MSA containing domains. The same workers demonstrated that this temperature-induce morphological transition could be used to modulate the macroscopic conductivity of the bulk sample [82].

Morphological transitions controlled by temperature have also been reported very recently by Wiesner and coworkers for diblock copolymers containing a polyether dendron linked to a PEO linear block (see Fig. 5.17) [83]. In contrast to most previous combinations of coil–dendron systems, the interface of these linearly extended dendrons is modeled in the middle of the dendritic structure rather than at the focal point. The hydrophilic part is thus composed of linear PEO and a PEO-like dendritic core, whereas the hydrophobic block consists of eight docosyl chains. Two samples, 1 and 2 (Fig. 5.17), differing in their composition were analyzed. The morphological transitions occurring upon heating of the samples were monitored by SAXS. At room temperature, the amphiphilic dendrons self-assemble into a crystalline lamellar organization (see for example Fig. 5.17B). After PEO and hydrophobic block melting but before isotropization, the SAXS pattern of 1 showed reflections (Fig. 5.17A), which can be indexed as a cubic micellar structure with $Pm3n$ symmetry. After PEO and hydrophobic periphery

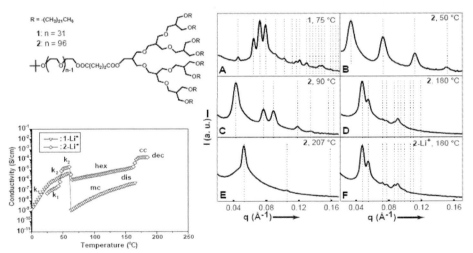

Fig. 5.17 Chemical structure (top left) of the two extended dendrons 1 and 2. Selected SAXS patterns of 1 and 2 at different temperatures (right). Ionic conductivity measured as a function of temperature for samples 1 and 2 doped with Li$^+$ ions (bottom left): k, crystalline; mc, micellar; hex, hexagonal columnar; lam, lamellar; cc, continuous cubic; dis, disordered; dec, decomposition (reprinted with permission from [83]).

melting, **2** shows three distinct mesophases as a function of temperature before isotropization. For the first mesophase, the SAXS pattern showed reflections consistent with a hexagonal mesophase (Fig. 5.17C). As the temperature increases, a cubic structure of $Ia3d$ symmetry was observed (Fig. 5.17D). Upon further heating, **2** displayed two reflections, in agreement with a lamellar mesophase (Fig. 5.17E). This sequence of transitions cannot be understood on the basis of simple diblock copolymer phase diagrams and is probably due to a larger expansion parallel to the interface as a function of temperature of the linear PEO/branched (dendritic) section of the molecules versus the hydrophobic docosyl parts.

Furthermore, it was demonstrated that the temperature-induced morphological transition was strongly correlated with mechanical properties and ionic conductivity of ion-doped samples at the macroscopic scale. Indeed, the different mesophases observed for this system have allowed the study of charge transport within a nanostructured material in which the conducting medium is confined to either micelles (zero-dimensional), cylinders (1D) or lamellae (2D), or is a continuous (3D) network throughout the entire macroscopic sample (Fig. 5.17).

5.7
Conclusions and Outlook

The recent advances in the field of stimuli-responsive block copolymers in bulk samples, thin films and in a selective solvent for one of the blocks have been reviewed briefly. Experimental investigation on these "smart block copolymers" has undergone tremendous progress in the past few years because they may find useful applications as versatile bio-related intelligent systems, such as gene or drug delivery, chromatography, microfiltration, actuators, sensors, injectable polymeric matrixes and artificial tissues or organs.

The progress achieved to date in the field of controlled or living polymerization techniques, especially controlled radical polymerization, has meant that a wide variety of stimuli-responsive block copolymers with well-defined architectures and molecular characteristic features are available. These new materials are now being investigated systematically with respect to their stimuli-responsive properties.

This research field is, however, still in its infancy and important breakthroughs are expected to occur within the next few years. Most of them are foreseen for bio-related applications provided that the following issues are considered.

Biocompatibility and biodegradability should be taken into account whenever these polymeric systems are applied in a physiological environment such as drug delivery.

Depending on the application, the physical form of the stimuli-responsive block copolymer has to be adjusted. In addition to micelles, bulk samples and thin films, hydrogels, functional surfaces and bioconjugates have to be considered.

The response time of such materials is also a very important factor for bio-applications but has scarcely been investigated. It could be adjusted in principle by appropriate design of the molecular structures.

The combination of other functionalities with stimuli-responsiveness could broaden the use of stimuli-responsive block copolymers in bio-applications. For example, functional groups could be incorporated into the stimuli-responsive copolymer in order to allow the coupling with specific drugs or recognition units. Coupling of enzymes to stimuli responsive block polymers could provide on/off functionality depending on the corresponding stimulus change.

Some interesting applications for bioseparation purposes could also be envisaged.

Abbreviations

AFM	atomic force microscopy
CMC	critical micelle concentration
CMT	critical micelle temperature
HABA	2-(4′-hydroxybenzeneazo)benzoic acid
HEMA	2-hydroxyethyl methacrylate
IPEC	interpolyelectrolyte complex
LCST	lower critical solubility temperature
MSA	methane sulfonic acid
MW	molecular weight
PAA	poly(acrylic acid)
PAMPS	poly(2-acrylamido)2-methyl propane sulfonate)
PBD	polybutadiene
PBO	poly(butylene oxide)
PCEMA	poly(cinnamoyl ethyl methacrylate)
PDEAEMA	poly(N,N-diethylaminoethyl methacrylate)
PDMAEMA	poly(N,N-dimethylaminoethyl methacrylate)
PDP	pentadecyl phenol
PEO	poly(ethylene oxide)
PEOVE	poly(2-ethoxyethyl vinyl ether)
PEOEOVE	poly[2-(2-ethoxy)ethoxyethyl vinyl ether)
PGA	poly(L-glutamic acid)
PLys	poly(L-lysine)
PMAA	poly(methacrylic acid)
PMMA	poly(methyl methacrylate)
PMOVE	poly(2-methoxyethyl vinyl ether)
PMTGEVE	poly[methyl tri(ethyleneglycol) vinyl ether]
PMVE	poly(methyl vinyl ether)
PNaMA	poly(sodium methacrylate)
PNIPAM	poly(N-isopropyl acrylamide)
PPO	poly(propylene oxide)
PS	polystyrene
PSCNa	poly(sodium 4-styrene carboxylate)
PSS	poly(styrene sulfonate)

PVA	poly(vinyl alcohol)
P2VP	poly(2-vinylpyridine)
P4VP	poly(4-vinylpyridine)
SAXS	small-angle X-ray scattering
SANS	small-angle neutron scattering
TEM	transmission electron microscopy
UCST	upper critical solubility temperature

References

1 Hamley IW (ed.) (1998) *The physics of block copolymers*. Oxford Science Publication, Oxford.

2 See for example: (a) Okano T (ed.) (1998) *Biorelated polymers and gels*. Academic Press, San Diego, CA; (b) Jeong B, Gutowska A (2002) *Trends Biotechnol.* 20, 305; (c) Galaev LY, Mattiasson B (2000) *Trends Biotechnol.* 17, 335. (d) Yokoyama M (2002) *Drug Discov. Today* 7, 426. (e) Chilkoti A, Dreher MR, Meyer DE, Raucher D, (2002) *Adv. Drug Deliv. Rev.* 54, 613. (f) Kataoka K, Harada A, Nagasaki Y (2001) *Adv. Drug Deliv. Rev.* 47, 113.

3 Kurisawa M, Yui N (1998) *J. Control. Release* 54, 191.

4 Hoffman AS, Stayton PS, Bulmus V, Chen G, Chen J, Cheung C, Chilkoti A, Ding Z, Dong L, Fong R, Lackey CA, Long CJ, Miura M, Morris JE, Murthy N, Nabeshima Y, Park TG, Press OW, Shimoboji T, Shoemaker S, Yang HJ, Monji N, Nowinski RC, Cole CA, Priest JH, Milton Harris J, Nakamae K, Nishino T, Miyata T (2000) *J. Biomed. Mater. Res.* 52, 577.

5 Gil ES, Hudson SM (2004) *Prog. Polym. Sci.* 29, 1173.

6 Rodriguez-Hernandez J, Chécot F, Gnanou Y, Lecommandoux S (2005) *Prog. Polym. Sci.* 30, 691.

7 Cölfen H (2001) *Macromol. Rapid Commun.* 22, 219.

8 Chu B, Zhou Z (1996) Physical chemistry of polyoxyalkylene block copolymer surfactants, in Nace VN ed. *Nonionic surfactants: polyoxyalkylene block copolymers*, (vol. 60), Marcel Dekker, New York.

9 Almgren M, Brown W, Hvidt S (1995) *Colloid Polym. Sci.* 273, 2.

10 Booth C, Attwood D (2000) *Macromol. Rapid Commun.* 21, 501.

11 Wanka G, Hoffmann H, Ulbricht W (1990) *Colloid Polym. Sci.* 268, 101.

12 Li H, Yu G, Price C, Booth C, Hecht E, Hoffmann H (1997) *Macromolecules* 30, 1347.

13 Topp MDC, Dijkstra PJ, Talsma H, Feijen J (1997) *Macromolecules* 30, 8518.

14 Virtanen J, Holappa S, Lemmetyinen H, Tenhu H (2002) *Macromolecules* 35, 4763.

15 Neradovic D, Hinrich WLJ, Kettenes-van den Bosch JJ, Hennink WE (1999) *Macromol. Rapid Commun.* 20, 577.

16 Neradovic D, van Nostrum CF, Hennink WE (2001) *Macromolecules* 34, 7589.

17 Aoshima S, Kobayashi E (1995) *Makromol. Chem. Makromol. Symp.* 95, 91.

18 Forder C, Patrickios CS, Billingham NC, Armes SP (1996) *Chem. Commun.* 883.

19 Forder C, Patrickios CS, Armes SP, Billingham NC (1996) *Macromolecules* 29, 8160.

20 Sugihara S, Kanaoka S, Aoshima S (2004) *J. Polym. Sci. Part A: Polym. Chem.* 42, 2601.

21 Martin TJ, Prochazka K, Munk P, Webber SE (1996) *Macromolecules* 29, 6071.

22 Reddy NK, Fordham PJ, Attwood D, Booth C (1990) *J. Chem. Soc., Faraday Trans.* 86, 1569.

23 Gohy JF, Mores S, Varshney SK, Zhang JX, Jérôme R (2002) *e-Polymers* 21.

24. FÖRSTER S, ABETZ V, MÜLLER AHE (2004) *Adv. Polym. Sci.* 166, 173.
25. BRONSTEIN LM, SIDOROV SN, VALETSKY PM, HARTMANN J, CÖLFEN H, ANTONIETTI M (1999) *Langmuir* 15, 6256.
26. SIDOROV SN, BRONSTEIN LM, KABACHII YA, VALETSKY PM, SOO PL, MAYSINGER D, EISENBERG A (2004) *Langmuir* 20, 3543.
27. MATHUR AM, DRESCHER B, SCRANTON AB, KLIER J (1998) *Nature (London)* 392, 367.
28. GOHY JF, VARSHNEY SK, JÉRÔME R (2001) *Macromolecules* 34, 3361.
29. HOLAPPA S, KARESOJA M, SHAN J, TENHU H (2002) *Macromolecules* 35, 4733.
30. KHOUSAKHOUN E, GOHY JF, JÉRÔME R (2004) *Polymer* 45, 8303.
31. GABASTON LI, FURLONG SA, JACKSON RA, ARMES SP (1999) *Polymer* 40, 4505.
32. BÜTÜN V, BILLINGHAM NC, ARMES SP (1997) *Chem. Commun.* 671.
33. GOHY JF, ANTOUN S, JÉRÔME R (2001) *Macromolecules* 24, 7435.
34. SUMERLIN BS, LOWE AB, THOMAS DB, MCCORMICK CL (2003) *Macromolecules* 36, 598.
35. MCCORMICK CL, LOWE AB (2004) *Acc. Chem. Res.* 37, 312.
36. YUSA S, SHIMADA Y, MITSUKAMI Y, YAMAMOTO T, MORISHIMA Y (2003) *Macromolecules* 36, 4208.
37. LIU S, WEAVER JVM, TANG Y, BILLINGHAM NC, ARMES SP (2000) *Macromolecules* 35, 6121.
38. TANG YQ, LIU SY, ARMES SP, BILLINGHAM NC (2003) *Biomacromolecules* 4, 1636.
39. GOHY JF, CREUTZ S, GARCIA M, MAHLTIG B, STAMM M, JÉRÔME R (2000) *Macromolecules* 33, 6378.
40. GOLUB T, DE KEIZER A, COHEN-STUART MA (2000) *Macromolecules* 33, 7672.
41. LOWE AB, BILLINGHAM NC, ARMES SP (1997) *Chem. Commun.* 1035.
42. TEOH SK, RAVI P, DAI S, TAM KC (2005) *J. Phys. Chem. B* 109, 4431.
43. GOHY JF, VARSHNEY SK, ANTOUN S, JÉRÔME R (2000) *Macromolecules* 33, 9298.
44. PATRICKIOS CS, FORDER C, ARMES SP, BILLINGHAM NC (1997) *J. Polym. Sci. Part A: Polym. Chem.* 35, 1181.
45. CHUNG JE, YOKOYAMA M AND OKANO T (2000) *J. Control. Release* 65, 103.
46. KOHORI F, SAKAI K, AOYAGI T, YOKOYAMA M, SAKURAI Y, OKANO T (1998) *J. Control. Release* 55, 87.
47. VERDONCK B, GOHY JF, KHOUSAKOUN E, JÉRÔME R, DU PREZ F (2005) *Polymer* 46, 9899.
48. GOHY JF, WILLET N, VARSHNEY S, ZHANG J-X, JÉRÔME R (2001) *Angew. Chem., Int. Ed. Engl.* 40, 3214.
49. GOHY JF, WILLET N, VARSHNEY SK, ZHANG J-X, JÉRÔME R (2002) *e-Polymers* 35.
50. GOHY JF, LOHMEIJER BGG, VARSHNEY SK, DÉCAMPS B, LEROY E, BOILEAU S, SCHUBERT US (2002) *Macromolecules* 35, 9748.
51. GOHY JF, LOHMEIJER BGG, SCHUBERT US (2003) *Chem. Eur. J.* 9, 3472.
52. GIEBELER E, STADLER R (1997) *Macromol. Chem. Phys.* 198, 3815.
53. WANKA G, HOFFMANN H, ULBRICHT W (1994) *Macromolecules* 27, 4145.
54. ALEXANDRIDIS P, HOLMQVIST P, LINDMAN B (1997) *Colloids Surf. A* 130, 3.
55. ALEXANDRIDIS P, OLSSON U, LINDMAN B (1998) *Langmuir* 14, 262.
56. SHEN H, EISENBERG A (1999) *J. Phys. Chem. B* 103, 9473.
57. ZHANG L, EISENBERG A (1995) *Science* 268, 1728.
58. LIU S, BILLINGHAM NC, ARMES SP (2001) *Angew. Chem., Int. Ed. Engl.* 40, 2328.
59. AROTÇARENA M, HEISE B, ISHAYA S, LASCHEWSKY A (2002) *J. Am. Chem. Soc.* 124, 3787.
60. SCHILLI CM, ZHANG M, RIZZARDO E, THANG SH, CHONG YK, EDWARDS K, KARLSSON G, MUELLER AHE (2004) *Macromolecules* 37, 7861.
61. CHÉCOT F, LECOMMANDOUX S, GNANOU Y, KLOK H-A (2002) *Angew. Chem., Int. Ed. Engl.* 4, 1339.
62. CHÉCOT F, LECOMMANDOUX S, KLOK H-A, GNANOU Y (2003) *Eur. Phys. J. Part E* 10, 25.
63. LECOMMANDOUX S, SANDRE O, CHÉCOT F, RODRIGUEZ-HERNANDEZ J, PERZINSKY R (2005) *Adv. Mater.* 17, 712.
64. RODRIGUEZ-HERNANDEZ J, LECOMMANDOUX S (2005) *J. Am. Chem. Soc.* 127, 2026.
65. SZCZUBIALKA K, NOWAKOWSKA M (2003) *Polymer* 44, 5269.

66 Rehse N, Knoll A, Magerle R, Krausch G (2003) *Macromolecules* 36, 3261.
67 Thurn-Albrecht T, Schotter J, Kästle GA, Emley N, Shibauchi T, Krusin-Elbaum L, Guarini K, Black CT, Tuominen MT, Russell TP (2000) *Science* 290, 2126.
68 Park M, Harrison C, Chaikin PM, Register RA, Adamson DH (1997) *Science* 276, 1401.
69 Morkved TL, Lu M, Urbas AM, Ehrichs EE, Jaeger HM, Mansky P, Russell TP (1996) *Science* 273, 931.
70 Huang E, Russell TP, Harrison C, Chaikin PM, Register RA, Hawker CJ, Mays J (1998) *Macromolecules* 31, 7641.
71 Elbs H, Drummer C, Abetz V, Krausch G (2002) *Macromolecules* 35, 5570.
72 Kim SH, Misner MJ, Russell TP (2004) *Adv. Mater.* 16, 2119.
73 Fukunaga K, Elbs H, Magerle R, Krausch G (2000) *Macromolecules* 33, 947.
74 Thurn-Albrecht T, Steiner R, DeRouchey J, Stafford CM, Huang E, Bal M, Tuominen M, Hawker CJ, Russell TP (2000) *Adv. Mater.* 12, 787.
75 Zalusky AS, Olayo-Valles R, Wolf JH, Hillmyer MA (2002) *J. Am. Chem. Soc.* 124, 12761.
76 Ndoni S, Vigild ME, Berg RH (2003) *J. Am. Chem. Soc.* 125, 13366.
77 Mäki-Ontto R, de Moel K, de Odorico W, Ruokolainen J, Stamm M, ten Brinke G, Ikkala O (2001) *Adv. Mater.* 13, 117.
78 Sidorenko A, Tokarev I, Minko S, Stamm M (2003) *J. Am. Chem. Soc.* 125, 12211.
79 Gohy JF, Lohmeijer BGG, Schubert US (2002) *Macromolecules* 35, 4560.
80 Lohmeijer BGG, Schubert US (2002) *Angew. Chem., Int. Ed. Engl.* 41, 3825.
81 Fustin CA, Lohmeijer BGG, Duwez AS, Jonas AM, Schubert US, Gohy JF (2005) *Adv. Mater.* 17, 1162.
82 Ruokolainen R, Mäkinen R, Torkelli M, Mäkelä T, Serimaa R, ten Brinke G, Ikkala O (1998) *Science* 280, 557.
83 Cho BK, Jain A, Gruner SM, Wiesner U (2004) *Science* 305, 1598.

6
Self-assembly of Linear Polypeptide-based Block Copolymers

Sébastien Lecommandoux, Harm-Anton Klok, and Helmut Schlaad

6.1
Introduction

Now that the fundamental principles that underlie the self-assembly of block copolymers have been addressed in numerous theoretical and experimental studies, these materials are finding increasing interest in several nanotechnology applications, such as nanostructured membranes, templates for nanoparticle synthesis and high-density information storage [1–3]. The self-assembly of block copolymers composed of two (A and B) or more (A, B, C, ...) chemically incompatible, amorphous segments is determined by the interplay of two competitive processes [4, 5]. On the one hand, in order to avoid unfavorable monomer contacts, the blocks segregate and try to minimize the interfacial area. Minimization of the interfacial area, however, involves chain stretching, which is entropically unfavorable. It is the interplay between these two processes that determines the final block copolymer morphology. Depending on the volume fractions of the respective blocks, amorphous AB diblock copolymers can form lamellar, hexagonal, spherical and gyroid structures. More complex morphologies can also be generated, but these require alternative strategies. One possibility is to increase the number of different blocks [6]. Other strategies to create more complex block copolymer nanostructures include the introduction of more rigid or (liquid) crystalline blocks [7, 8]. Often, such conformationally restricted segments introduce additional secondary interactions, such as electrostatic, hydrogen bonding or π–π interactions, which have an impact on the block copolymer self-assembly. In particular, rod–coil type block copolymers composed of a rigid (crystalline) block and a flexible (amorphous) segment have attracted increased interest over the past years [9, 10]. Morphological studies on rod–coil block copolymers have revealed several unconventional nanoscale structures, which were previously unknown for purely amorphous block copolymers. These findings underline the potential of manipulating chain conformation and interchain interactions to further engineer block copolymer self-assembly.

The self-assembly of amphiphilic block copolymers in solution, driven by the incompatibility of constituents, into ordered structures in the sub-micrometer range

---

Block Copolymers in Nanoscience. Massimo Lazzari, Guojun Liu, Sébastien Lecommandoux
Copyright © 2006 WILEY-VCH Verlag GmbH & Co. KGaA, Weinheim
ISBN: 3-527-31309-5

is a current topic in colloid and materials science [11–15]. The basic structures of diblock copolymers in solution are, in the order of decreasing curvature, spherical and cylindrical micelles and vesicles; the curvature of the core–corona interface is essentially determined by the volume fractions of comonomers and environmental factors (solvent, ionic strength, etc.) [16]. Polymer vesicles, often also referred to as "polymersomes", are particularly interesting as mimetics for biological membranes [17–20]. Deviation from this conventional aggregation behavior and appearance of more complex superstructures occur, as in biological systems, when specific non-covalent interactions, chirality and secondary structure effects come into play [21, 22]. Particularly interesting are block copolymers that combine advantageous features of synthetic polymers (solubility, processability, rubber elasticity, etc.) with those of polypeptides or polysaccharides (secondary structure, functionality, biocompatibility, etc.).

This chapter discusses the solid-state and solution structures, organization and properties of polypeptide-based block copolymers. Most of the block copolymers studied so far are composed of a synthetic block and a peptide segment and are an interesting class of materials, both from a structural and a functional point of view [23, 24]. Peptide sequences can adopt ordered conformations, such as α-helices or β-strands (Fig. 6.1). In the former case, this leads to block copolymers with rod–coil character. Peptide sequences with a β-strand conformation can undergo intermolecular hydrogen bonding, which also offers additional means to direct nanoscale structure formation compared with purely amorphous block copolymers. Combining peptide sequences and synthetic polymers, however, is not only interesting to enhance control over nanoscale structure formation, but can also result in materials that can interface with biology. Such biomimetic hybrid polymers or molecular chimeras [25] may produce sophisticated superstructures with new materials properties. "Smart" materials based on polypeptides may reversibly change the conformation and, along with it, properties in response to an environ-

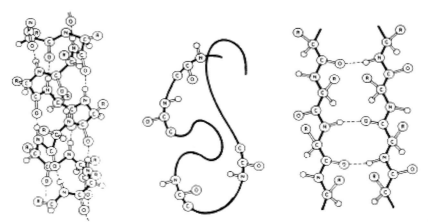

Fig. 6.1 Illustration of the basic secondary structure motifs of polypeptides: (a) α-helix, (b) random coil and (c) antiparallel β-sheet.

mental stimulus, such as a change in pH or temperature [26]. Also, polypeptide block copolymers may be used as model systems to study generic self-assembly processes in natural proteins. Obviously, such materials would be of great potential interest for a variety of biomedical and bioanalytical applications.

6.2
Solution Self-assembly of Polypeptide-based Block Copolymers

6.2.1
Aggregation of Polypeptide-based Block Copolymers

6.2.1.1 Polypeptide Hybrid Block Copolymers

Corona-forming Polypeptides Block copolymers with soluble and thus corona-forming polypeptide segments include linear diblock and triblock copolymer samples. In most cases, studies on the aggregation behavior were carried out in aqueous solutions with samples consisting of a soft polybutadiene (PB) or polyisoprene (PI) (exhibiting a glass transition temperature, T_g, below the freezing point of water) and an α-helical poly(L-glutamate) (PLGlu) or poly(L-lysine) (PLLys) segment.

The first study, reported in 1979 by Nakajima et al. [27], dealt with the structure of aggregates of symmetric triblock copolymers consisting of a coiled *trans*-1,4-PB middle-block and two α-helical poly(γ-benzyl L-glutamate) outer-blocks, $PBLGlu_{53-188}$-b-PB_{64}-b-$PBLGlu_{53-188}$ (subscripts denote number averages of repeat units, P_n), so-called "once-broken rods", in chloroform. The shape and dimensions of the aggregates in solution were calculated on the basis of simple thermodynamic considerations by taking into account chain conformation and the interfacial free energy. Predictions were found to be in good agreement with the structures (of solvent-cast films) observed by transmission electron microscopy (TEM). Depending on the composition of the copolymer, aggregates had a spherical, cylindrical or lamellar structure with a characteristic size of about 25–45 nm. Similar data were also obtained for block copolymers based on poly(γ-methyl L-glutamate) (PMLGlu) and poly(N^ε-benzyloxycarbonyl L-lysine) (PZLLys) [28]. These results suggest that the aggregation of rod–coil block copolymers might be treated in the same way as conformationally isotropic samples [29], provided that the rigid segment is dissolved in the continuous phase.

Schlaad and coworkers [30] and Lecommandoux and coworkers [31] investigated the aggregates of PB_{27-119}-b-$PLGlu_{20-175}$ in aqueous saline solution by dynamic and static light scattering (DLS/SLS), small-angle neutron scattering (SANS) and TEM. Copolymers were found to form spherical micelles with a hydrodynamic radius of $R_h < 40$ nm (70–75 mol-% glutamate) or unilamellar vesicles with $R_h = 50–90$ nm (17–54 mol-% glutamate); cylindrical micelles have not been observed so far. Against all expectations, however, Klok and Lecommandoux and coworkers [32, 33] earlier reported not micelles but vesicles for a PB_{40}-b-$PLGlu_{100}$ containing 71 mol-% glutamate.

DLS and SANS showed that any pH-induced changes of the secondary structure of poly(L-glutamate) from a random coil (pH > 6) to an α-helix (pH < 5) (CD spectroscopy) did not have a severe impact on the morphology (curvature) of the aggregates. SANS further suggested that aggregation numbers remained the same [31], despite equilibration of the sample and a dynamic exchange of polymer chains between aggregates [30]. Coiled and α-helical polypeptide chains seem to have similar spatial requirements at the core–corona interface (see also [27]). However, as the contour length of an all-*trans* polypeptide chain is more than twice that of an α-helix, in particular the hydrodynamic size of the aggregates might decrease when decreasing the pH of the solution. A decrease of the hydrodynamic radius by 20% or less could be observed for PB_{48}-*b*-$PLGlu_{56-145}$ [31] but, however, not for PB_{27-119}-*b*-$PLGlu_{24-64}$ [30]. PI_{49}-*b*-$PLLys_{123}$ micelles in saline, also reported by Lecommandoux and coworkers [34], exhibited a hydrodynamic radius of $R_h \approx 44$ nm at pH 6 (coil) and of 23 nm at pH 11 (helix) (DLS), which corresponds to a decrease in size by almost 50%. Interestingly, although the polypeptide segment is of nearly the same length, the effect seen for PI_{49}-*b*-$PLLys_{123}$ micelles is much larger than that for PB_{48}-*b*-$PLGlu_{114}$ ($\approx 8\%$). A possible explanation might be that the PLGlu helices are disrupted [28] and/or folded and hence less stretched as PLLys helices.

It is worth noting that the coronae of micelles of PB_{48}-*b*-$PLGlu_{114}$ and PI_{49}-*b*-$PLLys_{178}$ could be stabilized using 2,2′-(ethylenedioxy)bisethylamine and glutaric dialdehyde as cross-linking agents, respectively (Lecommandoux et al. [35]). The size and morphology of the aggregates were not affected by the chemical modification reaction.

Ouchi and coworkers [36] studied the aggregation of poly(L-lactide)-*block*-poly(aspartic acid), $PLL_{95, 270}$-*b*-$PAsp_{47-270}$, in pure water and in 0.2 M phosphate buffer solution at pH 4.4–8.6. Irrespective of the chemical composition of the copolymer (21–74 mol-% aspartic acid), however, only spherical aggregates with $R_h = 10$–80 nm could be observed [DLS and scanning force microscopy (SFM)]. The aggregation behavior of the samples was rationalized in terms of a balance between hydrophobic interactions in the PLL core and electrostatic repulsion and hydrogen-bridging interactions in the PAsp corona. That the aggregates might be non-equilibrium structures being kinetically trapped in a "frozen" state due to semi-crystallinity of PLL chains was not considered. Metastability could be an explanation for the exclusive formation of spherical micelles in addition to the seemingly arbitrary changes of the size of aggregates in buffer solutions at different pH.

Klok and coworkers [37] used DLS/SLS, SANS and analytical ultracentrifugation (AUC) for the analysis of aggregates formed by $PS_{8, 10}$-*b*-$PLLys_{9-72}$ (PS = polystyrene) in dilute aqueous solution at neutral pH; at this pH, PLLys was in a random coil conformation. They observed, however, cylindrical micelles regardless of the length of the PLLys segment. A conclusive explanation of this unexpected aggregation behavior could not be given.

Core-forming polypeptides Aggregates with an insoluble polypeptide core have been prepared with block or random copolymers having linear or branched architecture. Most studies focused on aqueous systems (designated for use as drug carriers in biomedical applications), and the only inverse systems investigated to date are PB-b-PLGlu in dilute tetrahydrofuran (THF) and CH_2Cl_2 solution and PS-b-PZLLys in CCl_4.

Harada and Kataoka [38–40] were the first to investigate the formation of polyion complex (PIC) micelles in an aqueous milieu from a pair of oppositely charged linear polypeptide block copolymers, namely of PEG_{113}-b-$PAsp_{18, 78}$ polyanions and PEG_{113}-b-$PLLys_{18, 78}$ polycations [PEG = poly(ethylene glycol)]. Complexation studies were carried out at pH 7.29, where both block copolymers had the same degree of ionization (α = 0.967) and were thus double-hydrophilic in nature and did not form aggregates in water. Mixing of the copolymers at a 1:1 ratio of amino acid residues resulted in the formation of stable and monodispersed spherical core-shell assemblies of 30 nm in diameter (DLS). Another interesting feature connected with PIC micelles is that of "chain-length recognition" [40]. PIC micelles are exclusively formed by matched pairs of chains with the same block lengths of polyanions and polycations, even in mixtures with different block lengths. The key determinants in this recognition process are considered to be the strict phase separation between the PIC core and the PEO corona, requiring regular alignment of the molecular junctions at the core–corona interface, and the charge stoichiometry (neutralization).

Yonese and coworkers [41] studied the aggregation behavior of PEG_{113}-b-$PMLGlu_{20, 50}$ and lactose-modified PEG_{75}-b-$PMLGlu_{32}$ in water. As shown by DLS, the copolymers formed large aggregates with a hydrodynamic radius of $R_h \approx 250$ nm. Contrary to what was claimed by these workers, it seems more likely that these aggregates were vesicles rather than spherical micelles. Key in the aggregation behavior might be the association of α-helical PMLGlu segments, as evidenced by CD spectroscopy, promoting the formation of plane bi-layers which then close into vesicles [15]. Further systematic studies on this system and detailed analysis of structures are lacking.

Closely related to this system are PEO_{272}-b-$PBLGlu_{38-418}$ and $PNIPAAm_{203}$-b-$PBLGlu_{39-123}$ [PEO = poly(ethylene oxide), PNIPAAm = poly(N-isopropylacrylamide)] described by Cho and coworkers [42, 43]. The aqueous polymer solutions, prepared by the dialysis of organic solutions against water, contained large spherical aggregates ($R_h \approx 250$ nm) with a broad size distribution (DLS). Although the size suggested a vesicular structure of the aggregates, aggregation numbers ($Z < 100$, method of determination not specified) were far below the values of several thousands typically being reported for polymer vesicles [18]. It is also worth noting that the PNIPAAm chains exhibit LCST (lower critical solution temperature) behavior. However, raising the temperature to the LCST (≈ 34 °C) had no serious impact on the size of the aggregates.

Dong and coworkers described symmetric triblock copolymers with a glyco methacrylate middle block and two outer poly(L-alanine) (PLAla) or PBLGlu blocks [44, 45]. The aggregates formed in dilute aqueous solution were spherical in shape and were 200–700 nm in diameter (TEM). TEM further revealed a compact structure of the aggregates as with multi-lamellar vesicles. The dimensions of the

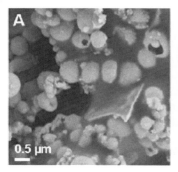 

Fig. 6.2 SEM images of the aggregates formed by (A) PS_{258}-b-$PZLLys_{57}$ and (B) PS_{258}-b-$PZLLys_{109}$ in dilute CCl_4 solution; specimens were prepared by shock-freezing a 0.2 wt-% polymer solution with liquid nitrogen and subsequent freeze-drying [47].

particles, however, were found to decrease with increasing concentration of the copolymer.

Naka et al. [46] studied the aggregation behavior of poly(acetyliminoethylene)-*block*-poly(L-phenylalanine), $PAEI_{41}$-b-$PLPhe_{4,8}$, in a 0.05 M phosphate buffer at pH 7. Aggregates were observed despite the very low number of hydrophobic L-phenylalanine units, and the size of which was in the order of R_h = 425 nm (DLS). The seemingly high tendency of these polymers to form aggregates was attributed to the establishment of hydrogen bridges between the amino acid units, as shown by IR spectroscopy, in addition to hydrophobic interactions. Visualization of the aggregates with TEM strongly suggested the presence of coacervates or large clusters of small micelles but no vesicles.

The existence of vesicles could be demonstrated for PS_{258}-b-$PZLLys_{57}$ in dilute CCl_4 solution (Losik and Schlaad [47]). Scanning electron microscopy (SEM) showed collapsed hollow spheres of about 300–600 nm in diameter, indicative of vesicles, and also sheet-like structures, supposedly bi-layers that are not yet closed to vesicles (Fig. 6.2A) [15]. The preference for a lamellar structure might be, as in the previous examples, attributed to a stiffening of the core by the 2D-arrangement of crystallizable PZLLys α-helices. PS_{258}-b-$PZLLys_{109}$, on the other hand, was found to form large compact fibrils being hundreds of nanometers in diameter and several tens of microns in length (Fig. 6.2B); these aggregates might be cylindrical multi-lamellar vesicles. However, the processes involved in the formation of these structures are not yet known.

Likewise, Lecommandoux and coworkers [31] found vesicles for PB_{48}-b-$PLGlu_{20}$ in THF and in CH_2Cl_2 solution (R_h = 106–108 nm, DLS/SLS). The formation of vesicles rather than micelles was attributed to the α-helical rod-like secondary structure of the insoluble PLGlu that forms a planar interface.

6.2.1.2 Block Copolypeptides

Besides the polypeptide hybrid block copolymers described earlier, a few purely peptide-based amphiphiles and block/random copolymers (copolypeptides) exist. In the latter case, both the core and corona of aggregates consist of a polypeptide. All of the studies reported so far have dealt with aggregation in aqueous media.

Doi and coworkers [48, 49] observed the spontaneous formation of aggregates of PMLGlu$_{10}$ with a phosphate (P) head group in water. Immediately after sonication, the freshly prepared solution contained globular assemblies (diameter: 50–100 nm, TEM), which after 1 h transformed into fibrous aggregates, promoted by intermolecular hydrogen bonding between peptide chains. After one day, these fibrils assembled into a twisted ribbon-like aggregate (TEM). As its thickness was ≈4 nm, which is close to the contour length of PMLGlu$_{10}$-P in a β-sheet conformation (CD and FT-IR spectroscopy), these workers concluded that the formation of the ribbon was driven by a stacking of anti-parallel β-sheets via hydrophobic interactions (see above) [22].

Lecommandoux and coworkers [50] showed that zwitterionic PLGlu$_{15}$-b-PLLys$_{15}$ in water can self-assemble into unilamellar vesicles with a hydrodynamic radius of greater than 100 nm (Fig. 6.3). A change in the pH from 3 to 12 induced an inversion of the structure of the membrane (NMR) and was accompanied by an increase in the size of vesicles from 110 to 175 nm (DLS).

Non-ionic block copolypeptides made of L-leucine and ethylene glycol modified L-lysine residues, PLLeu$_{10-75}$-b-PELLys$_{60-200}$, were described by Deming and coworkers [51]. The copolymers adopted a rod-like conformation, due to the strong tendency of both segments to form α-helices, as confirmed by CD spectroscopy. The self-assembled structures observed in aqueous solutions included (sub-)micrometer vesicles, sheet-like membranes and irregular aggregates. Here again, it was shown that the vesicle formation is related to the systematic presence of the polypeptide in a rod-like conformation in the hydrophobic part of the membrane, inducing a low interfacial curvature and as a result a hollow structure.

Meyrueix and coworkers [52] performed the selective precipitation of PLLeu$_{180}$-b-PLGlu$_{180}$ and obtained nanoparticles, which could be purified and further suspended in water or in a 0.15 M phosphate saline buffer at pH 7.4. The colloidal

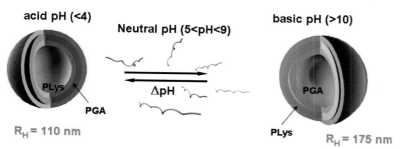

Fig. 6.3 Schematic representation of the self-assembly of zwitterionic PLGlu$_{15}$-b-PLLys$_{15}$ in water into unilamellar vesicles [50].

dispersions were stable, due to the electrosteric stabilization of the particles by poly(sodium L-glutamate) brushes, containing spherical or cylindrical micelles, besides the large hexagonally-shaped platelets with a diameter of about 200 nm (TEM). Different shapes of particles were due to the heterogeneity of copolymer chains with respect to chemical composition (NMR): glutamate-rich chains formed micelles and leucine-rich ones formed platelets. CD spectroscopy and X-ray diffraction suggested that the core of platelets consisted of crystalline, helical PLLeu segments, and the structural driving force was thus related to the formation of leucine zippers in a three-dimensional array.

6.2.2
Polypeptide-based Hydrogels

Protein-based hydrogels are used for many applications, ranging from food and cosmetic thickeners to support matrices for drug delivery and tissue replacement. These materials are usually prepared using proteins extracted from natural resources, which can give rise to inconsistent properties unsuitable for medical applications.

Recently, Deming and coworkers [53–56] designed and synthesized diblock copolypeptide amphiphiles containing charged and hydrophobic segments. It was found and demonstrated that gelation depends not only on the amphiphilic nature of the polypeptides, but also on chain conformation, meaning α-helix, β-strand or random coil. Specific rheological measurements were performed to evidence the self-assembly process responsible for gelation [55]: the rod-like helical secondary structure of enantiomerically pure PLLeu blocks is instrumental for gelation at polypeptide concentrations as low as 0.25 wt-%. The hydrophilic polyelectrolyte segments have stretched coil configurations and stabilize the twisted fibrillar assemblies by forming a corona around the hydrophobic cores (Fig. 6.4).

Interestingly, these hydrogels can retain their mechanical strength up to temperatures of about 90 °C and recover rapidly after stress. This new mode of assembly was found to give rise to polypeptide hydrogels with a unique combination of properties, such as heat stability and injectability, making them attractive for applications in foods, personal care products and medicine. In this context, their potential application as tissue engineering scaffolds has been recently studied [57].

6.2.3
Organic/Inorganic Hybrid Structures

Recently, polypeptide-based copolymers have also been used for the stabilization or synthesis of inorganic species. As a first example, Stucky and coworkers [58] used block copolypeptides to direct the self-assembly of silica into spherical and columnar morphologies at room temperature and neutral pH. Stucky, Deming and colleagues [59] also designed a double-hydrophilic block copolypeptide poly{N^{ε}-2[2-(2-methoxyethoxy)ethoxy]acetyl L-lysine}$_{100}$-b-poly(sodium L-aspartate)$_{30}$ (PELLys-b-PNaLAsp) that can direct the crystallization of calcium carbonate into micro-

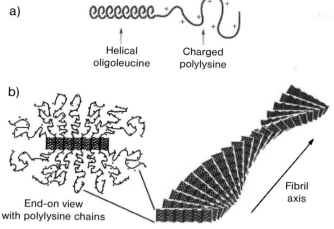

Fig. 6.4 Drawings showing (A) representation of a block copolypeptide chain and (B) proposed packing of block copolypeptide amphiphiles into twisted fibrillar tapes, with helices packed perpendicular to the fibril axes. Polylysine chains were omitted from the fibril drawing for clarity (reprinted from [55] with permission of The American Chemical Society).

spheres. They incorporated PLAsp in the diblock because domains of anionic aspartate residues are known to nucleate calcium carbonate crystallization. This effect is believed to be caused by matching interactions between aspartate and the atomic spacing of certain crystal faces in the growing mineral.

Also worth mentioning is the application of polypeptide-based copolymers in the production of magnetic nanocomposite materials. Lecommandoux et al. [60] obtained stable dispersions of super-paramagnetic micelles and vesicles by combining an aqueous solution of PB_{48}-b-$PLGlu_{56-145}$ with a ferrofluid consisting of

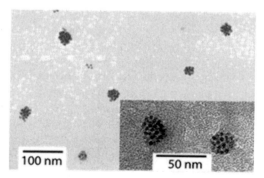

Fig. 6.5 TEM images of clusters of maghemite nanoparticles deposited from dispersions in water in the presence of $PELLys_{100}$-b-$PAsp_{30}$ (reprinted from [61] with permission of The American Chemical Society).

maghemite (γ-Fe$_2$O$_3$) nanoparticles. Incorporation of one mass equivalent of ferrofluid into the hydrophobic core of aggregates did not alter their morphology, as deduced from SLS and SANS data, but caused a substantial increase in the outer diameter by a factor of 6 (DLS). Interestingly, the hybrid vesicles underwent deformation under a magnetic field, as shown by 2D-SANS experiments. Held and coworkers [61] earlier reported that monodisperse, highly crystalline maghemite nanoparticles in organic solvents could be transferred into an aqueous medium using tetramethylammonium hydroxide stabilized at neutral pH. Combination of the aqueous maghemite solution with PELLys$_{100}$-b-PAsp$_{30}$ led to the formation of uniform clusters comprising approximately 20 nanoparticles (Fig. 6.5).

6.3
Solid-state Structures of Polypeptide-based Block Copolymers

6.3.1
Diblock Copolymers

6.3.1.1 Polydiene-based Diblock Copolymers

One of the first reports on the nanoscale solid-state structure of peptide–synthetic hybrid block copolymers was published by Gallot and coworkers [62] in 1976. In this publication, the solid-state structure of a series of PB-b-PBLGlu and PB-b-PHLGln diblock copolymers was investigated using a combination of techniques, including infrared and CD spectroscopy, X-ray scattering and TEM. The block copolymers covered a broad composition range with peptide contents ranging from 19 to 75 %. Interestingly, SAXS revealed a well-ordered lamellar superstructure characterized by up to four higher-order Bragg spacings for all of the investigated samples. The lamellar superstructure was confirmed by electron microscopy experiments, which were carried out on OsO$_4$-stained specimens. The intersheet spacings determined from the electron micrographs were in good agreement with the diffraction data. Wide-angle X-ray scattering (WAXS) experiments indicated that the α-helical peptide blocks were assembled in a hexagonal array. For a number of block copolymer samples it was found that the calculated length of the peptide helix was larger than the thickness of the polypeptide layer. To accommodate this difference, it was proposed that the peptide helices were folded in the peptide layer. The lamellar structure consists of plane, parallel equidistant sheets. Each sheet is obtained by superposition of two layers: (1) the PB chains in a more or less random coil conformation and (2) the α-helical polypeptide blocks in a hexagonal array of folded chains. The hexagonal-in-lamellar structure was also found for PB-b-PZLLys and PB-b-PLLys block copolymers by the same workers [63–66]. In the case of the PB-PLLys block copolymers, no periodic arrangement of the PLLys chains in the peptide layer was found. This is due to the fact that the polypeptide segments in these block copolymers are not exclusively α-helical but are composed roughly of 50 % random coil, 35 % α-helix and 15 % β-strand domains [63].

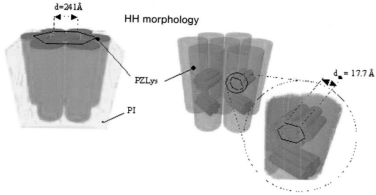

Fig. 6.6 Schematic model representation of the hexagonal in hexagonal (HH) morphology obtained for PI$_{49}$-b-PZLLys$_{178}$ copolymer cast from dioxane solution.

More recently, the solid-state nanoscale structure of PI-b-PZLLys diblock copolymers was reported [34]. Diblock copolymers composed of a PI block with a number-average degree of polymerization of 49 and a PZLLys block containing 61–178 amino acid residues were investigated with dynamic mechanical analysis and X-ray scattering. For the PI$_{49}$-b-PZLLys$_{35}$, PI$_{49}$-b-PZLLys$_{61}$ and PI$_{49}$-b-PZLLys$_{92}$, the X-ray scattering data were in agreement with a hexagonal-in-lamellar morphology. Interestingly, for PI$_{49}$-b-PZLLys$_{92}$ the lamellar spacing was found to decrease when the samples were prepared from dioxane instead of THF/N,N-dimethylformamide (DMF) and suggested folding of the peptide helices. For PI$_{49}$-b-PZLLys$_{123}$ and PI$_{49}$-b-PZLLys$_{178}$ a hexagonal-in-hexagonal structure was found. This morphology is illustrated in Fig. 6.6. This structure is unprecedented for polydiene-based peptide hybrid block copolymers, but has also been found for low molecular weight PS-b-PBLGlu copolymers [73].

6.3.1.2 Polystyrene-based Diblock Copolymers

In an early study, Gallot and coworkers [64] reported on the bulk nanoscale structure of PS-b-PZLLys diblock copolymers, which were based on a PS block with a

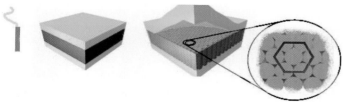

Fig. 6.7 Schematic model representation of the hexagonal-in-lamellar (HL) morphology obtained for polypeptide-based rod–coil diblock copolymers cast from solution.

number-average molecular weight of M_n = 37 kg mol^{-1} and had peptide contents ranging from 18 to 80 mol-%. X-ray scattering patterns of dry samples that had been evaporated from dioxane showed two sets of signals, characteristic of a hexagonal-in-lamellar superstructure. At very low angles, Bragg spacings characteristic of a layered superstructure were found, whereas at somewhat larger angles, there was a second set of reflections pointing towards a hexagonal arrangement of peptide helices. For several samples, the calculated length of the peptide helix was larger than the peptide layer thickness as determined from the X-ray data. In these cases, it was proposed that the helical PZLLys chains were folded in the peptide layer. Thus, the bulk nanoscale structure of the PS-*b*-PZLLys copolymers can be described in terms of the same hexagonal-in-lamellar model as was also proposed for the PB-based block copolymer described earlier (Fig. 6.7).

Removal of the of the side-chain protective groups of the peptide segment resulted in PS-*b*-PLLys diblock copolymers [63]. These copolymers were not water soluble, but formed mesomorphic gels at water contents of less than 50%. The X-ray scattering patterns indicated a lamellar superstructure, both in the gel state and the dry samples. In contrast to the side-chain protected block copolymers, no evidence for a periodic arrangement of the peptide chains was found. This is not too surprising considering that IR spectra indicated that roughly 50% of the peptide blocks have a random coil conformation, 35% an α-helical secondary structure and 15% a β-strand conformation.

Along the same lines, Douy and Gallot [66] also studied the bulk nanoscale organization of PS-*b*-PBLGlu. For block copolymers composed of a PS block with M_n = 25 kg mol^{-1} and containing 31–94 mol-% peptide, the same hexagonal-in-lamellar morphology as described above for the PS-*b*-PZLLys was found. The biocompatibility of PS-*b*-PBLGlu copolymers has been discussed in two publications [67, 68]. Mori et al. [68] studied diblock copolymers composed of a PS block with a number-average degree of polymerization of 87 and PBLGlu segments with number-average degrees of polymerization of 23, 52 or 83. Thrombus formation was assessed by exposing films of the diblock copolymers and the corresponding homopolypeptides to fresh canine blood. It was found that thrombus formation on the diblock copolymer films was reduced compared with the corresponding homopolymers. For the block copolymers, thrombus formation decreased with decreasing PBLGlu block length. Also, adsorption of plasma proteins such as bovine serum albumine, bovine γ-globulin and bovine plasma fibrinogen was reduced on the block copolymers compared with PS homopolymer.

The characterization of the solid-state nanoscale organization of PS-polypeptide hybrid block copolymers has recently been refined in a series of publications by Schlaad and coworkers [69–71]. In a first report, three PS-*b*-PZLLys diblock copolymers with peptide volume fractions of 0.48, 0.74 and 0.82 were investigated [69]. SAXS patterns recorded from DMF cast films confirmed the hexagonal-in-lamellar morphology published earlier by Gallot and coworkers [64]. In their paper, Schlaad and coworkers went a step further and analyzed their SAXS data using the interface-distribution concept and the curvature-interface formalism. These evaluation techniques suggested that the bulk nanoscale structure of the

PS-b-PZLLys diblock copolymers does not consist of plain but of undulated lamellae. The concept of the interface-distribution function and the curvature-interface formalism were also applied to compare the solid-state structures of two virtually identical PS based diblock copolymers; PS$_{52}$-b-PZLLys$_{111}$ ($\phi_{peptide}$ = 0.82) and PS$_{52}$-b-PBLGlu$_{104}$ ($\phi_{peptide}$ = 0.79) [70]. Analysis of the SAXS data obtained on DMF-cast films indicated a hexagonal-in-undulated (or zigzag) lamellar morphology for both block copolymers. However, the X-ray data also revealed two striking differences between the samples. The first difference concerns the thickness of the layers, which are a factor of three smaller for PS$_{52}$-b-PBLGlu$_{104}$ as compared with PS$_{52}$-b-PZLLys$_{111}$. Whereas the PZLLys helices are fully stretched, the PBLGlu helices are folded twice in the layers. As peptide folding increases the area per chain at the PS-PBLGlu interface, the thickness of the PS layers also has to decrease in order to cover the increased interfacial area. The second difference concerns the packing of the peptide helices. For the PZLLys-based diblock copolymer it was estimated that about 180 peptide helices form an ordered domain. The level of ordering, however, was considerably lower for the peptide blocks of PS$_{52}$-b-PBLGlu$_{104}$ and only ≈80 helices were estimated to form a single hexagonally ordered domain.

In addition, the influence of the polydispersity of the polypeptide block on the solid-state morphology of PS-b-PZLLys diblock copolymers has also been studied [71]. To this end, a series of five diblock copolymers was prepared from an identical ω-amino-polystyrene macroinitiator (P_n = 52; polydispersity index, PDI = 1.03). The peptide content in these diblock copolymers varied between 0.43 and 0.68 and the PDI ranged from 1.03 to 1.64. Evaluation of the SAXS data with the interface-distribution function and the curvature-interface formalism confirmed, as expected, the hexagonal-in-undulated (or zigzag) lamellar solid-state morphology. Fractionation of the peptide helices according to their length leads (locally) to the formation of an almost plane, parallel lamellar interface, which is disrupted by kinks (undulations). The curvature at the PS-PZLLys interface, however, was found to be strongly dependent on the chain length distribution of the peptide block. Block copolymers with the smallest molecular weight distribution produced lamellar structures with the least curvature. Increasing the chain length distribution of the peptide block (block copolymers with PDI ≈ 1.25) leads to larger fluctuations in the thickness of the PZLLys layers, which increases the number of kinks and the curvature at the lamellar interface. At even larger polydispersities (PDI ≈ 1.64), however, the number of kinks decreases again. With increasing polydispersity of the peptide block, the thickness fluctuations become larger and larger, as does the interfacial area. At a certain point, at sufficiently high polydispersity, the system tries to compensate for the increased interfacial tension and minimizes the number of kinks (Fig. 6.8).

Ludwigs et al. [72] used SFM to investigate the formation of hierarchical structures of PS$_{52}$-b-PBLGlu$_{104}$ in thin films. Thin films with a thickness of ≈4 and 40 nm were prepared by spin-coating of dilute polymer solutions on silicon substrates and were subsequently annealed in saturated THF vapor to achieve a controlled crystallization of the α-helical PBLGlu. On the smallest length-scale, the structure was found to be built of short ribbons or lamellae of interdigitated poly-

130 | *6 Self-assembly of Linear Polypeptide-based Block Copolymers*

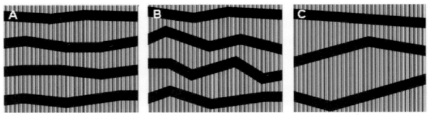

Fig. 6.8 Schematic representation of the disordered zigzag lamellar morphology formed by polypeptide-based diblock copolymers with low (A), moderate (B) and high polydispersity (C) with respect to the length of helices. Polypeptide helices are represented as cylinders, and polyvinyl sheets are depicted in black (reprinted from [71] with permission of The American Chemical Society).

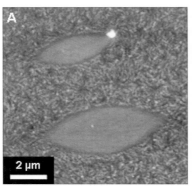

Height, $\Delta z = 60$ nm

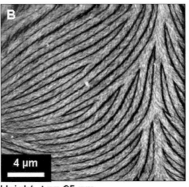

Height, $\Delta z = 25$ nm

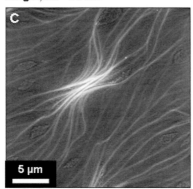

Height, $\Delta z = 80$ nm

Fig. 6.9 SFM height images of a film of PS_{52}-b-$PBLGlu_{104}$ obtained by spin-coating from a 5 mg mL^{-1} THF solution and subsequent exposure to saturated THF vapor for 3.5 (A), 22.5 (B) and 42 h (C) (reprinted from [72] with permission of The American Chemical Society).

mer chains. PBLGlu helices were fully stretched in thin films, in contrast to what has been observed in the 3D organized bulk mesophase (see above). Depending on the time of solvent annealing, different ordered structures on the micrometer length-scale could be observed (Fig. 6.9).

The examples discussed so far have all involved relatively high molecular weight diblock copolymers. In these cases, the molecular weight of the polypeptide block is usually sufficiently high so that it forms a stable α-helix and the common hexagonal-in-lamellar morphology is found. The situation changes, however, when the molecular weight of the block copolymers is significantly decreased. The influence of molecular weight on the solid-state organization of polystyrene-based peptide–synthetic hybrid block copolymers has been studied for a series of low-molecular weight PS-b-PBLGlu and PS-b-PZLLys [73, 74]. These diblock copolymers consisted of a short PS block with $P_n \approx 10$, a polypeptide block containing ≈10 to 80 amino acid repeat units and were characterized by means of variable temperature FT-IR spectroscopy and X-ray scattering. These experiments allowed the construction of "phase diagrams", which are shown in Fig. 6.10. The phase diagrams reveal a number of interesting features. At temperatures below 200 °C and for sufficiently long polypeptide blocks, a hexagonal arrangement of the diblock copolymers was found, analogous to the hexagonal-in-lamellar morphology of the high-molecular weight analogues. Upon decreasing the length of the peptide block, however, several novel solid-state structures were discovered. For very short peptide block lengths (PS$_{10}$-b-PBLGlu$_{10}$, PS$_{10}$-b-PZLLys$_{20}$, PS$_{10}$-b-PZLLys$_{40}$ and PS$_{10}$-b-PZLLys$_{60}$) a lamellar supramolecular structure was found. This is due to the fact that for such short peptide block lengths, a substantial fraction of the peptide blocks adopts a β-strand secondary structure. Self-assembly of these diblock copolymers in a β-sheet type fashion results in the lamellar structures observed by SAXS. For PS$_{10}$-b-PBLGlu$_{20}$ a peculiar and until then unprecedented structure was found. This structure that consisted of hexagonally packed diblock copolymer molecules, which are organized in a hexagonal superlattice, has been referred to as the double hexagonal or hexagonal-in-hexagonal morphology. Apart from several unconventional solid-state nanoscale structures, another factor that distinguishes the phase diagrams in Fig. 6.10 from those of most conventional, conformationally isotropic block copolymers is the influence of temperature. For a number of diblock copolymers, increasing the temperature above 200 °C results in a change from a hexagonal-in-hexagonal (PS$_{10}$-b-PBLGlu$_{20}$) or hexagonal (PS$_{10}$-b-PBLGlu$_{40}$, PS$_{10}$-b-PZLLys$_{80}$) to a lamellar morphology. FT-IR spectroscopy experiments suggested that these morphological transitions are induced by an increase in the fraction of peptide blocks that have a β-strand conformation.

6.3.1.3 Polyether-based Diblock Copolymers

PEG–polypeptide block copolymers are of particular interest, from both a structural and a functional point of view. Unlike the hybrid block copolymers discussed in the previous paragraphs, which were based on amorphous synthetic polymers, PEG is a semi-crystalline polymer. In addition to microphase separation and the tendency of the peptide blocks towards aggregation, crystallization of PEG intro-

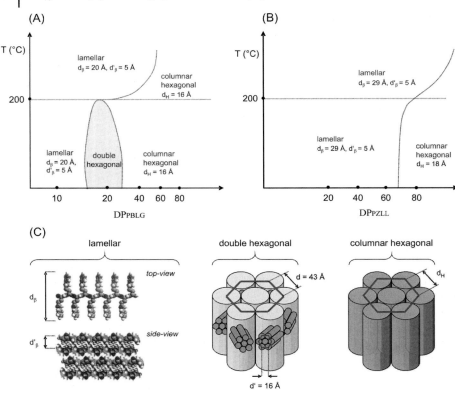

Fig. 6.10 Phase diagrams describing the solid-state nanoscale structure of (A) PS-b-PBLGlu and (B) PS-b-PZLLys diblock copolymers; (C) illustration of the lamellar, double hexagonal and hexagonal morphologies found for the low molecular weight hybrid block copolymers (reprinted from [73, 74] with permission of The American Chemical Society).

duces an additional factor that can influence the structure formation of these hybrid block copolymers. Ma and coworkers [75] have investigated the solid-state structure and properties of three PEG-b-PAla copolymers that were prepared from a PEG macroinitiator with M_n = 2 kg mol^{-1}. The diblock copolymers contained 39.8, 49.6 and 65.5 mol-% alanine. From FT-IR spectra and DSC measurements, these workers proposed a microphase-separated bulk structure.

AB diblock and ABA triblock copolymers composed of PEG as the A block and random coil segments of poly(D,L-valine-co-D,L-leucine) as the B block(s) were investigated by Cho and coworkers [76]. DSC experiments revealed PEG crystallization and showed that the PEG melting temperature was decreased compared with that of the PEG homopolymer. TEM micrographs suggested a lamellar microphase-separated structure for one of the triblock copolymer samples.

6.3.1.4 Polyester-based Diblock Copolymers

One of the first studies focusing on the solid-state properties of polypeptide–polyester synthetic hybrid block copolymers was reported by Jérôme and coworkers [77]. DSC experiments on a poly(ε-caprolactone)$_{50}$-*block*-poly(γ-benzyl L-glutamate)$_{40}$-(PCL$_{50}$-*b*-PBLGlu$_{40}$) diblock copolymer revealed two endotherms. The first endotherm was found at 60 °C and is due to the melting of the PCL. The second endotherm, which was located at 110 °C, was, mistakenly, interpreted as the melting transition of PBLGlu. This transition, however, is not a melting transition, but instead reflects the conformational transition of the PBLGlu helix from a 7/2 to an 18/5 helical structure. Although no further structural investigations were carried out, the observation of two separate endotherms occurring at temperatures identical to the transitions found for the respective homopolymers was a first indication for the existence of a microphase-separated structure. Similar results were reported by Chen and coworkers [78] who investigated the thermal properties of a series of PCL-*b*-PBLGlu copolymers composed of PCL blocks containing 13–51 repeat peptide segment units with and 22–52 amino acid repeat units.

Caillol et al. [79] have studied the solid-state structure and properties of a series of PLL-*b*-PBLGlu [PLL = poly(L-lactide)] copolymers. The PLL block in these copolymers contained 10–40 repeat units and the peptide segments were composed of 20–100 repeat units. DSC thermograms of the block copolymers revealed three transitions corresponding to the T_g of PLL ($\approx$50 °C), the 7/2 to 18/5 helix transition of PBLGlu ($\approx$100 °C) and the melting temperature of PLL ($\approx$160 °C), respectively. This observation was already providing a first hint towards a microphase-separated bulk morphology. SAXS experiments, which were performed at 100 °C, indicated the existence of hexagonally ordered assemblies of α-helical PBLGlu chains. With decreasing glutamate content, the peaks corresponding to this hexagonal organization decreased in intensity and another scattering peak appeared, which was ascribed to a lamellar assembly of PBLGlu chains with a β-strand secondary structure. Increasing the temperature to 200 °C not only resulted in melting of PLL, but also led to a decrease in intensity of the diffraction peaks corresponding to the hexagonally ordered α-helical PBLGlu segments and an increase in the fraction of PBLGlu segments that are ordered in a lamellar β-strand fashion.

6.3.1.5 Diblock Copolypeptides

A step forward in the design of hierarchically ordered structures with biofunctionality has been the subject of recent reports on the synthesis of block copolymers based on polypeptides. In the first such report [80], organo-nickel initiators rather than amines were used to avoid the unwanted α-amino acid N-carboxyanhydrides (NCA) side reactions, which had, for more than 50 years, hampered the formation of well-defined copolypeptides. This approach gave rise to various peptidic-based block copolymers that have mainly been studied in solution (see previous section). A second approach addressed the side reaction problem directly by using amines in combination with high-vacuum techniques, to ensure the necessary conditions for the living polymerization of NCAs: PBLGlu-*b*-PGly (PGly = polyglycine) were prepared for the first time with this methodology [81].

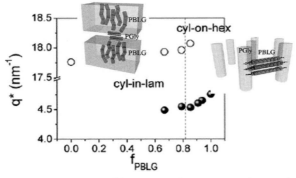

Fig. 6.11 Dependence of the WAXS peak positions on the PBLGlu (PBLG) volume fraction, corresponding to the distance between PBLGlu α-helices (filled circles) and PGly β-sheets (open circles). The interhelix (inter-sheet) distance increases (decreases) with increasing polypeptide volume fraction. The vertical line separates the two nanodomain morphologies.

Despite these important synthetic efforts, the solid-state morphology of purely peptidic block copolymers is largely unexplored. Hadjichristidis and coworkers [81] recently investigated the self-assembly of a series of narrow polydispersity PBLGlu-b-PGly diblock copolymers within the composition range $0.67 < f_{BLGlu} < 0.97$ and the temperature range $303 < T < 433$ K. SAXS and WAXS, ^{13}C NMR and DSC were used for the structure investigation coupled with dielectric spectroscopy for both the peptide secondary structure and the associated dynamics. These techniques not only provided insight into the nanophase morphology but also gave information about the type and persistence of peptide secondary structures. Particular evidence has been found for hexagonal-in-lamellar and cylinder-on-hexagonal nanostructures (Fig. 6.11). The thermodynamic confinement of the blocks within the nanodomains and the disparity in their packing efficiency results in multiple chain folding of the PGly secondary structure that effectively stabilizes a lamellar morphology for high f_{BLGlu}. Nanoscale confinement proved to be important in controlling the persistence length of secondary peptide motifs.

6.3.2
Triblock Copolymers

6.3.2.1 Polydiene-based Triblock Copolymers

Whereas Gallot and coworkers have mainly studied the solid-state organization of PB-based diblock copolymers, Nakajima et al. concentrated on ABA-type hybrid block copolymers containing PB as the B component ("once-broken rods"). In a first series of publications, the structure and properties of PBLGlu-b-PB-b-PBLGlu triblock copolymers containing 7.5–32.5 mol-% (= 3.0–14.3 vol-%) PB were investigated [82–84]. Infrared spectroscopy and WAXS experiments on films of the triblock copolymers indicated that the PBLGlu blocks were predominantly α-helical. From the WAXS experiments, it was concluded that the PBLGlu blocks assembled into different structures, depending on the type of solvent that was used to cast the films. In benzene cast films, the peptide helices were relatively poorly ordered, similar to the so-called form A morphology of PBLGlu [85]. In contrast, the PBLGlu segments in films cast from $CHCl_3$ were well ordered and contained paracrystalline and mesomorphic regions. Based on TEM, a cylindrical microstructure was proposed for a triblock copolymer containing 8 vol-% PB. Electron micrographs for other samples were not reported, but based on volume fraction considerations it was predicted that triblock copolymers containing 12 and 14 vol-% PB would form either cylindrical or lamellar superstructures [83]. Interestingly, copolymers having the same composition but polypeptide segments made of either enantiomerically pure or racemic γ-benzyl glutamate exhibited not only different secondary structures (α-helix or random coil, respectively; FTIR and WAXS) but also different superstructures (TEM). A cylindrical or lamellar morphology was proposed in the first case and a more spherical superstructure in the second [86].

Further support for the microphase-separated structure of the PBLGlu-b-PB-b-PBLGlu triblock copolymers was obtained from dynamic mechanical spectroscopy and water permeability experiments [84]. The temperature dependence of the dynamic modulus and the loss modulus could be explained well by assuming a microphase-separated structure. Furthermore, the hydraulic permeability of water through membranes prepared from the copolymers was approximately three orders of magnitude larger compared with a pure PBLGlu membrane. The hydraulic water permeability was found to increase with increasing PB content in the block copolymers. This was explained in terms of microphase-separated structure and the presence of an interfacial zone that separates the ordered domains formed by the α-helical PBLGlu chains from the unordered PB phase (Fig. 6.12). The interfacial zone consists of amino acid residues that are located close to the N-terminus of the peptide block and in the vicinity of the PB segment. The amino acid residues in the interfacial zone do not form regular secondary structures. As the amide groups of the peptide chains in the interfacial zone are not involved in intramolecular hydrogen bonding, they are able to bind water molecules. Consequently, increasing the interfacial zone, e.g., by increasing the PB content, leads to an increase in the water permeability.

The bulk and surface structure of solvent-cast films from a series of PBLGlu-b-PB-b-PBLGlu triblock copolymers with much higher PB contents (50–80 mol-%) than the samples discussed above have been described by Gallot and coworkers [87]. The organization of these copolymers was compared with that of three other triblock copolymers with approximately the same PB content ($\approx$50 mol-%) but which were composed of poly(N^ε-trifluoroacetyl L-lysine) (PTLLys), poly(N^5-hydroxyethyl L-glutamine) (PHLGln) or polysarcosine (PSar) as the peptide block. For any sample investigated, X-ray scattering experiments indicated a hexagonal-in-lamellar bulk morphology. X-ray photoelectron spectroscopy (XPS) measurements revealed that for the triblock copolymers with hydrophobic peptide blocks, i.e., PBLGlu or PTLLys, the surface composition was identical with that in the bulk of the sample. In contrast, the surfaces of films prepared from the triblock copolymers with the

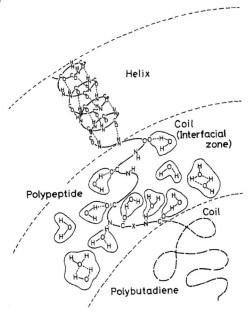

Fig. 6.12 Hydrogen-bonded water and water clusters in an interfacial zone formed by unordered peptide chains that separate the PB domains from the helical PBLGlu phase in films of ABA triblock copolymers.

more hydrophilic peptide segments, i. e., PHLGln or PSar, were PB enriched. Furthermore, the XPS data suggested that the lamellar superstructures formed by the triblock copolymers were perpendicular to the air–polymer interface.

In addition, the solid-state organization and properties of PZLLys-b-PB-b-PZLLys triblock copolymers have been investigated. Nakajima and coworkers [88, 89] have studied copolymers composed of a central PB block with M_n = 3.6 kg mol^{-1} and PB contents ranging from 12 to 52 mol-%. WAXS patterns obtained from solution-cast triblock copolymer films were in agreement with the α-helical secondary structure of the peptide blocks. The bulk microphase-separated structure of the five different block copolymer samples could be successfully characterized by means of TEM. For the samples with the largest PB volume fraction (56 and 65 vol-%), a lamellar superstructure was found. However, the electron micrographs suggested cylindrical and spherical microphase-separated structures for triblock copolymers with smaller PB volume fractions.

Other polybutadiene-based ABA type triblock copolymers that have been investigated include PMLGlu-b-PB-b-PMLGlu and PMGlu-b-PB-b-PMGlu [PMGlu = poly(γ-methyl D,L-glutamate)] [86, 90]. Infrared spectroscopy experiments on solvent-cast films indicated that the incorporation of 50% of the D-isomer disrupts the α-helical secondary structure and induces a random coil conformation in significant portions of the peptide blocks. From the infrared spectra and WAXS experiments, it was estimated that the helix content of a PMGlu homopolypeptide was about 60% of that of the corresponding PMLGlu. TEM images of OsO$_4$ stained samples provided evidence for the microphase-separated solid-state structure. Interestingly, different morphologies were observed when comparing the images of PMLGlu-b-PB-b-PMLGlu and PMGlu-b-PB-b-PMGlu samples with the same PB content ($\approx$30 mol-%). A cylindrical morphology was proposed for the first and a spherical structure for the second [86] (see above). The difference in morphology was ascribed to the less regular secondary structure of the peptide block in the case of the D,L-triblock copolymer, which prevents a highly ordered organization of the peptide domains and facilitates the formation of spherical PB domains. The ATR-IR spectra further showed that adsorption of bovine serum albumine (BSA) and bovine fibrinogen (BF) did not lead to denaturation. From these observations, these workers concluded that the surfaces of the PMGlu-b-PB-b-PMGlu membranes interact only weakly or reversibly with these plasma proteins and it was predicted that this may also lead to a good overall biocompatibility.

The solid-state structure and properties of PELGlu-b-PB-b-PELGlu [PELGlu = poly(γ-ethyl L-glutamate)] triblock copolymers containing 31.5–94.5 mol-% (= 17–88 vol-%) PELGlu have been studied using the same techniques as described above for the other triblock copolymers [91, 92]. The secondary structure of the PELGlu blocks was found to be predominantly α-helical and the helix content in the triblock copolymers decreased from 95 to 60% upon decreasing the peptide content from 95 to 61%. Interestingly, the WAXS data suggested that the PELGlu helices were packed in a pseudohexagonal, i.e., monoclinic, arrangement instead of the hexagonal structure observed for most of the other investigated peptide–synthetic hybrid block copolymers. TEM experiments on OsO$_4$ stained films indicated a microphase-separated structure. Based on the electron micrographs, a spherical microphase-separated structure was proposed for the copolymer containing 17 vol-% PB, while cylindrical and lamellar morphologies were suggested for triblock copolymers containing 28 and 44 vol-%, respectively, 68 and 88 vol-% PB. The biocompatibility of the PELGlu-b-PB-b-PELGlu triblock copolymers was assessed by coating the samples onto a polyester mesh fiber cloth, which was subsequently subcutaneously implanted in mongrel dogs for four weeks. It was found that the foreign body reaction and degradation of the PELGlu-b-PB-b-PELGlu samples were less pronounced as compared with PMLGlu-b-PB-b-PMLGlu, PBLGlu-b-PB-b-PBLGlu and PZLLys-b-PB-b-PZLLys triblock copolymers.

The bulk nanoscale structure of a series of PBLGlu-b-PI-b-PBLGlu copolymers containing 37.4–81.1 mol-% PBLGlu was studied by means of infrared spectroscopy, WAXS, dynamic mechanical analysis and electron microscopy [93]. Based on the electron micrographs, a cylindrical morphology was proposed for triblock copolymers containing 74.6 and 81.1 mol-% PBLGlu. Water permeability measurements also supported the microphase-separated bulk morphology [94]. Further insight into the bulk morphology of the PBLGlu-b-PI-b-PBLGlu triblock copolymers was obtained from pulsed proton NMR experiments [95]. The NMR signals of the block copolymers were composed of three components with different spin–spin relaxation times (T_2). The three different T_2 values were attributed to the microphase-separated structure, which consists of three regions (the ordered helical peptide domains, the unordered interfacial peptide region and the rubbery PI phase) with different molecular mobility. The spin–lattice relaxation times (T_1) that were obtained provided insight into the domain sizes, which were in good agreement with the results from TEM. The surface structure of $CHCl_3$-cast films was studied by XPS and contact angle measurements [96]. It was found that the chemical composition of the microphase-separated films at the surface was different from that in the bulk. The PI content at the film surface was higher than that in the bulk. Water contact angle measurements indicated that the block copolymer films were wetted easier than the respective homopolymers for the same reasons as the previous samples.

Treatment of a PBLGlu-b-PI-b-PBLGlu film with a mixture of 3-amino-1-propanol and 1,8-octamethylenediamine led to the formation of hydrophilic, cross-linked PHLGln-b-PI-b-PHLGln membranes being obtained [97]. The swelling ratio of these membranes in pseudoextracellular fluid (PECF) was found to decrease with increasing PI content and increasing cross-link density. Tensile tests in PECF revealed that the triblock copolymer membranes had a larger Young's modulus, increased tensile strength and elongation at breaking compared with membranes prepared from PBLGlu homopolymer. Enzymatic degradation experiments using papain showed that the triblock copolymer films were more resistant towards degradation than the corresponding homopolypeptide membranes. The half-times for sample degradation increased with decreasing peptide content, which was in agreement with the swelling behavior of the membranes.

6.3.2.2 Polystyrene-based Triblock Copolymers

Tanaka and coworkers studied ABA type triblock copolymers composed of a central PS block flanked by two polypeptide segments (PBLGlu, PZLLys or PSar) [98]. TEM of a $CHCl_3$-cast film of $PBLGlu_{25}$-b-PS_{165}-b-$PBLGlu_{25}$ that was stained with phosphotungstic acid revealed a lamellar phase separated structure. In contrast, no microphase separation was observed in a film of $PSar_{73}$-b-PS_{421}-b-$PSar_{73}$. These workers proposed that the different block copolymer morphologies could be related to the different secondary structure of the peptide block; while the PBLGlu segments are predominantly helical, the PSar may not form any regular secondary structure. Fibrinogen adsorption on the block copolymer films was studied with ATR-IR spectroscopy and compared with that on the corresponding homopolymer films [98]. It was found that fibrinogen adsorption on PS and PSar homopolymer films and on PSar-b-PS-b-PSar triblock copolymer films led to denaturation of the protein. In contrast, protein adsorption on the microphase-separated PBLGlu-b-PS-b-PBLGlu surfaces was reported to stabilize the protein's secondary structure. Blood clotting tests suggested that thrombus formation was retarded compared with the respective homopolymers.

Samyn and coworkers [99] extended the investigations of ABA triblock copolymers and studied the solid-state organization of three different PBLGlu-b-PS-b-PBLGlu copolymers containing 34, 55 and 92 wt-% PBLGlu. TEM micrographs of ultramicrotomed and RuO_4 stained specimens and SAXS experiments indicated a lamellar morphology for the copolymers with 34 and 55 wt-% PBLGlu. The sample containing 92 wt-% PBLGlu did not form a lamellar structure. WAXS patterns yielded d-spacings reflecting the intermolecular distance between neighboring peptide α-helices. Ion permeability measurements on dioxane-cast films indicated that the bulk morphology influences the membrane properties [100]. The membranes prepared from the lamellae forming 34 and 55 wt-% PBLGlu containing triblock copolymers showed cation selectivity. In contrast, the membrane prepared from the triblock copolymer containing 92 wt-% PBLGlu did not show such selectivity. It was proposed that uptake of cations into the triblock copolymer membranes was facilitated by the interactions between the cations and the ester functions in the block copolymers. The difference in selectivity was explained in terms of the interfacial zone (as discussed earlier), which separates the PS and PBLGlu domains only in the films generated by the former two triblock copolymers.

6.3.2.3 Polysiloxane-based Triblock Copolymers

Imanishi and coworkers [101] have studied the structure, antithrombogenicity and oxygen permeability of ABA triblock copolymers composed of poly(dimethylsiloxane) (PDMS) as the B block and PBLGlu, PBGlu [poly(γ-benzyl D,L-glutamate)], PZLLys or PSar as the A block. Several series of triblock copolymers were prepared using bifunctional PDMS macroinitiators and targeting various peptide block lengths. TEM images of DMF-cast films provided evidence for a microphase-separated morphology for PZLLys$_{49}$-b-PDMS$_{400}$-b-PZLLys$_{49}$ and PZLLys$_{91, 160}$-PDMS$_{256}$-PZLLys$_{91, 160}$. The images revealed a spherical morphology composed of PDMS islands in a PZLLys matrix. The formation of these spherical domains was attributed to the solvent that was used for sample preparation. While DMF is a good solvent for PZLLys, it is a poor solvent for PDMS. In a separate publication, the same workers also described non-spherical microphase-separated structures [102]. In CH_2Cl_2-cast films of a triblock copolymer with a very high PDMS content (PBLGlu$_{48}$-b-PDMS$_{508}$-b-PBLGlu$_{48}$, 83 mol-% PDMS) more extended, rod-like PBLGlu aggregates in a matrix of PDMS were observed. The TEM experiments also provided insight into the effects of peptide secondary structure and the nature of the casting solvent on the thin film morphology [101].

Thin films of PBLGlu$_{42}$-b-PDMS$_{148}$-b-PBLGlu$_{42}$ prepared from DMF showed a spherical morphology. Changing the solvent from DMF (a good solvent for PBLGlu) to CH_2Cl_2 (a fairly non-selective solvent) resulted in coarsening of the microphase-separated structures. PBGlu$_{42}$-b-PDMS$_{148}$-b-PBGlu$_{42}$ films prepared from CH_2Cl_2 also showed a microphase-separated structure in which spherical PDMS domains were embedded in a PBGlu matrix. The dimensions of the spherical domains, however, were much smaller than those observed by TEM. These different morphologies reflect the influence of the peptide secondary structure on the block copolymer self-assembly; PBLGlu$_{42}$ adopts an α-helical conformation and PBGlu$_{42}$ a random coil conformation. Studies on adsorption/denaturation of proteins and oxygen permeation measurements from these triblock copolymers also tend to describe the relationship between the film morphology and these properties. In addition, a detailed study of the gas permeation properties of PBLGlu-b-PDMS-b-PBLGlu films cast from CH_2Cl_2 and DMF solution with

PDMS contents ranging from 46 to 83 mol-% has been reported [103] and revealed that the oxygen permeability of the triblock copolymer films in water was found to increase exponentially with increasing PDMS content, in agreement with a microphase-separated morphology of the membranes. Similar results were reported by Kugo et al., who studied oxygen and nitrogen transport across PBLGlu-*b*-PDMS-*b*-PBLGlu triblock copolymers containing 63–81 mol-% PBLGlu [104].

6.3.2.4 Polyether-based Triblock Copolymers

Inoue and coworkers [105, 106] studied the adhesion behavior of rat lymphocytes on solvent-cast films of PBLGlu-*b*-PEG-*b*-PBLGlu triblock copolymers. The triblock copolymers were prepared from α,ω-bis-amino functionalized PEG macroinitiators with molecular weights of 1.0 and 4.0 kg mol^{-1} and had PEG contents varying from 11 to 33 wt-%. Rat lymphocyte adhesivity was found to decrease with increasing PEG content. At the same PEG content, the adhesivity of the triblock copolymers based on the macroinitiator with a molecular weight of 4 kg mol^{-1} was lower than that of samples based on the macroinitiator with 1 kg mol^{-1}. In addition to overall lymphocyte adhesivity, these workers also studied the adhesion of specific subpopulations: B-cells and T-cells. All triblock copolymers showed a preference towards B-cells. These experiments, however, revealed that the observed differences in cell adhesion behavior were neither due to differences in the conformation of the peptide blocks, nor could they be attributed to differences in surface hydrophilicity. It was therefore proposed that the observed effects were caused by differences in the higher order surface structures, i.e., in terms of the microphase-separated morphology and/or PEG crystallinity.

Kugo et al. [107] studied the solid-state conformation of the peptide segment of a series of PBLGlu-*b*-PEG-*b*-PBLGlu copolymers containing a PEG segment with a molecular weight of 4 kg mol^{-1} and 36–86 mol-% PBLGlu. FT-IR spectroscopy experiments on CHCl$_3$-cast films revealed that the PBLGlu blocks, which had degrees of polymerization of 25–276, had an α-helical secondary structure. The helix content of the triblock copolymer containing PBLGlu$_{276}$ blocks was found to be similar to that of the PBLGlu homopolymer. Swelling the triblock copolymer films with water resulted in a decrease in helix content, as indicated by the CD spectra. This decrease in helicity was attributed to competition of water clusters to form hydrogen bonds with the peptide backbone. The effect was even more pronounced when pseudo-extracellular fluid was used instead of water.

A first detailed study of the solid-state nanoscale structure of peptide–PEG hybrid block copolymers was published by Cho et al. [108]. They investigated thin, CHCl$_3$-cast films of PBLGlu-*b*-PEG-*b*-PBLGlu copolymers, which were composed of a PEG block of 2 kg mol^{-1} and contained 25–76 mol-% PBLGlu. TEM micrographs of RuO$_4$-stained specimens revealed a lamellar morphology for triblock copolymers containing 25–64 mol-% PBLGlu. The microphase-separated structure was proposed to consist of chain folded, crystalline PEG domains and helical PBLGlu domains (IR). WAXS patterns were consistent with the ordered, crystalline-like solid-state modification C of PBLGlu. In contrast, in films cast from benzene, the peptide blocks only formed poorly ordered arrays. The sensitivity of the organization of the PBLGlu blocks towards the nature of the casting solvent is identical with the behavior of the PBLGlu homopolymer.

In a separate study, the enzymatic degradation behavior of PBLGlu-b-PEG-b-PBLGlu triblock copolymers was investigated [109]. The rate of degradation was found to increase with increasing PEG content in the triblock copolymers from 1.4 to 3.1 to 13.6 mol-%. A similar dependence on PEG content was observed for the level of swelling. Exposure of the triblock copolymer samples to a PBS solution without the enzyme did not result in measurable weight loss, indicating that hydrolytic degradation did not take place.

While the data reported by Cho et al. described the structure and organization of thin solvent cast films of PBLGlu-b-PEG-b-PBLGlu, Floudas et al. have extensively studied the bulk nanoscale organization of these materials [110]. To this end, a series of triblock copolymers with PBLGlu volume fractions (f_{PBLGlu}) ranging from 0.07 to 0.89 was investigated using SAXS/WAXS, polarizing optical microscopy (POM), DSC and FT-IR spectroscopy. For triblock copolymers with $f_{PBLGlu} \leq 0.25$, PEG crystallization was observed, however, with significant undercooling. Triblock copolymers with $f_{PBLGlu} \geq 0.43$ did not show PEG crystallization. SAXS experiments, which were carried out at 373 K, i.e., above the melting point of PEG, also revealed a different behavior for triblock copolymers with small and large PBLGlu volume fractions. For triblock copolymers with $f_{PBLGlu} \geq 0.43$ only a weakly phase separated structure was found, whereas for samples with $f_{PBLGlu} \leq 0.25$ the SAXS data clearly indicated a microphase-separated structure. WAXS patterns showed that in the microphase-separated state the PEG phase was semi-crystalline and the peptide phase consisted of hexagonally ordered assemblies of PBLGlu α-helices that coexisted with β-sheet structures. For triblock copolymers with $f_{PBLGlu} \geq 0.43$, PEG is amorphous and interspersed with aggregates of α-helical PBLGlu segments and unordered peptide chains. These different bulk structures are illustrated schematically in Fig. 6.13. This figure illustrates how the competing interactions that promote the bulk self-assembly of the PBLGlu-b-PEG-b-PBLGlu triblock copolymers lead to the formation of hexagonally ordered structures covering different length scales. At the smallest length scale, hydrogen-bonding interactions stabilize peptide secondary structures (α-helices and β-strands) and PEG chain folding occurs. On the next higher level, peptide α-helices and β-strands form hexagonal assemblies and β-sheet structures, respectively. Finally, the mutual incompatibility of the peptide and PEG block leads to microphase separation.

Additional insight into the solid-state nanoscale organization of the triblock copolymers just discussed was obtained by combining SAXS/WAXS with various microscopic techniques (TEM and AFM) [111]. A "broken lamellar" morphology was observed in the TEM micrographs of PBLGlu$_{58}$-b-PEG$_{90}$-b-PBLGlu$_{58}$ ($f_{PBLGlu} = 0.58$). Annealing converted this metastable structure into a non-uniform microphase-separated pattern, which was proposed to consist of "puck-like" PEG domains in a PBLGlu matrix. For PBLGlu$_{105}$-b-PEG$_{90}$-b-PBLGlu$_{105}$ ($f_{PBLGlu} = 0.67$), a lamellar morphology was found in the as-cast film, which was

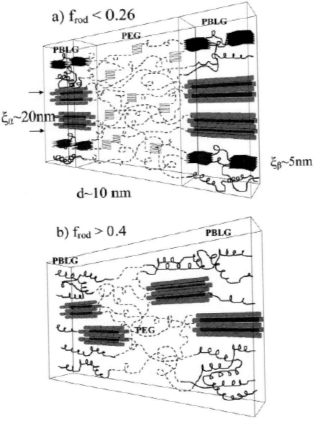

Fig. 6.13 Highly schematic model of the phase state in the PBLGlu-b-PEG-b-PBLGlu triblock copolymers. (a) Phase state corresponding to low peptide volume fractions depicting a microphase-separated copolymer consisting of all the peptide and PEG secondary structures. (b) Phase state corresponding to $f_{rod} > 0.4$ depicting phase mixing resulting in the appearance of only one (α-helical) secondary structure (reprinted from [110] with permission of The American Chemical Society).

transformed into a "broken lamellar" structure upon annealing. Based on these results, a morphology map was constructed.

In addition to PBLGlu-b-PEG-b-PBLGlu, PZLLys-b-PEG-b-PZLLys triblock copolymers have also been studied. Cho et al. [112] reported on the solid-state structure of a series of PZLLys-b-PEG-b-PZLLys composed of a PEG block with $M_n = 2$ kg mol^{-1} and PZLLys contents of 25.2, 49.9 and 83.0 mol-% (= 68, 86 and 98 vol-%). Infrared spectra of CHCl$_3$-cast films were in agreement with a helical secondary structure of the peptide blocks. DSC experiments provided a first hint for the existence of a microphase-separated structure and revealed two T_g values for all samples. The higher T_g was very close to that of the PZLLys homopolymer and the lower T_g approximately 20 °C higher than that of PEG homopolymer. A PEG melting transition was not observed. These results were interpreted in terms of a microphase-separated structure with hard, crystalline PZLLys domains and soft, amorphous PEG segments. The presence of a microphase-separated structure was confirmed by TEM micrographs of RuO_4-stained thin films.

Akashi and coworkers [113] reported on the solid-state nanoscale structure of ABA type triblock copolymers composed of a central PEG block flanked by two poly(β-benzyl L-aspartate) (PBLAsp) blocks. The molecular weight of the central PEG block was 11 or 20 kg mol^{-1} and the degrees of polymerization of the peptide blocks ranged from 12 to 32. WAXS and POM studies on CH_2Cl_2-cast films showed PEG crystallization in all samples. The intensity of the crystalline PEG reflection peak, however, was found to decrease with increasing length of the PBLAsp block. The observation of PEG crystallization was interpreted as a first indication for microphase separation. In addition to the PEG signal, the WAXS patterns also contained reflections at $2\Theta = 5.9°$ (= 15 Å), which were assigned to a hexagonally packed array of PBLAsp helices, a result confirmed by FT-IR spectroscopy. In the SAXS patterns of PBLAsp$_{25}$-b-PEG$_{250}$-b-PBLAsp$_{25}$, PBLAsp$_{25}$-b-PEG$_{454}$-b-PBLAsp$_{25}$ and PBLAsp$_{32}$-b-PEG$_{454}$-b-PBLAsp$_{32}$ broad and weak diffraction peaks were observed, indicating the formation phase separated structures. Thermal analysis of the triblock copolymers, however, revealed several interesting properties. DSC experiments showed that the melting temperature of the crystalline PEG domains decreased linearly with increasing PBLAsp content, reflecting the strong influence of the peptide segments on PEG crystallization. More interestingly, these workers found that heating the as-cast films above 333 K and cooling down to 303 K, converted a certain fraction of the α-helical PBLAsp chains into β-strands and was accompanied by a decrease in PEG crystallinity. On a macroscopic level, this led to increased strength and elasticity of the films.

Cho et al. [114] have studied triblock copolymers composed of a middle block of poly(propylene glycol) (PPG) with a molecular weight of 2 kg mol^{-1} flanked by two PBLGlu segments. Three triblock copolymer samples were investigated with PPG contents of 17.0, 26.0 and 60.0 mol-%, respectively. According to infrared spectra that were recorded from $CHCl_3$-cast films, the PBLGlu blocks possessed an α-helical secondary structure. WAXS patterns revealed a 12.5 Å interhelical spacing and were in agreement with a solid-state modification C of PBLGlu. No further details on the solid-state nanoscale structure and the possibility of microphase separation were reported. Platelet adhesion on glass beads coated with block copolymers containing 33 or 47 mol-% PPG was reduced compared with beads modified with PMLGlu homopolymer or block copolymers with 70.0 mol-% PPG. These differences were attributed to differences in surface composition and morphology, which, unfortunately, were not discussed further. Finally, Hayashi et al. [115] have reported on PMLGlu-b-PTHF-b-PMLGlu [PTHF = poly(tetrahydrofuran)] triblock copolymers that were prepared from an amino-functionalized PTHF macroinitiator with a molecular weight of 9.6 kg mol^{-1}. Three different triblock copolymers were studied with PMLGlu contents and degrees of polymerization of 86.8 mol-%/

460, 89.7 mol-%/605 and 91.3 mol-%/730, respectively. With respect to the solid-state structure and organization, only infrared spectra and WAXS data were discussed.

6.3.2.5 Miscellaneous

The structure and properties of an ABA triblock copolymer composed of a poly(ether urethane urea) (PEUU) B block with M_n = 15.8 kg mol^{-1} and two PBLGlu$_{29}$ B blocks were described by Ito et al. [116]. DSC thermograms showed a single endotherm located between the T_g of the PEUU and the 7/2 to 18/5 helix transition of PBLGlu, suggesting that there was no phase separation. Platelet adhesion on DMF-cast films of the triblock copolymer was significantly reduced compared with the respective homopolymers and a PEUU/PBLGlu blend. Thrombus formation on block copolymer films was found to be ≈50 % less compared with a glass surface. However, no significant difference in antithrombogenicity between PEUU, PBLGlu, their blend and the block copolymer was observed.

Another, very early study focused on two PBLGlu-b-PBAN-b-PBLGlu [PBAN = poly(butadiene-co-acrylonitrile)] triblock copolymers [117]. These block copolymers were composed of a PBAN block with M_n = 3.4 kg mol^{-1} and two PBLGlu segments containing either 80 or 160 repeat units. TEM micrographs of OsO$_4$-stained films cast from dioxane, which is a selective solvent for PBLGlu, revealed a lamellar morphology. When the non-selective solvent CHCl$_3$ was used for the preparation of the TEM specimens, the images were more homogeneous and phase separation was less distinct. This suggests that, depending on the solvent conditions, the PBAN block can affect PBLGlu secondary structure.

Electro- and photoactive peptide–synthetic hybrid triblock copolymers have been prepared using bis(benzyl amine)-terminated poly(9,9-dihexylfluorene-2,7-diyl) (PHF) as a macroinitiator for the ring-opening polymerization of BLGlu-NCA [118]. The electroactive and photoactive properties of the triblock copolymers were similar to those of the PHF homopolymer, indicating that the introduction of the PBLGlu segments did not interfere with charge injection and transport and other material properties. FT-IR spectra of CHCl$_3$-cast films of PBLGlu$_{23}$-b-PHF$_{15}$-b-PBLGlu$_{23}$ indicated an α-helical secondary structure. The FT-IR spectra of PBLGlu$_{16}$-b-PHF$_{28}$-b-PBLGlu$_{16}$ displayed, however, an additional peak at 1630 cm^{-1}, indicating the coexistence of α-helical and β-strand conformations. The thin film morphologies of the triblock copolymers were investigated with AFM using different casting solvents (Fig. 6.14). When 2,2,2-trifluoroacetic acid (TFA)–CHCl$_3$ 30/70 (v/v) was used for the preparation of samples, globular or spherical aggregates with diameters of about 32 and 40 nm were observed for PBLGlu$_{23}$-b-PHF$_{15}$-b-PBLGlu$_{23}$ and PBLGlu$_{16}$-b-PHF$_{28}$-b-PBLGlu$_{16}$, respectively. In this solvent mixture, the PBLGlu chains have a random coil conformation and the triblock copolymers were proposed to form spherical nanostructures composed of a PHF core and a PBLGlu shell. When the solvent was changed to TFA–CHCl$_3$ 3/97 (v/v), the SFM images revealed parallel fibrillar structures being 79 ± 25 nm (PBLGlu$_{23}$-b-PHF$_{15}$-b-PBLGlu$_{23}$) and 83 ± 17 nm (PBLGlu$_{16}$-b-PHF$_{28}$-b-PBLGlu$_{16}$) in width and 4–10 μm in length. Under these conditions,

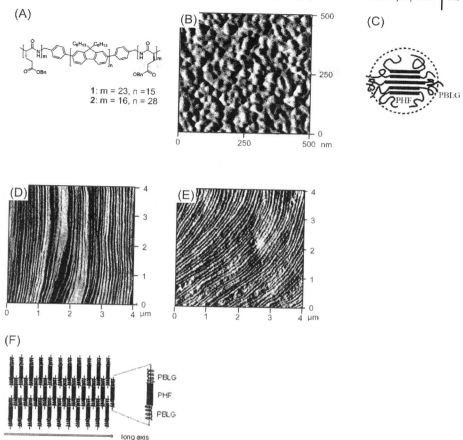

Fig. 6.14 (A) Chemical structure of PBLGlu-b-PHF-b-PBLGlu triblock copolymers **1** and **2**; (B) AFM image of a thin film of **1** cast from TFA–CHCl$_3$ 30/70 (v/v); (C) schematic representation of the spherical nanostructures that can be observed in (B); (D) and (E) AFM images of triblock copolymers **1** (D) and **2** (E) cast from TFA–CHCl$_3$ 3/97 (v/v); (F) model proposed for the self-assembly of **1** and **2** in the fibrillar structures shown in (D) and (E) (reprinted from [118] with permission of The American Chemical Society).

PBLGlu adopts an α-helical secondary structure. As the widths of the fibrils were much larger than the extended length of the triblock copolymers, a side-by-side antiparallel stacking was proposed to explain the fibril formation.

6.4
Summary and Outlook

In summary, the first part of this chapter described and discussed the phase behavior of biomimetic polypeptide-based copolymers in solution with respect to the occurrence of secondary structure effects. Evidently, incorporation of crystallizable polypeptide segments inside the core of an aggregate has an impact on the curvature of the core–corona interface and promotes the formation of fibrils or vesicles, or other flat superstructures. Spherical micelles are not usually observed. Copolymers with soluble polypeptide segments, on the other hand, seem to behave as conventional block copolymers. A pH-induced change of the conformation of coronal polypeptide chains only affects the size of aggregates, not their shape. It is evident that the biomimetic approach using polypeptide hybrid polymers is very successful in the creation of novel superstructures with hierarchical order. However, although begun about 30 years ago in the mid-1970s, this field is still in a premature state. Most systematic studies on aggregation in solution have been reported only during the last five to ten years. A comprehensive picture of the processes involved in the formation of hierarchical structures is still lacking. The application potential of polypeptide copolymers has also not been exhausted. Most studies deal with ordinary micelles for the controlled delivery of drugs or genes. Not much attention has been, for whatever reason, paid to gel structures and other colloidal systems, such as emulsions, polymer latexes and inorganic–organic hybrid nanoparticles.

The solid-state structure, organization and properties of peptide–synthetic hybrid block copolymers were discussed in the second part. The most notable difference between peptide-based block copolymers and their fully synthetic and amorphous analogues is their hierarchical solid-state organization. In contrast to most synthetic amorphous block copolymers, which typically exhibit structural order only over a single length scale, peptide-based block copolymers can form hierarchically organized nanoscale structures that cover several different length scales. At the smallest length scale, peptide sequences fold into regular secondary structures, such as α-helices or β-strands. On the next higher level, peptide α-helices and β-strands can assemble into hexagonal superstructures and β-sheets, respectively. Finally, phase separation between the peptide and synthetic blocks leads to the formation of ordered domains with the largest characteristic length scales. For a large number of peptide-based block copolymers, lamellar phase separated morphologies have been observed. These lamellar structures, however, are often found over a much broader range of compositions compared with regular, fully amorphous diblock copolymers. This behavior, as in solution, can be explained easily considering the flat interface that is generated from the rod–rod packing. In addition to these more conventional morphologies, structural investigations on peptide-based hybrid block copolymers have also led to the discovery of various novel phase separated structures, which were not previously known for fully amorphous diblock copolymers. Both observations reflect the fact that the solid-state structure formation of peptide hybrid block copolymers is not solely dictated by phase separation, as is the case for amorphous diblock copolymers, but is also influenced by other factors such as intra- and intermolecular hydrogen bonding and chain conformation. While much of the early interest in peptide–synthetic hybrid block copolymers was driven by their potential use as membrane materials or for the development of antithrombogenic surfaces, more recent studies revealed that these materials can also have interesting mechanical properties [113].

The major drawback of most of the block copolymers discussed in this chapter is that they have been prepared via the conventional amine-initiated NCA polymerization. The polymerization of NCAs under these conditions does not allow very accurate control over polymer chain length, results in rather broad molecular weight distributions and is also not very useful for preparing defined block copolypeptides [119, 120]. It is obvious that these limitations possibly restrict further engineering of the structure and organization of peptide–synthetic hybrid block copolymers and could also hamper the exploration of their full practical potential. Over recent years, however, a number of alternative NCA polymerization strategies have been developed, which provide enhanced control over polypeptide chain length and chain length distribution and also allow access to defined block copolypeptides [121–123]. In the last 5 years, a great deal of effort has been focused on controlling the synthesis and understanding the structural behavior of these polypeptide-based block copolymers. These systems, which have been known for a long time, are currently gaining more and more attention due to the possibilities of making highly ordered materials on the nano- to micrometer-length scale and bio-compatible aggregates that respond to external stimuli. The door to new innovations in materials science is now open.

References

1 Park C, Yoon J, Thomas EL (2003) *Polymer* 44, 6725.
2 Hamley IW (2003) *Nanotechnology* 14, R39.
3 Lazzari M, López-Quintela MA (2003) *Adv. Mater.* 15, 1583.
4 Bates FS, Fredrickson GH (1990) *Annu. Rev. Phys. Chem.* 41, 525.
5 Hamley IW (1998) *The physics of block copolymers*, Oxford University Press, Oxford, New York, Tokyo.
6 Bates FS, Fredrickson GH (1999) *Phys. Today* 52, 32.
7 Walther M, Finkelmann H (1996) *Progr. Polym. Sci.* 21, 951.
8 Mao G, Ober CK (1997) *Acta Polym.* 48, 405.
9 Klok H-A, Lecommandoux S (2001) *Adv. Mater.* 13, 1217.
10 Lee M, Cho BK, Zin WC (2001) *Chem. Rev.* 101, 3869.
11 Förster S, Antonietti M (1998) *Adv. Mater.* 10, 195.
12 Bates FS, Fredrickson GH (1999) *Phys. Today* 52, 32.
13 Cölfen H (2001) *Macromol. Rapid. Commun.* 22, 219.
14 Förster S, Konrad M (2003) *J. Mater. Chem.* 13, 2671.
15 Antonietti M, Förster S (2003) *Adv. Mater.* 15, 1323.
16 Choucair A, Eisenberg A (2003) *Eur. Phys. J. E* 10, 37.
17 Discher BM, Won Y-Y, Ege DS, Lee JC-M, Bates FS, Discher DE, Hammer DA (1999) *Science* 284, 1143.
18 Discher DE, Eisenberg A (2002) *Science* 297, 967.

19 Taubert A, Napoli A, Meier W (2004) Curr. Opin. Chem. Biol. 8, 598.
20 Kita-Tokarczyk K, Grumelard J, Haefele T, Meier W (2005) Polymer 46, 3540.
21 Cornelissen JJLM, Rowan AE, Nolte RJM, Sommerdijk NAJM (2001) Chem. Rev. 101, 4039.
22 Löwik DWPM, van Hest JCM (2004) Chem. Soc. Rev. 33, 234.
23 Vandermeulen GWM, Klok H-A (2004) Macromol. Biosci. 4, 383.
24 Klok, H-A (2005) J. Polym. Sci. Part A Polym. Chem. 43, 1.
25 Schlaad H, Antonietti M (2003) Eur. Phys. J. E 10, 17.
26 Rodriguez-Hernandez J, Chécot F, Gnanou Y, Lecommandoux S (2005) Progr. Polym. Sci. 30, 691.
27 Nakajima A, Kugo K, Hayashi T (1979) Macromolecules 12, 844.
28 Hayashi T (1985), in *Developments in block copolymers*, ed. Goodman I, Elsevier Applied Science Publishers, London, p. 109.
29 Förster S, Zisenis M, Wenz E, Antonietti M (1996) J. Chem. Phys. 104, 9956.
30 Kukula H, Schlaad H, Antonietti M, Förster S (2002) J. Am. Chem. Soc. 124, 1658.
31 Chécot F, Brûlet A, Oberdisse J, Gnanou Y, Mondain-Monval O, Lecommandoux S (2005) Langmuir 21, 4308.
32 Chécot F, Lecommandoux S, Gnanou Y, Klok H-A (2002) Angew. Chem., Int. Ed. Engl. 41, 1340.
33 Chécot F, Lecommandoux S, Klok H-A, Gnanou Y (2003) Eur. Phys. J. E 10, 25.
34 Babin J, Rodríguez-Hernández J, Lecommandoux S, Klok H-A, Achard M-F (2005) Faraday Discuss. 128, 179.
35 Rodríguez-Hernández J, Babin J, Zappone B, Lecommandoux S (2005) Biomacromolecules 6, 2213.
36 Arimura H, Ohya Y, Ouchi T (2005) Biomacromolecules 6, 720.
37 Lübbert A, Castelletto V, Hamley IW, Nuhn H, Scholl M, Bourdillon L, Wandrey C, Klok H-A (2005) Langmuir 21, 6582.
38 Harada A, Kataoka K (1995) Macromolecules 28, 5294.
39 Harada A, Kataoka K (1997) J. Macromol. Sci.-Pure Appl. Chem. A34, 2119.
40 Harada A, Kataoka K (1999) Science 283, 65.
41 Toyotama A, Kugimiya S-i, Yamanaka J, Yonese M (2001) Chem. Pharm. Bull. 49, 169.
42 Cheon J-B, Jeong Y-I, Cho C-S (1998) Korea Polym. J. 6, 34.
43 Cheon J-B, Jeong Y-I, Cho C-S (1999) Polymer 40, 2041.
44 Dong C-M, Sun X-L, Faucher KM, Apkarian RP, Chaikof EL (2004) Biomacromolecules 5, 224.
45 Dong C-M, Faucher KM, Chaikof EL (2004) J. Polym. Sci., Part A: Polym. Chem. 42, 5754.
46 Naka K, Yamashita R, Nakamura T, Ohki A, Maeda S (1997) Macromol. Chem. Phys. 198, 89.
47 Losik M, Schlaad H, unpublished results.
48 Doi T, Kinoshita T, Kamiya H, Tsujita Y, Yoshimizu H (2000) Chem. Lett. 262.
49 Doi T, Kinoshita T, Kamiya H, Washizu S, Tsujita Y, Yoshimizu H (2001) Polym. J. 33, 160.
50 Rodríguez-Hernández J, Lecommandoux S (2005) J. Am. Chem. Soc. 127, 2026.
51 Bellomo EG, Wyrsta MD, Pakstis L, Pochan DJ, Deming TJ (2004) Nat. Mater. 3, 244.
52 Constancis A, Meyrueix R, Bryson N, Huille S, Grosselin J-M, Gulik-Krzywicki T, Soula G (1999) J. Colloid. Interf. Sci. 217, 357.
53 Deming TJ (2005) Soft Matter 1, 28.
54 Nowak AP, Breedveld V, Pakstis L, Ozbas B, Pine DJ, Pochan D, Deming TJ (2002) Nature (London) 417, 424.
55 Breedveld V, Nowak AP, Sato J, Deming TJ, Pine DJ (2004) Macromolecules 37, 3943.
56 Pochan DJ, Pakstis L, Ozbas B, Nowak AP, Deming TJ (2002) Macromolecules 35, 5358.
57 Pakstis LM, Ozbas B, Hales KD, Nowak AP, Deming TJ, Pochan D (2004) Biomacromolecules 5, 312.

58 Cha JN, Stucky GD, Morse DE, Deming TJ (2000) *Nature (London)* 403, 289.
59 Euliss LE, Trnka TM, Deming TJ, Stucky GD (2004) *Chem. Commun.* 1736.
60 Lecommandoux S, Sandre O, Chécot F, Rodríguez-Hernández J, Perzynski R (2005) *Adv. Mater.* 17, 712.
61 Euliss LE, Grancharov SG, O'Brien S, Deming TJ, Stucky GD, Murray CB, Held GA (2003) *Nano Lett.* 3, 1489.
62 Perly B, Douy A, Gallot B (1976) *Makromol. Chem.* 177, 2569.
63 Billot J-P, Douy A, Gallot B (1976) *Makromol. Chem.* 177, 1889.
64 Billot J-P, Douy A, Gallot B (1977) *Makromol. Chem.* 178, 1641.
65 Douy A, Gallot B (1977) *Polym. Eng. Sci.* 17, 523.
66 Douy A, Gallot B (1982) *Polymer* 23, 1039.
67 Gallot B, Douy A, Hayany H, Vigneron C (1983) *Polym. Sci. Technol.* 23, 247.
68 Mori A, Ito Y, Sisido M, Imanishi Y (1986) *Biomaterials* 7, 386.
69 Schlaad H, Kukula H, Smarsly B, Antonietti M, Pakula T (2002) *Polymer* 43, 5321.
70 Losik M, Kubowicz S, Smarsly B, Schlaad H (2004) *Eur. Phys. J. E* 15, 407.
71 Schlaad H, Smarsly B, Losik M (2004) *Macromolecules* 37, 2210.
72 Ludwigs S, Krausch G, Reiter G, Losik M, Antonietti M, Schlaad H (2005) *Macromolecules* 38, 7532.
73 Klok H-A, Langenwalter JF, Lecommandoux S (2000) *Macromolecules* 33, 7819.
74 Lecommandoux S, Achard M-F, Langenwalter JF, Klok H-A (2001) *Macromolecules* 34, 9100.
75 Zhang G, Ma J, Li Y, Wang Y (2003) *J. Biomater. Sci. Polym. Edn.* 14, 1389.
76 Cho I, Kim J-B, Jung H-J (2003) *Polymer* 44, 5497.
77 Degée P, Dubois P, Jérôme R, Theyssié P (1993) *J. Polym. Sci. Part A Polym. Chem.* 31, 275.
78 Rong G, Deng M, Deng C, Tang Z, Piao L, Chen X, Jing X (2003) *Biomacromolecules* 4, 1800.
79 Caillol S, Lecommandoux S, Mingotaud A-F, Schappacher M, Soum A, Bryson N, Meyrueix R (2003) *Macromolecules* 36, 1118.
80 Deming TJ (1997) *J. Am. Chem. Soc.* 119, 2759.
81 Papadopoulos P, Floudas G, Schnell I, Aliferis T, Iatrou H, Hadjichristidis N (2005) *Biomacromolecules* 6, 2352.
82 Nakajima A, Hayashi T, Kugo K, Shinoda K (1979) *Macromolecules* 12, 840.
83 Nakajima A, Kugo K, Hayashi T (1979) *Macromolecules* 12, 844.
84 Nakajima A, Kugo K, Hayashi T (1979) *Polymer J.* 11, 995.
85 McKinnon AJ, Tobolsky AV (1968) *J. Phys. Chem.* 72, 1157.
86 Hayashi T, Chen G-W, Nakajima A (1984) *Polymer J.* 16, 739.
87 Gervais M, Douy A, Gallot B, Erre R (1988) *Polymer* 29, 1779.
88 Kugo K, Hayashi T, Nakajima A (1982) *Polymer J.* 14, 391.
89 Kugo K, Hata Y, Hayashi T, Nakajima A (1982) *Polymer J.* 14, 401.
90 Kugo K, Murashima M, Hayashi T, Nakajima A (1983) *Polymer J.* 15, 267.
91 Chen G-W, Hayashi T, Nakajima A (1981) *Polymer J.* 13, 433.
92 Sato H, Nakajima A, Hayashi T, Chen G-W, Noishiki Y (1985) *J. Biomed. Mater. Res.* 19, 1135.
93 Yoda R, Komatsuzaki S, Nakanishi E, Hayashi T (1995) *Eur. Polym. J.* 31, 335.
94 Yoda R, Komatsuzaki S, Hayashi T (1996) *Eur. Polym. J.* 32, 233.
95 Yoda R, Shimoda M, Komatsuzaki S, Hayashi T, Nishi T (1997) *Eur. Polym. J.* 33, 815.
96 Yoda R, Komatsuzaki S, Hayashi T (1995) *Biomaterials* 16, 1203.
97 Yoda R, Komatsuzaki S, Nakanishi E, Kawaguchi H, Hayashi T (1994) *Biomaterials* 15, 944.
98 Imanishi Y, Tanaka M, Bamford CH (1985) *Int. J. Biol. Macromol.* 7, 89.
99 Janssen K, Van Beylen M, Samyn C, Scherrenberg R, Reynaers H (1990) *Makromol. Chem.* 191, 2777.
100 Janssen K, Van Beylen M, Samyn C, Van Driessche W (1989) *Makromol. Chem. Rapid. Commun.* 10, 457.

101 Kumaki T, Sisido M, Imanishi Y (1985) *J. Biomed. Mater. Res.* 19, 785.
102 Kang I-K, Ito Y, Sisido M, Imanishi Y (1988) *Biomaterials* 9, 138.
103 Kang I-K, Ito Y, Sisido M, Imanishi Y (1988) *Biomaterials* 9, 349.
104 Kugo K, Nishioka H, Nishino J (1987) *Chem. Express* 2, 21.
105 Nishimura T, Sato Y, Yokoyama M, Okuya M, Inoue S, Kataoka K, Okano T, Sakurai Y (1984) *Makromol. Chem.* 185, 2109.
106 Yokoyama M, Nakahashi T, Nishimura T, Maeda M, Inoue S, Kataoka K, Sakurai Y (1986) *J. Biomed. Mater. Res.* 20, 867.
107 Kugo K, Ohji A, Uno T, Nishino J (1987) *Polymer J.* 19, 375.
108 Cho C-S, Kim S-W, Komoto T (1990) *Makromol. Chem.* 191, 981.
109 Cho C-S, Kim SU (1988) *J. Control. Release* 7, 283.
110 Floudas G, Papadopoulos P, Klok H-A, Vandermeulen GWM, Rodríguez-Hernandez J (2003) *Macromolecules* 36, 3673.
111 Parras P, Castelletto V, Hamley IW, Klok H-A (2005) *Soft Matter* 1, 284.
112 Cho CS, Jo B-W, Kwon J-K, Komoto T (1994) *Macromol. Chem. Phys.* 195, 2195.
113 Tanaka S, Ogura A, Kaneko T, Murata Y, Akashi M (2004) *Macromolecules* 37, 1370.
114 Cho C-S, Kim S-W, Sung Y-K, Kim K-Y (1988) *Makromol. Chem.* 189, 1505.
115 Hayashi T, Kugo K, Nakajima A (1984) *Cont. Topics Polym. Sci.* 4, 685.
116 Ito Y, Miyashita K, Kashiwagi T, Imanishi Y (1993) *Biomat. Artif. Cells Immob. Biotechnol.* 21, 571.
117 Barenberg S, Anderson JM, Geil PH (1981) *Int. J. Biol. Macromol.* 3, 82.
118 Kong X, Jenekhe SA (2004) *Macromolecules* 37, 8180.
119 Kricheldorf HR (1987) *α-Aminoacid-N-carboxyanhydrides and related heterocycles,* Springer-Verlag, Berlin, Heidelberg, New York.
120 Deming TJ (2000) *J. Polym. Sci. A Polym. Chem.* 38, 3011.
121 Deming TJ (1997) *Nature (London)* 390, 386.
122 Dimitrov I, Schlaad H (2003) *Chem. Commun.* 2944.
123 Aliferis T, Iatrou H, Hadjichristidis N (2004) *Biomacromolecules* 5, 1653.

7
Synthesis, Self-assembly and Applications of Polyferrocenylsilane (PFS) Block Copolymers

Xiaosong Wang, Mitchell A. Winnik, and Ian Manners

7.1
Introduction

High molecular weight polyferrocenylsilane (PFS) macromolecules were discovered in the early 1990s and represent an interesting class of metal-containing polymers [1]. Depending on the substituent groups on silicon, PFS materials can be semicrystalline or amorphous [2]. In this chapter, PFS will refer to semicrystalline dimethyl substituted polyferrocenylsilane (see Scheme 7.1) unless indicated otherwise. Owing to the presence of iron, PFSs exhibit a range of intriguing properties that traditional organic polymers do not possess, including redox activity [associated with the reversible Fe(II)/Fe(III) couple], charge dissipativity and an ability to act as a magnetic ceramic or catalyst precursor [3, 4].

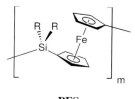

PFS

Scheme 7.1

The incorporation of PFS segments into self-organizing motifs, such as block copolymers, provides further possibilities for supramolecular chemistry and the development of functional nanomaterials [5].

This chapter summarizes recent developments in the synthesis and self-assembly of PFS block copolymers and their applications in material science.

Block Copolymers in Nanoscience. Massimo Lazzari, Guojun Liu, Sébastien Lecommandoux
Copyright © 2006 WILEY-VCH Verlag GmbH & Co. KGaA, Weinheim
ISBN: 3-527-31309-5

7.2
Synthesis of PFS Block Copolymers

The discovery that silicon-bridged [1]ferrocenophanes (**1**) could undergo living anionic ring-opening polymerization (ROP) at room temperature using initiators such as BuLi was reported in the mid-1990s (see Scheme 7.2) [6], and permitted the synthesis of PFS with controlled molecular weights and narrow molecular weight distributions (polydispersities <1.2). The comparative reactivity of PFS anionic end-groups follows the sequence of PS > PFS > PDMS [PS = polystyrene, PDMS = poly(dimethylsiloxane)] [6]. Therefore, by sequential addition of monomers in an order of decreasing end-group reactivity, block copolymers with skeletal transition metal atoms were successfully prepared. The first "prototypical" block copolymers that we synthesized by consecutive addition of monomers in a one-pot anionic polymerization were organic–organometallic (PS-*b*-PFS) (see Scheme 7.3) and organometallic–inorganic (PFS-*b*-PDMS) block copolymers (see Scheme 7.4) [6].

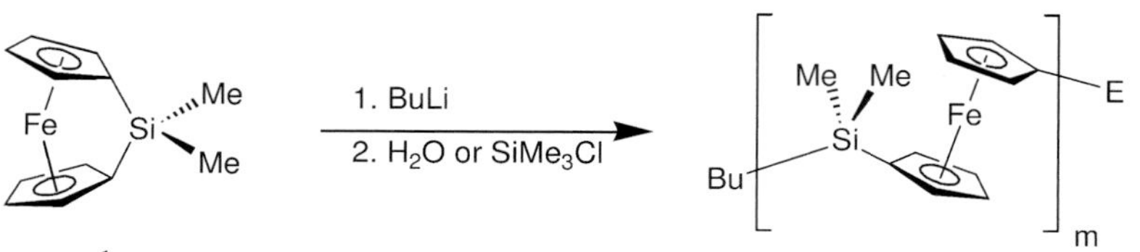

Scheme 7.2 Synthesis of PFS Block Copolymers

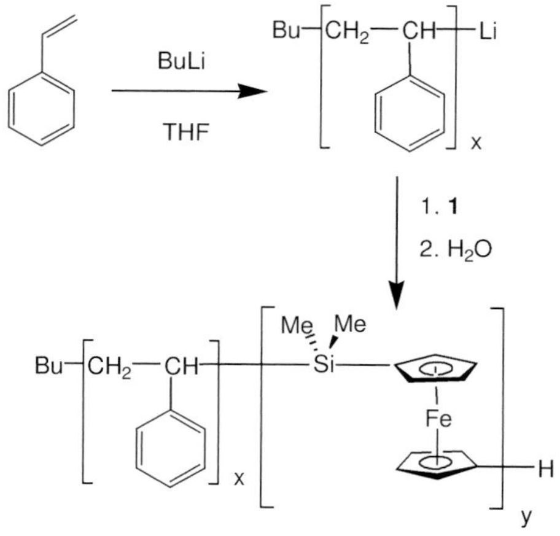

PS-*b*-PFS

Scheme 7.3

Simply extending this method to other monomers has allowed the synthesis of PI-b-PFS [7] (PI = polyisoprene), PFS-b-PMVS [8] [PMVS = poly(methylvinylsiloxane)] and PFP-b-PFS-b-PDMS [PFP = poly(ferrocenyl phosphine)] triblock copolymers [9]. In principle, this sequential anionic polymerization technique can be applied to integrate a PFS block with any polymers compatible with the anionic mechanism. One exception is a methacrylate type of polymer, because side reactions such as nucleophilic attack of the carbanionic chain ends to the carbonyl groups of methacrylates may occur if the monomers are added directly at the anionic chain end derived from **1**.

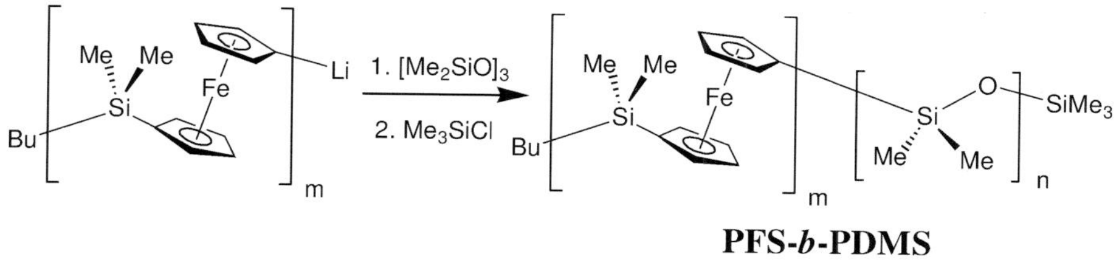

PFS-b-PDMS

Scheme 7.4

To circumvent nucleophilic side reactions, we synthesized the first PFS–methacrylate block copolymer by a two-step anionic polymerization (see Scheme 7.5) [10]. Hydroxy-terminated PFS chains (PFS-OH) were first synthesized using *tert*-butyldimethylsilyloxy-1-propyllithium as an OH-protected initiator. Deprotonation

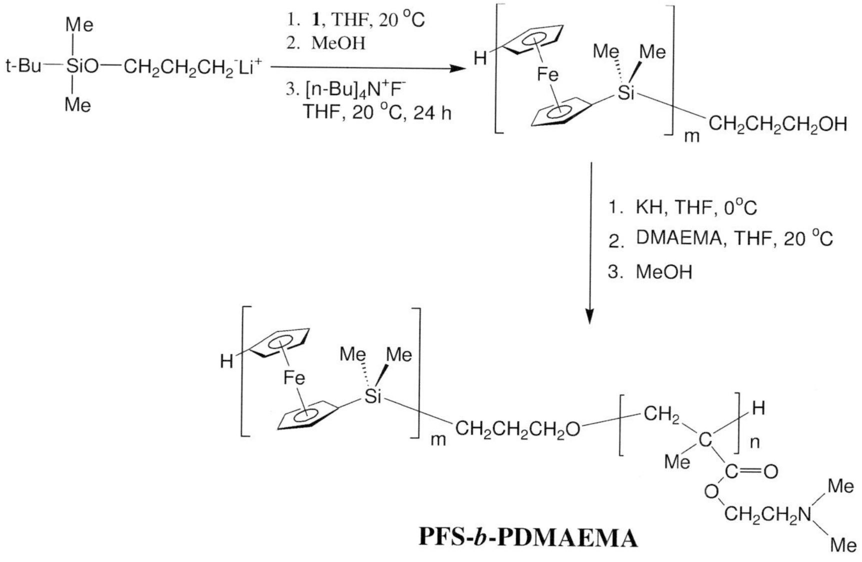

PFS-b-PDMAEMA

Scheme 7.5

of PFS-OH using potassium hydride gives an oxanionic chain end, which initiates the growth of dimethylaminoethyl methacrylate (DMAEMA) as a second block.

Kloninger and Rehahn reported a method to synthesize PFS-*b*-PMMA block copolymers (see Scheme 7.6) [11]. They transferred living PFS chain ends to DPE anionic species (DPE = 1,1-diphenylethylene) by sequentially adding 1,1-dimethylsilacyclobutane (DMSB) and DPE at the end of the living anionic chain end derived from **1**, followed by the polymerization of MMA to subsequently grow the second block of PMMA. Under optimum conditions, PFS-*b*-PMMA block copolymers can be obtained in >90% yield (in addition to <10% PFS homopolymer) with polydispersity indices below 1.1.

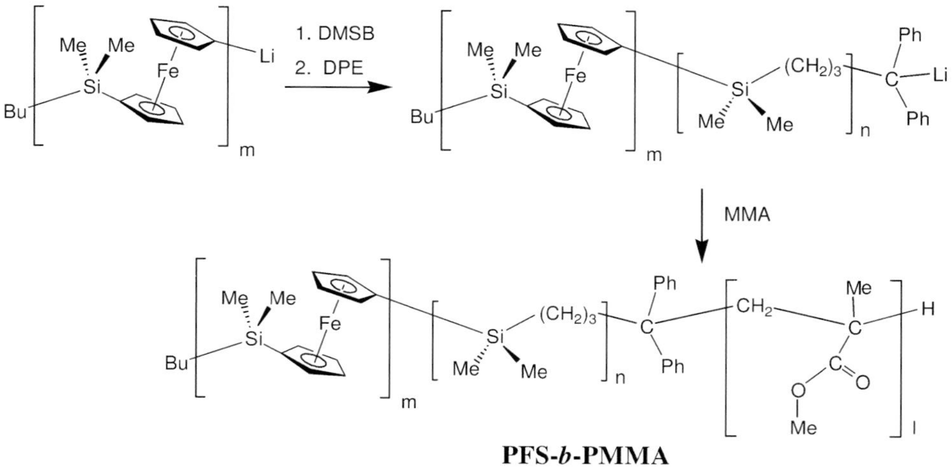

Scheme 7.6

We observed that, at elevated temperature (50 °C), DPE could relatively effectively cap the living anionic polyferrocenylsilane directly [12]. Therefore, the addition of DMSB to "pump up" the activity of PFS anionic chain ends can be omitted but the yield of the isolated PFS-*b*-PMMA is significantly lower [13]. The living DPE-capped PFS polymers can also react with chloromethyl functionalities of poly(styrene-co-chloromethylstyrene) (PS-co-PCMS), leading to the first PS-*g*-PFS graft copolymers (see Scheme 7.7) [12].

The synthesis of PFS block copolymers by the combination of the ROP of **1** and a living free radical polymerization has also been described. PFS-*b*-PMMA was synthesized as shown in Scheme 7.8, which involved a step of Ru-catalyzed living free radical polymerization [14]. The molecular weights of the block copolymers are well controlled with narrow polydispersities, although the conversion of MMA is incomplete. In view of the versatility of living free radical polymerization in terms of reaction conditions, the choice of monomers and the compatibility with functional groups, further studies in this methodology may provide a route to access a range of novel PFS block copolymers under mild conditions.

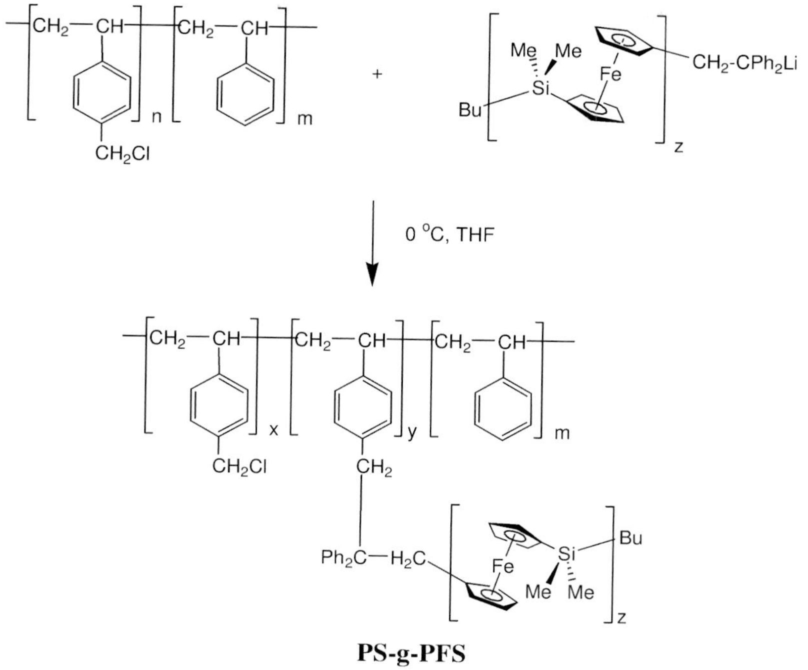

7.2 Synthesis of PFS Block Copolymers | 155

Scheme 7.7

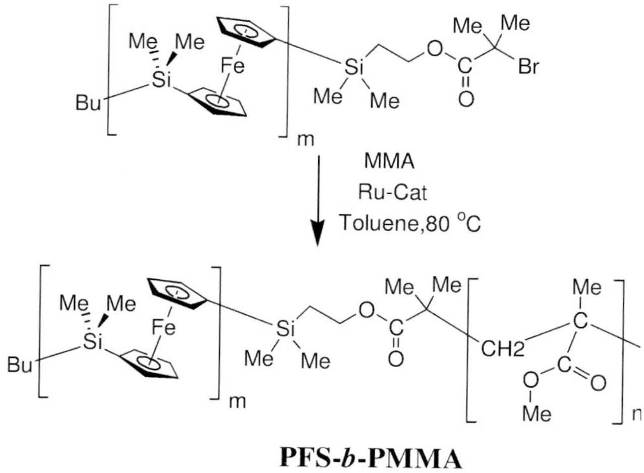

PS-g-PFS

Scheme 7.8

PFS-b-PMMA

The first organometallic miktoarm star copolymer, PFS(PI)$_3$, with a PDI of 1.04 was synthesized through an anionic polymerization by using SiCl$_4$ as a coupling agent, as shown in Scheme 7.9 [15]. The well-defined structure was confirmed by the characterizations of GPC and ^{1}H NMR spectroscopy.

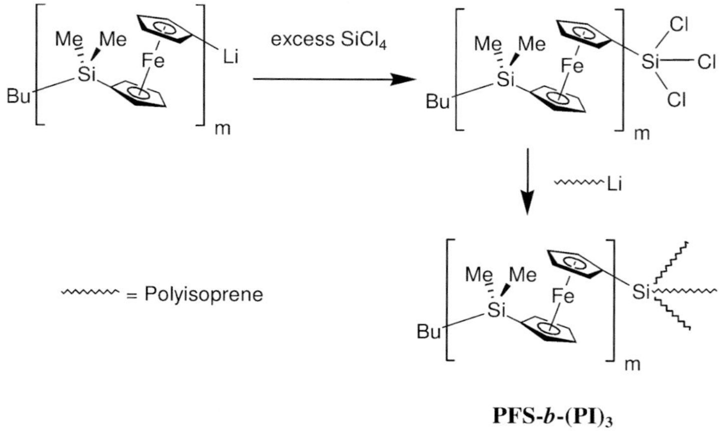

Scheme 7.9

In this context, it is worthwhile noting a recently reported photolytic living anionic ROP of **1**. The polymerization was initiated by UV-VIS irradiation in the presence of anionic promoter Na[C$_5$H$_5$] and propagated via the cleavage of the Fe–Cp bonds in the monomer (see Scheme 7.10) [16]. This method differs fundamentally from the previously reported living anionic ROP in the presence of organolithium initiators as illustrated in Scheme 7.2. Organolithium initiated anionic polymerization proceeds in the absence of UV irradiation and involves Si–Cp bond cleavage. Significantly, the propagating centers for the new photolytic methodology are silyl-substituted cyclopentadienyl anions, which are less basic than iron coordinated Cp anions. This method not only provides a new concept for living anionic polymer-

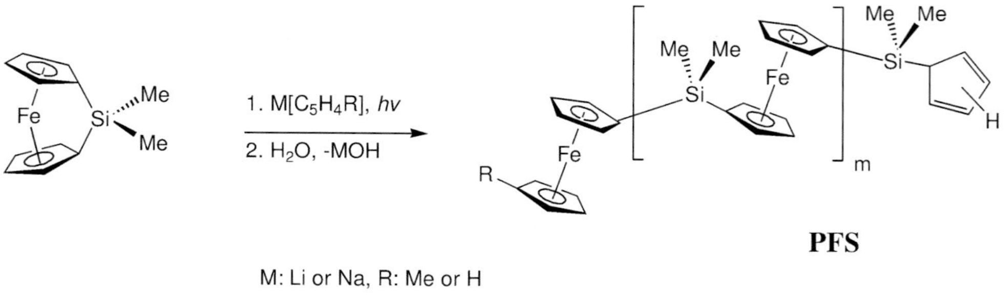

Scheme 7.10

ization but also offers opportunities to synthesize PFS block copolymers under mild conditions.

The hybridization of PFS with polypeptide will not only produce a new type of materials but also offer an opportunity to explore biological applications of PFS. Very recently, we reported our first efforts in this regard [17]. As shown in Scheme 7.11, PFS-*b*-PBLG [PBLG = poly(γ-benzyl-L-glutamate)] block copolymer was synthesized in two steps: preparing amino terminated PFS (PFS-NH$_2$) through anionic ROP of ferrocenylsilane, followed by an ROP of γ-benzyl-L-glutamate-N-carboxyanhydrides initiated by the PFS-NH$_2$ macroinitiator.

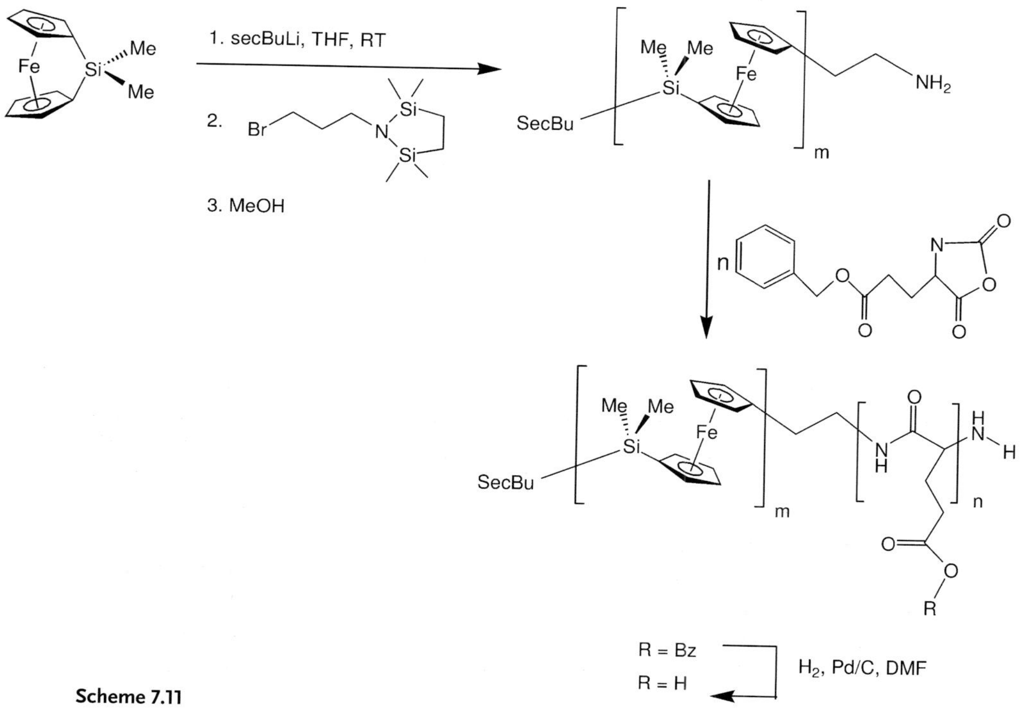

Scheme 7.11

7.3
Solution Self-assembly of PFS Block Copolymers

Solution self-assembly of PFS block copolymers allows the generation of discrete supramolecular organometallic nanomaterials with a range of one-dimensional morphologies, including cylinders [18], tubes [19], fibers [20] and tapes [7]. When a bulk polymer sample of $PFS_{50}\text{-}b\text{-}PDMS_{300}$ (block ratio of PFS/PDMS = 1/6) was heated with n-hexane, a selective solvent for PDMS, in a sealed vial at 80 °C, the cylinders were solubilized and could be visualized by TEM (see Fig. 7.1a), in which the high electron density of iron-rich PFS blocks provides the contrast in the image [18]. As illustrated schematically in Fig. 7.1b, the cylinders possess an iron-rich, organometallic core of PFS surrounded by an insulating sheath of PDMS.

Apart from cylinders, PFS-b-PDMS block copolymers can also self-assemble into apparently hollow tubular structures when the block ratios of PFS/PDMS reaches 1/12 [19]. As shown in Fig. 7.2, $PFS_{40}\text{-}b\text{-}PDMS_{480}$ (PFS/PDMS = 1/12) were self-organized in hexane into nanotubes, in which the PFS blocks aggregate to form

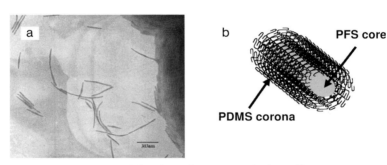

Fig. 7.1 (a) TEM image and (b) scheme for PFS nanocylinders self-assembled from $PFS_{50}\text{-}b\text{-}PDMS_{300}$ (PFS/PDMS = 1/6) in hexane (reproduced from [5a] with permission).

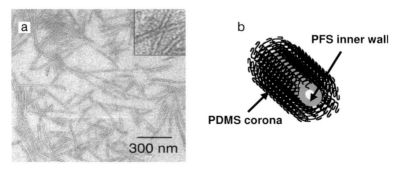

Fig. 7.2 (a) TEM image and (b) scheme for PFS nanotubes self-assembled from $PFS_{40}\text{-}b\text{-}PDMS_{480}$ (PFS/PDMS = 1/12) in hexane (reproduced from [17] with permission).

a shell with a cavity in the middle of the tube, while the PDMS blocks form the corona (see Fig. 7.2b) [19]. Several additional TEM techniques including dark-field TEM and energy-filtered TEM also supported the nanotube structures [21]. The presence of the hollow cavities was further supported by trapping tetrabutyllead in the voids and performing energy-dispersive X-ray measurements on the resulting structure [19].

The essential role of the block ratio in determining the micellar morphologies of PFS block copolymers was also discerned in our studies on the PI-b-PFS system [7]. PI_{30}-b-PFS_{60} and PI_{320}-b-PFS_{53} have similar PFS length but differ significantly in PI length, with block ratios (PI/PFS) of 1/2 and 6/1, respectively. As shown in Fig. 7.3, for the block copolymers of PI_{320}-b-PFS_{53} with longer PI chains, cylinders were observed (see Fig. 7.3a), while the tape-like micelles as shown in Fig. 7.3b self-assembled from PI_{30}-b-PFS_{60}.

Under certain circumstances, spherical aggregates were also found. For example, when PFS cylindrical micelles were heated above the T_m of PFS blocks (ca. 120–145 °C) [19] or an amorphous poly(ferrocenylmethylphenylsilane) block was used to replace semicrystalline PFS in the block copolymers [22], spheres became the only morphology observed by TEM. These experiments suggested that the crystallization of PFS chains could be a possible reason for the formation of observed one-dimensional supramolecular assemblies. PFS crystallization was indeed detected in both PFS nanocylinders [22] and nanotubes [19] by Wide-Angle X-ray Diffraction (WAXD) experiments on the micelles solid samples dried from the corresponding solutions.

PFS block copolymers are also able to self-assemble in water, providing hydrophilic blocks are incorporated [23, 24]. Following the successful synthesis of PFS_9-b-$PDMAEMA_{50}$ (see Scheme 7.5), we performed a reaction of the block copolymer with methyl iodide obtaining a corresponding quaternized PFS_9-b-$qPDMAEMA_{50}$ [24]. We further investigated the micellization behavior of these two block copolymers. A range of intermediate morphologies, such as vesicles, rods and cylinders, were captured in the case of pristine PFS_9-b-$PDMAEMA_{50}$ during

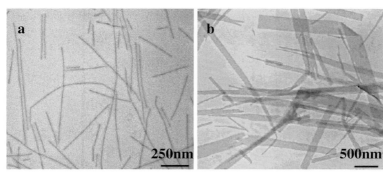

Fig. 7.3 TEM images for the micelles self-assembled from (a) PI_{320}-b-PFS_{53} in THF–hexane (2/8, v/v) and (b) PI_{30}-b-PFS_{60} THF–hexane (3/7, v/v) (reproduced from [7] with permission).

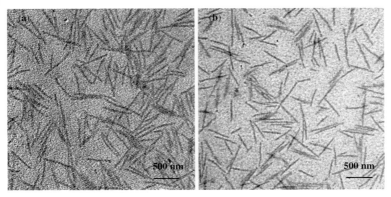

Fig. 7.4 TEM for the micelles prepared by (a) mixing PFS$_9$-b-PDMAEMA$_{50}$ with ethanol and (b) replacing ethanol with water by dialysis (reproduced from [24] with permission).

the course of our investigation, though spheres appear to be the thermodynamically stable structures for the aggregates of both polymers in water. Nevertheless, alcohol solvents such as EtOH, ^{i}PrOH are able to induce PFS$_9$-b-PDMAEMA$_{50}$ to self-assemble into cylindrical micelles by simply dissolving the block copolymers in the solvent. Therefore, water-soluble cylindrical micelles could then be prepared via the dialysis of PFS$_9$-b-PDMAEMA$_{50}$ in ethanol against water [24].

One potential application of these well-defined aggregates may be as etching resists for semiconducting substrates, such as GaAs or Si, and offer potential access to magnetic or semiconducting nanoscopic patterns on various substrates [25]. In a collaboration with Spatz and Moller at the University of Ulm, Germany, the PFS cylinders have been positioned on the surface of a GaAs resist by capillary forces along grooves, which were previously formed from electron beam etching of the surface. Subsequent reactive plasma etching generated connected ceramic lines of reduced size (see Fig. 7.5) [26].

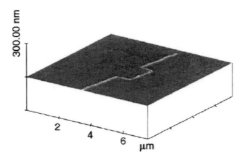

Fig. 7.5 Scanning force micrograph for the ceramic line dervied from aligned PFS-b-PDMS cylindrical micelles by H$_2$ plasma etching (reproduced from [26] with permission).

Despite the success in the fabrication of nanoceramic lines by plasma etching of aligned micellar precursors, our further exploitations of PFS nanostructures were often frustrated by the labile nature of the self-assembled micelles, in that they easily undergo dissociation in a common solvent or at an elevated temperature.

7.4 Shell Cross-linked Nanocylinders and Nanotubes

Several groups have demonstrated that the stability of self-assembled micelles can be improved by performing a shell cross-linking reaction [27]. The synthesis of PFS-b-PMVS and PI-b-PFS block copolymers through the sequential anionic polymerization, followed by the self-assembly in hexane, allowed us to access either nanocylinders or nanotubes with pendent vinyl groups attached to the corona chains (Scheme 7.12). By taking advantage of pendent vinyl groups, we performed a Pt-catalyzed hydrosilylation cross-linking reaction with tetramethyldisiloxane as a cross-linker, leading to shell cross-linked nanocylinders and nanotubes [8, 28].

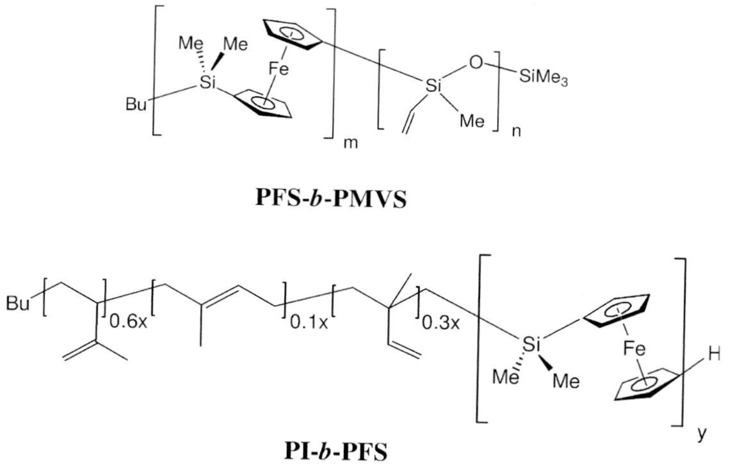

PFS-b-PMVS

PI-b-PFS

Scheme 7.12

As a result of shell cross-linking, the one-dimensional micellar structures were locked-in and preserved even if transferred from hexane to a common solvent [8, 28]. Figure 7.6 illustrates a TEM image for PI_{320}-b-PFS_{53} shell cross-linked micelles. The TEM sample was prepared from a micellar solution in THF, a common solvent for both PFS and PI blocks.

We have further demonstrated that the shell cross-linking reaction is essential in making PFS nanoceramics. Upon heating up to 600 °C with a temperature ramp of 1 °C min^{-1} under N_2, we pyrolyzed the shell cross-linked micelles. The resulting ceramics, with excellent shape retention, were characterized by TEM (see

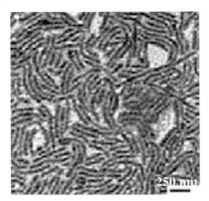

Fig. 7.6 TEM image for shell cross-linked PI_{320}-b-PFS_{53} cylindrical micelle solid samples prepared by drying a drop of THF solution on a carbon coated copper grid. THF is a common solvent for both blocks (reproduced from [28] with permission).

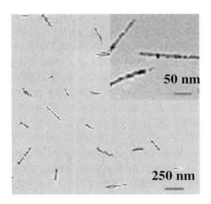

Fig. 7.7 TEM image for cylindrical nanoceramic replica derived from PI_{320}-b-PFS_{53} shell cross-linked micelles through a pyrolysis process under N_2 with a temperature ramp of 1 °C min^{-1} up to 600 °C (reproduced from [28] with permission).

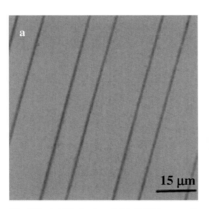 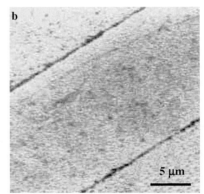

Fig. 7.8 (a) Optical and (b) SEM images for microfluidically aligned shell cross-linked PI_{320}-b-PFS_{53} nanocylinders (reproduced from [28] with permission).

Fig. 7.7). In a control experiment, pyrolysis of uncross-linked PI$_{320}$-b-PFS$_{53}$ cylindrical micelles was found to lead to the destruction of the structure. This comparison indicated that the shell cross-linking plays an essential role in the shape retention and permits the formation of a ceramic replica, presumably due to high ceramic yield resulting from the cross-linked structure [28].

We have also aligned and patterned shell cross-linked micelles on a flat silicon substrate by a microfluidic technique (see Fig. 7.8). Ordered magnetic nanoceramic arrays derived from a pyrolysis process, which may be of interest as magnetic memory materials or as catalysts for the growth of carbon nanotubes [4].

As for shell cross-linked nanotubes (see Fig. 7.9a), we proved that a redox-activity due to PFS chains is conserved by performing cyclic voltammetry experiments (see Fig. 7.9b) [8]. These nanotubes represent an extraordinary new type of reactive nanostructure and provide an opportunity to encapsulate guest compounds or particles in the cavities through an *in situ* redox reaction using PFS as a reductant [29].

As illustrated in Scheme 7.13, Ag nanoparticles were prepared within the tubes via a reaction with Ag[PF$_6$]. We found that partial pre-oxidation of the PFS domains with an organic oxidant, such as tris(4-bromophenyl)aminium hexachloroanti-

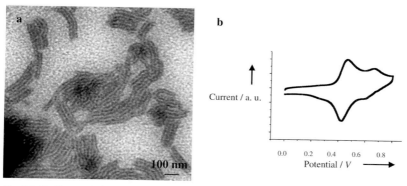

Fig. 7.9 Shell cross-linked nanotubes of PFS$_{48}$-b-PMVS$_{300}$ block copolymers. (a) TEM image for the sample prepared for THF solution. (b) Cyclic voltammetry in dichloromethane–benzonitrile (2/1) with 0.1 M [Bu$_4$N][PF$_6$] as supporting electrolyte (reproduced from [8] with permission).

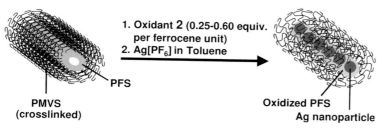

Scheme 7.13

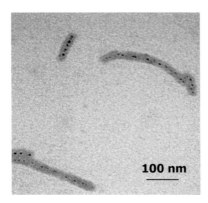

Fig. 7.10 TEM images for shell cross-linked nanotubes of PFS_{48}-b-$PMVS_{300}$ block copolymers encapsulated with Ag nanoparticles (reproduced from [29] with permission).

monate, is a key step for the efficient formation of one-dimensional arrays of silver nanoparticles confined within the nanotubes (see Fig. 7.10) [29]. Further attempts to synthesize metal nanowires through this encapsulation method are currently in progress.

7.5
Self-assembly of PFS Block Copolymers in the Solid State

Polyferrocenylsilane block copolymers also phase separate in the solid state to generate periodic, nanoscopic iron-rich domains that can be observed by TEM without resorting to staining techniques [6]. This bulk self-assembly behavior has been studied on various block copolymers such as PS-b-PFS [30], PI-b-PFS [5b], PFS-b-PMMA [5e]. Oxidation of the PFS block was reported as having influence on the order–disorder transition temperature [5d], and the high etch resistance of the PFS nanodomain has been used to nanopattern surfaces [5f]. The periodic PFS domains can be converted into iron nanoclusters that can be used as heterogeneous catalysts [4, 31]. As shown in Fig. 7.11a, perpendicular ordering of PFS microdomains traversing the film thickness was observed throughout the sectioned sample from a PS_{374}-b-PFS_{45} thin film. This thin film was pyrolyzed above 600 °C, leading to iron-rich ceramic nanodomain replicas as shown in Fig. 7.11b [30]. In a collaboration with Lu at Agilent Laboratories, Russell at the University of Massachusetts, Liu at Duke University and Ajayan and Ryu at Rensselaer Polytechnic Institute, we have used the iron-rich ceramics as catalysts for the growth of single-walled carbon nanotubes (SWCN) on the substrate by a simple one step chemical vapor deposition (CVD). The resulting SWCN lying on the surface were characterized by SEM (Fig. 7.12) [4]. Multi-walled CN have also been reported by Hinderling et al. using a similar procedure but involving a prior plasma treatment [31].

Most studies [5, 32] indicated that the self-assembly of PFS block copolymers in the bulk state follows the theory that the longer block forms the continuous phase,

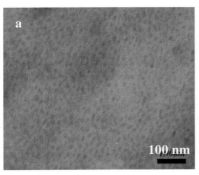 

Fig. 7.11 Self-assembly of PFS block copolymers in the solid state. (a) TEM image for PS_{374}-b-PFS_{45} thin film microtomed parallel to the substrate, (b) AFM image (2 × 2 µm) of pyrolyzed sample (600 °C, 2 h, N_2) of a PS_{374}-b-PFS_{45} thin film in which the PS matrix has been cross-linked by UV radiation (reproduced from [30] with permission).

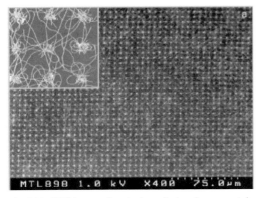

Fig. 7.12 SEM images for single-walled carbon nanotubes (SWCN) synthesized by chemical vapor deposition (CVD) with iron-rich ceramic particles derived from patterned PS-b-PFS islands as catalyst (reproduced from [4] with permission).

whereas the shorter block forms the imbedded structures. In a recent study, we discovered an interesting exception [33].

As illustrated in Fig. 7.13, the morphology of PFS_{90}-b-$PDMS_{900}$ film casting from toluene shows somewhat disordered hexagonal packed objects that are formed by the PDMS blocks (white areas), while the continuous phase appears gray (Fig. 7.13). Scanning TEM (STEM) in the dark field mode to view an image of the electrons that are elastically scattered by the PFS domains also confirmed this morphology. As shown in the inset of Fig. 7.11, we observed the presence of PFS rings (white) in the midst of the relatively electron-poor PDMS domains (black and gray).

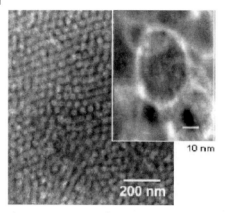

Fig. 7.13 TEM image for a cryomicrotomed PFS$_{90}$-b-PDMS$_{900}$ film cast from toluene. Inset: high-resolution STEM image of this sample obtained in the dark field mode (reproduced from [33] with permission).

The small-angle X-ray scattering (SAXS) pattern of the film using a Cu-K$_\alpha$ source shows broad peaks with relative positions of $3^{1/2}$, $7^{1/2}$ and $21^{1/2}$, suggesting hexagonally packed cylinders.

Therefore, based on our preliminary experiments, it was confirmed that an asymmetric PFS$_{90}$-b-PDMS$_{900}$ copolymer undergoes phase segregation where the longer block forms isolated cylinders surrounded by a shell of PFS, with the remaining PDMS filling the interstitial spaces.

7.6
Summary

Living anionic ROP of **1** represents an unprecedented breakthrough in the controlled synthesis of metal-containing polymers, which allows access to a range of well-defined block copolymers with predicted molecular weights, narrow polydispersities and designed architectures.

Self-assembly of PFS block copolymers either in solution or in bulk generated a number of interesting nanostructures with iron-rich, redox-active PFS domains. These nanomaterials have proven useful for various applications, including fabrication of nanoceramics and catalysts for carbon nanotube growth.

References

1 FOUCHER, D. A., TANG, B. Z., MANNERS, I. *J. Am. Chem. Soc.* **1992**, *114*, 6246.
2 RULKENS, R., PERRY, R., LOUGH, A. J., MANNERS, I., LOVELACE, S. R., GRANT, C., GEIGER, W. E. *J. Am. Chem. Soc.* **1996**, *118*, 12683.
3 a) MANNERS, I. *Science* **2001**, *294*, 1664. b) KULBABA, K., MANNERS, I. *Macromol. Rapid Commun.* **2001**, *22*, 711.
4 a) LU, J. Q., KOPLEY, T. E., MOLL, N., ROITMAN, D., CHAMBERLIN, D., FU, Q., LIU, J., RUSSELL, T. P., RIDER, D. A., MANNERS, I., WINNIK, M. A. *Chem. Mater.* **2005**, *17*, 2227; b) LASTELLA, S., JUNG, Y. J., YANG, H., VAJTAI, R., AJAYAN, P. M., RYU, C. Y., RIDER, D. A., MANNERS, I. *J. Mater. Chem.* **2004**, *14*, 1791.
5 a) MASSEY, J., POWER, K. N., WINNIK, M. A., MANNERS, I. *Adv. Mater.* **1998**, *10*, 1559; b) LAMMERTINK, R. G. H., HEMPENIUS, M. A., THOMAS, E. L., VANCSO, G. J. *J. Poly. Sci. B-Poly. Phys.* **1999**, *37*, 1009;. c) LI, W., SHELLER, N., FOSTER, M. D. BALAISHIS, D., MANNERS, I., ANNIS, B., LIN, J. S. *Polymer* **2000**, *41*, 719; d) EITOUNI http://pubs.acs.org/cgi-bin/article.cgi/jacsat/2004/126/i24/html/#ja048570cAF2, H. B., BALSARA, N. P. *J. Am. Chem. Soc.* **2004**, *126*, 7446; e) KLONINGER, C, REHAHN M. *Macromolecules* **2004**, *37*, 8319; f) LAMMERTINK, R. G. H., HEMPENIUS, M. A., VAN DEN ENK, J. E., CHAN, V. Z. H., THOMAS, E. L., VANCSO, G. J. *Adv. Mater.* **2000**, *12*, 98.
6 NI, Y., RULKENS, R., MANNERS, I. *J. Am. Chem. Soc.* **1996**, *118*, 4102.
7 CAO, L., MANNERS, I, WINNIK, M. A. *Macromolecules* **2002**, *35*, 8258.
8 WANG, X. S., WINNIK, M. A., MANNERS, I. *Angew. Chem., Int. Ed. Engl.* **2004**, *43*, 8258.
9 WANG, X. S., WINNIK, M. A., MANNERS, I. *Macromolecules* **2002**, *35*, 9146.
10 WANG, X. S., WINNIK, M. A., MANNERS, I. *Macromol. Rapid Commun.* **2002**, *23*, 210.
11 KLONINGER, C, REHAHN M. *Macromolecules* **2004**, *37*, 1720.
12 POWER-BILLARD, K. N., WIELAND, P., SCHAFER, M., NUYKEN, O., MANNERS, I. *Macromolecules* **2004**, *37*, 2090.
13 VANDERARK, L., MANNERS, I. unpublished results.
14 WANG, X. S., WINNIK, M. A., MANNERS, I. *Macromol. Rapid Commun.* **2003**, *24*, 403.
15 KORCZAGIN, I., HEMPENIUS, M. A., VANCSO, G. J. *Macromolecules* **2004**, *37*, 1686.
16 TANABE, M., MANNERS, I. *J. Am. Chem. Soc.* **2004**, *126*, 11434.
17 KIM, K. T., VANDERMEULEN, G. W. M., WINNIK, M. A., MANNERS, I. *Macromolecules* **2005**, *38*, 4958.
18 MASSEY, J. A., POWER, K. N., MANNERS, I., WINNIK, M. A. *J. Am. Chem. Soc.* **1998**, *120*, 9533.
19 RAEZ, J., MANNERS, I., WINNIK, M. A. *J. Am. Chem. Soc.* **2002**, *124*, 10381.
20 RAEZ, J., MANNERS, I., WINNIK, M. A. *Langmuir* **2002**, *18*, 7229.
21 FRANKOWSKI, D. J., RAEZ, J., MANNERS, I., WINNIK, M. A., KHAN, S. A., SPONTAK, R. J. *Langmuir* **2004**, *20*, 9304.
22 MASSEY, J. A., TEMPLE, K., CAO, L., RHARBI, Y., RAEZ, J., WINNIK, M. A., MANNERS, I. *J. Am. Chem. Soc.* **2000**, *122*, 11577.
23 GOHY, J. F., LOHMEIJER, B. G. G., ALEXEEV. A., WANG, X. S., MANNERS, I., WINNIK, M. A., SCHUBERT, U. S. *Chem. Eur. J.* **2004**, *10*, 4315.
24 WANG, X. S., WINNIK, M. A., MANNERS, I. *Macromolecules*, **2005**, *38*, 1928.
25 CAO, L., MASSEY, J. A., WINNIK, M. A., MANNERS, I., RIETHMULLER, S., BANHART F., SPATZ, J. P., MÖLLER, M. *Adv. Funct. Mater.* **2003**, *13*, 271.
26 MASSEY, J. A., WINNIK, M. A., MANNERS, I., CHAN, V. Z-H., OSTERMANN, J. M., ENCHELMAIER, R., SPATZ, J. P., MÖLLER, M. *J. Am. Chem. Soc.* **2001**, *123*, 3147.
27 a) THURMOND, K. B., KOWALEWSKI, T., WOOLEY, K. L. *J. Am. Chem. Soc.* **1996**, *118*, 7239; b) THURMOND, K. B., KOWALEWSKI, T., WOOLEY, K. L. *J. Am. Chem. Soc.* **1997**, *119*, 6656; c) DING, J.,

Liu, G. *Macromolecules* **1998**, *31*, 6554; d) Bütün, V., Lowe, A. B., Billingham, N. C., Armes, S. P. *J. Am. Chem. Soc.* **1999**, *121*, 4288.

28 Wang, X. S., Arsenault, A., Ozin, G. A., Winnik, M. A., Manners, I. *J. Am. Chem. Soc.* **2003**, *125*, 12686.

29 Wang, X. S., Wang, H., Winnik, M. A., Manners, I. *J. Am. Chem. Soc.* **2005**, *127*, 8924.

30 Temple, K., Kulbaba, K., Power-Billard, K. N., Manners, I., Leach, K. A., Xu, T., Russell, T. P., Hawker, C. J. *Adv. Mater.* **2003**, *15*, 297.

31 Hinderling, C., Keles, Y., Stockli, T., Knapp, H. E., de los Arcos, T., Oelhafen, P., Korczagin, I., Hempenius, M. A., Vancso, G. J., Pugin, R. L., Heinzelmann, H. *Adv. Mater.* **2004**, *16*, 876.

32 Lammertink, R. G. H., Hempenius, M. A., Vancso, G. J., Shin, K., Rafailovich, M. H., Sokolov, J. *Macromolecules* **2001**, *34*, 942.

33 Raez, J., Zhang, Y., Cao, L., Petrov, S., Görgl, R., Wiesner, U., Manners, I., Winnik, M. A. *J. Am. Chem. Soc.* **2003**, *125*, 6010.

8
Supramolecular Block Copolymers Containing Metal–Ligand Binding Sites: From Synthesis to Properties

Khaled A. Aamer, Raja Shunmugan, and Gregory N. Tew

8.1
Introduction

Through control of both constituents and interactions, supramolecular science offers an important route to large and complex structures unattainable by other methods. Polymeric materials have been used rarely as important components in the synthesis of supramolecular materials, despite their many advantages. Most importantly, they increase organizational length scales compared with small molecules, opening a range extending from the atomic (0.1 nm) to the mesoscopic (2000 nm). Here we describe current methods to create supramolecular materials based on copolymer molecules and metal-driven assembly. A majority of the work has so far focused on using metal–ligand interactions to drive polymer formation. We will highlight some of the advances in this area, but our primary focus is on the incorporation of metal–ligands into the side-chains of macromolecules. Specifically, a heavy emphasis is on terpyridine (terpy) but many of the principles outlined should also apply to other ligands. Studies in this field are of both fundamental and practical importance, yielding insights into the process of metal-driven self-assembly, while at the same time delivering a wide variety of unique material properties.

Synthesis of new materials in which supramolecular bonds or interactions are critical to organization and structure will lead to novel architectures and interesting properties. There are many interactions used to self-assemble molecules into supramolecular materials, including hydrogen and metal bonds, π–π and donor–acceptor associations, electrostatics, hydrophilic–hydrophobic and van der Waals forces [1]. Supramolecular design allows access to larger and much more complex molecules than traditional methods of synthesis [1]. However, the use of polymeric substrates as basic elements in supramolecular materials has not received much attention [2]. Recent advances in polymer synthesis, including the discovery of living free radical polymerization or controlled radical polymerization (CRP) [3, 4], allows access to a wider variety of well-defined polymer architectures with increased functionality. Incorporation of supramolecular interactions into polymeric materials will increase the complexity of structure already found in synthetic polymer

Block Copolymers in Nanoscience. Massimo Lazzari, Guojun Liu, Sébastien Lecommandoux
Copyright © 2006 WILEY-VCH Verlag GmbH & Co. KGaA, Weinheim
ISBN: 3-527-31309-5

molecules. However, supramolecular design allows more than just the building of complex architectures, it enables the simultaneous integration of increased functionality into the material [5]. When appropriately designed, a supramolecular architecture will direct the organization of constituent molecules and express their functionality in the final material. Such materials are referred to as supramolecular materials as the supramolecular interaction is critical to their overall architecture, order, properties and function [6]. The noncovalent bond must directly influence the materials structure or property to be considered a supramolecular material [6]. In fact, when a macromolecule plays a critical role in the overall assembly process, this may be called supramolecular polymer science, which is clearly different from supramolecular chemistry, as the constituents are macromolecules as opposed to small molecules.

Many natural materials have unique and complex architectures that directly control their properties. Gaining a better understanding of the fundamental principles governing this assembly and their properties is one of the goals of supramolecular science [7]. This field has evolved rapidly in the last 20 years, from two molecules associating in solution, to the promise of new materials for the next century. Research in this area attempts to incorporate intermolecular forces to build novel materials with superior properties than those that can be obtained through traditional routes. These new materials are likely to find target applications in areas ranging from nanotechnology and biology to polymer composites and electronics [1]. Supramolecular methods allow control of structure at length scales that are currently difficult to achieve by other methods. Moreover, supramolecular design allows an increased level of complexity in relation to other techniques, and this is sure to generate new materials with tailored properties for the next several decades [2]. These design principles are critical in the nanotechnology revolution. The study of covalent macromolecules containing supramolecular functionality for further assembly promises to yield unique materials. The likely outcome of such research will be new polymer processing options [2], control over defect structure,

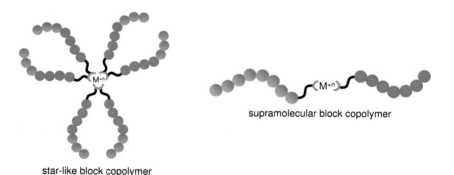

Fig. 8.1 Schematic representation (left) of star-like polymer containing the metal–complex at the core and (right) two homo-blocks connected through a metal–complex to generate a supramolecular diblock copolymer.

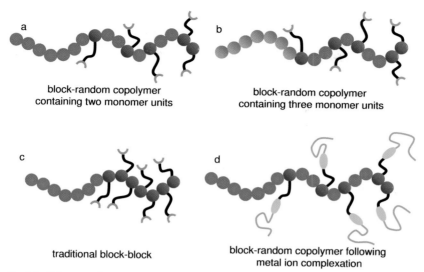

Fig. 8.2 Schematic diagram of block-random (a, b) and traditional diblock copolymers (c). (d) Represents a block-random copolymer functionalized through the supramolecular metal–ligand complex to generate a new macromolecule.

new phase-separated morphologies, novel properties (including electronic and optical) and materials able to respond to external stimuli.

Eventually, the essential features of many biological systems will be successfully mimicked by materials created through supramolecular design principles. In fact, the properties of these supramolecular systems will probably outperform natural systems because they are not limited to the building blocks of life but to the toolbox of organic and materials synthesis. In this chapter, we confine our discussion to block copolymers in which metal–ligands are an important component of the system. An entire survey of the literature is impossible and we will confine our examples to those specifically focused on block copolymers with potential applications toward self-assembling materials. A further constraint is a focus on poly-pyridine ligands. The ability to generate block copolymer architectures with metal–ligands confined to at least one block will have important applications in the field of supramolecular polymer science. The chapter is divided into two sections: (1) block copolymers in which a metal–ligand resides at the central junction between the two blocks (Fig. 8.1) and (2) block copolymers in which metal–ligands are appended to the side-chain of one or more blocks (Fig. 8.2).

8.2
Block Copolymers with Chain-end Containing Metal Complexes

A significant contribution to this area has grown out of the supramolecular chemistry field, in which small molecules are used to form larger macromolecules [1, 8]. Although many groups have assembled α,ω-functionalized metal–ligand small molecules and demonstrated their assembly into larger structures, we are focused here on the production of block copolymers via the use of metal–ligand interactions [9–12]. Two groups have been especially active in this area, but with significantly different approaches. Fraser focused on metal–ligand cores to generate a host of star-like copolymers, while Schubert used end-functionalized homopolymers to create amphiphilic block copolymers. More recently, diblocks were also reported by reversible addition-fragmentation chain transfer polymerization (RAFT) [13, 14].

8.2.1
Metal Complexes in the Center of Star Polymers

Linear and star polymers containing the metal–ligand complex in the center have been studied by Fraser [15–25]. Using either the convergent or the divergent approach, multiarm star homo- and diblock copolymers were constructed. In the divergent approach, multiarm metalloinitiators based on tris(bipyridine) containing Fe(II) and Ru(II) were used to initiate the ring-opening polymerization (ROP) of oxazolines, poly(lactic acid) (PLA) and poly(ε-caprolactone) (PCL). It was also demonstrated that atom transfer radical polymerization (ATRP) could be carried out to generate blocks of polystyrene (PS) and poly(methyl methacrylate) (PMMA). The convergent approach involves growth of the polymer chain from functionalized bipyridine (bpy) initiators followed by assembly of the star polymer via complexation of the metal ion. Figure 8.3 illustrates the two approaches and different star polymers synthesized [21, 22].

By combining dissimilar blocks with metal complexes at the core, Fraser showed that metal-centered heteroarm stars could be produced with luminescent lanthanide ions at the interface of the microphase separated domains. This block copolymer was prepared from end-functionalized homopolymers containing dibenzoylmethane (dbm) and bpy [18]. Dbm (Fig. 8.4a) was used to chelate a lanthanide ion more strongly and the polymers were based on PLA and PCL. Complexation with europium (Eu^{3+}) produced a new block copolymer, $Eu(dbmPLA)_3(bpyPCL_2)$. Europium-based systems are of special interest as they exhibit high luminescence quantum efficiencies. Casting of this macromolecule from chloroform generated phase-separated materials, which were characterized by transmission electron microscopy (TEM). A schematic diagram of the resulting lamellar structures is shown in Fig. 8.4b with the metal ions confined to the interface between lamellae. Upon heating, morphological changes consistent with thermally induced ligand dissociation and fragmentation of the block copolymer at the block junction were observed. This is a good example of supramolecular polymer science. In addition, the use of lanthanide ions has not attracted much

8.2 Block Copolymers with Chain-end Containing Metal Complexes | 173

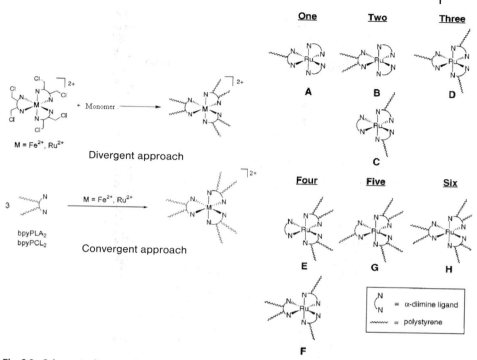

Fig. 8.3 Schematic diagram showing the divergent and convergent approaches (left). A–F are the various star polymers synthesized with different arms.

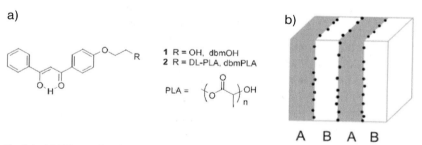

Fig. 8.4 (a) Dibenzoylmethane initiator and PLA macroligand along with (b) a schematic representation of the lamellar morphology (A = PCL, B = PLA, ● = Eu^{3+} center).

attention as compared with transition metal ions, despite their many attractive qualities, including weaker binding constants and excellent emission. The weaker binding constants could be used to tune different segments of the macromolecular block architecture.

8.2.2
Supramolecular Diblock Copolymers Connected by Metal Complexes

The generation of amphiphilic block copolymers remains of great interest due to their unique solution and bulk properties. In addition, various applications have been described. Applying principles of supramolecular chemistry to generate novel amphiphilic block copolymers was outlined by Schubert and coworkers [26–32]. It was suggested that these supramolecular block copolymers offer a number of advantages over classical covalent block copolymers. Generating a collection of building blocks would allow a library of block copolymers to be formed through self-organization of different complementary blocks via the metal ions. The metal–ligand complex introduces electrochemical, photochemical and redox properties that are new to block copolymer systems. In addition, the metal complex can be highly charged, generating chemical disparity at the junction point. The metal–ligand chemistry can be tuned to introduce reversibility into the system, which is an essential property for controlling the nanostructured assembly of the system.

Outlined in Fig. 8.5 is the synthesis of a polystyrene-*block*-poly(ethylene oxide) copolymer, designated as PS_{20}-[Ru]-PEO_{70} to describe the PS and PEO chains with an average degree of polymerization of 20 and 70, respectively, and [Ru] to highlight the bis(terpy) Ru(II) linker at the block junction [29, 30, 32]. The polymeric precursors of PEO-terpyRu(III) and PS-terpy were synthesized by living anionic polymerization processes and have a polydispersity index (PDI) lower than 1.1.

The self-assembly of this amphiphilic PS_{20}-[Ru]-PEO_{70} block copolymer was studied in water. The system forms micelles with the glassy PS in the core and the PEO chains in the corona. Kinetically frozen micelles were obtained by dissolving the copolymer in a good solvent (DMF) followed by dialysis against water to remove the DMF and freeze the PS core. Micelle size and polydispersity were studied by DLS, AFM and TEM. DLS showed two micelle populations with a hydrodynamic radius (D_h) of 65 ± 4 nm and 202 ± 37 nm. The data were attributed to a first population (65 nm) of block copolymer micelles with spherical morphology, while the second population (202 nm) represented another type of aggregate. Dilution experiments were performed to investigate the equilibrium between the two micelle populations but showed no significant change in the two populations, consistent with a very slow, or even nonexistent, exchange process between primary micelles and the larger aggregates due to the glassy styrene block. Micellar morphology was studied by TEM and suggested spherical micelles with a high tendency to aggregate into larger objects, as shown in Fig. 8.6, with average diameters in good agreement with the DLS data. Diluting the original micellar solution 100 times enabled the observation of individual micelles via AFM. These individual micelle sizes agreed well with those obtained from DLS and TEM.

In order to understand the influence of the central metal complex on micellar self-assembly, a classical PS-*b*-PEO copolymer was synthesized and studied under similar conditions. Factors believed to affect micellar size and shape are mainly controlled by three independent parameters: the stretching of the core-forming chains, the interfacial tension between the micellar core and the solvent and the

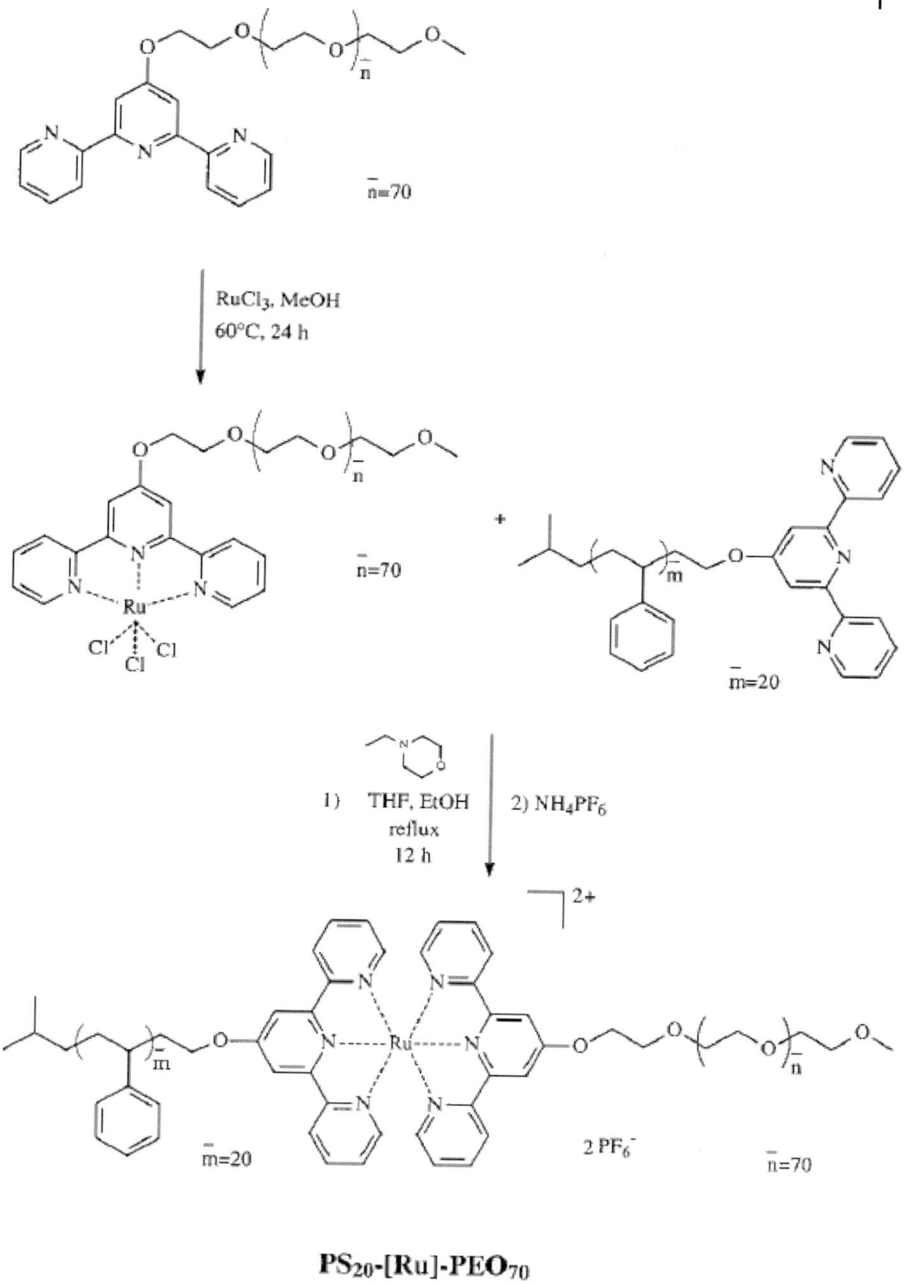

Fig. 8.5 Synthetic pathway to the metallo-supramolecular PS$_{20}$-[Ru]-PEO$_{70}$ diblock copolymer.

(a)

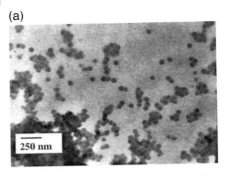

(b)

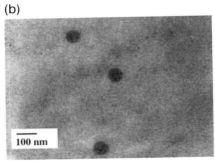

Fig. 8.6 TEM micrographs of the PS_{20}-[Ru]-PEO_{70} micelles in water (without staining).

repulsions among chains in the corona. Experimental variables that can be easily controlled are ionic strength, salt concentration, salt type, pH and temperature. Addition of salt to the PS-[Ru]-PEO during micelle formation had a profound effect on micelle size. DLS studies showed a steady decrease of the 65 nm population as the ionic strength of the medium increased. The final micellar size at high ionic strength (salt concentration of 1 molar) was comparable in size to that of PS-b-PEO in the absence of salt; although the D_h of PS-[Ru]-PEO micelles were always bigger than those of PS-b-PEO. This was attributed to the steric affect created by the bis(terpy)Ru(II) complex at the interface. The further addition of salt results in an increase in the number of aggregates, in agreement with a decreased solubility of the PEO blocks in aqueous salt solution and a subsequent clustering of the less stable micelles into aggregates. Such behavior was confirmed by DLS as more salt was added. It was observed that the negative counterion played a role in reducing the micelle size, an effect which is not currently well understood. The salt effect on D_h was reversible as the primary micelles increased in size when the added salt was eliminated by dialysis. A summary of the salt behavior is shown in Fig. 8.7.

Increasing the temperature from 20 and 65 °C decreased the primary micelle size from 65 to 40 nm, with no effect on the size of the larger aggregate population. In sharp contrast, the D_h of the covalent block copolymer, PS_{22}-b-PEO_{70}, micelles were insensitive to temperature in the investigated range. This effect was attributed

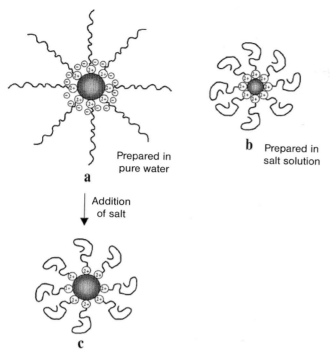

Fig. 8.7 Schematic representation of the PS$_{20}$-[Ru]-PEO$_{70}$ micelles prepared in water (a) and in a salt solution (b), and the influence of the addition of salt (c). PS core in gray, open circles with 2+ sign for bis-(terpyridine)ruthenium(II) complexes, open circles with minus sign for PF$_6^-$ anions; PF$_6^-$ anions have been only drawn in (a).

to the presence of stretched "polyelectrolyte-like" PEO segments in the metallo-supramolecular micelles. The role of solvated water around the metal complex could be another significant influence.

In another demonstration of the use of supramolecular block copolymers, Schubert reported the pH-sensitive core-shell corona aqueous micelles made from polystyrene-*block*-poly(2-vinylpyridine)-*block*-poly(ethylene oxide) ABC triblock copolymer PS-*b*-P2VP-[Ru]-PEO [33]. The pH sensitivity of the P2VP shell was used to tune the size of this system from a hydrodynamic diameter of 75 nm at pH > 5 to 135 nm at pH < 5. This effect was attributed to electrostatic repulsion between the charged P2VP blocks at low pH.

8.3
Block Copolymers with Side-chain Metal Complexes in One Block

8.3.1
Polymerizing Pre-formed Metal Complexes

In recent years, olefin metathesis reaction, and specifically ring opening metathesis polymerization (ROMP), has attracted widespread attention as a living polymerization. The technique is remarkably tolerable to many functional groups. Sleiman and coworkers used ROMP to construct homopolymers and block copolymers containing metal complexes as pendant groups from the polymer backbone [34, 35]. The polymerization of the Ru(II) tris(bpy) functionalized monomer is very popular due to the luminescent properties and high stability [36–42]. Unfortunately, the incorporation of this pre-formed metal complex does not allow the supramolecular functionality to be used to connect addition groups. Nevertheless, Sleiman's work is included here as it is a recent addition to this larger area and represents an interesting series of block copolymers. Further, the "polyelectrolyte", or charged nature, of the metal-containing block is exploited in their work.

The polymer system is based on oxynorbornene functionalized Ru(II) tris(bipyridine) metal complexes. Using "Grubbs" catalyst Gen. III, the polymerization kinetics, as monitored by ^{1}H NMR experiments, was linear first-order in monomer concentration and fast, suggesting living polymerization. The reaction was extended to synthesize block copolymers through sequential addition of monomers. A homopolymer of n-butyl functionalized oxynorbornene was extended with Ru(II) tris(bipyridine) functionalized oxynorbornene monomer to give a diblock copolymer, Fig. 8.8 [34]. The copolymer molecular weight varied with the monomer ratio, as expected, and molecular weights (MW) as high as 57.2 kDa were prepared.

Characterizing the monomer, homopolymer and the diblock with UV/VIS absorbance and luminescence experiments illustrated that the UV/VIS spectra of the homopolymer and copolymer in acetonitrile, including the molar absorptivity values for the metal to ligand charge transfer (MLCT) bands, are fairly similar to those

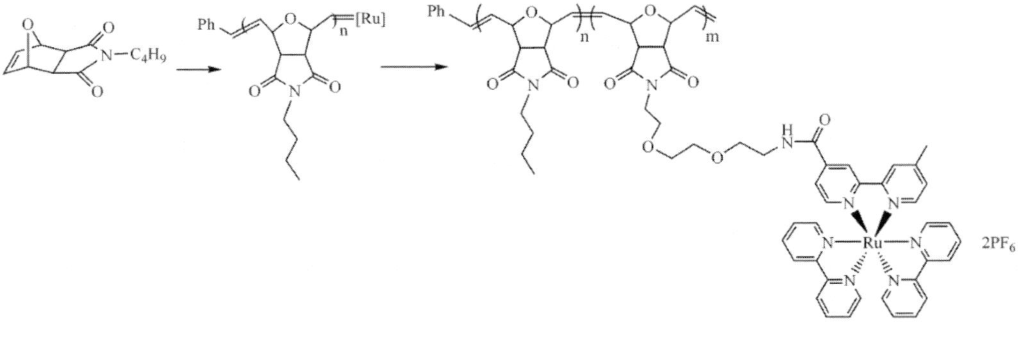

Fig. 8.8 Synthesis of block copolymers containing tris(bpy) via ROMP.

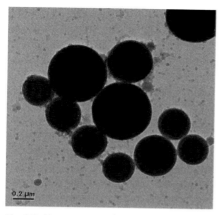

Fig. 8.9 Transmission electron micrographs of block copolymer in acetonitrile–toluene (1/2.5).

of the monomer. Such an observation is consistent with the absence of a strong electronic interaction between the chromophores in the multinuclear homopolymer and copolymer. The emission spectra in acetonitrile of the homopolymer and block copolymer show nearly identical emission maxima and quantum yield, consistent with mutually independent Ru(II) centers. The emissive ^{3}MLCT excited states thus probably lie at the same energy for all three complexes (monomer, homopolymer and block copolymer). Cyclic voltammetry experiments provided similar results. Based on these observations, it was concluded that the attachment of Ru(II) complexes to the polymer backbone had a negligible impact on the electronic and electrochemical properties of the Ru(II) complex.

Self organization of the block copolymer into micellar aggregates was performed by adding toluene slowly to a solution of the copolymer in acetonitrile, until the observed turbidity indicated the onset of aggregation (acetonitrile–toluene, 1/2.5). Figure 8.9 shows the TEM of the dried turbid solution with spherical particles of large diameter (200–600 nm). The large micellar size was attributed to the aggregation of several micelles. The micellar structure was visualized to have a core of the Ru(II) complex containing block and a corona of the more hydrophobic block in the toluene rich medium. Emission spectra of the micellar aggregate in acetonitrile–toluene at 1/500 revealed that the luminescence intensity of the Ru(II) complex containing block was enhanced in acetonitrile–toluene solution as compared with the luminescence intensity in acetonitrile solution. The enhancement was argued to be a result of both the exposure of the Ru(II) centers to the toluene solvent and the effective shielding and protection of the Ru(II) centers from oxygen quenching.

8.3.2
Block Copolymers Containing Free Metal–Ligand Side-chains

8.3.2.1 Polymerization of Metal–Ligand Containing Monomers

Ikkala, ten Brinke and coworkers have been extremely active in the area of supramolecular polymer science [2]. The majority of their work has focused on "ionic", or hydrogen bonded, complexation between poly(vinyl pyridine) and acidic surfactants. Extension of these principles to include metal ions has recently been reported [43–45]. These materials have a plethora of applications that can be imagined from the unique combination of metal ion functionality coupled to the block copolymer self-assembly. Mesoporous materials are a direct application of such materials.

Their system of zinc dodecylbenzene sulfonate [Zn(DBS)$_2$] and [poly(styrene-b-(4-vinyl pyridine)] (PS-block-P4VP) was used to demonstrate this concept [44]. Owing to the strong attraction between Zn(II) ions in Zn(DBS)$_2$ and the nitrogen lone pair of electrons of P4VP, a system of PS-b-P4VP[Zn(DBS)$_2$]$_y$ was produced. The architecture of this system is a linear block connected to a comb-shaped block, which acts as a supramolecular template for mesoporous material. PS-block-P4VP[Zn(DBS)$_2$]$_y$ self-assembled into hierarchical morphologies of ordered lamellar domains of P4VP[Zn(DBS)$_2$]$_y$ blocks in a matrix of glassy PS. Depending on the overall architecture (total molecular weight and weight ratio of each block), structure-within-structure morphologies were achieved. In the system of PS-block-P4VP[Zn(DBS)$_2$]$_{0.9}$ with $M_{n,PS}$ = 238.1 kDa and $M_{n,P4VP}$ = 49.5 kDa, a lamellar-within-lamellar structure was found in which P4VP[Zn(DBS)$_2$]$_{0.9}$ forms a lamellar structure perpendicular to the larger lamellar microphase separation, which had dimensions of 100 nm. This is shown schematically in Fig. 8.10.

Extraction of the coordinated Zn(DBS)$_2$ by a selective solvent (methanol) leads to porous lamellar structures. Small angle X-ray scattering data (SAXS) indicates that the lamellar structure was retained after removal of Zn(DBS)$_2$. Such porous lamellar would be stabilized by the glassy PS phase, different lamellar orientation and the relativity high number of defects. TEM mapping of the morphology before and after solvent extraction confirmed the lamellar structure as shown in Fig. 8.11.

Porous material with different pore sizes was achieved by varying the copolymer molecular weight. A system of PS-block-P4VP[Zn(DBS)$_2$]$_{0.8}$ with $M_{n,PS}$ = 41.4 kDa and $M_{n,P4VP}$ = 1.9 kDa formed a lamellar structure with the long period of 25 nm. Lamellar morphology is also observed even if the weight fraction of P4VP[Zn(DBS)$_2$]$_{0.8}$ domain is 0.23, although a cylindrical morphology would be expected on the basis of simple diblock copolymer considerations. This indicates an asymmetric phase diagram, as was also observed in PS-block-P4VP/alkylphenol bonded systems. The thicknesses of the PS and P4VP[Zn(DBS)$_2$]$_{0.8}$ lamellae are 20 and 5 nm, which qualitatively agrees with the P4VP[Zn(DBS)$_2$]$_{0.8}$ weight fraction.

This system allowed structurally well-defined mesoporous materials with a narrow distribution of pore sizes, a very high density of pores and high surface area per volume. The key factors are the ability to extract Zn(DBS)$_2$ from the system, which is governed by the coordination strength of the system components and

8.3 Block Copolymers with Side-chain Metal Complexes in One Block

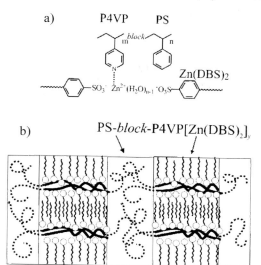

Fig. 8.10 Bonding between PS-*block*-P4VP and Zn(DBS)$_2$ and lamellar-within-lamellar structure of PS-*block*-P4VP[Zn(DBS)$_2$]$_{0.9}$. Note that Zn(DBS)$_2$ can be coordinated to more than one pyridine ring, however, still counterbalanced by the steric hindrances due to the polymer chains. Therefore, the actual coordination geometry is not known and is not addressed here.

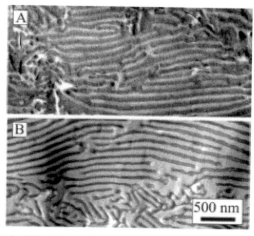

Fig. 8.11 (A) Lamellar structure for the methanol extracted and unstained sample, where the electron deficient P4VP/air domain is now white. (B) Extracted and I$_2$-stained sample where the P4VP domain shows dark due to the increased electron density in that domain. It should be noted that if the amphiphiles were not removed, the P4VP[Zn(DBS$_2$)]$_{0.9}$ domain would always be dark in the images, with or without staining.

ability of the PS glassy phase. In addition, it was speculated that the different orientations of the non-ligand lamellar and defects help prevent the structure from collapsing. The mesoporous material could be obtained in other morphologies upon proper selection of the block copolymer and composition. This work is inspiring and demonstrates the potential of this area. At the same time, it represents a critically important first step that clearly shows the tremendous potential. One only has to stop and look briefly to imagine the possibilities. The road ahead is sure to generate novel and stimulating results by combining new, sophisticated chemistry with detailed physical chemistry characterization tools.

In another demonstration of the use of ROMP, Weck and coworkers synthesized homopolymers and diblock copolymers possessing both palladated pincer complexes and diaminopyridine moieties (hydrogen-bonding entities), Fig. 8.12 [46, 47]. The polymers possess the elements of hydrogen bonding and metal recognition, a good combination for orthogonal functionalization. The side-chain functionalized polymers possessing a terminal palladated SCS pincer complex function as both well-defined Heck catalysts and recognition units capable of quantitative self-assembly of pyridine-containing molecules. Although no self-assembly studies on the diblock copolymer have been reported so far, the concept was elegantly demonstrated using the corresponding random copolymer architecture. It is expected that block copolymer studies will yield novel materials for application in the area of supramolecular polymer science.

The incorporation of metal–ligands, beyond simple pyridine and carboxylic acids, into polymeric architectures has fewer examples due to the complexity required to incorporate such functionality into various copolymer architectures. Early work focused on the incorporation of bipy or terpy into polymer side-chains of polyoxazolines [48–50] and copolymers based on 4'-vinyl-2,2':6',2''-terpyridinyl [51–54]. Pyridine-based metal–ligands have been used by a few groups to design and synthesize polymeric structures with different architectures [55, 56]. The advancement

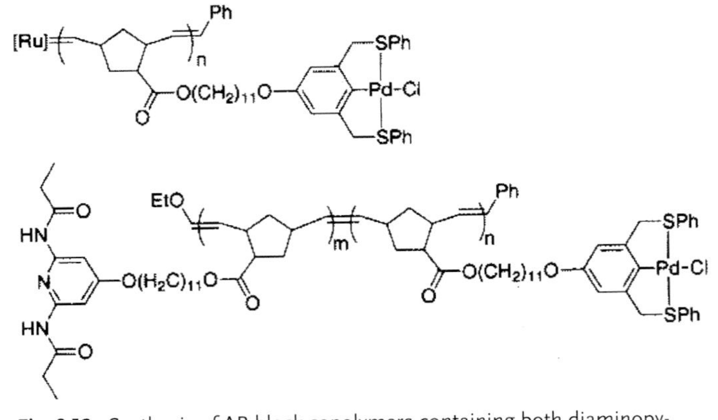

Fig. 8.12 Synthesis of AB block copolymers containing both diaminopyridine and Pd(II) pincer receptor units.

of CRP; however, is already making an impact here and is sure to continue as an important synthetic tool for this field. This was seen for Fraser's studies; however, this work still contained one metal–ligand complex at the core. The ability to synthesize a homo-block containing something such as terpy is a much more difficult task.

The ability to localize the metal–ligand to at least one segment would provide block copolymer architectures for use in supramolecular polymer science. As a result, we initiated a large program to develop a broad synthetic platform that would enable the synthesis of various monomers leading to macromolecules containing a high density of terpy in the side-chain. Initial work focused on the direct polymerization of terpy functionalized vinyl monomers; however, owing to the difficulty of using standard size-exclusion chromatography (SEC) for the routine characterization of these materials, an indirect approach was pursued. Highlights of these approaches are outlined below. CRP including ATRP, nitroxide mediated radical polymerization (NMP) and RAFT are valuable techniques to control polymer architecture and are tolerant to a broad range of functionalities. The ability to polymerize vinyl functionalized terpy monomers directly by both NMP and RAFT has been successful [57, 58]. Not surprisingly, ATRP has been unsuccessful to date, we believe mostly due to the large molar ratio of terpy in the reaction.

Knowing that NMP produced random copolymers with low PDI, block copolymer architectures were pursued. Figure 8.13 shows an example of a block-random copolymer prepared by NMP. This so-called block-random copolymer has become more popular with the advancement of CRP due to the polymerization mechanism. Starting from a polystyrene macroinitiator, a solution of 90 mol-% styrene and 10 mol-% Sty$_{terpy}$ was polymerized to afford the copolymer. Overlaid SEC chromatograms showed excellent initiation of the macroinitiator and growth of the second random block [55]. This SEC chromatogram, containing less than 10 mol-% terpy, already began to show peak broadening due to adhesion of terpy to the sta-

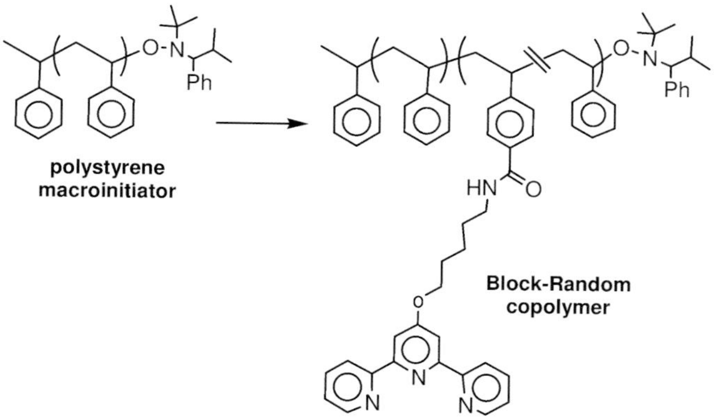

Fig. 8.13 Synthesis of poly[styrene-b-(styrene-ran-Sty$_{terpy}$)] by NMP.

tionary phase. These results showed clearly that NMP allows block architectures, based on styrene, to be prepared in which the terpy unit is confined to one segment.

The pursuit of traditional block copolymers remains attractive because of their remarkable microphase separation properties, which promote ordering of the two dissimilar blocks into different morphologies. The incorporation of terpy into one of these blocks is very tempting, as this supramolecular functionality can be further used to direct assembly or provide additional functionality. Following similar synthetic methods for NMP, but polymerizing only Sty_{terpy}, yielded a new material with 27 mol-% of Sty_{terpy}, based on the nitrogen content from elemental analysis. Along with NMR and the disappearance of the polystyrene macroinitiator from the SEC curve, we concluded that initiation of the second block was successful. To our dismay; however, we were unable to obtain an SEC curve supporting block copolymer formation. It is clear that the new sample adheres tenuously to the stationary phase of the SEC column and no solvent–salt combination we examined was able to prevent this adhesion. All attempts to remove the terpy function including treatment with refluxing methanol–NaOH and transamidation with benzylamine in the presence of $Sc(OTf)_3$ and toluene at 90 °C failed. In addition to the NMR confirmation that Sty_{terpy} polymerized and the absence of the poly(S) macroinitiator in the SEC, the best evidence for diblock copolymer formation was the appearance of discrete colored domains in the optical microscopy techniques. Inspection of the annealed film showed a clear material supporting the absence of macrophase separation that would be expected if two homopolymers were present and demixed. Although it seems reasonable that formation of the block copolymer was successful, the evidence is less than satisfying. Therefore, we turned our attention to a synthetic approach, which does not include the direct polymerization of terpy functionalized monomers.

8.3.2.2 Post-polymerization Attachment of the Metal–Ligand

Given the difficulty in characterizing this simple diblock copolymer prepared from the direct approach, we considered an alternative strategy for generating these copolymers [57, 59]. A post-polymerization approach might allow easy characterization of the pre-polymer by traditional methods such as SEC, while convenient monitoring of the reaction to incorporate terpy could be followed by another technique. Generating reactive polymers for subsequent modification has been studied extensively [60], but the use of activated ester monomers has gained favor recently due to their chemical versatility [61, 62]. We have focused on *N*-methacryloxysuccinimide (OSu) as these esters are more hydrolytically stable than other commonly used active esters and the conversion can easily be followed by IR and NMR. Optimized ATRP conditions for the homopolymerization of OSu in non-polar solvent were developed [59]. These conditions were then used to generate a series of block copolymers outlined below in Fig. 8.14. An alternative approach also worked well in which block copolymers containing one or more blocks of *t*-butyl acrylate were prepared. Selective deprotection of these blocks provides carboxylic functions that are then reacted with amine functionalized terpy, via coupling chemistry, to generate terpy containing block copolymers.

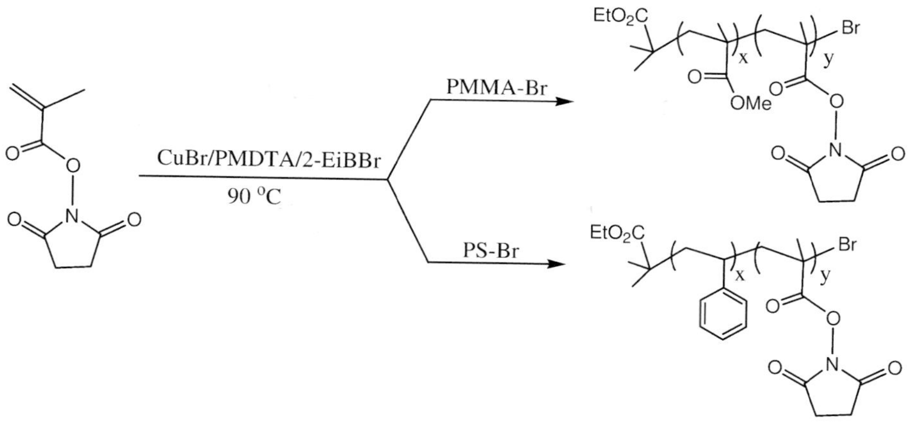

8.3 Block Copolymers with Side-chain Metal Complexes in One Block

Fig. 8.14 ATRP synthesis of poly(MMA-b-OSu) and poly(S-b-OSu).

Both poly(MMA-b-OSu) and poly(S-b-OSu) block copolymers were readily obtained from macroinitiators prepared by ATRP. Formation of the diblock copolymers was confirmed by overlaying SEC traces of the macroinitiator and the resulting copolymer. A typical example is shown in Fig. 8.15. All block copolymers showed monomodal SEC peaks suggesting very good chain extension from the macroinitiator. These results show that the polydispersity remains low in the transition from one block to the other and the process appears to work well regardless of whether the macroinitiator belongs to the same monomer class (methacrylate) or not. In all cases, a macroinitiator was used to polymerize the active ester monomer leading to very good block copolymer formation. Block copolymers of various com-

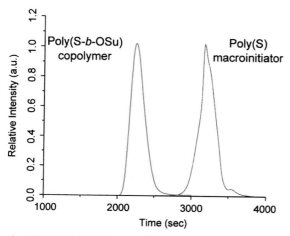

Fig. 8.15 Overlaid GPC traces of macroinitiator and block copolymer showing very good initiation.

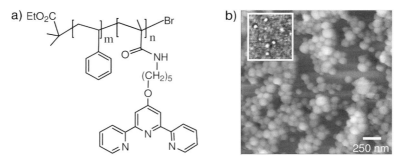

Fig. 8.16 (a) Metal–ligand containing block copolymer and (b) self-assembled structures formed upon the addition of metal ions such as Ru^{3+}.

positions were prepared with total MW ranging from 13 000 to 22 000 and PDI between 1.19 and 1.27.

Although we are just initiating structural investigations that exploit the supramolecular function, the results are very promising. Attachment of terpy groups to every monomer unit within the block generates a dense array of ligand functionality. One of the outcomes is that the backbone becomes sterically constrained. As a result, the incorporation of bis-metal complexes leads to the formation of rigid macromolecules that self-organize into a hexagonal lattice of polymeric cylinders [63]. Unlike studies reported by Sleiman on Ru(II) tris(bpy), incorporation of Ir(III) bis(terpy) complexes into the polymer architecture causes red-shifted emission compared with the monomer, or isolated, complex [63]. We recently reported lanthanide alloys that generate a unique yellow emission due to the polymeric architecture [64]. Lanthanide ions are likely to continue gaining favor because of their unique properties, including emission and binding associations.

DLS indicates that the block copolymer shown in Fig. 8.16a exists as single chains in a good solvent. Addition of metal ions to this solution creates a block copolymer which is more amphiphilic and assembles spontaneously into polymer micelles (ca. 50 nm). Owing to the polyelectrolytic nature of the metal–ligand centers, the size of the micelles depends on the dielectric of the solvent media. In addition, when these micelles are spun from solution and examined by atomic force microscopy, various nanostructures are observed. Shown in Fig. 8.16b is an AFM image obtained by spinning a concentrated solution of these micelles onto a gold surface. This image shows spherical objects in the order of ca. 150 nm with a relatively large polydispersity. When the concentration is further diluted hollow rings self-assemble, which are composed of micelles [65]. Continued dilution yields individual micelles on the gold surface as shown in the inset of Fig. 8.16b. This is just a highlight of the types of self-assembly we are finding in these systems. The future prospects are bright for these and other similar systems.

8.4
Conclusion

This chapter highlights recent activity in the area of supramolecular polymer chemistry. The incorporation of metal–ligand functionality into block copolymer architectures for use in supramolecular organization is just emerging as a powerful methodology to create materials with hierarchical order and increased functionality for applications in nanoscience. Continued improvement of synthetic tools is needed to create these rather complex macromolecules. A deeper understanding of metal–ligand interactions will lead to more opportunities in this arena. The use of lanthanides will provide more dimensionality to the landscape of copolymers containing metal–ligand side-chains. It is clear that many possibilities remain and the continued growth of new research activity in this area is promising for the future. Much basic research remains to be accomplished. At the same time, advanced materials are easily envisaged but will require continued creativity and focused effort. We look forward to the evolution of this area with great anticipation.

Acknowledgments

We thank the ARO Young Investigator and PECASE programs for generous support of this work. We also thank NIH, ARL Center of Excellence, NSF-MRSEC and the ONR Young Investigator, NSF-CAREER, 3M Nontenured faculty grant and Dupont Young Faculty Award programs for supporting research in our laboratory. My conversations with our friends and colleagues have been tremendously insightful.

References

1 J.-M. Lehn, *Supramolecular chemistry — Concepts and prospectives*, VCH, Weinheim, **1995**.
2 O. Ikkala, G. ten Brinke, *Science* **2002**, *295*, 2407–2409.
3 K. Matyjaszewski, J. Xia, *Chem. Rev.* **2001**, *101*, 2921–2990.
4 C. J. Hawker, A. W. Bosman, E. Harth, *Chem. Rev.* **2001**, *101*, 3661–3688.
5 J.-M. Lehn, *Science* **2002**, *295*, 2400–2403.
6 S. I. Stupp, M. U. Pralle, G. N. Tew, L. M. Li, M. Sayar, E. R. Zubarev, *MRS Bulletin* **2000**, *25*, 42–48.
7 G. M. Whitesides, B. Grzybowski, *Science* **2002**, *295*, 2418–2421.
8 R. P. Sijbesma, F. H. Beijer, L. Brunsveld, B. J. B. Folmer, J. H. K. K. Hirschberg, R. F. M. Lange, J. K. L. Lowe, E. W. Meijer, *Science* **1997**, *278*, 1601–1604.
9 D. G. Kurth, P. Lehmann, M. Schutte, *Proc. Natl. Acad. Sci.* **2000**, *97*, 5704–5707.
10 N. Severin, J. P. Rabe, D. G. Kurth, *J. Am. Chem. Soc.* **2004**, *126*, 3696–3697.
11 J. B. Beck, S. J. Rowan, *J. Am. Chem. Soc.* **2003**, *125*, 13922–13923.
12 Y. Q. Zhao, J. B. Beck, S. J. Rowan, A. M. Jamieson, *Macromolecules* **2004**, *37*, 3529–3531.
13 G. C. Zhou, I. I. Harruna, *Macromolecules* **2004**, *37*, 7132–7139.
14 G. C. Zhou, I. I. Harruna, *Macromolecules* **2005**, *38*, 4114–4123.
15 R. M. Johnson, C. L. Fraser, *Macromolecules* **2004**, *37*, 2718–2727.

16 A. P. SMITH, C. L. FRASER, *Macromolecules* **2003**, *36*, 2654–2660.

17 A. P. SMITH, C. L. FRASER, *J. Polym. Sci. Polym. Chem.* **2002**, *40*, 4250–4255.

18 J. L. BENDER, P. S. CORBIN, C. L. FRASER, D. H. METCALF, F. S. RICHARDSON, E. L. THOMAS, A. M. URBAS, *J. Am. Chem. Soc.* **2002**, *124*, 8526–8527.

19 A. P. SMITH, C. L. FRASER, *Macromolecules* **2002**, *35*, 594–596.

20 C. L. FRASER, A. P. SMITH, X. F. WU, *J. Am. Chem. Soc.* **2000**, *122*, 9026–9027.

21 J. E. MCALVIN, C. L. FRASER, *Macromolecules* **1999**, *32*, 1341–1347.

22 X. F. WU, C. L. FRASER, *Macromolecules* **2000**, *33*, 4053–4060.

23 C. M. PARK, J. E. MCALVIN, C. L. FRASER, E. L. THOMAS, *Chem. Mater.* **2002**, *14*, 1225–1230.

24 R. M. JOHNSON, C. L. FRASER, *Biomacromolecules* **2004**, *5*, 580–588.

25 J. E. MCALVIN, C. L. FRASER, *Macromolecules* **1999**, *32*, 6925–6932.

26 V. VOGEL, J. F. GOHY, B. G. G. LOHMEIJER, J. A. VAN DEN BROEK, W. HAASE, U. S. SCHUBERT, D. SCHUBERT, *J. Polym. Sci. Polym. Chem.* **2003**, *41*, 3159–3168.

27 G. MAYER, V. VOGEL, B. G. G. LOHMEIJER, J. F. GOHY, J. A. VAN DEN BROEK, W. HAASE, U. S. SCHUBERT, D. SCHUBERT, *J. Polym. Sci. Polym. Chem.* **2004**, *42*, 4458–4465.

28 J. F. GOHY, B. G. G. LOHMEIJER, U. S. SCHUBERT, *Macromol. Rapid Commun.* **2002**, *23*, 555–560.

29 J. F. GOHY, B. G. G. LOHMEIJER, S. K. VARSHNEY, U. S. SCHUBERT, *Macromolecules* **2002**, *35*, 7427–7435.

30 J. F. GOHY, B. G. G. LOHMEIJER, U. S. SCHUBERT, *Adv. Mater.* **2003**, *9*, 3472–3479.

31 J. F. GOHY, H. HOFMEIER, A. ALEXEEV, U. S. SCHUBERT, *Macromol. Chem. Phys.* **2003**, *204*, 1524–1530.

32 J.-F. O. GOHY, B. G. G. LOHMEIJER, U. S. SCHUBERT, *Macromolecules* **2002**, *35*, 4560–4563.

33 J. F. GOHY, B. G. G. LOHMEIJER, S. K. VARSHNEY, B. DECAMPS, E. LEROY, S. BOILEAU, U. S. SCHUBERT, *Macromolecules* **2002**, *35*, 9748–9755.

34 B. CHEN, H. F. SLEIMAN, *Macromolecules* **2004**, *37*, 5866–5872.

35 B. CHEN, K. METERA, H. F. SLEIMAN, *Macromolecules* **2005**, *38*, 1084–1090.

36 E. HOLDER, V. MARIN, A. ALEXEEV, U. S. SCHUBERT, *J. Polym. Sci. Polym. Chem.* **2005**, *43*, 2765–2776.

37 V. MARIN, E. HOLDER, R. HOOGENBOOM, U. S. SCHUBERT, *J. Polym. Sci. Polym. Chem.* **2004**, *42*, 4153–4160.

38 J. M. J. FRECHET, *J. Polym. Sci. Polym. Chem.* **2003**, *41*, 3713–3725.

39 M. CHENG, W. B. EULER, *Inorg. Chem.* **2003**, *42*, 5384–5391.

40 J. SERIN, X. SCHULTZE, A. ADRONOV, J. M. J. FRECHET, *Macromolecules* **2002**, *35*, 5396–5404.

41 C. R. RIVAROLA, S. G. BERTOLOTTI, C. M. PREVITALI, *J. Polym. Sci. Poyml. Chem.* **2001**, *39*, 4265–4273.

42 S. C. PAULSON, S. A. SAPP, C. M. ELLIOTT, *J. Phys. Chem. B* **2001**, *105*, 8718–8724.

43 S. VALKAMA, R. MAKI-ONTTO, M. STAMM, G. TEN BRINKE, O. IKKALA, in *Nanoporous Materials III*, Vol. 141, **2002**, pp. 371–378.

44 S. VALKAMA, T. RUOTSALAINEN, H. KOSONEN, J. RUOKOLAINEN, M. TORKKELI, R. SERIMAA, G. TEN BRINKE, O. IKKALA, *Macromolecules* **2003**, *36*, 3986–3991.

45 T. RUOTSALAINEN, J. TURKU, P. HEIKKILA, J. RUOKOLAINEN, A. NYKANEN, T. LAITINEN, M. TORKKELI, R. SERIMAA, G. TEN BRINKE, A. HARLIN, O. IKKALA, *Adv. Mater.* **2005**, *17*, 1048–1052.

46 J. M. POLLINO, M. WECK, *Org. Lett.* **2002**, *4*, 753–756.

47 J. M. POLLINO, L. P. STUBBS, M. WECK, *Macromolecules* **2003**, *36*, 2230–2234.

48 Y. CHUJO, K. SADA, T. SAEGUSA, *Macromolecules* **1993**, *26*, 6315–6319.

49 Y. CHUJO, K. SADA, T. SAEGUSA, *Macromolecules* **1993**, *26*, 6320–6323.

50 Y. CHUJO, K. SADA, T. SAEGUSA, *Polym. J.* **1993**, *25*, 599–608.

51 K. T. POTTS, D. A. USIFER, *Macromolecules* **1988**, *21*, 1985–1991.

52 K. T. POTTS, D. A. USIFER, A. GUADALUPE, H. D. ABRUNA, *J. Am. Chem. Soc.* **1987**, *109*, 3961–3967.

53 K. HANABUSA, A. NAKAMURA, T. KOYAMA, H. SHIRAI, *Makromol. Chem.* **1992**, *193*, 1309–1319.

54 K. HANABUSA, K. NAKANO, T. KOYAMA, H. SHIRAI, N. HOJO, A. KUROSE, *Makromol. Chem.* **1990**, *191*, 391–394.

55 K. J. CALZIA, G. N. TEW, *Macromolecules* **2002**, *35*, 6090–6093.

56 U. S. Schubert, H. Hofmeier, *Macromol. Rapid Comm.* **2002**, *23*, 561–566.
57 G. N. Tew, K. Aamer, R. Shunmugam, *Polymer* **2005**, *46*, 8440–8447.
58 K. Aamer, G. N. Tew, *Macromolecules* **2004**, *37*, 1990–1993.
59 R. Shunmugam, G. N. Tew, *J. Polym. Sci. Polym. Chem.* **2006**, *43*, 5831–5843.
60 S. M. Heilmann, J. K. Rasmussen, L. R. Krepski, *J. Polym. Sci. Polym. Chem.* **2001**, *39*, 3655.
61 S. Monge, D. M. Haddleton, *Eur. Polym. J.* **2004**, *40*, 37.
62 A. Godwin, M. Hartenstein, A. H. E. Muller, S. Brocchini, *Angew. Chem., Int. Ed. Engl.* **2001**, *40*, 595.
63 K. Aamer, G. N. Tew, **2005**, unpublished work.
64 R. Shunmugam, G. N. Tew, *J. Am. Chem. Soc.* **2006**, *127*, 13567–13572.
65 R. Shunmugam, G. N. Tew, **2005**, unpublished work.

9
Methods for the Alignment and the Large-scale Ordering of Block Copolymer Morphologies

Massimo Lazzari and Claudio De Rosa

9.1
Introduction

9.1.1
Motivation

Self-assembly is an enormously powerful concept in modern material science, which was first associated with the use of synthetic strategies for the preparation of nanostructures only about 15 years ago [1–3]. Since then, a large variety of carefully designed building blocks have been proposed and employed for working "from atoms up" with the aim of fabricating 2D- and 3D-structures. In particular, within the last decade, interest in a potentially ideal nanoscale tool has been growing exponentially, for example, phase-separated block copolymers (BCs) [4–17].

Immiscibility among the BC constituents is common, and phase separation results in the series of morphologies introduced in Chapter 1, e.g., lamellar, gyroid, hexagonal and body-centered cubic for diblock copolymers [4, 7, 9], the size and shape of which may be conveniently tuned by changing the molecular weights and compositions of the BCs [5, 18–20]. Part of the enormous potential for nanomaterial fabrication of these self-organized patterns of chemically distinct domains that have periodicity in the mesoscale, i. e., between a few tens and hundreds of nanometers, has already been demonstrated [21–31]. However, in our opinion their development into practical routes suitable for industrial applications will probably only be fully exploited when a few key limitations have been efficiently overcome. As an example, writing and replication processes in microelectronics require spatial and orientational control of patterns, which in the case of BCs entails the solution of the non-trivial problem of large-area ordering and precise orientation of domains.

Regular structures can be generated by controlling the typical disadvantages of spontaneous phase separation. Firstly, similar to polycrystalline materials, the self-assembly of BCs in bulk is prone to form grains with a high level of local order but a very short persistence length, especially in the case of cylinders and lamellae, which on a larger scale correspond to bulk materials with isotropic properties. Although

---

Block Copolymers in Nanoscience. Massimo Lazzari, Guojun Liu, Sébastien Lecommandoux
Copyright © 2006 WILEY-VCH Verlag GmbH & Co. KGaA, Weinheim
ISBN: 3-527-31309-5

annealing permits a consistent annihilation of the corresponding grain boundary defects, in addition to other defects such as dislocation, disclination, terracing and asymmetry defects [32–37], the formation of defect-free structures points to more efficient strategies based on the application of external forces or spatial constraints. In the same way, long-range oriented patterns, e. g., cylindrical or lamellae domains parallel to the substrate, with all of the axes aligned in a single direction rather than randomly in the parallel plane, may only be induced through a careful engineering of surface effects or by well-oriented fields.

It could be stated that for many technological applications such an appealing spontaneous organization has to be directed by some form of templating (in this context considered in its broader sense), where the BC components could not only interact with each other but also take advantage of external controlling interactions. Numerous methods to induce and control BC domain orientation have been explored, particularly for substrate-supported films to create perfectly periodic 2D-patterns, ranging from the first demonstration of the effectiveness of the application of electric fields [38] to the most recent application of chemically nanopatterned templates [31]. Most methods proposed to date will be critically presented in this chapter, and attention will also be given to the combination of techniques to yield 3D-control.

9.1.2
Organization of the Chapter

The scope of this chapter is to give an overview of the variety of methods that can induce long-range ordered BC morphologies, focusing on the most promising areas of ongoing research. The main strategies applied so far may be classified into three different approaches:

1. Control of orientation by applying external fields, such as electric, magnetic, thermal, mechanical and solvent evaporation. Depending on the field imposed, thin films and/or samples in bulk may be oriented.

2. Induction of large-area ordering by facilitating the self-assembly, generally of thin films, on templates either topographically or chemically nanopatterned.

3. Modulation of surface interactions as a result of:
   a) Preferential interaction of one block with the surface.
   b) Neutralization of attractions to the surface.
   c) Epitaxial organization of domains onto a crystalline substrate.
   d) Directional crystallization of a BC solvent.
   e) Graphoepitaxy.
   f) 2D-geometric confinements.

Before highlighting recent advances in such control strategies, a preliminary section specifically for all researchers that are new to this field will offer practical information on film preparation and on the methods used in practice to achieve thermodynamic equilibrium morphologies. Because one of the aims of this book is to try to offer solutions to experimental difficulties and research needs, we decided to highlight the practical aspects of phase separation and orientation with respect to the theoretical concepts.

9.2
How to Help Phase Separation

Except for some specific instances, e. g., direct use as photonic crystals or in the case of amphiphilic BC for drug and genetic delivery (see specific chapters in this book), applications of BCs require the preparation of thin films with thicknesses ranging from a few tens of nanometers (in some cases down to thicknesses of less than the corresponding equilibrium period of the BC) to several micrometers. Films with low surface roughness may be produced by spin-coating or dip-coating from relatively dilute solutions, i. e., approximately 1–5 % by weight, onto solid substrates with uniform flatness. The thickness and the uniformity of the film surface mainly depend on the concentration of the solution, the volatility of the solvent and the specific instrumental speed, i. e., spin speed or withdrawal speed, respectively. Silicon wafers, eventually coated with metals, semiconductors, carbon or polymers are often utilized as the support. During dip-coating processes, and also for films from direct casting, the solvent may evaporate slowly thus allowing a stable organization of macromolecules close to thermodynamic equilibrium, whereas in the case of spun-cast films, the solvent is driven off so quickly that non-equilibrium structures could be observed. Moreover, the concentration gradient due to fast solvent evaporation can in fact have a significant effect on domain orientation, as discussed in the appropriate section.

In thin films the self-assembled BC morphologies are influenced not only by molecular weights, polydispersity and composition but also by other variables such as the selectivity of the solvent for one block, surface–interfacial interactions and the interplay between structure periodicity and film thickness, which can cause significant deviations from the predicted phases in the bulk state. An elegant demonstration and a didactic example of the influence of surface effects in thin films has been reported by Krausch and coworkers [39]. Their experimental data for a triblock copolymer of ABA type are compared with those of the simulated film in Fig. 9.1. Several more studies have demonstrated the importance of confinement effects and will be treated in detail later.

Independently of the casting techniques, even for a film prepared taking all the precautions, it is not possible to obtain a perfectly ordered morphology over a large area. In fact, an optimization of the nanostructures is almost indispensable and can be carried out by different annealing processes, with the double objective of obtaining the equilibrium morphology and eliminating the defects. Annealing is usually performed before any further manipulation, even though some approaches include the alignment of domains within the orientation process. In principle, thermal annealing is the simplest option and the most commonly used method, which consists of controlled heating at a temperature above that of the glass transition

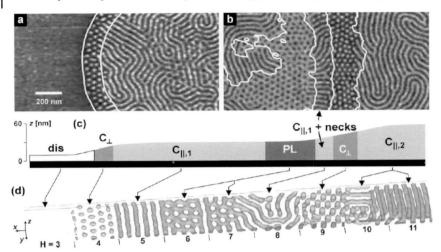

Fig. 9.1 The tapping-mode SFM phase images (a and b) of thin films of a cylinder-forming polystyrene-*block*-polybutadiene-*block*-polystyrene (PS-*b*-PB-*b*-PS) after solvent annealing showing a sequence of phases as a function of the height profile (c), which is in good agreement with the simulation of an $A_3B_{12}A_3$ copolymer film (d). Owing to preferential wetting, the silicon substrate is covered with a homogeneous 10-nm thick PB layer, which, with increasing thickness from left to right, is followed by isolated domains of PS, parallel oriented PS cylinders, perforated lamella, parallel oriented PS cylinders again and finally two layers of parallel oriented cylinders (reproduced from Knoll et al. [39], copyright (2002) with permission from The American Physical Society).

temperatures of the constituent blocks, preferably in an inert atmosphere or under vacuum, for a specific time. For polymers with high molecular weights and complex architectures, the high degree of chain entanglements and the difficult diffusion of one polymer block through the domains of the other blocks pose large kinetic barriers to equilibrium. In such cases, and for partially crystalline BCs with high fusion temperatures, the annealing window for the relevant conditions of temperature and time that are theoretically necessary and those of the order–disorder transition or polymer decomposition, may be insufficient to reach equilibrium. On the other hand, degradation of one or more blocks has been used to stabilize the morphologies for further applications, through decomposition, partial oxidation or cross-linking [23, 26, 27, 40–43], or even to induce a hierarchical transition of morphologies [44].

Alternative annealing approaches entail an increase in chain mobility by the addition of some type of plasticizer, either transient, such as a solvent, or low molecular weight homopolymers [45], which also swell each domain. Exposing thin films to vapors of a good solvent potentially allows long-range equilibrium morphologies to be produced without any thermal treatment. However, as genuinely neutral solvents are rare, the use of solvents selective for one block may induce structures far removed from the equilibrium situation [39, 46–51], which remain frozen after solvent removal and are sometimes difficult to obtain by other methods [49]. As a conjecture, a fully selective solvent should lead to a micellar-like microdomain phase. Once more Krausch and coworkers first demonstrated for a polystyrene-*block*-poly(2-vinylpyridine)-*block*-poly(*t*-butyl acrylate) (PS-*b*-P2VP-*b*-P*t*BA) triblock copolymer that the choice of the annealing solvent strongly influences the types of metastable structures observed, and also investigated the time development of the microdomain structure [49]. More recently, the morphology evolution has been followed in other BCs, and particularly in polystyrene-*block*-poly(methyl methacrylate) (PS-*b*-PMMA) thin films with different thicknesses exposed to PMMA-selective solvent vapors (Fig. 9.2) [52, 53]. For related topics, the reader is also invited to refer to Chapter 12.

An in-depth understanding of the formation of either equilibrium structures or kinetically trapped morphologies during solvent annealing require not only an accurate choice of the solvent but also the proper control of experimental parameters, such as vapor pressure, treatment time and solvent extraction rate. In practice, this knowledge facilitates the preparation of reproducible nanopatterns in thin films, which also, in the case of metastable assemblies, show a long-term stability, at least for glass transition temperatures of all the components that are well above room temperature.

Another plasticizing agent that presents a modest equilibrium sorption and can be easily removed is supercritical CO_2. Although its rapid diffusion in most polymers allows an equally rapid equilibrium distribution in thin films, CO_2 annealing has so far been limited to a surprisingly low number of BCs, essentially PS-*b*-PMMAs, having molecular weights up to 300 000 g mol^{-1} [54, 55].

9.3
Orientation by External Fields

Since the first macroscopic alignment of cylindrical domains of an industrial triblock copolymer (a PS-*b*-PB-*b*-PS, Kraton 102) by extrusion carried out by the Keller group in the early 1970s [56, 57], many more mechanical flow fields have been proposed to control BC alignment. Following the Keller belief that BC microstructuring and orientation were exciting research subjects, but with very limited applicability [58], and after the shear flow experiments on the same copolymer by Skolios and coworkers [59–61], in the 1990s research interests returned strongly to alignment by relatively weak external fields. Another reason for this was the greater availability of BCs with different architectures and chemical compositions [5, 18–20]. Particularly in recent years, most investigations moved from bulk materials to thin films, due to their nanotechnological potential, with a special focus on the use of electric fields. Active external fields will be covered later, while those that are passive, such as with interfacial interactions, are considered in the next sections.

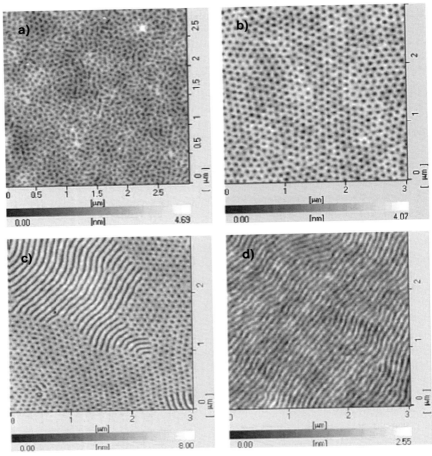

Fig. 9.2 AFM topographic images of PS-b-PMMA ($M_{n\,PS}$ = 133 000, $M_{n\,PMMA}$ = 130 000) thin films cast onto an Si substrate with an SiO$_x$ surface layer exposed to chloroform vapor. After 40 h (a) the initially featureless structure shows a disordered microstructure that evolves, 60 h (b), into a hexagonally packed nanocylinder structure. Further treatment, 80 h (c), promotes the evolution after 100 h (d) to a mixed morphology, which completely develops into stripes having the repeat spacing (L_0), of 90 nm (reproduced from Xuan et al., [52], copyright (2004) with permission from The American Chemical Society). For longer annealing times, the film returns to a flat surface. It has also been shown that only when the film thickness is less than ½L_0 can the packed nanocylinder structures form.

9.3.1
Mechanical Flow Fields

Extrusion [56, 57], compression [62–65], flows involving oscillatory shearing [59–61, 66–73] and other steady shearing [74–77], and techniques that combine different flow fields [78–82] have been successfully applied to induce alignment in BCs, either in a microphase separated molten state or, to a lesser extent, in gels of solvent-swollen microdomains [83–90], with the objective of forming single crystal structures. Large amplitude oscillatory shear (LAOS), first proposed at the end of the 1980s to orient commercial triblock copolymers [66, 67], is the most widely employed technique, also due to the easy characterization of the shear field with respect to other methods. As an example, common parallel plate type rheometers enable the modulation of valuable experimental variables to be carried out, such as shear rate, frequency, strain amplitude and temperature.

Early studies on the effect of mechanical flow fields on the orientation of BC domains have already been summarized in many excellent publications and books, for example, by Fasolka and Mayes [9], Hamley [4, 91], the group working with Thomas [14, 92] and others [68, 93, 94], so need not be reviewed here. Di-, tri- and multiblock copolymers in bulk with different microstructures, such as lamellar [57, 62, 63, 68–73], cylindrical [64–66, 76, 77, 80, 82], spherical [74, 75] and gyroid [81], can be preferentially oriented parallel or, where applicable, perpendicular or mixed-perpendicular to the flow direction. Most experimental observations have been supported and even anticipated by theory, with several good theoretical works being published in recent years [95–101]. Further developments concerned the extension of the shearing techniques to thin films. Albalak and Thomas [78, 79] proposed a novel casting method, termed roll-casting, in which a BC solution is subjected to a flow between two counter-rotating rolls (Fig. 9.3a). The solution is compressed and, as a result of solvent evaporation and shearing, the microphase separates into a film with a high degree of orientation and close to single-crystal characteristics. Only cylinder morphology from triblock commercial polymers has been oriented in this way [80, 81], with a film thickness intrinsically limited by the geometry of the device to hundreds of nanometers. Thinner films of sphere- and cylinder-forming polystyrene-*b*-poly(ethylene propylene) (PS-*b*-PEP) have been shear-aligned by simple flowing techniques [75, 76], and extended to arbitrarily large areas. Films of 30–50 nm spread onto a silicon substrate are covered by an elastomer pad, which is then slowly pulled forward by a constant force (Fig. 9.3b). The steady shear has to be applied at temperatures between the T_g and the T_{ODT} of the copolymer, thus providing an essentially infinite orientational order in all directions without the typical limitations of sheared bulk samples, related to the presence of multigrains.

Unconventional methods, which to some extent are based on shearing and a stretching field, such as electrospinning [102–104] and orientation by spin coating [105, 106], have been proposed in the last few years, but still need to be investigated in depth. Electrospinning in particular [107–109] seems to offer great potential for the fabrication of nanofibers with internal oriented structuring from cylinder-forming copolymers with at least one partially crystalline block.

9.3.2
Electric and Magnetic Fields

After the pioneering work by Amundson et al. on lamellar diblock copolymer thin films [110–112], static electric fields have been widely used in copolymer melts to macroscopically orient lamellar or cylindrical morphologies [38, 113–119], with

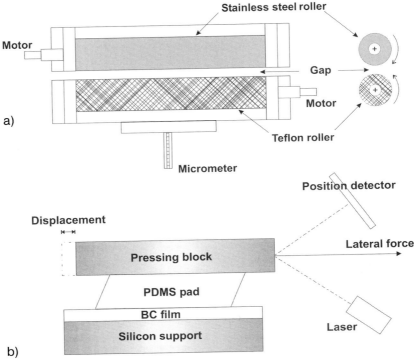

Fig. 9.3 Schematic of apparatus used to induce alignment in BC thin films: (a) In the roll-caster (upper view and section) two independent rollers, one of stainless steel and the other of Teflon, are driven at the same angular velocity by two independent motors and separated by a micrometer-controlled gap. (b) In the shear-alignment set-up the PDMS pad is pressed with a constant weight against the film; the support is heated at the annealing temperature while the displacement of the pressing block is continuously monitored with a laser–mirror–optical sensor assembly.

most of the investigations being focused on PS-*b*-PMMA. Although morphologies parallel to the substrate could be obtained using an in-plane field, a uniaxial orientation along the field and perpendicular to the substrate is preferably induced between the electrodes compressing the film. As initially proposed by Russell and coworkers, a convenient procedure consists in the preparation of films between two Kapton sheets coated on one side with Al, with the film thickness controlled through Kapton spacers. Kapton is a commercially available polyimide with excellent mechanical properties that allows further manipulation after alignment of the domains [23] and facilitates sample characterization, such as for microtoming. To avoid electric shorting, one of the electrodes has to be placed in contact with the polymer, while the second is inverted with the Kapton side facing the polymer.

The general statement that the orientation of BC microdomains is possible if the applied field is high enough for a given difference in the dielectric constant between blocks, can be expressed by Eq. (9.1) [120,121]:

$$E_c = \Delta\gamma^{1/2} \frac{2(\varepsilon_A + \varepsilon_B)^{1/2}}{\varepsilon_A - \varepsilon_B} t^{-1/2} \qquad (9.1)$$

where E_c is the critical electric field strength, ε_A and ε_B are the dielectric constants of blocks and t is the film thickness, which also takes into account the surface interactions through the difference between the interfacial energies of each block with the substrate, $\Delta\gamma$. In the absence of preferential interactions with the substrate, i.e., $\Delta\gamma = 0$, the domains do not need any external field to orient normal to the substrate (see Section 9.5.1). From a practical point of view, Eq. (9.1) permits the evaluation of the critical parameter for a given set of the different parameters, eventually imposed by the same experimental design. The thickness of the region to be oriented is limited to a few millimeters by the electric field strengths, in the range of approximately 1–100 V μm^{-1}, considered as the upper limit that prevents dielectric breakdown. Theoretical works have investigated various aspects of the dynamics of mesophase formation and orientation, mainly in diblock copolymer melts [90–92, 100–105], with a special mention required for the capacitor analogy proposed by Pereira and Williams for symmetric BCs [120]. The appeal of this model actually lies in its simplicity, which, at a more qualitative level, also permits justification of the alignment of lamellae perpendicular to the electrodes on the basis of preferential travel of the electric field lines along easy paths, though each and every block has the highest polarizability. In the same way, parallel orientation is unfavorable, as the electric field lines would be forced to cross both blocks evenly. Tsori et al. have recently proposed that in these systems, alignment may also occur by other strong orienting forces, such as by the presence of a conductivity or mobility contrast [121], postulating an important role for the ions remaining in the polymers after synthesis in the reorientation mechanism [122].

The alignment pathways in thin films of BCs with various morphologies have been recently followed using *in situ* small-angle X-ray scattering [115, 118]. Application of an electric field (perpendicular to the plane of the film) to the disordered melt of an asymmetric BC induced the progressive growth of micro-phase separated structures oriented parallel to the field only at temperatures below the order–disorder transition temperature. This occurs up to a stationary, equilibrium state, and also for a copolymer already having ordered domains parallel to the substrate, either a cylindrical PS-*b*-PMMA [115] or a lamellar polystyrene–polyisoprene (PS-PI) system [118], alignment parallel to the field could be achieved within minutes. Analysis of the scattering patterns suggested that during the reorientation, the initial large grains are broken up into smaller pieces by the amplification of interfacial fluctuations, which are able to rotate in the field. The final state of the symmetric BCs of PS and PI consists of many small grains with a high degree of orientation of the lamellae parallel to the electric field, but with a random orientation in the perpendicular plan. This observation also finds some theoretical support in the prediction by Tsori [121–126]. The limits imposed by the film thickness and the interfacial energy effects, as predicted in Eq. (9.1), have been faced experimentally by the Russell group for symmetric and asymmetric BC thin films between sub-

Fig. 9.4 Cross-sectional TEM of ≈300 nm (a) and ≈700 nm (b) symmetric PS-b-PMMA films between PDMS coated Kapton electrodes annealed at 170 °C under ≈40 V μm^{-1} electric field for 6 and 16 h, respectively. The scale bar represents 100 nm (reproduced from Xu et al. [117], copyright (2004) with permission from The American Chemical Society).

strates with modulated interfacial interactions [117, 127, 128]. In the presence of a preferential interaction of one block with the substrate (see Section 9.5.1) such effects are dominant for film thicknesses up to approximately 10 L_0, regardless of the electric field applied normal to the surface (see, for example, Fig. 9.4a). For thicker films, the effects dissipate with distance from the surface and the domains orient in the direction of the electric filed, at least in the interior of the film (Fig. 9.4b), with a distance of propagation inversely proportional to the strength of the interfacial interactions [128].

All of these melt-based procedures suffer severe limitations because of the melt viscosity itself, which is directly dependent on the molecular weight and architecture of BCs, and the thickness of the region to be oriented. In addition, the achievement of high degrees of orientation eventually requires high temperatures and electric field strengths close to the decomposition conditions. An alternative approach to circumvent such restrictions consists in the addition of a neutral solvent with the effect of increasing chain mobility, as already discussed above for thermal annealing, thus facilitating large scale domain alignment. Concentrated diblock (PS-b-PI) [129] and triblock [polystyrene-$block$-poly(hydroxyethyl methacrylate)-$block$-poly(methyl methacrylate), PS-b-PHEMA-b-PMMA] [130] copolymer solutions have been aligned under a dc electric field during solvent evaporation, leading to highly anisotropic bulk samples (Fig. 9.5). Orientation kinetics and mechanisms of alignment, i.e., grain orientation and nucleation and growth of domains, have been corroborated by simulations [129] based on dynamic density functional theory [131].

In summary, several groups have demonstrated that the application of unidirectional electric fields coupled with an appropriate choice of materials and exper-

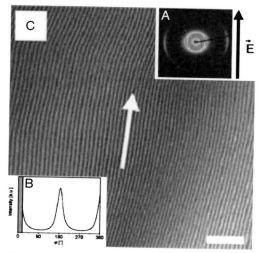

Fig. 9.5 2D-SAXS pattern (a), azimuthal intensity distribution for a first-order reflection (b) and TEM micrograph (c) of a bulk sample prepared from a 40 wt-% solution in toluene of symmetric PS-b-PI (M_n = 80 000). The arrows indicate the direction of the electric field and the scale bar represents 400 nm (reproduced from Böker et al. [129], copyright (2003) with permission from The American Chemical Society).

imental conditions allow an effective fabrication of patterns with a high degree of orientation and high aspect ratio, suitable for nanotechnological applications. However, orientation of domains is azimuthally degenerate and a further improvement to achieve morphologies with orientation controlled in the three dimensions logically requires the application of a second, orthogonal, external field. The first sequential combination towards long-range, 3-D ordered thin films was shown to control the orientation of lamellar microdomains through the application of an elongational flow field to obtain an in-plane orientation in ≈30 μm thick films, and an electric field normal to the surface [132]. The concurrent application of the orthogonal field, e. g., a mechanical shearing, is expected to reduce the presence of defects and grain boundaries in an even more effective fashion.

A further alignment approach is available for materials that exhibit anisotropic susceptibility due to an anisotropic molecular structure. Magnetic field induced orientation has been achieved for liquid crystalline diblock copolymers with a dielectric diamagnetic isotropy, possibly through the magnetic alignment of LC mesogens [133, 134], and also for BCs with a crystallizable block through an accurate control of the crystallization process [135]. The latter investigation is particularly likely to open a new fruitful research field and, at the same time, pave the way to an alternative method of relatively wide applicability, which only has the prerequisite of a high nucleation density during crystallization. Magnetic fields also offer the ability to apply very high fields without the risks of electric fields, associated with the danger and limit of electric breakdown.

9.3.3
Solvent Evaporation and Thermal Gradient

The observation of lamellar and cylindrical microdomains in thin films perpendicular to the surface as a result of solvent evaporation was first reported by Turturro and coworkers [136] and then investigated in more detail by Kim and Libera for a similar triblock copolymer [137, 138]. On the basis of these and other studies on either spun-cast or solution-cast films from solutions in a good solvent for all the blocks, a reasonable mechanism of orientation could be proposed (Fig. 9.6). In the first moments after deposition the T_g of the swollen film is still well below room temperature, thus allowing free chain mobility. With the decrease in the solvent concentration, the BC undergoes a transition from the disordered to the ordered state and, as in thin films the diffusion of the solvent produces a gradient of concentration, the ordering front rapidly propagates from the air surface to the substrate. The consequent decrease of T_g below room temperature, for at least one block, locks in the structures, which, due to the high directionality of the solvent gradient, are highly oriented normal to the surface. This behavior has been reported so far for films of less than one-half micron thick, e. g., on PS-PB systems [136–139], polystyrene-*block*-poly(ethylene oxide) (PS-*b*-PEO) [140, 141] and polystyrene-*block*-polyferrocenyldimethylsilane (PS-*b*-PFS) [142], whereas the applicability of these conditions can be extended to any BCs having the T_g of one block above room temperature.

Notwithstanding the excellent results obtained by the Hashimoto group [143, 144], the use of temperature gradient apparatus to produce highly oriented single crystals is limited to bulk samples and presumably it is too difficult to suppose that this approach could be suitable of general use.

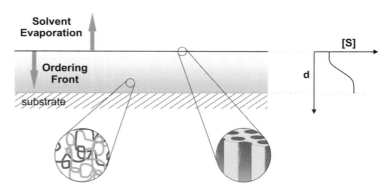

Fig. 9.6 Schematic of the solvent evaporation in a diblock copolymer thin film (adapted from [139]). The diffusion produces a gradient in the concentration of the solvent, [S], as a function of depth, *d*, which induces an ordering front from the film surface to the substrate.

9.4 Templated Self-assembly on Nanopatterned Surfaces

The combination of top-down strategies to fabricate patterns that direct the bottom-up organization of organic or inorganic building blocks is a strategy often used in the micrometer and, to a minor extent, in the nanometer regime. Also, in the case of BCs, long-range order and orientation may be induced if self-assembly is forced to occur into/onto a guide, either topographically or chemically patterned, or in other 2D- or 3D-confinements. In this section, only the effect of surfaces with features having periodicity commensurable with those of the BC morphologies is considered. The study of the control of the registration and orientation of domains by topographic confinements (in the form of surface relief or groove structures) with a length scale much larger than the BC lattice parameters had a somehow different theoretical and historical development, and is better known as graphoepitaxy. It will therefore be discussed later as it follows such different nomenclature.

Rockford and coworkers decorated miscuit silicon wafers with gold to fabricate striped and essentially flat surface of nonpolar gold and polar silicon oxide, with periodicity ranging from 40 to 70 nm [145, 146]. Films of symmetric PS-*b*-PMMA solution cast onto such substrates showed, after thermal annealing, lamellar orientation normal to the plane and parallel to the striping, with the greatest ordering for patterns commensurable with the bulk lamellar period of the BC. A mismatch in the length scale of 10 % and 25 % in thick and ultrathin films, respectively, appeared

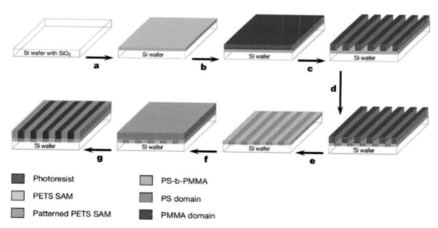

Fig. 9.7 Schematic representation of the fabrication process of chemically nanopatterned surfaces that template the self-assembly of symmetric PS-*b*-PMMA (M_n = 104 000). SAM of phenylethyltrichlorosilane is deposited on a silicon wafer (a). Photoresist is then spin-coated (b) and patterned with alternating lines and spaces by ultraviolet interferometric lithography (c). The topographic pattern is converted into a chemical pattern by irradiation with soft X-rays in the presence of oxygen (d). After the photoresist removal (e), a toluene solution of PS-*b*-PMMA is spin-coated onto the patterned SAM (f). Thermal annealing (g) facilitate surface-directed self-assembly (reproduced from Kim et al. [148], copyright (2003) with permission from Nature Publishing Group).

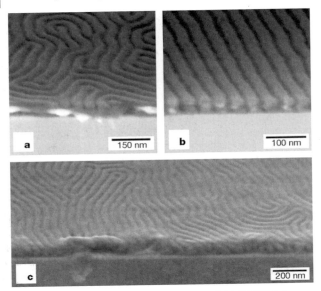

Fig. 9.8 Cross-sectional images of symmetric PS-b-PMMA (M_n = 104 000, L_0 = 48 nm) films on unpatterned surfaces (a) and chemically nanopatterned surfaces with 47.5 nm periodicity (b), prepared as reported in Fig. 9.7. When the surface pattern is greater than L_0 (c), lamellae exhibit an imperfect ordering (reproduced from Kim et al. [148], copyright (2003) with permission from Nature Publishing Group).

to be sufficient to lose control over orientation. A novel integrated fabrication strategy that makes use of advanced lithographic techniques has been tuned, by Nealey and coworkers, to produce perfect periodic domain ordering [147, 148]. Although Nealey actually discussed epitaxial self-assembly, we prefer to reserve that term for a different methodology to control spatial and orientation order, as introduced by De Rosa et al. [149] (for a definition of epitaxy see the next section). As schematically reported in Fig. 9.7, in the first essential step of this procedure a self-assembled monolayer (SAM) is precisely patterned throughout the photoresist using extreme ultraviolet interferometric lithography [150, 151]. Following the conversion of the topographic pattern to a chemical pattern and the photoresist removal, a symmetric PS-b-PMMA is spin-coated onto the chemical pattern and, as the modified regions present polar groups that preferentially wet the PMMA block, the self-assembly results in lamellae oriented perpendicular to the substrate. High quality defect-free BC films require pretty much the same period of the chemical pattern and the BC, with commensurability within just a few percent. For values within approximately 10 %, the morphology can still be surface directed but it is not perfect (Fig. 9.8). An extension of this method, through the use of ternary blends of diblock copolymers and the corresponding homopolymers, demonstrated that is possible to pattern chemical templates with periodicities substantially deviating from those of the BC [31]. Redistribution of homopolymer macromolecules during thermal annealing facilitates defect-free assembly of the blend as a whole, to form 50–80 nm periodic arrays in addition to pattern structures with different bend geometries. A further optimization of blend composition, polymer chemistry and lithographic techniques will make it possible to fabricate nonregular shaped structures of different types, which implicitly afford the production of nanodevices requiring patterns more complex than simple periodic arrays.

9.5
Epitaxy and Surface Interactions

The control of the orientation of the microdomains in the microstructure of the BCs can be obtained through a bias field induced by surface interactions. Different types of interactions can be established depending on the nature of the surface and of the BC.

Specific interactions, such as epitaxy and directional crystallization, are involved in semicrystalline or amorphous BCs and crystalline substrates. It will be shown in the following sections that epitaxy and/or graphoepitaxy can be used in combination with the directional crystallization providing powerful methods for creating regular surface patterns in BC thin films.

9.5.1
Preferential Wetting and Homogeneous Surface Interactions

The simplest interaction of a BC film deposited on a substrate is the preferential wetting of one block at an interface to minimize interfacial and surface energies. As a consequence, a parallel orientation of microdomains, lamellae and cylinders is often induced at the interface and this orientation tends to propagate throughout the entire film [9, 152–163].

The microstructure can be altered by variation of the film thickness on the substrate and preferential interactions of blocks with the substrate [48, 161, 162, 164–166]. Symmetric boundary conditions are established when one of the blocks preferentially interacts with both the substrate and the air surface [156], while asymmetric conditions pertain when one block is preferentially wetted by the substrate and the other block by the superstrate [160].

The control of orientation of the microdomains can also be achieved by confining a BC between two surfaces, that is, adding a superstrate to a BC film supported on a substrate [167–170]. Strong or weak interactions of BCs with the surfaces can be created by coating the surface walls with a homopolymer or a random copolymer, respectively, containing the same chemical species as the confined BC [169]. In the case of a neutral surface, e.g., by using a random copolymer, the lamellar microdomains rearrange themselves so that the direction of periodicity is parallel to the substrate [169]. Moreover, decreasing the confined film thickness, that is, creating a large incompatibility strain of the natural domain period of the BC and the film thickness, induces a heterogeneous in-plane structure where both parallel and perpendicular lamellae are located near the confining substrate [170]. Various theoretical studies have predicted the structural behavior of BC thin films in a confined geometry [171–180] and are basically consistent with experimental results.

Great influences of both film thickness and surface interaction on the orientations of the microdomains have also been found for BCs forming cylinder [153, 181–189] and sphere [190, 191] morphologies.

Alignment of microdomains in both substrate supported and confined films has also been achieved by tuning the specific interaction between the BC and the substrate through the modification of the surface [27, 114, 192–198]. An example is the application of a random copolymer brush anchored on the substrate, which has allowed changing the orientation of the domains from parallel to perpendicular by altering the composition of the random copolymer [198].

9.5.2
Epitaxy

Epitaxy is defined as the oriented growth of a crystal on the surface of a crystal of another substance (the substrate). The growth of the crystals occurs in one or more strictly defined crystallographic orientations defined by the crystal lattice of the crystalline substrate [199–201]. The resulting mutual orientation is due to a 2D- or, less frequently, a 1D-structural analogy, with the lattice matching in the plane of contact of the two species [199]. The term epitaxy, literally meaning "on surface arrangement", was introduced in the early theory of organized crystal growth based on structural matching [199–201]. Discrepancy between atomic or molecular spacings is measured by the quantity $100(d - d_0)/d_0$, where d and d_0 are the lattice periodicities of the adsorbed phase and the substrate, respectively. In general 10–15 % discrepancies are considered as an upper limit for epitaxy to occur in polymers [201].

Inorganic substrates were first used for the epitaxial crystallization of polymers [202–216]. Successive studies have demonstrated the epitaxial crystallizations of polyethylene (PE) and linear polyesters onto crystals of organic substrates, such as condensed aromatic hydrocarbons (naphthalene, anthracene, phenanthrene, etc.), linear polyphenyls and aromatic carboxylic acids [217–221]. In the case of PE, a unique orientation of the crystals grown on all the substrates is observed, with different contact planes depending on the substrate [217–219].

An example of an epitaxial relationship between PE crystals and an organic substrate is shown in Fig. 9.9, with the substrate constituted by a crystal of benzoic acid [218] (monoclinic structure with $a = 5.52$ Å, $b = 5.14$ Å, $c = 21.9$ Å and $\beta = 97°$, with a melting temperature of 123 °C). A clear match between the PE interchain distance (the b-axis of PE equal to 4.95 Å) and the b-axis periodicity of benzoic acid crystal (5.14 Å), and between the c-axis periodicity of PE (2.5 Å) and the a-axis of benzoic acid crystal (5.52 Å), produces the crystallization of PE onto preformed crystals of benzoic acid with lamellae standing edge-on, that is, normal to the surface of benzoic acid crystal (Fig. 9.9). The PE polymer chains lie flat on the substrate surface with their chain axis parallel to the substrate surface and parallel to the a-axis of

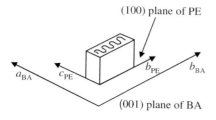

$(100)_{PE} \parallel (001)_{BA}$ $b_{PE} \parallel b_{BA}$ and $c_{PE} \parallel a_{BA}$

Fig. 9.9 PE lamella oriented edge-on on the (001) face of a BA crystal substrate after epitaxial crystallization. The (100) plane of PE is in contact with the (001) plane of BA, and b- and c-axes of PE are parallel to b- and a-axes of BA, respectively [218].

benzoic acid crystal, and the b-axis of PE parallel to the b-axis of BA crystal [218]. The (100) plane of PE is in contact with the (001) exposed face of benzoic acid.

In order to apply epitaxial control, BCs need to have at least one crystalline block, which can interact with a crystalline surface. Although the versatility of epitaxy in the case of crystallization of polymers has clearly been shown, the method of using organic substrates to control the crystallization and the morphology of semicrystalline BCs was introduced only in 2000, when De Rosa and coworkers [149, 222] first used epitaxy to control the molecular and microdomain orientation of PE-b-PEP-b-PE and PS-b-PE semicrystalline BC thin films. They used benzoic acid (BA) [149, 222, 223] or anthracene (AN) [223, 224] and obtained precise control of the molecular orientation of the crystalline block and subsequent overall long-range order of the BC microdomains.

The selected area electron diffraction pattern and the TEM bright-field image of a thin film of PE-b-PEP-b-PE epitaxially crystallized onto a BA crystal are reported in Fig. 9.10A and B, respectively. The diffraction pattern essentially presents only the 0kl reflections of PE, therefore, it corresponds to the b^*c^* section of the reciprocal lattice of PE (Fig. 9.10A). This indicates that the chain axis of the crystalline PE lies flat on the substrate surface and is oriented parallel to the a-axis of the BA crystals, as in the case of the PE homopolymer (Fig. 9.9) [149]. The (100) plane of PE is in contact with the (001) plane of BA; therefore, the crystalline PE lamellae stands edge-on on the substrate surface, with the b-and c-axes of PE oriented parallel to the b- and a-axes of BA, respectively. The bright-field TEM image (Fig. 9.10B) indicates that epitaxy has produced, instead of a spherulitic structure, a highly aligned lamellar structure with long, thin crystalline PE lamellae with a thickness of 10–15 nm, evidenced as dark regions, oriented along the $[010]_{PE} \parallel [010]_{BA}$ direction. The dark-field image created by using the strongest 110 reflection, shown in Fig. 9.10C, reveals the same parallel array of crystalline edge-on PE lamellae oriented along the b-axis of BA crystals. This can be seen as bright regions due to the lamellae all being arranged in Bragg diffraction conditions determined by the 2D-epitaxy [149]. A scheme of the orientation of the PE-b-PEP-b-PE obtained on BA

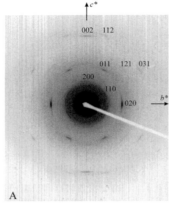

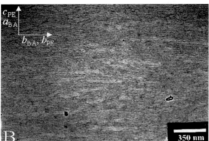 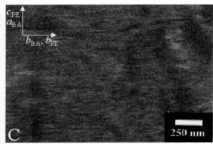

Fig. 9.10 Selected area electron diffraction pattern (A), TEM bright-field image (B) and TEM (110) dark-field image (C) of a thin film of PE-*b*-PEP-*b*-PE epitaxially crystallized onto a BA crystal. In the TEM bright-field image, the dark regions correspond to the denser crystalline PE phase, which form long lamellae standing edge-on on the substrate surface and preferentially oriented with the *b*-axis of PE parallel to the *b*-axis of the BA (B). In the dark-field image, the bright regions correspond to the crystalline PE lamellae in the Bragg condition (C) (reproduced from De Rosa et al. [149], copyright (2000) with permission from The American Chemical Society).

crystals is shown in Fig. 9.11 [149]. The biaxial matching of the BA and PE lattices creates a highly ordered lamellar BC microdomain state. The widths of the crystalline PE lamellae are highly uniform, the PE crystals and the intervening noncrystalline PEP are all parallel, and the orientations of both the *c*- and *b*-axes of PE crystals over many micron sized regions are very high [149].

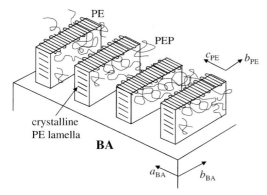

Fig. 9.11 Schematic model of the crystalline and amorphous microdomains in the PE-b-PEP-b-PE triblock copolymer epitaxially crystallized on the BA crystal. The epitaxy shows the relative orientation of crystalline PE lamellae on the BA crystal: $(100)_{PE} \| (001)_{BA}$ and $c_{PE} \| a_{BA}$, $b_{PE} \| b_{BA}$ (reproduced from De Rosa et al. [149], copyright (2000) with permission from The American Chemical Society).

9.5.3
Directional Crystallization

In polymer–diluent binary mixtures, if both polymer and the solvent are crystallizable above room temperature, then eutectic-like behavior can be observed. In fact, both melting temperatures of the polymer and the solvent are depressed up to the eutectic composition [225–230].

The presence of solvent in a non-crystalline BC depresses the order–disorder temperature of the BC depending on the solvent quality. Several theories predict the phase behavior of BC–organic solvent mixtures [231–233]. Therefore, it can be expected that mixtures of BCs with crystallizable solvents exhibit eutectic behavior.

Park et al. [234] used this hypothesis of eutectic formation to globally organize the eutectic mixture for the first time and induce alignment of microdomains in the solid BC. They used organic diluents, such as benzoic acid (BA) and anthracene (AN), and amorphous asymmetric PS-b-PI and symmetric PS-b-PMMA diblock copolymers [234]. In this process the first requirement is that the diluent must dissolve the BC above the melting temperature of the solvent. Secondly, the diluent must have a tendency to crystallize directionally into large, plate-like crystals. When the organic diluent, which initially was a solvent for the BC, directionally crystallizes along its fast growth direction under a temperature gradient, the BC undergoes microphase separation, due to a rapid decrease in the solvent concentration. At the same time, the orientation of the microdomains nucleated from the eutectic transforming solution is determined by the fast directional growth of the organic diluent. In the case of BA and AN, the fast growth directions are both b-axis and consequently the inter-material dividing surface of the microdomains of the BC tends to orient along this direction, which also corresponds to the temperature

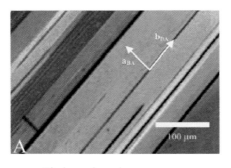

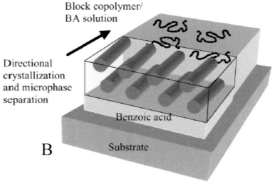

Fig. 9.12 (A) Polarized optical microscopy image of directionally crystallized BA crystals. The large, flat and elongated BA crystals are aligned with the b-axis parallel to the growth front direction (reproduced from Park et al. [234], copyright (2001) with permission from The American Chemical Society). (B) Schematic model of the microstructure formation of a cylinder forming BC under directional crystallization of the BA from homogeneous solution (reproduced from Park et al. [14], copyright (2003) with permission from Elsevier).

gradient direction [234]. A polarized optical microscope image of directionally crystallized BA crystal is shown in Fig. 9.12A [234]. The BA crystal is elongated along the fast growth b-axis and presents a flat (001) surface. The different thicknesses of the BA crystals lead to the different colors under the microscope. A schematic model of cylindrical microdomains well aligned from the directional eutectic transformation of the homogeneous BC solution is shown in Fig. 9.12B [14]. This process occurs within a few seconds (the growth velocity of the BA crystal is nearly 1 cm s^{-1}) and the microstructure is formed and then kinetically trapped.

Examples of the ordered microstructures obtained by this process are shown in Fig. 9.13 for symmetric PS-b-PMMA (Fig. 9.13A) and asymmetric PS-b-PI (Fig. 9.13B and C) diblock copolymers prepared by directional solidification with BA crystals [234]. In the case of PS-b-PMMA, the darker regions correspond to the RuO$_4$-stained PS microdomains (Fig. 9.13A). Edge-on parallel alternating lamellae of PS and PMMA are well aligned along the fast growth direction of the BA crystals,

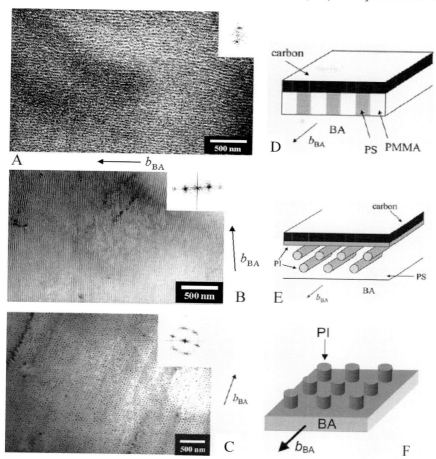

Fig. 9.13 TEM bright-field images of thin films directionally solidified with BA of the symmetric PS-b-PMMA (20-nm thick) stained with RuO₄ (A), the asymmetric PS-b-PI stained with OsO₄ with a thickness of 50 nm (B) and 20 nm (C). The dark regions correspond to the stained PS lamellae (A), parallel PI cylinders (B) and vertically oriented PI cylinders (C) well aligned along the fast growth direction of the BA crystals (b-axis). Insets show the FFT power spectrum of the TEM micrographs. Schematic models of the microstructures of PS-b-PMMA (D), PS-b-PI (E) and ultrathin film of PS/PI (F) processed with BA (reproduced from Park et al. [234], copyright (2001) with permission from The American Chemical Society, and from Park et al. [14], copyright (2003) with permission from Elsevier).

as shown in the schematic model of Fig. 9.13D [80]. The bright-field TEM image of a film of the asymmetric PS-b-PI with a thickness of approximately 50 nm, directionally solidified with BA (Fig. 9.13B) shows OsO₄-stained darker PI cylinders, lying in-plane and well oriented along the crystallographic b-axis of the BA, as reported in the scheme of Fig. 9.13E [234]. The average diameter of the PI cylindrical microdomains is approximately 20 nm, while the average distance between the cylinders is 40–50 nm.

This process has also been used in combination with thickness effects to achieve vertically aligned cylinders employing ultrathin films [234]. Cylindrical microdomains, indeed, orient vertically for relatively thin films due to incommensurability effects, which is similar to what was observed for lamellar diblock copolymers [235]. A bright-field TEM image of a thinner film (of approximately 20-nm thickness) of PS-b-PI, prepared from a more dilute solution, directionally solidified with BA and stained with OsO_4, is shown in Fig. 9.13C [234]. The dark OsO_4-stained vertically aligned cylindrical PI microdomains are oriented into rows along the b-axis of the BA crystal and packed in an approximate hexagonal lattice, as shown in the schematic model of the microstructure reported in Fig. 9.13F [234].

9.5.4
Graphoepitaxy and Other Confining Geometries

Graphoepitaxy is a process whereby an artificial surface topography of a crystalline or amorphous substrate influences and controls the orientation of the crystal growth in thin film [236–239]. Graphoepitaxy has been used, often in combination with epitaxy, to obtain high orientation of polymeric crystals onto substrates constituted by films of other polymers [240–244]. For instance various polymers, such as PE, nylons, polyesters, liquid crystalline polymers, have been grown in highly oriented form on the oriented surface of poly(tetrafluoroethylene) (PTFE) crystals [240]. The alignment of materials on PTFE films can occur by graphoepitaxial or epitaxial mechanisms, depending on the materials. In some instances both the lattice structure and the surface topography of the PTFE films promote the alignment so that the two mechanisms can operate in conjunction [244].

Graphoepitaxy was utilized by Segalman and coworkers [245, 246] to control the orientation of BC microdomains. The procedure is an example of the combined top-down and bottom-up approaches to patterning. They used a topographically alternating mesa and well patterns fabricated by conventional photolithography and chemical etching techniques to align spherical P2VP microdomains of a PS-b-P2VP diblock copolymer. A large area single crystal of the P2VP spheres was obtained on the mesas and wells [245].

Nanostructures with long-range order were also developed using graphoepitaxy in combination with BC lithography [247, 248]. Cheng et al. employed a diblock copolymer of PS-b-PFS, in which the organometallic PFS block provides excellent (10:1) etching contrast in an oxygen plasma [247]. A topographically patterned silica substrate was fabricated by interference lithography. The substrate was then used for a templating BC self-assembly. Monolayer films of the BC were deposited by spin-casting and annealed onto the patterned silica substrate. The film was then oxidized by etching in an oxygen plasma. SEM images of annealed and etched PS-b-PFS films, reported in Fig. 9.14, revealed ordered arrays of PFS spheres in grooves of different widths [247]. The microdomain patterns were transferred onto the un-

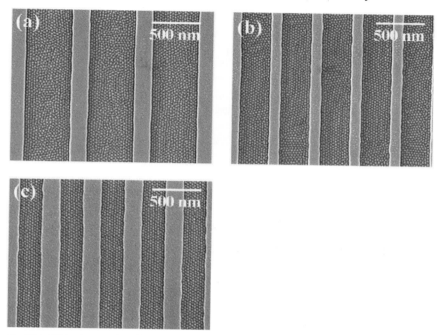

Fig. 9.14 SEM micrographs of annealed and oxygen plasma treated PS-b-PFS films on silica gratings with (a) 500-nm wide grooves, (b) 320-nm wide grooves, (c) 240-nm wide grooves (reproduced from Park et al. [14], copyright (2003) with permission from Elsevier).

derlying silica substrate using an RIE (reactive ion etcher) process with a CHF_3 plasma and well-ordered arrays of silica posts were created [247]. The oxidized PFS domains are perfectly ordered by the guided self-assembly of the BC in the pre-patterned substrate and can be potentially useful for fabricating 2D-photonic crystal waveguide structures.

Unlike graphoepitaxy, which can be considered to approximate a 1D-confinement of a thin film between (often parallel) solid walls, 2D-confinement of BCs has been predicted to induce more complicated effects on phase-separation and morphology orientation [249, 250]. Recent investigations confirmed for various PS-based diblock copolymers that in the case of non-planar geometry of confinements, such as in cylindrical pores, confining dimensions incommensurable with the copolymer period and the imposed curvature produces unusual morphologies [251–253]. Self-assembly of asymmetric diblock copolymers introduced as melt into nanoporous alumina templates occurs with an alignment of the cylinders along the pore axis, favored by the preferential wetting of the pore with the majority block [252]. Furthermore, for symmetric BCs, concentric lamellae oriented parallel to the long axes of the pores have been observed, with an overall number of lamellae and forbidden segregations directly dependent on the pore diameter and the preferential wetting of one block, respectively [252, 253].

9.5.5
Combination of Directional Crystallization and Graphoepitaxy

The idea of combining graphoepitaxy from a topographical pattern fabricated by conventional photolithography procedures and the directional crystallization of a solvent was first introduced by Park et al. [254]. The thin film of the BC was confined between crystals of benzoic acid, obtained from directional eutectic crystallization of the homogeneous BC solution, and the topographic substrate pattern. The confinement induces a thickness variation of the BC film, resulting in the two different orientations of the microdomains [254]. The patterned substrate was produced via standard lithographic techniques and consists of 30-nm high, 2 µm × 2 µm square mesas arranged in a square array with a 4-µm spacing. The square-shaped mesas are made by selective etching of a thermally grown silicon oxide on a 4-in wafer. A PS-b-PI diblock copolymer was directionally solidified with benzoic acid on the patterned substrate [254]. The pre-patterned substrate structure produces thickness variations in the directionally solidified BC thin films; the films are thinner on the mesas and thicker in the plateau regions. As shown in Section 9.5.3 (Fig. 9.13B and C), the directional eutectic solidification produces orientation of PI cylinders along the b-axis of the benzoic acid crystal or normal to the substrate depending on the thickness [234, 254].

The BC film directionally solidified on the patterned substrate was then subjected to O_2-RIE to selectively remove the cylindrical PI microdomains [254]. A higher magnification AFM image after O_2-RIE, shown in Fig. 9.15A, reveals the double orientation of the cylindrical PI microdomains. In the thicker film regions, approximately 50-nm thick, cylindrical PI microdomains are well aligned along the b-axis of the BA crystal with the cylinder axis parallel to the substrate and to the walls of the mesas. In the thinner film regions, approximately 20 nm, the PI cylinders are hexagonally packed with their cylinder axes perpendicular to the substrate and with the a_1 hexagonal lattice direction parallel to b-axis of BA, that is along the [10] direction of the square substrate pattern [254]. The SEM image of Fig. 9.15B shows that the microdomains successively transform from parallel to vertical when they go from the plateau into the mesa regions [254]. The cylindrical hole structures on the mesa areas after O_2-RIE are also shown in Fig. 9.15C in height contrast AFM. The vertically ordered PI cylindrical domains, which appear dark, on the mesa regions are essentially empty. A scheme of this process of directional solidification of the BC, combined with topographically mediated substrate patterning, is shown in Fig. 9.15D. The BC films confined between the top BA crystal and the bottom pre-patterned substrate undergo thickness variation (h_1 and h_2), leading to two different microdomain orientations.

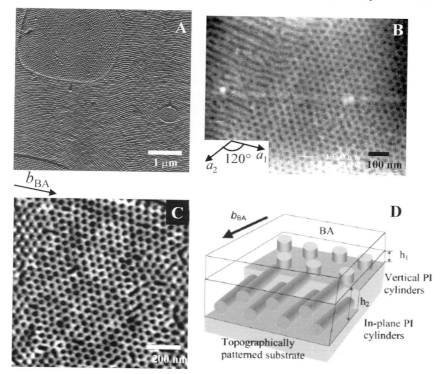

Fig. 9.15 AFM (A, C) and SEM (B) images of a thin film of asymmetric PS-b-PI, directionally solidified with BA on the pre-patterned substrate and subsequently etched by O_2-RIE. The tapping mode AFM image (A) shows that the cylindrical PI microdomains, with two different orientations with respect to the substrate, are well aligned along the fast growth direction of the BA crystals, indicated by the arrow. The square-shaped mesa regions exhibit vertically oriented, hexagonally packed PI cylinders. The thicker matrix regions show the in-plane PI cylinders. The SEM image reveals that aligned PI cylinders transform their in-plane to vertical orientation with respect to the substrate. The two hexagonal lattice directions of the vertically ordered PI cylinders are shown in inset (B). The height mode AFM image on a mesa region (C) indicates that vertically ordered cylindrical PI microdomains are selectively removed by O_2-RIE, which appear dark (reproduced from Park et al. [254], copyright (2001) with permission from The American Institute of Physics). Scheme of the orientation of microdomains of PS-b-PI between the top BA crystal and the bottom pre-patterned substrate (D) (reproduced from Park et al. [14], copyright (2003) with permission from Elsevier).

9.5.6
Combination of Epitaxy and Directional Crystallization

The method of combination of directional crystallization and epitaxy was introduced by De Rosa et al. [222]. The process is based on the directional solidification of the eutectic solution of a semicrystalline BC in a crystallizable organic solvent and the following epitaxial crystallization of the crystalline block onto the organic crystalline substrate [222]. The organic substrate, such as benzoic acid or anthracene, at a temperature higher than its melting temperature is a solvent for the BC, and becomes a substrate, when it crystallizes at lower temperatures, for the epitaxial crystallization of the BC [222–224, 255].

Examples of the ordered microstructures obtained with this method are provided by PS-b-PEP-b-PE PS-b-PE semicrystalline BCs containing crystallizable PE blocks directionally solidified with BA or anthracene [222–224].

The electron diffraction pattern of PS-b-PEP-b-PE triblock copolymer film directionally solidified and epitaxially crystallized with BA is shown in Fig. 9.16A [223]. As in the case of the melt-compatible PE-b-PEP-b-PE (Fig. 9.10A) [149], the pattern of Fig. 9.16A essentially presents only the 0kl reflections of PE, indicating orientation of the PE crystals similar to that found for the PE-b-PEP-b-PE triblock copolymer [149], and for the homopolymer, shown in Figs. 9.9 and 9.11. The crystalline PE lamellae are oriented edge-on on the substrate surface with the (100) plane of PE in contact with the (001) plane of BA. The b- and c-axes of PE are parallel to the b- and a-axes of benzoic acid, respectively.

The bright-field TEM image of the unstained film directionally solidified and epitaxially crystallized onto BA, reported in Fig. 9.16B [223], shows the highly oriented lamellar structure with long thin crystalline lamellae oriented along the $[010]_{PE} \| [010]_{BA}$ direction. The bright-field image of the same sample after staining with RuO_4, shown in Fig. 9.16C, indicates that the PS blocks, corresponding to the darker stained regions, are organized in parallel cylinders whose axes are generally oriented along the elongated direction $[010]_{PE} \| [010]_{BA}$ of the crystalline lamellae [223].

As the order–disorder temperature of PS-b-PEP-b-PE (>250°C) is much higher than the crystallization temperature of the PE block (almost 65 °C) [256, 257], the directional crystallization of BA induces the orientation of the cylindrical PS microdomains along the fast growth direction of the BA crystal (b-axis) [223], as discussed in the Section 9.5.3 for amorphous BCs [234]. Subsequently the epitaxial crystallization of the crystalline PE block onto the BA crystalline substrate induces the molecular orientation of the PE chains, leading to the oriented crystalline PE lamellae in the presence of the pre-existing ordered PS cylinders [223]. A schematic model shown in Fig. 9.16D displays the final microstructure generated by combination of the directional crystallization and epitaxy. The process induces alignment of the PS cylinders and PE lamellae along the same direction, resulting in multilayered ordering of the PS cylinders [223].

A different microstructure of the same PS-b-PEP-b-PE is obtained when anthracene (AN) is used as the crystallizable solvent [224], due to a different epitaxial relationship between PE and anthracene crystals [217]. The SAD electron diffraction pattern of the PS-b-PEP-b-PE film epitaxially crystallized onto AN is shown in Fig. 9.17A [224]. The pattern exhibits 0kl reflections of two sets of PE crystals having two different orientations [224], corresponding to the b^*c^* sections of the PE reciprocal lattice symmetrically rotated by an angle of almost 70° with respect to the a^*-axis of the AN. This angle of 70° corresponds to the angle made by the [110] and [1$\bar{1}$0] directions in the ab-plane of the AN crystal (a_{AN} = 8.65 Å, b_{AN} =

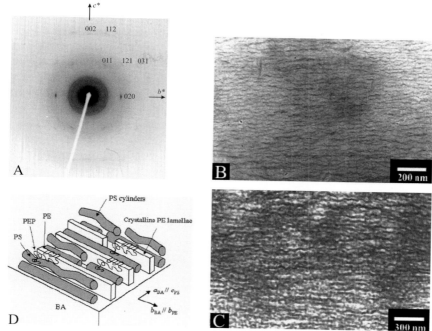

Fig. 9.16 SAD electron diffraction pattern (A), TEM bright-field image of unstained (B) and stained with RuO$_4$ (C) thin film of PS-*b*-PEP-*b*-PE triblock copolymer directionally solidified and epitaxially crystallized onto BA. The dark regions in the TEM bright-field image of the unstained film (B) correspond to the crystalline PE phase, which form long lamellae oriented edge-on, and aligned with the *b*-axis parallel to the *b*-axis of BA. The dark regions in the TEM bright-field image of the film stained with RuO$_4$ (C) correspond to the PS phase, which form cylinders aligned parallel to the crystalline lamellae. Schematic model of the microstructure in thin films of the PS-*b*-PEP-*b*-PE terpolymer directionally solidified and epitaxially crystallized onto BA (D). The directional solidification induces in-plane cylindrical PS microdomains aligned along the fast growth direction of BA crystals. The following epitaxial crystallization of PE blocks onto the BA crystals induces the formation of long, thin crystalline PE lamellae in between the PS cylinders. The PE crystals have their (100) planes in contact with the (001) plane of the BA crystal with $a_{BA} \| c_{PE}$ and $b_{BA} \| b_{PE}$ (reproduced from Park et al. [223], copyright (2003) with permission from Wiley-VCH).

6.04 Å, c_{AN} = 11.18 Å, β = 124.7°, melting temperature of 216 °C). The *c*-axis of PE is therefore oriented parallel to two equivalent [110] and [1$\bar{1}$0] directions of the AN crystal, and two sets of PE lamellae oriented edge-on on the AN crystals are produced, with the (100) plane of PE in contact with the (001) plane of AN (Fig. 9.17B) [70]. The epitaxial relationship between PE and AN crystals can be readily explained in terms of matching between the *b*-axis of PE and the inter-row spacing of the (110) plane of AN and a second matching of 2*c* of PE with the Bragg distance of the (110) plane (d_{110}) of AN [217]. Similar lattice matching occurs between PE and other aromatic substances [217].

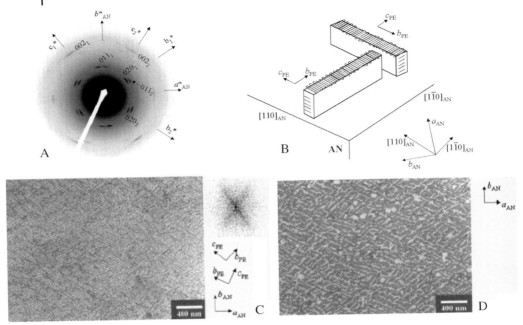

Fig. 9.17 (A) Selected area electron diffraction pattern of a thin film of PS-b-PEP-b-PE block terpolymer epitaxially crystallized onto AN. (B) Scheme showing the two orientations of edge-on PE lamellae onto the (001) exposed face of the AN crystals. The (100) plane of PE is normal to the electron beam and parallel to the (001) face of the AN crystals. The c-axes of the two sets of PE crystals are parallel to the [110] and [1$\bar{1}$0] direction of the AN crystal. (C) TEM bright-field image of an unstained thin film of PS-b-PEP-b-PE epitaxially crystallized on AN. The dark regions correspond to the crystalline PE phase, which forms cross-oriented PE lamellae standing edge-on. (D) TEM bright-field image of the film of PS-b-PEP-b-PE epitaxially crystallized on AN and stained with RuO$_4$. The dark regions correspond to the PS phase, which forms a double-oriented structure similar to that of the PE microdomains (reproduced from Park et al. [224], copyright (2001) with permission from Wiley-VCH).

The bright-field TEM image of the unstained film of the PS-b-PEP-b-PE epitaxially crystallized onto AN, shown in Fig. 9.17C, provides a direct visualization of the crystalline PE lamellae that presents a cross-oriented texture. The lamellae having a thickness of 10–20 nm are oriented along the two different [110] and [1 $\bar{1}$ 0] directions of the AN crystals [224]. A bright-field image of a different region of the same sample after staining with RuO$_4$, shown in Fig. 9.17D, indicates that the PS blocks (the darker regions) are organized into a similar double oriented pattern [224]. The directional solidification of the mixture of the terpolymer and AN develops the orientation of the cylindrical PS microdomains along the fast growth direction of the AN crystals (b-axis direction) before crystallization of the PE. The epitaxial crystallization induces a high orientation of the molecular chains of the crystalline phase, resulting in PE lamellae standing edge-on on the substrate surface, oriented in two directions due to the crystallographic degeneracy. This structure in turn induces reorganization and reorientation of the pre-existing PS cylinders [224].

This thin-film process suggests the possibility of creating new 2D-microdomain textures via self-assembled semicrystalline BCs using a combination of one or more crystallizable blocks and various crystalline organic substrates [224].

The asymmetric PS-*b*-PE diblock copolymer is an interesting example of crystallization of PE confined in PE cylinders formed by self-assembly in the melt [222, 223]. The electron diffraction pattern of an unstained film of PS-*b*-PE film epitaxially crystallized onto BA is shown in Fig. 9.18A. The pattern is similar to those obtained for the PS-*b*-PEP-*b*-PE (Fig. 9.16A) [223] and PE-*b*-PEP-*b*-PE (Fig. 9.10A) [245] epitaxially crystallized onto BA. In addition, in this instance the pattern presents only the 0*kl* reflections of PE. This indicates that the PE lamellae are oriented edge-on on the (001) surface of BA, with the (100) plane of PE in contact with the (001) plane of BA and the *b*- and *c*-axes of PE parallel to the *b*- and *a*-axes of BA, respectively, as shown in Figs. 9.9 and 9.11. A bright-field image of a thin film of PS-*b*-PE directionally solidified and epitaxially crystallized onto BA and stained with RuO_4, is shown in Fig. 9.18B [222]. The presence of a very well ordered array of light-unstained PE cylinders in the dark-stained PS matrix is apparent [222]. The cylinders are vertically aligned and are packed on a hexagonal lattice, having an average lattice constant of nearly 40 nm. The order extends to large distances (>20 μm).

The PE crystals have been visualized with dark-field imaging, employing the strong 110 reflection of PE (Fig. 9.18C). Small rectangular PE crystals are observed in the dark-field image of Fig. 9.18C (the light regions), well aligned along the *b*-axis direction of the BA crystals and packed on a pseudo-hexagonal lattice, the size and orientation of which are the same as seen in the bright-field image of Fig. 9.18B [222]. The PE crystals are 7-nm thick and 20-nm long, with their longest dimension parallel to the [110]* reciprocal lattice direction [222]. The size and spacing of PE microdomains observed in bright- and dark-field images show that there is precisely one crystalline PE lamella centered in each cylinder, with the *b*-axis of the lamella oriented parallel to the *b*-axis of the BA, confirming the epitaxy of PE on BA demonstrated by the electron-diffraction pattern (Fig. 9.16A). A scheme of the nanostructure obtained in PS/PE thin film is shown in Fig. 9.18D [222, 223].

The same PE-*b*-PS was also processed with AN to manipulate both molecular and microstructure orientation [223]. The electron diffraction pattern of an unstained film is shown in Fig. 9.19A. As in the case of Fig. 9.17A for the PS-*b*-PEP-*b*-PE triblock copolymer, the pattern exhibits the 0*kl* reflections of two sets of PE crystals in two different orientations, indicating that PE lamellae stand edge-on on the substrate surface with the *c*-axes of PE oriented parallel to the [110] or [1 $\bar{1}$0] directions (Fig. 9.17B) [223]. A RuO_4 stained bright-field TEM image of the PS-*b*-PE diblock copolymer directionally solidified and epitaxially crystallized onto AN is shown in Fig. 9.19B. Also, in this case the light-unstained PE cylinders are vertically aligned but the order in the lateral packing is lower than that observed for the same BC crystallized onto BA (Fig. 9.18B) [223]. The lattice has, indeed, many defects, resulting in very small grain size with short-range order. This is probably because

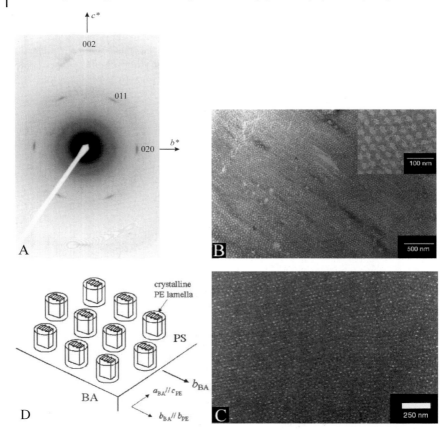

Fig. 9.18 SAD electron diffraction pattern (A), TEM bright-field image (B) and TEM dark-field image of a thin film of PS-*b*-PE directionally solidified and epitaxially crystallized onto BA. In the TEM bright-field image the film of PS-*b*-PE was stained with RuO_4, therefore, the darker regions correspond to the stained PS matrix, while the light regions correspond to the crystalline PE microdomains, which form hexagonally packed cylinders oriented perpendicular to the substrate surface. Inset, magnified region of (B) showing the non-circular shape of the PS-PE interface. The dark-field image (C) of the unstained film was created using the (110) diffraction reflection. Small rectangular PE crystals are observed, well aligned along the *b*-axis direction of the BA crystals and packed on a pseudo-hexagonal lattice whose size and orientation is the same as seen in (B) (reproduced from De Rosa et al. [222], copyright (2000) with permission from Nature Publishing Group). (D) Schematic models of the microstructures in thin films of PS-*b*-PE directionally solidified and epitaxially crystallized onto BA. Vertically oriented PE cylinders are packed on a hexagonal lattice and contain one crystalline PE lamella oriented edge-on on the substrate with the *b*-axis parallel to the *b*-axis of the BA crystal (reproduced from Park et al. [223], copyright (2003) with permission from Wiley-VCH).

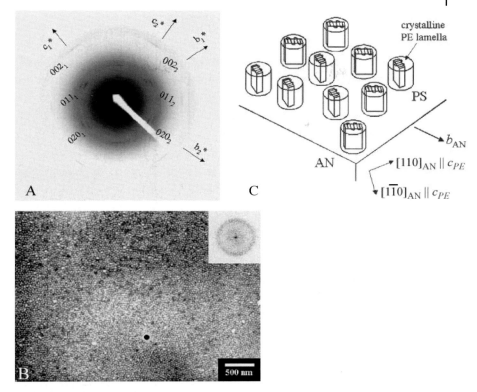

Fig. 9.19 SAD electron diffraction pattern (A) and TEM bright-field image (B) of a thin film of PS-b-PE directionally solidified and epitaxially crystallized onto AN. In the TEM bright-field image the film of PS-b-PE was stained with RuO_4, therefore, the darker regions correspond to the stained PS matrix, while the light regions correspond to the crystalline PE microdomains, which form hexagonally packed cylinders oriented perpendicular to the substrate surface. (C) Schematic model of the microstructure in thin films of PS-b-PE, directionally solidified and epitaxially crystallized onto AN. Vertically oriented PE cylinders are packed on a hexagonal lattice and contain one crystalline PE lamella oriented edge-on on the substrate with two different orientations, in which the c-axes of PE lamellae are parallel to the [110] or [1$\bar{1}$0] directions of AN crystals (reproduced from Park et al. [223], copyright (2003) with permission from Wiley-VCH).

the shape anisotropy of AN crystals in the growth process is lower than that of BA crystals and the epitaxial crystallization of the PE block on AN, which causes the double texture of the crystalline PE lamellae, may disturb the lateral ordering of the PE cylinders [223]. A scheme of the microstructure obtained is shown in Fig. 9.19C. The PE cylinders are vertically aligned and contain one PE crystalline lamella oriented along two equivalent crossed directions [223].

It has been suggested by De Rosa et al. [222] that the mechanism of the pattern formation of PS-b-PE on BA or AN is based on the combination of epitaxy and directional crystallization and is basically similar to that for PS-*b*-PEP-*b*-PE (Fig. 9.16) [222, 223]. The directional crystallization of the organic crystal (BA or AN) produces the formation of the globally oriented PE cylindrical microdomains due to the polymer first undergoing the order–disorder transition. However, in this instance the subsequent epitaxial crystallization of the PE block influences the final microstructure, where vertically oriented cylindrical PE microdomains [222] (Fig. 9.18D) instead of in-plane PE cylinders are formed (Fig. 9.16D) [223]. The morphological evolution of the system is depicted schematically in Fig. 9.20. The initial homogeneous solution confined between the glass substrates (Fig. 9.20A) transforms, due to the imposed directional eutectic solidification, into large crystals of BA having (001) surfaces coexisting with a thin liquid layer near the eutectic composition (Fig. 9.20B). Decreasing the temperature also causes this layer to solidify directionally, by thickening the pre-existing BA crystal and the formation of a thin, metastable vertically oriented lamellar microdomain film (Fig. 9.20C). The PE microdomains that result aligned along the fast growth direction of substrate crystals, as is shown in Fig. 9.12B. Many sequences of events are possible, depending on the affinity of the respective blocks for the substrate surfaces, the polymer film thickness, microdomain period and the type of subsequent epitaxy and orientation of chains in the crystallizable block with respect to the crystalline substrate,. In the case of the PS-*b*-PE/BA system, the structure transforms due to the instability of the flat interface at this composition and from the in-plane PE *c*-axis orientation induced by the epitaxy, so that vertical cylinders readily form (Fig. 9.20D). The depression of the glass transition temperature of PS by plasticization from the BA allows the intermediate structure to reorganize when the crystallization of PE occurs, even at 60 °C. The domains evolve from an aligned precursor state (Fig. 9.20D), into the final structure with vertical cylinders containing one PE crystalline lamella (Fig. 9.20E) [222].

In the case of PS-*b*-PEP-*b*-PE, due to the presence of the PEP mid-block and the appropriate condition for the epitaxy of the PE block, in-plane crystalline PE lamellae were produced without hindering the pre-existing PS cylinders (Fig. 9.16D) [223]. However, in the case of PS-*b*-PE, the fact that the epitaxial crystallization of the PE block must take place inside the cylindrical microdomains and the *c*-axis of the PE block must contact with the crystalline BA or AN surface, makes the two controlling forces of directional crystallization and epitaxy determine the final microstructure orientation in a more cooperative manner. Indirect evidence of the contribution of the PE epitaxy to the vertically ordered microstructure is given by the results reported by Park et al. [234], who introduced the method of directional crystallization (see Section 9.5.3) using a PS-*b*-PI that has a volume fraction and molecular weight similar to those of the PS-*b*-PE. They found in-plane PI cylinder orientation with the same experimental conditions due to the absence of the epitaxy (Fig. 9.13B) [234]. The plasticization of the PS matrix from BA or AN enables the subsequent epitaxial crystallization to transform the microstructure into vertical cylinders [222, 223].

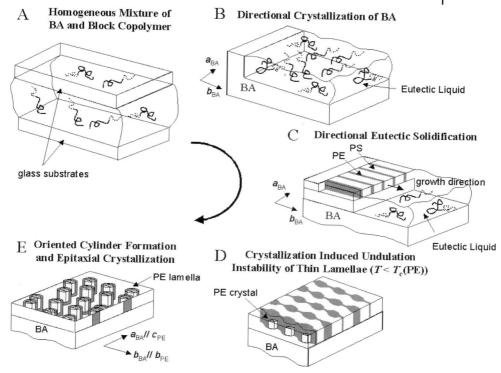

Fig. 9.20 Structural evolution during the directional eutectic solidification and epitaxial crystallization of the BC from the crystallizable solvent. (A) Homogeneous solution of PS-b-PE in BA between two glass substrates. (B) Directional solidification forms crystals of BA coexisting with a liquid layer of more concentrated polymer. (C) Second directional solidification, showing the eutectic liquid layer transforming into a BA crystal (which grows on the pre-eutectic BA crystal) and an ordered lamellar BC. (D) Because of the highly asymmetric composition of the BC and the epitaxial crystallization of the PE in contact with the BA substrate, the flat interfaces of vertically oriented lamellae are unstable, and spontaneously deform in order to achieve a more preferred interfacial curvature and allow epitaxial growth of PE. (E) The layers transform into an array of vertically oriented, pseudohexagonally packed semicrystalline PE cylinders. A single, chain-folded PE lamella is formed in each cylinder. The PE crystals have their (100) planes contacting the (001) plane of the BA crystal with $a_{BA} \| c_{PE}$ and $b_{BA} \| b_{PE}$ (reproduced from De Rosa et al. [222], copyright (2000) with permission from Nature Publishing Group).

9.6
Summary and Outlook

The importance of developing methods which allow control over the alignment of domains arises from the opportunities they provide for the exploitation of BCs in nanotechnological applications, such as in nanolithography or for the fabrication of photonic materials and other optical and electronic devices. In this chapter we have attempted to cover most of techniques proposed essentially during the last decade to induce long-range ordered and preferentially oriented morphologies, considering as seminal work the partial alignment of cylindrical domains in BC melts obtained in the early 1970s by the application of a mechanical field [56].

Macroscopically oriented thin films have been achieved by methods of general applicability, including roll-casting, shearing and through the use of unidirectional electric fields. Less direct approaches have established a control of surface interactions or entail a fine tuning of surface topography and geometric confinements, eventually through lithographically patterned substrates. Moreover, when BCs fulfill specific structural or molecular requirements, other strategies are available: (a) semicrystalline BCs or copolymers having the T_g of one block above room temperature may undergo morphology alignment by epitaxy or solvent evaporation, respectively; (b) BCs soluble in a crystallizable solvent are suitable for alignment by directional crystallization, eventually in combination with epitaxy or graphoepitaxy. Other approaches, essentially proposed for bulk BC systems, only achieve moderate orientation with multigrain structures, e. g., through the application of different mechanical flow fields, or appear to be too complicated to assume nimble developments (e. g., for orientation by temperature gradients).

Finally, a scenario could be foreseen in which only a few of the techniques proposed so far will allow the development of processes for the preparation of defect-free nanostructured materials suitable for technological applications. This is despite the fact that most of them are of remarkable interest as model studies, which, in the best case, could assist researchers working in this field and on the development of other complementary technologies, or they may simply be intended for academic exercises. In particular, there is a considerable interest in the combination of different methods of alignment for the production of "truly" defect free 3D-nanostructures, with the goal of overcoming the azimuthal degeneration of domains, intrinsic for unidirectional interactions. The best potential is logically offered by the simultaneous application or orthogonal directing interactions. However, in our opinion, it is worth devoting additional effort to further explore the integration of ultraviolet interferometric lithography and, more generally, the implementation of the combination of top-down and bottom-up approaches.

Acknowledgements

M. L. and C. D. R. gratefully acknowledge financial support from Ministerio de Educación y Ciencia (MAT2005-07554-C02-01 and the program Ramón y Cajal) and MIUR (Prin 2004 project), respectively. M. L. has also benefited from useful discussions with Prof. M. A. Lopez-Quintela, University of Santiago, Spain. Finally, we thank M. Geoghegan for critically reading the manuscript.

References

1. C. T. Seto, G. M. Whitesides, *J. Am. Chem. Soc.* **1990**, *112*, 6409.
2. J. S. Manka, D. S. Lawrence, *J. Am. Chem. Soc.* **1990**, *112*, 2440.
3. U. Koert, M. M. Harding, J.-M. Lehn, *Nature (London)* **1990**, *346*, 339.
4. I. W. Hamley, *The Physics of Block Copolymers*, Oxford University Press, Oxford, **1998**.
5. N. Hadjichristidis, S. Pispas, G. Floudas, *Block Copolymers. Synthetic Strategies, Physical Properties and Applications*, Wiley, New York, **2003**.
6. I. W. Hamley (ed.), *Developments in Block Copolymers Science and Technology*, Wiley, Chichester, **2003**.
7. F. S. Bates, G. H. Fredrickson, *Phys. Today* **1999**, *52*, 32.
8. H.-A. Klok, S. Lecommandoux, *Adv. Mater.* **2001**, *13*, 1217.
9. M. J. Fasolka, A. M. Mayes, *Annu. Rev. Mater. Res.* **2001**, *31*, 323.
10. G. Krausch, R. Magerle, *Adv. Mater.* **2002**, *14*, 1579.
11. D. Y. Ryu, U. Jeong, J. K. Kim, T. P. Russell, *Nature Mater.* **2002**, *1*, 114.
12. I. W. Hamley, *Angew. Chem., Int. Ed. Engl.* **2003**, *42*, 1692.
13. I. W. Hamley, *Nanotechnology* **2003**, *14*, R39.
14. C. Park, J. Yoon, E. L. Thomas, *Polymer*, **2003**, *44*, 6725.
15. M. Lazzari, M. A. Lopez-Quintela, *Adv. Mater.* **2003**, *15*, 1583.
16. R. A. Segalman, *Mat. Sci. Eng. R* **2005**, *48*, 191.
17. A.-V. Ruzette, L. Leibler, *Nature Mater.* **2005**, *4*, 19.
18. K. Hatada, T. Kitayama, O. Vogl (eds.), *Macromolecular Design of Polymeric Materials*, Marcel Dekker, New York, **1997**.
19. N. Hadjichristidis, M. Pitzikalis, S. Pispas, H. Iatrou, *Chem. Rev.* **2001**, *101*, 3747.
20. K. Matyjaszewski, T. P. Davis (eds.), *Handbook of Radical Polymerization*, Wiley, New York, **2002**.
21. M. Park, C. Harrison, P. M. Chaikin, R. A. Register, D. H. Adamson, *Science* **1997**, *276*, 1401.
22. R. R. Li, P. D. Dapkus, M. E. Thompson, W. G. Jeong, C. Harrison, P. M. Chaikin, R. A. Register, D. H. Adamson, *Appl. Phys. Lett.* **2000**, *76*, 1689.
23. T. Thurn-Albrecht, J. Schotter, G. A. Kästle, N. Emley, T. Shibauchi, L. Krusin-Elbaum, K. Guarini, C. T. Black, M. T. Tuominen, T. P. Russell, *Science* **2000**, *290*, 2126.
24. M. Park, P. M. Chaikin, R. A. Register, D. H. Adamson, *Appl. Phys. Lett.* **2001**, *79*, 257.
25. J. Y. Cheng, C. A. Ross, V. Z.-H. Chan, E. L. Thomas, R. G. H. Lammertink, G. J. Vancso, *Adv. Mater.* **2001**, *13*, 1174.
26. K. W. Guarini, C. T. Black, K. R. Milkove, R. L. Sandstrom, *J. Vac. Sci. Technol. B* **2001**, *19*, 2784.
27. H.-C. Kim, X. Jia, C. M. Stafford, D. H. Kim, T. J. McCarthy, M. T. Tuominen, C. J. Hawcker, T. P. Russell, *Adv. Mater.* **2001**, *13*, 795.
28. K. Naito, H. Hieda, M. Sakurai, Y. Kamata, K. Asakawa, *IEEE Trans. Magn.* **2002**, *38*, 1949.
29. J. Y. Cheng, A. M. Meyes, C. A. Ross, *Nature Mater.* **2004**, *3*, 823.
30. J. Yoon, W. Lee, E. L. Thomas, *MRS Bull.* **2005**, *30*, 721.
31. M. P. Stokykovich, M. Müller, S. O. Kim, H. H. Solak, E. W. Edwards, J. J. de Pablo, P. F. Nealey, *Science* **2005**, *308*, 1442.
32. E. L. Thomas, D. M. Anderson, C. S. Henkee, D. Hoffman, *Nature (London)* **1988**, *334*, 598.
33. S. P. Gido, E. L. Thomas, *Macromolecules* **1994**, *27*, 6137.
34. S. P. Gido, E. L. Thomas, *Macromolecules* **1997**, *30*, 3739.
35. J. Hahm, W. A. Lopes, H. M. Jaeger, S. J. Sibener, *J. Chem. Phys.* **1998**, *109*, 10111.
36. T. Goldacker, V. Abetz, R. Stadler, I. Erukhimovich, L. Leibler, *Nature (London)* **1999**, *398*, 138.
37. J. Hahm, S. J. Sibener, *J. Chem. Phys.* **2001**, *114*, 4730.
38. T. L. Morkved, M. Lu, A. M. Urbas, E. E. Ehrichs, H. M. Jaeger,

P. Mansky, T. P. Russell, *Science* **1996**, *273*, 931.

39 A. Knoll, A. Horvat, K. S. Lyakhova, G. Krausch, G. J. A. Sevink, A. V. Zvelindovsky, R. Magerle, *Phys. Rev. Lett.* **2002**, *89*, 035501.

40 T. Hashimoto, K. Tsutsumi, Y. Funaki, *Langmuir* **1997**, *13*, 6869.

41 G. Liu, J. Ding, A. Guo, M. Hertfort, D. Bazett-Jones, *Macromolecules* **1997**, *30*, 1851.

42 G. Liu, J. Ding, T. Hashimoto, K. Kimishima, F. M. Winnik, S. Nigam, *Chem. Mater.* **1999**, *11*, 2233.

43 J. M. Leiston-Belanger, T. P. Russell, E. Drockenmuller, C. J. Hawker, *Macromolecules* **2005**, *38*, 7676.

44 Y.-H. La, E. W. Edwards, S.-M. Park, P. F. Nealey, *Nano Lett.* **2005**, *5*, 1379.

45 U. Jeong, D. Y. Ryu, D. H. Kho, J. K. Kim, J. T. Goldbach, D. H. Kim, T. P. Russell, *Adv. Mater.* **2004**, *16*, 533.

46 J.-C. Meiners, A. Quintel-Ritzi, J. Mlinek, H. Elbs, G. Krausch, *Macromolecules* **1997**, *30*, 4945.

47 G. Kim, M. Libera, *Macromolecules* **1998**, *31*, 2569.

48 M. J. Fasolka, P. Banerjee, A. M. Mayes, G. Pickett, A. C. Balazs, *Macromolecules* **2000**, *33*, 5702.

49 H. Elbs, C. Drummer, V. Abetz, G. Krausch, *Macromolecules* **2002**, *35*, 5570.

50 E. Buck, J. Fuhrmann, *Macromolecules* **2001**, *34*, 2172.

51 J. Peng, Y. Xuan, H. F. Wang, B. Y. Li, Y. C. Han, *Polymer* **2005**, *46*, 5767.

52 Y. Xuan, J. Peng, L. Cui, H. Wang, B. Li, Y. Han, *Macromolecules* **2004**, *37*, 7301.

53 Y. Chen, H. Huang, Z. Hu, T. He, *Langmuir* **2004**, *20*, 3805.

54 V. S. RamachandraRao, R. R. Gupta, T. P. Russell, J. J. Watkins, *Macromolecules* **2001**, *34*, 7923.

55 B. D. Vogt, V. S. RamachandraRao, R. R. Gupta, K. A. Lavery, T. J. Francis, T. P. Russell, J. J. Watkins, *Macromolecules* **2003**, *36*, 4029.

56 A. Keller, E. Pedemonte, F. M. Willmouth, *Nature (London)* **1970**, *225*, 538.

57 M. J. Folkes, A. Keller, F. P. Scalisi, *Colloid. Polym. Sci.* **1973**, *251*, 1.

58 E. Pedemonte, personal communication.

59 A. Skoulios, *J. Polym. Sci. Polym. Symp.* **1977**, *58*, 369.

60 G. Hadziioannou, A. Mathis, A. Skoulios, *Colloid. Polym. Sci.* **1979**, *257*, 15.

61 G. Hadziioannou, A. Mathis, A. Skoulios, *Colloid. Polym. Sci.* **1979**, *257*, 136.

62 P. Kofinas, R. E. Cohen, *Macromolecules* **1995**, *28*, 336l.

63 P. L. Drzal, J. D. Barnes, P. Kofinas, *Polymer* **2001**, *42*, 5633.

64 D. J. Quiram, R. A. Register, G. R. Marchand, D. H. Adamson, *Macromolecules* **1998**, *31*, 4891.

65 O. L. J. van Asselen, I. A. van Casteren, J. G. P. Goossens, H. E. H. Meijer, *Macromol. Symp.* **2004**, *205*, 85.

66 F. A. Morrison, H. H. Winter, *Macromolecules* **1989**, *22*, 3533.

67 F. A. Morrison, H. H. Winter, W. Gronski, J. D. Barnes, *Macromolecules* **1990**, *23*, 7200.

68 U. Wiesner, *Macromol. Chem. Phys.* **1997**, *198*, 3319.

69 B. Scott Pinheiro, K. I. Winey, *Macromolecules* **1998**, *31*, 4447.

70 H. Leist, D. Maring, T. Thurn-Albrecht, U. Wiesner, *J. Chem. Phys.* **1999**, *110*, 8225.

71 T. J. Hermel, L. F. Wu, S. F. Hahn, T. P. Lodge, F. S. Bates, *Macromolecules* **2002**, *35*, 4685.

72 S. Stangler, V. Abetz, *Rheol. Acta* **2003**, *42*, 569.

73 L. Wu, T. P. Lodge, F. S. Bates, *Macromolecules* **2004**, *37*, 8184.

74 J. M. Sebastian, W. W. Graessley, R. A. Register, *J. Rheol.* **2002**, *46*, 863.

75 D. E. Angelescu, J. H. Waller, R. A. Register, P. M. Chaikin, *Adv. Mater.* **2005**, *17*, 1878.

76 D. E. Angelescu, J. H. Waller, D. H. Adamson, P. Deshpande, S. Y. Chou, R. A. Register, P. M. Chaikin, *Adv. Mater.* **2004**, *16*, 1736.

77 K. F. Luo, Y. L. Yang, *Polymer* **2004**, *45*, 6745.

78 R. J. Albalak, E. L. Thomas, *J. Polym. Sci. Polym. Phys.* **1993**, *31*, 37.

79 R. J. Albalak, E. L. Thomas, *J. Polym. Sci. Polym. Phys.* **1994**, *32*, 341.

80 C. C. Honeker, E. L. Thomas, R. J. Albalak, D. A. Hajduk, S. M. Gruner, M. C. Capel, *Macromolecules* **2000**, *33*, 9395.
81 B. J. Dair, A. Avgeropoulos, N. Hadjichristidis, *Polymer* **2000**, *41*, 6231.
82 M. A. Villar, D. R. Rueda, F. Ania, E. L. Thomas, *Polymer* **2002**, *43*, 5139.
83 K. Mortensen, W. Brown, B. Norden, *Phys. Rev. Lett.* **1992**, *68*, 2340.
84 I. W. Hamley, K. A. Koppi, J. H. Rosedale, F. S. Bates, K. Almdal, K. Mortensen, *Macromolecules* **1993**, *26*, 5959.
85 I. W. Hamley, J. A. Pople, J. P. A. Fairclough, A. J. Ryan, C. Booth, Y. W. Yang, *Macromolecules* **1998**, *31*, 3906.
86 C. Daniel, I. W. Hamley, W. Ming-vanish, C. Booth, *Macromolecules* **2000**, *33*, 2163.
87 I. W. Hamley, *Curr. Opin. Colloid Interf. Sci.* **2000**, *5*, 342.
88 I. W. Hamley, *Philos. Trans. R. Soc. London* **2001**, *359*, 1017.
89 K. Mortensen, E. Theunissen, R. Kleppinger, K. Almdal, H. Reynaers, *Macromolecules* **2002**, *35*, 7773.
90 V. Castelletto, I. W. Hamley, M. Crothers, D. Attwood, Z. Yang, C. Booth, *Macromol. Sci. Phy. B* **2004**, *43*, 13.
91 I. W. Hamley, *J. Phys.-Condens. Mat.* **2001**, *13*, R643.
92 C. C. Honeker, E. L. Thomas, *Chem. Mater.* **1996**, *8*, 1702.
93 G. H. Fredikson, F. S. Bates, *Annu. Rev. Mater. Sci.* **1996**, *26*, 501.
94 H. Watanabe, *Acta Polym.* **1997**, *48*, 215.
95 A. V. Morozov, Zvelindovsky, J. G. E. M. Fraaije, *Phys. Rev. E* **2001**, *64*, 051803.
96 F. Corberi, G. Gonnella, A. Lamura, *Phys. Rev. E* **2002**, *66*, 016114.
97 A. V. M. Zvelindovsky, G. J. A. Sevink, *Europhys. Lett.* **2003**, *62*, 370.
98 B. Fraser, C. Denniston, M. H. Muser, *J. Polym. Sci. Polym. Phys.* **2005**, *43*, 970.
99 A. V. Zvelindovsky, G. J. A. Sevink, *J. Chem. Phys.* **2005**, *123*, 074903.
100 P. L. Chen, *Phys. Rev. E* **2005**, *71*, 061503.
101 I. Rychkov, *Macrom. Theory Simul.* **2005**, *14*, 207.
102 H. Fong, D. H. Reneker, *J. Polym. Sci. Polym. Phys.* **1999**, *37*, 3488.
103 T. Ruotsalainen, J. Turku, P. Heikkilä, J. Ruokolainen, A. Nykänen, T. Laitinen, M. Torkkeli, R. Serimaa, G. ten Brinke, A. Harlin, O, Ikkala, *Adv. Mater.* **2005**, *17*, 1048.
104 M. L. Ma, R. M. Hill, J. L. Lowery, S. V. Fridrikh, G. C. Rutledge, *Langmuir* **2005**, *21*, 5549.
105 J. Hahm, S. J. Sibener, *Langmuir* **2000**, *16*, 4766.
106 X. Li, Y. C. Han, L. J. An, *Langmuir* **2002**, *18*, 5293.
107 D. Li, Y. Xia, *Adv. Mater.* **2004**, *16*, 1151.
108 Y. Dzenis, *Science* **2004**, *304*, 1917.
109 M. L. Ma, R. M. Hill, J. L. Lowery, S. V. Fridrikh, G. C. Rutledge, *Langmuir* **2005**, *21*, 5549.
110 K. Amundson, E. Helfand, D. D. Davis, X. Quan, S. Patel, S. D. Smith, *Macromolecules* **1991**, *24*, 6546.
111 K. Amundson, E. Helfand, D. D. Davis, X. Quan, S. D. Smith, *Macromolecules* **1993**, *26*, 2698.
112 K. Amundson, E. Helfand, D. D. Davis, X. Quan, S. D. Hudson, S. D. Smith, *Macromolecules* **1994**, *27*, 6559.
113 A. Onuki, J. Fukuda, *Macromolecules* **1995**, *28*, 8788.
114 T. Thurn-Albrecht, R. Steiner, J. DeRouchey, C. M. Stafford, E. Huang, M. Bal, M. Tuominen, C. J. Hawker, T. P. Russell, *Adv. Mater.* **2000**, *12*, 787.
115 T. Thurn-Albrecht, J. DeRouchey, T. P. Russell, R. Kolb, *Macromolecules* **2002**, *35*, 8106.
116 S. Elhadj, J. W. Woody, V. S. Niu, R. F. Saraf, *Appl. Phys. Lett.* **2003**, *82*, 872.
117 T. Xu, Y. Zhu, S. P. Gido, T. P. Russell, *Macromolecules* **2004**, *37*, 2625.
118 J. DeRouchey, T. Thurn-Albrecht, T. P. Russell, R. Kolb, *Macromolecules* **2004**, *37*, 2538.
119 H. Xiang, Y. Lin, T. P. Russell, R. Kolb, *Macromolecules* **2004**, *37*, 5358.
120 G. G. Pereira, D. R. M. Williams, *Macromolecules* **1999**, *32*, 8115.
121 Y. Tsori, D. Andelman, *Macromolecules* **2002**, *35*, 5161.
122 B. Ashok, M. Muthukumar, T. P. Russell, *J. Chem. Phys.* **2001**, *115*, 1559.

123 A. V. Kyrylyuk, A. V. Zvelindovsky, G. J. A. Sevink, J. G. E. M. Fraaije, *Macromolecules* **2002**, *35*, 1473.
124 Y. Tsori, F. Tournilhac, L. Leibler, *Macromolecules* **2003**, *36*, 5873.
125 Y. Tsori, F. Tournilhac, D. Andelman, L. Leibler, *Phys. Rev. Lett.* **2003**, *90*, 145504.
126 O. Dürr, W. Dieterich, P. Maas, A. Nitzan, *J. Phys. Chem.* **2002**, *106*, 6149.
127 T. Thurn-Albrecht, J. DeRouchey, T. P. Russell, H. M. Jaeger, *Macromolecules* **2000**, *33*, 3250.
128 T. Xu, C. J. Hawker, T. P. Russell, *Macromolecules* **2003**, *36*, 6178.
129 A. Böker, H. Elbs, H. Hänsen, A. Knoll, S. Ludwigs, H. Zettl, A. V. Zvelindovski, G. J. A. Sevink, V. Urban, V. Abetz, A. H. E. Müller, G. Krausch, *Macromolecules* **2003**, *36*, 8078.
130 A. Böker, A. Knoll, H. Elbs, V. Abetz, A. H. E. Müller, G. Krausch, *Macromolecules* **2002**, *35*, 1319.
131 A. V. Zvelindovski, G. J. A. Sevink, *Phys. Rev. Lett.* **2003**, *90*, 049601.
132 T. Xu, J. T. Goldbach, T. P. Russell, *Macromolecules* **2003**, *36*, 7296.
133 C. Osuji, P. J. Ferreira, G. Mao, C. K. Ober, J. B. Vander Sande, E. L. Thomas, *Macromolecules* **2004**, *37*, 9903.
134 N. Tomikawa, Z. B. Lu, T. Itoh, C. T. Imrie, M. Adachi, M. Tokita, J. Watanabe, *Jpn. J. Appl. Phys. 2*, **2005**, *44*, L711.
135 T. Grigorova, S. Pispas, N. Hadjichristidis, T. Thurn-Albrecht, *Macromolecules* **2005**, *38*, 7430.
136 A. Turturro, E. Gattiglia, P. Vacca, G. T. Viola, *Polymer* **1995**, *21*, 3987.
137 G. Kim, M. Libera, *Macromolecules* **1998**, *31*, 2569.
138 G. Kim, M. Libera, *Macromolecules* **1998**, *31*, 2670.
139 S. H. Kim, M. J. Misner, T. Xu, M. Kimura, T. P. Russell, *Adv. Mater.* **2004**, *16*, 226.
140 M. Kimura, M. J. Mister, T. Xu, S. H. Kim, T. P. Russell, *Langmuir* **2003**, *19*, 9910.
141 Z. Lin, D. H. Kim, X. Wu, L. Boosahda, D. Stone, L. LaRose, T. P. Russell, *Adv. Mater.* **2002**, *14*, 1373.
142 K. Temple, K. Kulbaba, K. N. Power-Billard, I. Manners, K. A. Leach, T. Xu, T. P. Russell, C. J. Hawcker, *Adv. Mater.* **2003**, *15*, 297.
143 T. Hashimoto, J. Bodycomb, Y. Funaki, K. Kimishima, *Macromolecules* **1999**, *32*, 952.
144 T. Hashimoto, J. Bodycomb, Y. Funaki, K. Kimishima, *Macromolecules* **1999**, *32*, 2075.
145 L. Rockford, Y. Liu, P. Mansky, T. P. Russell, M. Yoon, S. G. J. Mochrie, *Phys. Rev. Lett.* **1999**, *82*, 2602.
146 L. Rockford, S. G. J. Mochrie, T. P. Russell, *Macromolecules* **2001**, *34*, 1487.
147 X. M. Yang, R. D. Peters, P. F. Nealey, H. H. Solak, F. Cerrina, *Macromolecules* **2000**, *33*, 9575.
148 S. O. Kim, H. H. Solak, M. P. Stoykovich, N. J. Ferrier, J. J. de Pablo, P. F. Nealey, *Nature (London)* **2003**, *424*, 411.
149 C. De Rosa, C. Park, B. Lotz, L. J. Fetters, J. C. Wittmann, E. L. Thomas, *Macromolecules* **2000**, *33*, 4871.
150 X. M. Yang, R. D. Peters, T. K. Kim, P. F. Nealey, *J. Vac. Sci. Technol. B*, **1999**, *17*, 3203.
151 H. H. Solak, C. David, J. Gobrecht, V. Golovkina, F. Cerrina, S. O. Kim, P. F. Nealey, *Microelectron. Eng.* **2003**, *67-68*, 56.
152 M. W. Matsen, *Curr. Opin. Colloid Interf. Sci.* **1998**, *3*, 40.
153 C. S. Henkee, E. L. Thomas, L. J. Fetters. *J. Mater. Sci.* **1988**, *23*, 1685.
154 G. Coulon, V. R. Deline, T. P. Russell, P. F. Green, *Macromolecules* **1989**, *22*, 2581.
155 S. H. Anastasiadis, T. P. Russell, S. K. Satija, C. F. Majkrzak. *Phys. Rev. Lett.* **1989**, *62*, 1852.
156 T. P. Russell, G. Coulon, V. R. Deline, D. C. Miller, *Macromolecules* **1989**, *22*, 4600.
157 S. H. Anastasiadis, T. P. Russell, S. K. Satija, C. F.Majkrzak, *J. Chem. Phys.* **1990**, *92*, 5677.
158 T. P. Russell, A. Menelle, S. H. Anastasiadis, S. K. Satija, C. F. Majkrzak, *Macromolecules* **1991**, *24*, 6269.

159 B. Collin, D. Chatenay, G. Coulon, D. Ausserre, Y. Gallot, *Macromolecules* **1992**, *25*, 1621.

160 G. Coulon, J. Dailant, B. Collin, J. J. Benattar, Y. Gallot, *Macromolecules* **1993**, *26*, 1582.

161 A. M. Mayes, T. P. Russell, P. Bassereau, S. M. Baker, G. S. Smith, *Macromolecules* **1994**, *27*, 749.

162 B. L. Carvalho, E. L. Thomas, *Phys. Rev. Lett.* **1994**, *73*, 3321.

163 S. Joly, D. Ausserre, G. Brotons, Y. Gallot, *Eur. Phys. J. E* **2002**, *8*, 355.

164 D. G. Walton, G. J. Kellogg, A. M. Mayes, P. Lambooy, T. P. Russell, *Macromolecules* **1994**, *27*, 6225.

165 A. P. Smith, J. F. Douglas, J. C. Meredith, E. J. Amis, A. Karim, *Phys. Rev. Lett.* **2001**, *87*, 015503.

166 A. P. Smith, J. F. Douglas, J. C. Meredith, E. J. Amis, A. Karim, *J. Polym. Sci. Part B Polym. Phys.* **2001**, *39*, 2141.

167 P. Lambooy, T. P. Russell, G. J. Kellogg, A. M. Mayes, P. D. Gallagher, S. K. Satija, *Phys. Rev. Lett.* **1994**, *72*, 2899.

168 N. Koneripalli, M. Singh, R. Levicky, F. S. Bates, P. D. Gallagher, S. K. Satija, *Macromolecules* **1995**, *28*, 2897.

169 G. J. Kellogg, D. G. Walton, A. M. Mayes, P. Lambooy, T. P. Russell, P. D. Gallagher, S. K. Satija, *Phys. Rev. Lett.* **1996**, *76*, 2503.

170 N. Koneripalli, R. Levicky, F. S. Bates, J. Ankner, H. Kaiser, S. K. Satija, *Langmuir* **1996**, *12*, 6681.

171 M. S. Turner, *Phys. Rev. Lett.* **1992**, *69*, 1788.

172 K. R. Shull, *Macromolecules* **1992**, *25*, 2122.

173 G. R. Pickett, T. A. Witten, S. R. Nagel, *Macromolecules* **1993**, *26*, 3194.

174 M. Kikuchi, K. Binder, *J. Chem. Phys.* **1994**, *101*, 3367.

175 G. Brown, A. Chakrabarti, *J. Chem. Phys.* **1995**, *102*, 1440.

176 G. T. Pickett, A. C. Balazs, *Macromolecules* **1997**, *30*, 3097.

177 M. W. Matsen, *J. Chem. Phys.* **1997**, *106*, 7781.

178 W. H. Tang, T. A. Witten, *Macromolecules* **1998**, *31*, 3130.

179 T. Geisinger, M. Muller, K. Binder, *J. Chem. Phys.* **1999**, *111*, 5251.

180 A. L. Frischknecht, J. G. Curro, L. J. D. Frink, *J. Chem. Phys.* **2002**, *117*, 10398.

181 Y. Liu, W. Zhao, X. Zheng, A. King, A. Singh, M. H. Rafailovich, J. Sokolov, K. H. Dai, E. J. Kramer, S. A. Schwarz, O. Gebizlioglu, S. K. Sinha, *Macromolecules* **1994**, *27*, 4000.

182 M. S. Turner, M. Rubinstein, C. M. Marques, *Macromolecules* **1994**, *27*, 4986.

183 M. A. van Dijk, R. van den Berg, *Macromolecules* **1995**, *28*, 6773.

184 L. H. Radzilowski, B. L. Carvalho, E. L. Thomas, *J. Polym. Sci. B Polym. Phys.* **1996**, *34*, 3081.

185 H. C. Kim, T. P. Russell *J. Polym. Sci. B Polym. Phys.* **2001**, *39*, 663.

186 K. Y. Suh, Y. S. Kim, H. H. Lee, *J. Chem. Phys.* **1998**, *108*, 1253.

187 H. P. Huinink, J. C. M. Brokken-Zijp, M. A. van Dijk, G. J. A. Sevink, *J. Chem. Phys.* **2000**, *112*, 2452.

188 M. Konrad, A. Knoll, G. Krausch, R. Magerle, *Macromolecules* **2000**, *33*, 5518.

189 A. Knoll, R. Magerle, G. Krausch, *J. Chem. Phys.* **2004**, *120*, 1105.

190 G. Szamel, M. Muller, *J. Chem. Phys.* **2003**, *118*, 905.

191 H. Yokoyama, T. E. Mates, E. J. Kramer, *Macromolecules* **2000**, *33*, 1888.

192 P. Mansky, T. P. Russell, C. J. Hawker, M. Pitsikalis, J. Mays, *Macromolecules* **1997**, *30*, 6810.

193 P. Mansky, Y. Liu, E. Huang, T. P. Russell, C. Hawker, *Science* **1997**, *275*, 1458.

194 E. Huang, T. P. Russell, C. Harrison, P. M. Chaikin, R. A. Register, C. J. Hawker, J. Mays, *Macromolecules* **1998**, *31*, 7641.

195 E. Huang, L. Rockford, T. P. Russell, C. J. Hawker, *Nature (London)* **1998**, *395*, 757.

196 E. Huang, S. Pruzinsky, T. P. Russell, J. Mays, C. J. Hawker, *Macromolecules* **1999**, *32*, 5299.

197 E. Huang, P. Mansky, T. P. Russell, C. Harrison, P. M. Chaikin, R. A. Register, C. J. Hawker, J. Mays, *Macromolecules* **2000**, *33*, 80.

198 C. Harrison, P. M. Chaikin, D. A. Huse, R. A. Register, D. H. Adamson, A. Daniel, E. Huang, P. Mansky, T. P. Russell, C. J. Hawker, D. A. Egolf, I. V. Melnikov, E. Bodenschatz, *Macromolecules* **2000**, *33*, 857.

199 L. Royer, *Bull. Soc. Fr. Mineral. Crystallogr.* **1928**, *51*, 7.

200 J. H. van deer Mere, *Discuss. Faraday Soc.* **1949**, *5*, 206.

201 G. S. Swei, J. B. Lando, S. E. Rickert, K. A. Mauritz, *Encyclopedia Polym. Sci. Eng.* **1986**, *6*, 209.

202 J. Willems, *Naturwissenschaften* **1955**, *42*, 176.

203 J. Willems, L. Willems, *Experientia* **1957**, *13*, 465.

204 J. Willems, *Discuss. Faraday Soc.* **1958**, *25*, 111.

205 E. W. Fischer, *Discuss. Faraday Soc.* **1958**, *25*, 204.

206 S. H. Wellinghoff, F. Rybnikar, E. Baer, *J. Macromol. Sci. (Phys.)* **1974**, *B10*, 1.

207 J. A. Koutsky, A. C. Walton, E. Baer, *J. Polym. Sci. Polym. Lett. Ed.* **1967**, *5*, 177.

208 J. A. Koutsky, A. C. Walton, E. Baer, *J. Polym. Sci. Polym. Lett. Ed.* **1967**, *5*, 185.

209 M. Ashida, Y. Uedn, T. Watanabe, *J. Polym. Sci. Polym. Phys. Ed.* **1978**, *16*, 179.

210 S. E. Rickert, E. Baer, *J. Mater. Sci. Lett.* **1978**, *13*, 451.

211 A. J. Lovinger, *Polym. Prep., Am. Chem. Soc.* **1980**, *21*, 253.

212 J. Martinez-Salazzas, P. J. Barham, A. Keller, *J. Polym. Sci. Polym. Phys. Ed.* **1984**, *22*, 1085.

213 J. C. Wittmann, B. Lotz, *J. Mater. Sci.* **1986**, *21*, 659.

214 F. Tuinstra, E. Baer, *J. Polym. Sci. Polym. Lett. Ed.* **1970**, *8*, 861.

215 A. J. Lovinger, *J. Polym. Sci. Polym. Phys. Ed.* **1983**, *21*, 97.

216 S. Y. Hobbs, *Nature Phys. Sci.* **1971**, *12*, 234.

217 J. C. Wittmann, B. Lotz, *J. Polym. Sci. Polym. Phys. Ed.* **1981**, *19*, 1837.

218 J. C. Wittmann, A. M. Hodge, B. Lotz, *J. Polym. Sci. Polym. Phys. Ed.* **1983**, *21*, 2495.

219 J. C. Wittmann, B. Lotz, *Polymer* **1989**, *30*, 27.

220 J. C. Wittmann, B. Lotz, *Prog. Polym. Sci.* **1990**, *15*, 909.

221 S. Kopp, J. C. Wittmann, B. Lotz, *Makromol. Chem. Macromol. Symp.* **1995**, *98*, 917.

222 C. De Rosa, C. Park, B. Lotz, E. L. Thomas, *Nature (London)* **2000**, *405*, 433.

223 C. Park, C. De Rosa, B. Lotz, L. J. Fetters, E. L. Thomas, *Macromol. Chem. Phys.* **2003**, *204*, 1514.

224 C. Park, C. De Rosa, B. Lotz, L. J. Fetters, E. L. Thomas, *Adv. Mater.* **2001**, *13*, 724.

225 P. Smith, A. J. Pennings, *Polymer* **1974**, *15*, 413.

226 P. Smith, A. J. Pennings, *J. Mater. Sci.* **1976**, *11*, 1450.

227 A. M. Hodge, G. Kiss, B. Lotz, J. C. Wittmann, *Polymer* **1982**, *23*, 985.

228 J. C. Wittmann, R. St. John Manley, *J. Polym. Sci. Polym. Phys. Ed.* **1977**, *15*, 1089.

229 J. C. Wittmann, R. St. John Manley, *J. Polym. Sci. Polym. Phys. Ed.* **1977**, *15*, 2277.

230 D. L. Dorset, J. Hanlon, G. Karet, *Macromolecules* **1989**, *22*, 2169.

231 G. H. Fredrickson, L. Leibler, *Macromolecules* **1989**, *22*, 1238.

232 T. P. Lodge, C. Pan, X. Jin, Z. Liu, J. Zhao, W. W. Maurer, F. S. Bates *J. Polym. Sci. Polym. Phys.* **1995**, *33*, 2289.

233 I. W. Hamley, J. P. A. Fairclough, A. J. Ryan, C. Y. Ryu, T. P. Lodge, A. J. Gleeson, J. S. Pedersen, *Macromolecules* **1998**, *31*, 1188.

234 C. Park, C. De Rosa, E. L. Thomas, *Macromolecules* **2001**, *34*, 2602.

235 M. A. Van Dijk, R. van den Berg, *Macromolecules* **1995**, *28*, 6773.

236 H. I. Smith, D. C. Flanders, *Appl. Phys. Lett.* **1978**, *32*, 349.

237 H. I. Smith, M. W. Geis, C. V. Thompson, H. A. Atwater, *J. Cryst. Growth* **1983**, *63*, 527.

238 T. Kobayashi, K. Takagi, *Appl. Phys. Lett.* **1984**, *45*, 44.

239 D. C. Flanders, D. C. Shaver, H. I. Smith, *Appl. Phys. Lett.* **1978**, *32*, 597.

240 J. C. Wittmann, P. Smith, *Nature (London)* **1991**, *352*, 414.

241 H. Hansma, F. Motamedi, P. Smith, P. Hansma, J. C. Wittmann, *Polymer Commun.* **1992**, *33*, 647.

242 P. Dietz, P. K. Hansma, K. J. Ihn, F. Motamedi, P. Smith, *J. Mater. Sci.* **1993**, *28*, 1372.

243 D. Fenwick, K. J. Ihn, F. Motamedi, J. C. Wittmann, P. Smith, *J. Appl. Polym. Sci.* **1993**, *50*, 1151.

244 D. Fenwick, P. Smith , J. C. Wittmann, *J. Mater. Sci.* **1996**, *31*, 128.

245 R. A. Segalman, H. Yokoyama, E. J. Kramer, *Adv. Mater.* **2001**, *13*, 1152.

246 R. A. Segalman, A. Hexemer, R. C. Hayward, E. J. Kramer, *Macromolecules* **2003**, *36*, 3272.

247 J. Y. Cheng, C. A. Ross, E. L. Thomas, H. I. Smith, G. J. Vancso, *Appl. Phys. Lett.* **2002**, *81*, 3657.

248 J. Y. Cheng, C. A. Ross, E. L. Thomas, H. I. Smith, G. J. Vancso, *Adv. Mater.* **2003**, *15*, 1599.

249 X.-H. He, M. Song, H.-J. Liang, C.-Y. Pan, *J. Chem. Phys.* **2001**, *114*, 10510.

250 G. J. A. Sevink, A. V. Zvelindovsky, J. G. E. M. Fraaije, H. P. Huinink, *J. Chem. Phys.* **2001**, *115*, 8226.

251 K. Shin, H. Xiang, S. I. Moon, T. Kim, T. J. McCarthy, T. P. Russell, *Science* **2004**, *306*, 76.

252 H. Xiang, K. Shin, T. Kim, S. I. Moon, T. J. McCarthy, T. P. Russell, *Macromolecules* **2004**, *37*, 5660.

253 Y. Sun, M. Steinhartm, D. Zschech, R. Adhikari, G. H. Michler, U. Gösele, *Macromol. Rapid Commun.* **2005**, *26*, 369.

254 C. Park, J. Y. Cheng, C. De Rosa, M. J. Fasolka, A. M. Mayes, C. A. Ross, E. L. Thomas *Appl. Phys. Lett.* **2001**, *79*, 848.

255 E. L. Thomas, C. De Rosa, C. Park, M. Fasolka, B. Lotz, A. M. Mayes, J. Yoon, *US Patent* N° 6,893,705 **2005** (MIT).

256 C. Park, S. Simmons, L. J. Fetters, B. Hsiao, F. Yeh, E. L. Thomas, *Polymer* **2000**, *41*, 2971.

257 C. Park, C. De Rosa, L. J. Fetters, E. L. Thomas, *Macromolecules* **2000**, *33*, 7931.

10
Block Copolymer Nanofibers and Nanotubes

Guojun Liu

10.1
Introduction

Block copolymer nanofibers in this chapter refer to cross-linked cylindrical structures that are made from block copolymers with diameters below 100 nm and lengths up to hundreds of micrometers. Figure 10.1 depicts the structures of nanofibers prepared from an A-B diblock and an A-B-C triblock copolymer, respectively. In the case of diblock nanofibers, either the core or the corona [1, 2] can be cross-linked. Once dried, nanofibers with cross-linked coronas may not re-disperse readily in solvents, the same as for block copolymer nanospheres with cross-linked shells [3]. Hence, our discussion in this chapter will be mainly on diblock nanofibers with cross-linked cores. For nanofibers with cross-linked cores, the soluble corona chains stretch into the solvent phase helping disperse the fibers, and the cross-linked core provides the structural stability. Although the nanofibers are depicted in Fig. 10.1 as being rigid and straight, in reality, they can bend or contain kinks.

Three scenarios can be differentiated for triblock nanofibers. In scenario one, the corona block is cross-linked. Again because of dispersibility considerations we have not prepared and studied such fibers. Our focus so far has been on scenario two where the middle layer of the nanofiber is cross-linked. In scenario three, the innermost block is cross-linked. The structure of such a nanofiber bears close resemblance to block copolymer cylindrical brushes [4] when the inner most block is short relative to the other blocks. Block copolymer cylindrical brushes can be obtained by polymerizing a diblock macromer. There have been a number of reports on the properties and applications of block copolymer cylindrical brushes [4]. While this chapter will not go beyond triblock nanofibers, except for one case when nanofibers of a tetrablock copolymer are discussed, the methodologies developed for di- and triblock copolymer nanofiber preparations should apply equally well to a preparation of nanofibers from tetra- and pentablock copolymers and to more complex copolymers.

Block Copolymers in Nanoscience. Massimo Lazzari, Guojun Liu, Sébastien Lecommandoux
Copyright © 2006 WILEY-VCH Verlag GmbH & Co. KGaA, Weinheim
ISBN: 3-527-31309-5

Fig. 10.1 Structural illustration of a diblock nanofiber (top) and a triblock nanofiber (bottom).

Fig. 10.2 Two configurations for a nanotube.

In principle, nanotubes can be prepared by cross-linking tubular micelles [5–8] of diblock copolymers. The focus of discussion here is, however, on nanotubes derived from triblock copolymers. Two possible configurations for such nanotubes are depicted in Fig. 10.2.

In case one, the tubular core is void. Such a nanotube is derived from a triblock nanofiber with a cross-linked intermediate layer by degrading the innermost block [9]. In case two, the tubular core is lined by a polymer. In this instance, a nanotube is derived from a triblock nanofiber with a cross-linked intermediate layer by cleaving pendant groups off the core block [10].

The preparation of block copolymer nanofiber [11] and nanotube [9] structures were only reported for the first time a few years ago. Over the years, there have been reports on the use of block copolymer nanofibers and nanotubes as vehicles for drug delivery [12], as scaffolds for cell growth [13, 14], as precursors for ceramic magnetic nanowires [15, 16] and as precursors for carbon nanofibers [17, 18], etc. While finding novel applications for such structures is of paramount importance, the emphasis of this chapter will be on research undertaken in my group aimed at achieving a fundamental understanding of the physical and chemical properties of these materials. In Section 10.2, the preparation of block copolymer nanofibers and nanotubes will be described and the solution properties of the nanofibers and nanotubes will be discussed in Section 10.3. The different reaction patterns of nanofibers and nanotubes will be examined in Section 10.4. In Section 10.5, some conclusions will be drawn and my perspectives on where block copolymer nanofiber and nanotubes research is going will be presented.

10.2
Preparation

10.2.1
Nanofiber Preparation

Nanofibers can be prepared from chemically processing either a block-segregated copolymer solid [19] or block copolymer cylindrical micelles [20] formed in a block-selective solvent. Figure 10.3 depicts the processes involved to obtain nanofibers from a block-segregated diblock solid.

The first step involves casting a film from a diblock with an appropriate composition so that the minority block segregates from the majority block forming hexagonally packed cylinders. This is followed by cross-linking the cylindrical domains. The cross-linked cylinders are then levitated from the film or separated from one another by stirring the film in a solvent that solubilizes the uncross-linked diblock.

Our literature search revealed that the first report on block copolymer nanofiber synthesis appeared in 1996 [11]. In that report, block-segregated solids of two polystyrene-*block*-poly(2-cinnamoyloxyethyl methacrylate), or PS-PCEMA, samples were used as the precursor. The two PS-PCEMA samples used had 1250 and 160 units of PS and PCEMA for polymer 1 and 780 and 110 for polymer 2, respectively. Thus, the volume fraction of PCEMA was ≈26% in both samples. After film formation from a diblock by evaporating a toluene solution slowly, the film was annealed at 90 °C for days to ensure clean phase segregation between PS and PCEMA. This film was then sectioned by ultra-microtoming to yield thin slices for transmission electron microscopic (TEM) examination. Shown in the left panel of Fig. 10.4 is a TEM image of a thin section for polymer 1 or PS_{1250}-$PCEMA_{160}$,

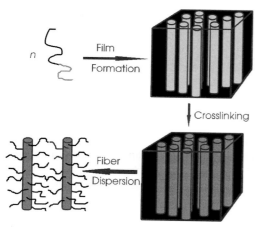

Fig. 10.3 Schematic illustration of the steps involved in the preparation of diblock copolymer nanofibers from a block-segregated solid.

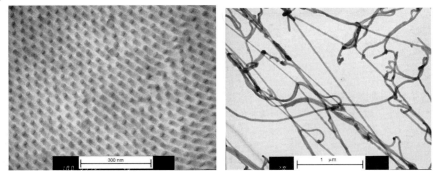

Fig. 10.4 Left: Thin-section TEM image of a solid PS_{1250}-$PCEMA_{160}$ sample. The dark PCEMA cylinders are slanted diagonally (from bottom right to top left). Right: TEM image of the PS_{1250}-$PCEMA_{160}$ nanofibers on a carbon-coated copper grid aspirated from THF.

where the subscripts denote the numbers of styrene and CEMA units, respectively (Scheme 10.1). It can be seen that the OsO_4-stained dark PCEMA phase consist of hexagonally packed cylinders dispersed in the PS matrix and that the cylinders slant, aligned along a diagonal direction of the image. A similar PCEMA block segregation pattern was found for polymer 2. Such films were then irradiated by UV light to cross-link the PCEMA cylindrical domains. Stirring the irradiated samples in THF helped levitate the cross-linked PCEMA cylinders from the films to yield solvent-dispersible nanofibers. In the right panel, a TEM image of the resultant nanofibers is shown. It can be observed that the fibers in this sample are still entangled.

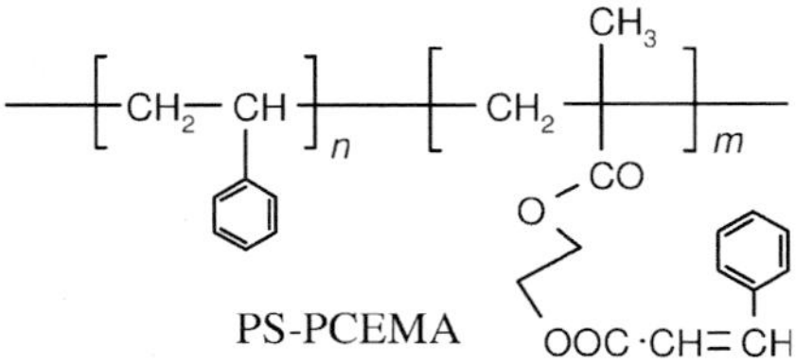

Scheme 10.1

The groups working with Wiesner [21], Ishizu [22], Muller [23] and my group [24], have also used chemical rather than photochemical methods to cross-link the cylindrical domains of block copolymer solids to prepare nanofibers. To prepare nanofibers from polystyrene-*block*-polyisoprene (PS-PI), we utilized a diblock that had 220 styrene and 140 isoprene units, respectively, which corresponded to a volume fraction of ≈30 % for PI. The cylindrical domains were then cross-linked by exposing the film to sulfur monochloride (Scheme 10.2).

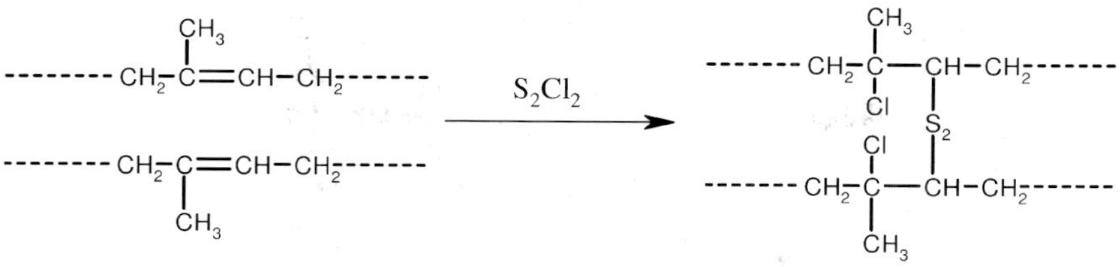

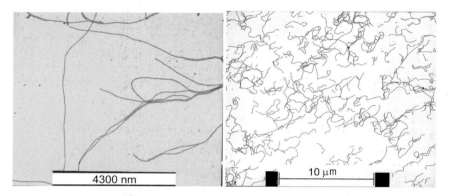

Scheme 10.2

Fig. 10.5 Left: TEM image of PS-PI nanofibers prepared from the cross-linking of block-segregated PS_{220}-PI_{140} solid. The length of the white box is 4300 nm. Right: TEM image of PS-PI nanofibers prepared from the cross-linking of PS_{130}-PI_{370} cylindrical micelles.

Isolated nanofibers were obtained by separating the cross-linked cylindrical domains in THF. The left panel of Fig. 10.5 shows a TEM image of such nanofibers.

A drawback with chemical cross-linking of block-segregated solids is the long diffusion time required for the cross-linker to penetrate the film. Insufficient reaction time leads to non-uniform cross-linking with higher degrees of cross-linking found close to the surfaces. Such non-uniform cross-linking can occur with photocross-linking for the short penetration distance of the light. These complications can be avoided by preparing nanofibers starting from block copolymer cylindrical micelles formed in a block-selective solvent.

Figure 10.6 depicts processes involved in preparation of diblock nanofibers starting from cylindrical micelles. Firstly, this requires the preparation of a diblock with an appropriate composition. Then, a selective solvent has to be found that solubilizes only one block of the diblock copolymer. In such a block-selective solvent, the insoluble blocks of different chains aggregate to form a cylindrical core stabilized by chains of the soluble block. Nanofibers are obtained by cross-linking the core chains.

A report on the preparation of block copolymer nanofibers from the cross-linking of cylindrical micelles first appeared in 1997 [25]. The cylindrical micelles were

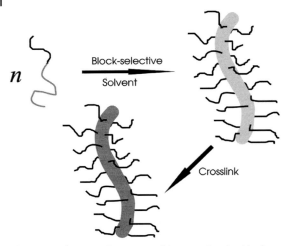

Fig. 10.6 Schematic illustration of the steps involved in the preparation of diblock copolymer nanofibers from cylindrical micelles formed in a block-selective solvent.

prepared also from PS-PCEMA in refluxing cyclopentane, which solubilized PS and not PCEMA. Nanofibers were obtained after the photocross-linking of the PCEMA cores. Aside from photocross-linking, the groups working with Bates [26], Discher [27], Manner [2], Wooley [28] and Stupp [13], in addition to my group [29, 30], have reported on the chemical cross-linking of cylindrical micelles to prepare nanofibers. For example, recently we prepared nanofibers by cross-linking PS-PI cylindrical micelles formed in a block-selective solvent, N,N-dimethyl acetamide, for PS [24, 31]. For cylindrical micelle formation with PI as the core, the PI weight fraction should be relatively high. We followed the recipe of Price [32] and used a sample consisting of 130 styrene units and 370 isoprene units. The cylindrical fibers were cross-linked using S_2Cl_2. The right-hand panel of Fig. 10.5 shows a TEM image of such nanofibers. Compared with fibers prepared from the solid-state syntheses described above, the fibers in the right-hand panel of Fig. 10.5 are shorter. A more quantitative analysis of the length distributions revealed that the fibers from the solution preparation approach are more monodisperse. Furthermore, the preparation yield can be high, approaching 100%.

10.2.2
Nanotube Preparation

Nanotubes have been prepared by my group mainly via the derivatization of triblock nanofibers. The first block copolymer nanotubes were prepared from PI_{130}-$PCEMA_{130}$-$PtBA_{800}$ [9], where PtBA denotes poly(tert-butyl acrylate). This involved firstly dispersing the triblock in methanol. As only the PtBA block was soluble in methanol, the triblock self-assembled into cylindrical micelles consisting of a PI core encapsulated in an insoluble PCEMA intermediate layer and a PtBA

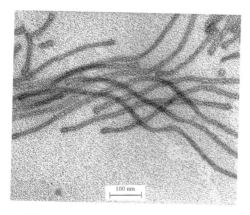

Fig. 10.7 TEM image of PCEMA-P*t*BA nanotubes where the PI core has been degraded.

corona. After photocross-linking the PCEMA intermediate layer, the PI core block was degraded by ozonolysis to yield nanotubes. The removal of the PI block was demonstrated by infrared absorption and TEM analyses. More importantly, Rhodamine B could be loaded into the tubular core. Figure 10.7 shows a TEM image of such nanotubes stained by OsO_4. The center of each tube appears lighter than the PCEMA intermediate layer because the PI block was decomposed.

To facilitate the incorporation of inorganic species into the tubular core, nanotubes containing PAA-lined cores were more desirable. Figure 10.8 depicts the steps involved in the preparation of PS-PCEMA-PAA nanotubes from PS_{690}-$PCEMA_{170}$-$PtBA_{200}$ via a solid-state precursor approach [10].

Step 1 (A→B in Fig. 10.8) involved casting films from the triblock containing concentric P*t*BA and PCEMA core-shell cylinders dispersed in the matrix of PS. This required the PS volume fraction to be ≈70 % [33] and was achieved by mixing

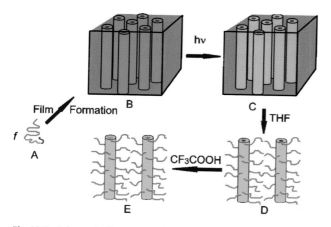

Fig. 10.8 Schematic illustration of processes involved in preparing PS-PCEMA-PAA nanotubes.

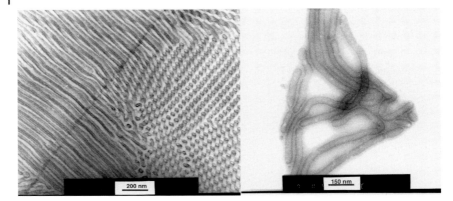

Fig. 10.9 Left: TEM image of a thin section of the PS_{690}-$PCEMA_{170}$-$PtBA_{200}$ solid. Right: TEM image of the PS-PCEMA-PAA nanotubes.

some PS homopolymers (hPS) with the triblock to increase the PS volume fraction. The left-hand panel of Fig. 10.9 shows a TEM image of a thin section of an hPS/PS-PCEMA-PtBA solid. On the right-hand side of this image we can see numerous concentric light and dark ellipses with short stems. These represent projections of cylinders with PCEMA shells and PtBA cores aligned slightly off the normal direction of the image. The PCEMA shells appear darker, because OsO_4 stained the PCEMA selectively. The diameter of the PtBA core is $\approx$20 nm. On the left of this image we see cylinders lying in the plane of the picture. Thus, the orientation of the cylindrical domains varied from grain to grain, in the micrometer size range, because we did not take special measures to effect their macroscopic alignment.

In step 2 (B→C, in Fig. 10.8), the block-segregated copolymer film was irradiated with UV light to cross-link the PCEMA shell cylinder. The cross-linked cylinders were levitated from the film by stirring in THF (C→D). PS-PCEMA nanotubes containing PAA-lined tubular cores were prepared by hydrolyzing the PtBA block in methylene chloride and trifluoroacetic acid (D→E). The right-hand panel of Fig. 10.9 shows a TEM image of the intestine-like nanotubes. The stained PCEMA layer does not have a uniform diameter across the nanotube length because of its uneven collapse during solvent evaporation. The presence of PAA groups inside the tubular core was demonstrated by our ability to carry out various aqueous reactions inside the tubular core, as will be discussed later.

10.3
Solution Properties

Figure 10.10 shows a comparison between the structures of a PS-PI nanofiber and a poly(*n*-hexyl isocyanate), PHIC, chain. In a PHIC chain, the backbone is made of imide units joined linearly and the hairs are the hexyl groups. Their counterparts in a PS-PI nanofiber are the cross-linked PI cylinder and PS chains, respectively.

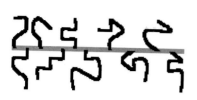 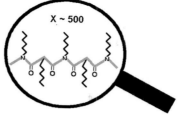

Fig. 10.10 Structural comparison between a PS-PCEMA nanofiber (left) and a PHIC chain (right) at different magnifications.

Other than a large size difference, a nanofiber bears remarkable structural resemblance to PHIC. Thus, block copolymer nanofibers can be viewed as a macroscopic counterpart of a polymer chain or a "suprapolymer" chain or "giant" polymer chain [34]. In this subsection, preliminary results showing the similarities and dissimilarities between the solution properties of polymer chains and diblock nanofibers will be reviewed.

To study the dilute solution properties of nanofibers and polymer chains, the fibers should be made sufficiently short, so that they remain dispersed in the solvent for a long, or even an infinitely long, period of time. The use of relatively short nanofibers also ensured their characterization by classic techniques, such as light scattering (LS) and viscometry. While we have studied nanofibers prepared from several block copolymer families, for clarity, the discussion will be restricted to PS-PI nanofibers obtained by cross-linking cylindrical micelles of PS_{130}-PI_{370} formed in N,N-dimethyl acetamide [24]. The preparation of such fibers has been discussed previously and the right-hand panel of Fig. 10.5 shows a TEM image of the nanofibers thus prepared in THF after aspiration onto a carbon-coated copper grid. As the magnification was known for such images, we were able to measure manually the lengths of more than 500 fibers for this sample. The data from such measurements allowed us to construct the length distribution function of this sample denoted as fraction 1 or F1 in Fig. 10.11. From the length distribution function, we obtained the weight- and number-average lengths L_w and L_n. The L_w and L_w/L_n values are 3490 nm and 1.35 for this sample.

While ultracentrifugation [34] or density gradient centrifugation could have been used, in principle, to separate the fibers into fractions of different lengths, we obtained nanofiber fractions with shorter lengths by breaking up the longer nanofibers by ultrasonication [26]. By adjusting the ultrasonication time, we produced fibers of different lengths. Also shown in Fig. 10.11 are the length distribution functions for samples denoted as F3 and F5, which were ultrasonicated for 4 and 20 h, respectively. As ultrasonication time increased, the distribution shifted to shorter lengths.

These fiber fractions were sufficiently short and allowed us to determine their weight-average molar, M_w, by light scattering. Figure 10.12 shows a Zimm plot for the light scattering data for sample F3 in the scattering angle, θ, range of 12 to 30°.

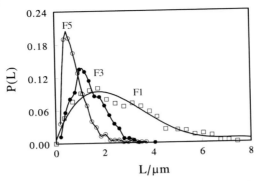

Fig. 10.11 Plot of fiber population density P(L) versus length L for PS_{130}-PI_{370} nanofiber fractions 1 (□), 3 (●) and 5 (○) generated from TEM image analysis.

The data quality appears high. Multiple runs of the same sample indicated that the data precision was high.

For the large-sized fibers, the $Kc/\Delta R_\theta$ data varied with $\sin^2(\theta/2)$ or the square of the scattering wave vector q non-linearly, despite the low angles used. We fitted the data using Eq. (10.1):

$$\frac{Kc}{\Delta R_\theta} = \frac{1}{M_w}\left[1 + (1/3)q^2 R_G^2 - kq^4 R_G^4\right] + 2A_2 c \qquad (10.1)$$

and obtained M_w, the radii of gyration R_G and the second Virial coefficient A_2 for the different fractions. Figure 10.13 plots the resultant M_w versus L_w, where the values for L_w were obtained from TEM length distribution functions $P(L)$. The linear increase in M_w with L_w suggests the validity of the M_w values determined. The validity of the M_w value for F3 was further confirmed recently by Professor Chi Wu's group at the Chinese University of Hong Kong, who performed a light scattering analysis of a nanofiber sample down to $\theta = 7°$. At such low angles, the $kq^4 R_G^4$ term in Eq. (10.1) was not required for curve fitting and data analysis by the Zimm method should yield accurate M_w and R_G values.

After nanofiber characterization, we then proceeded to check the dilute solution viscosity properties. Our experiments indicated that the nanofiber solutions were analogous to polymer solutions and were shear thinning, i.e., the viscosity of a sample decreased with increasing shear rate. This occurred for the alignment of the nanofibers along the shearing direction above a shear rate γ of ≈ 0.1 s^{-1} [35]. While both nanofiber and polymer solutions are shear thinning, the fields required for shear thinning are dramatically different. Polymers of ordinary molar mass, e.g. <10^6 g mol^{-1}, would experience shear thinning only if $\gamma \gtrsim 10^4$ s^{-1} [36]. The huge difference should be a direct consequence of the drastically different sizes between the two.

To minimize the shear-thinning effect, we measured the viscosities of dilute solutions of the nanofiber fractions in THF using a laboratory-built rotating cylinder viscometer at $\gamma = 0.082$ s^{-1} [37]. Figure 10.14 shows the $(\eta_r - 1)/c$ data plotted

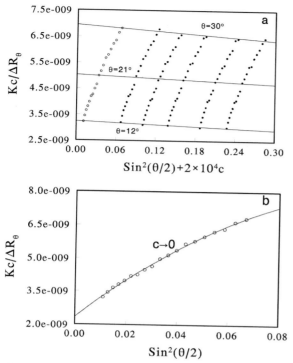

Fig. 10.12 Zimm plot for the light scattering data of F3 in the scattering angle range of 12 to 30°. The solid circles represent the experimental data. The hollow circles represent the extrapolated $Kc/\Delta R_\theta|_{c\to 0}$ data. (a) Linear extrapolation of data to zero concentration at the highest and lowest scattering angles of 30 and 12° is illustrated. (b) The result of curve fitting of the $Kc/\Delta R_\theta|_{c\to 0}$ data using Eq. (10.1).

against nanofiber concentrations c, where η_r, the relative viscosity, is defined as the ratio between the viscosities of the nanofiber solution and solvent THF. The solid lines represent the best fit to the experimental data by Eq. (10.2):

$$(\eta_r - 1)/c = [\eta] + k_h [\eta]^2 c \qquad (10.2)$$

where $[\eta]$ is the intrinsic viscosity and k_h is the Huggins coefficient. The linear dependence between $(\eta_r - 1)/c$ and c is in striking agreement with the behavior of polymer solutions. Even more interesting, k_h took values mostly between 0.20 and 0.60 in agreement with those found for polymers [36].

We further treated the $[\eta]$ data with the Yamakawa–Fujii–Yoshizaki (YFY) theory, originally developed for wormlike chains [38, 39]. According to Bohdanecky [40], the YFY theory could be cast in a much simpler form, Eq. (10.3):

$$\left(M_w^2/[\eta]\right)^{1/3} = A + B M_w^{1/2} \qquad (10.3)$$

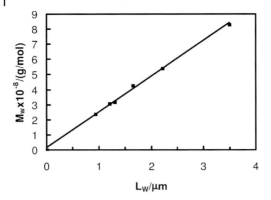

Fig. 10.13 Increase in LS M_w with TEM L_w for PS_{130}-PI_{370} nanofiber fractions.

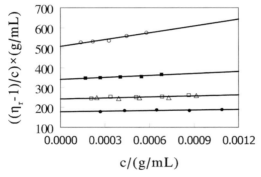

Fig. 10.14 From top to bottom, plot of $(\eta_r - 1)/c$ versus c for PS_{130}-PI_{370} nanofiber fractions 2, 3, 4 and 6 in THF. All the η_r data were obtained using the viscometer at a shear rate of 0.082 s^{-1} with the exception of those denoted by ($\triangle$), which were obtained at a shear rate of 0.047 s^{-1}.

for chains with a wide range of reduced chain lengths. In Eq. (10.3), A and B are fitting parameters that are related to the persistence length l_p and the hydrodynamic diameter d_h of the chains, respectively. Figure 10.15 shows the data that we obtained for the PS_{130}-PI_{370} nanofibers in THF plotted following Eq. (10.3). From the intercept A and slope B of the straight line, we calculated l_p and d_h for the nanofibers to be (1040 ± 150) and (69 ± 18) nm, respectively.

This procedure was repeated for the nanofibers in different solvents. Table 10.1 summarizes the l_p and d_h values that we determined in three different solvents for the PS_{130}-PI_{370} nanofibers. The d_h value in THF compares well with what we estimated from the sum between the diameter of the cross-linked PI core determined from TEM and the root-mean-square end-to-end distance of the PS coronal chains, and thus suggests the applicability of the FYF theory to the nanofiber solutions. What is more convincing is the decreasing trend for the determined d_h

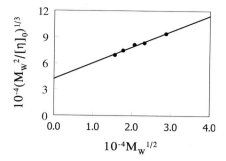

Fig. 10.15 Nanofiber viscosity data plotted following the Bohdanecky method.

values with increasing DMF content in THF–DMF mixtures. While both THF and DMF solubilize PS, d_h decreased with increasing DMF content because the extent of swelling for the cross-linked PI core decreased with increasing DMF content.

Table 10.1 Persistence length l_p and hydrodynamic diameter d_h of the nanofibers calculated from the viscosity data for PS_{130}-PI_{370} nanofiber fractions in various solvents.

Solvent	d_h/nm	l_p/nm
THF	69 ± 18	1040 ± 150
THF–DMF = 50/50	61±	850 ± 90
THF–DMF = 30/70	51 ± 12	830 ± 60

The l_p values reported in Table 10.1 are comparable to those reported by Discher and coworkers [41, 42] and by Bates and coworkers [43, 44] for PEO-PI cylindrical micelles with a core diameter of ≈20 nm prepared in water, where PEO denotes poly(ethylene oxide). While Bates and coworkers deduced the l_p values from small-angle neutron scattering, Discher and coworkers determined the l_p values from fluorescence microscopy. In the latter case, they compared the dynamic behavior of single cylindrical micelles before and after PI core cross-linking. After micelle cross-linking, the micelles became much more rigid dynamically, which means that the contour or conformation of the fibers, in contrast to the micelles, changed or flexed very little with time, despite their rotation in space as approximate rigid rotors. By performing a dynamic analysis of the flexion motion by subtracting off the spontaneously curved average shape of the fibers, they concluded that the *dynamic* l_p values of the fibers were about 50 times higher than those of the cylindrical micelles. From viscometry, one deduces the *static* l_p values of the nanofibers, which measure on average how much an ensemble of fibers bends. Therefore, one should not compare the viscometry l_p values determined by us with those determined by Discher and coworkers, who totally ignored the locked-in curvatures of the fibers in their analysis.

To get a clue to the static l_p values of the PEO-PI fibers studied by Discher and coworkers, from their fluorescence microscopy images we noticed that the kinks in the original cylindrical micelles were locked in after micelle cross-linking and the fibers assumed conformations similar to those before micelle cross-linking. Thus, the static l_p values of the nanofibers should be similar to those of the cylindrical micelles. The fact that the l_p values that we determined from viscometry are comparable to those of the PEO-PI cylindrical micelles with similar core diameters again suggests the validity of the YFY theory in treating the nanofiber viscosity data.

The above study demonstrates that block copolymer nanofibers have dilute solution properties similar to those of polymer chains. In an earlier report [45], we also demonstrated that block copolymer nanofibers have concentrated solution properties similar to those of polymer chains. According to the theories of Onsager [46] and Flory [47], polymer chains with $l_p/d_h > 6$ would form a liquid crystalline phase above a critical concentration. We did show the presence of such a liquid crystalline phase by polarized optical microscopy for PS-PCEMA nanofibers dissolved in bromoform at concentrations above ≈ 25 wt-% [45]. Furthermore, we observed that such liquid crystalline phases disappeared as the temperature was raised and the liquid crystalline to disorder transition was fairly sharp.

While block copolymer nanofibers behave similarly to polymer chains in many aspects, the drastic size difference between the two dictates that they have substantial property differences. Because of the large size of the nanofibers, they obviously move more sluggishly. Hence, we observed that a liquid crystalline phase was formed only after the PS-PCEMA nanofiber solution was sheared mechanically. Also, because of their sluggishness, the liquid crystalline phase could not reform spontaneously after cooling a system if it had been heated above the liquid crystalline to disorder transition temperature. Thus, we can predict, without performing any sophisticated experiments, that the analogy between nanofibers and polymer chains will fail after the molar mass or the size of the nanofibers exceeds a critical value. As the size of the nanofiber increases, the gravitational force driving the settling of the nanofiber increases and the dispersibility of the nanofiber decreases. Furthermore, the van der Waals forces between different nanofibers increase [48], which can cause different nanofibers to cluster and settle.

We recently examined the stability of nanofibers dispersed in THF prepared from PS_{130}-PI_{370}. This particular nanofiber sample had $L_w = 1650$ nm, $L_w/L_n = 1.21$ and $M_w = 4.3 \times 10^8$ g mol^{-1}, respectively. At a concentration of $\approx 8 \times 10^{-3}$ g mL^{-1} and under gentle stirring, no nanofiber settling was observed during 4 days of observation by light scattering. Without stirring, we noticed a 10% decrease in the light scattering intensity of the solution, which corresponded to ≈ 10 wt-% settling of the nanofibers in the first 4 days. No noticeable further settling was observed in another 8 days [24]. This could indicate that the longer fibers in this sample exceeded the critical size for settling. Our light scattering and centrifugation experiments suggested that the longer fibers first clustered and then settled. The fact that the clustering could be prevented by gentle stirring suggests that only a very shallow attraction potential existed between the fibers.

Although the critical length for settling was short for this sample, several micrometers, the critical length depends on many factors including the relative length between the core and soluble block and the absolute diameter of the cores. Methods of increasing the nanofiber dispersity may include increasing the length of the soluble block relative to the core block and decreasing the core diameter.

Because of the differences between the polymer chains and the nanofibers, we expect differences in the performances of these two classes of bulk materials. Unfortunately, the mechanical properties of block copolymer nanofibers or nanofiber composites have not been studied so far. We have not performed any detailed studies of solution properties of block copolymer nanotubes. As a result of the structural similarities between the two, we expect the nanofibers and nanotubes to have many similar solution properties.

10.4
Chemical Reactions

The similarities between the structure and the properties of the solutions between nanofibers and polymer chains prompted us to ask the question as to whether nanofibers and nanotubes would have chemical reaction patterns similar to those of polymer chains. A PI chain can be hydrogenated via "backbone modification" to yield a polyolefin chain. Through techniques such as anionic polymerization, etc., one can readily prepare "end-functionalized" polymers. The end-groups can be further derivatized or used for additional end-functionalization. This section will show that block copolymer nanotubes can also undergo backbone modification and end-functionalization.

10.4.1
Backbone Modification

Backbone modification has already been invoked to convert triblock nanofibers into nanotubes. Apart from the performance of organic reactions to the nanofibers and nanotubes, this sub-section discusses the performance of inorganic reactions in the cores of the nanotubes to convert them into polymer–inorganic hybrid nanofibers. Block copolymer nanofibers and nanotubes are soft materials. They will most probably find applications in bio-related disciplines, such as in the medical, pharmaceutical and cosmetic industries. For applications in nanoelectronic devices, polymer–inorganic hybrid nanofibers would be more desirable [49, 50]. The first report on the preparation of block copolymer–inorganic hybrid nanofibers appeared in 2001, which dealt with filling of the core of the PS-PCEMA-PAA nanotubes by γ-Fe_2O_3 [10].

The preparation first involved the equilibration between the nanotubes and $FeCl_2$ in THF. Fe(II) entered the nanotube core to bind with the core carboxyl groups. The extraneous $FeCl_2$ was then removed by precipitating the Fe(II)-containing nanotubes into methanol. Adding NaOH dissolved in THF containing 2 vol-% of water

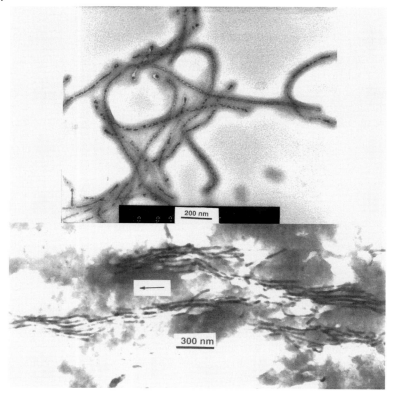

Fig. 10.16 Top: TEM image of PS-PCEMA-PAA/Fe_2O_3 hybrid nanofibers. Bottom: Bundling and alignment of the nanofibers in a magnetic field. The arrow indicates the magnetic field direction.

precipitated Fe(II) trapped in the nanotube core as ferrous oxide. The ferrous oxide was subsequently oxidized to γ-Fe_2O_3 via the addition of hydrogen peroxide [51]. The top panel in Fig. 10.16 shows a TEM image of the hybrid nanofibers. The γ-Fe_2O_3 particles can be seen to be produced exclusively inside the nanotube cores.

The production of γ-Fe_2O_3 in the confined space of the "nanotest-tubes" resulted in particles that were nanometer-sized. Hence the particles were superparamagnetic, as demonstrated by the results of our magnetic property measurement [10]. This meant that they were magnetized only in the presence of an external magnetic field and were demagnetized when the field was removed. To see how such fibers behaved in a solvent in a magnetic field, we dispersed the fibers in a solvent mixture consisting of THF, styrene, divinylbenzene, and a free radical initiator AIBN. The fiber dispersion was then dispensed into an NMR tube and mounted in the sample holder of an NMR instrument. In the 4.7-T magnetic field of the NMR, the solvent phase was gelled by raising the temperature to 70 °C to polymerize styrene and divinylbenzene. Thin sections were obtained from the gelled sample by ultramicrotoming. Shown in the bottom panel of Fig. 10.16 is a TEM image of nanofibers in a gelled sample. One consequence of the induced magnetization of the fibers is that they attracted one another and bundled in a magnetic field. Also clear from this image is that the fibers aligned along the magnetic field direction.

The bundling and alignment of the hybrid nanofibers in a magnetic field have important practical implications. For example, the controlled bundling of several nanofibers may form the basis of magnetic nanomechanical devices. For the construction of water-dispersible magnetic nanomechanical devices, the superparamagnetic nanofibers need to be water dispersible. We recently prepared water-dispersible polymer–Pd hybrid catalytic nanofibers from a tetrablock copolymer [52] and more recently polymer–Pd–Ni superparamagnetic nanofibers from a triblock copolymer [53]. The tetrablock that we used was PI-PtBA-P(CEMA-HEMA)-PGMA, where PGMA, being water soluble, denotes poly(glyceryl methacrylate) and P(CEMA-HEMA) denotes a random copolymer of CEMA and 2-hydroxyethyl methacrylate. The hydroxyl groups of the precursory PHEMA block was not fully cinnamated because P(CEMA-HEMA) facilitated the transportation of Pd^{2+} and Ni^{2+}.

We prepared the polymer–Pd hybrid nanofibers following the scheme depicted in Fig. 10.17 [52]. This involved first dispersing freshly-prepared PI-PtBA-P(CEMA-HEMA)-PGMA in water to yield cylindrical aggregates (A→B in Fig. 10.17). Such aggregates consisted of a PGMA corona and a PI core. Sandwiched between these two layers are a thin PtBA layer and a P(CEMA-HEMA) layer. Such cylindrical aggregates were then irradiated to cross-link the (CEMA-HEMA) layer (B→C). The PI core was degraded by ozonolysis (C→D). By controlling the ozonolysis time, we could control the degree of PI degradation. When not fully degraded, the residual double bonds of the PI fragments trapped inside the nanotubular core were able to sorb Pd(II), most probably via π-allyl complex formation, Scheme 10.3.

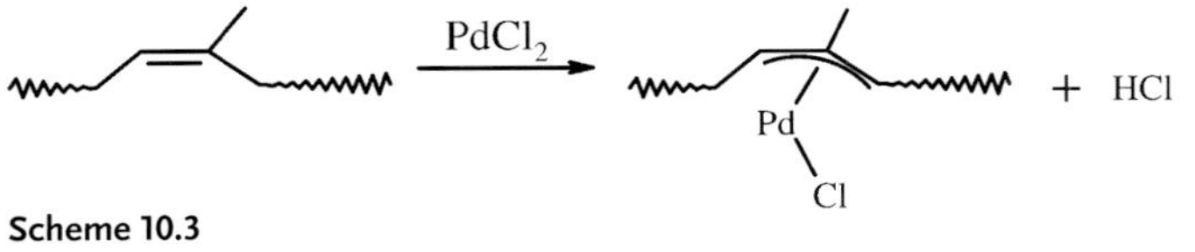

Scheme 10.3

The complexed Pd(II) was then reduced by $NaBH_4$ to Pd (D→E). The left panel of Fig. 10.18 is a TEM image for such nanotubes containing 4.0 wt-% reduced Pd nanoparticles. The Pd-loaded nanofibers were dispersible in water where many water-based electroless plating reactions occur. Thus, Pd could serve as a catalyst for the further electroless deposition of other metals. We, for example, loaded more Pd into the tubular core via electroless Pd plating onto the initially formed Pd nanoparticles to yield essentially continuous Pd nanowires (E→F, Fig. 10.17). The right-hand image in Fig. 10.18 shows a TEM image of such hybrid nanofibers after the incorporation of Pd to a total of 18.4 wt-%. In fact, we could tune the amount of Pd loaded into the nanotubes by adjusting the relative amounts of the Pd-loaded nanotubes and Pd^{2+} in a plating bath. As the Pd content increased,

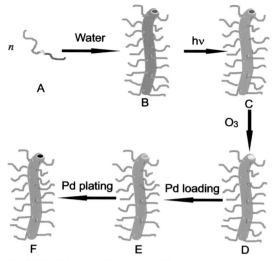

Fig. 10.17 Schematic illustration of the processes involved to produce water-dispersible polymer–Pd hybrid nanofibers.

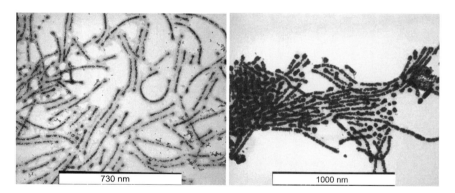

Fig. 10.18 TEM image of nanofibers containing 4 wt-% Pd (left); TEM image of nanofibers containing 18.4 wt-% Pd (right). The scale bars in the form of white boxes are 730 and 1000 nm long, respectively.

eventually the hybrid fibers could not be dispersed in water. In addition to Pd plating, we have also succeeded in the plating of Ni into the core of such nanotubes, as evidenced by our recent success in preparing water-dispersible triblock–Pd–Ni magnetic nanofibers [53].

The use of a tetrablock copolymer for the above project seems to be an over kill, as in the end the P*t*BA block was not used at all. This was, however, not by design. Our initial plan was to fully degrade the PI block and then to hydrolyze the P*t*BA block. We planned to introduce Pd(II) via its binding with the carboxyl groups of PAA. The binding between Pd(II) and the residual double bonds of PI was a surprise to us and should be useful in the future as a staining method for PI in the elucidation of complex segregation patterns of block copolymers.

10.4.2
End Functionalization

In the "supramolecular chemistry" of nanotubes and nanofibers, the end-functional groups should not be interpreted as the traditional carboxyl or amino groups, etc. Rather, they should be other nano "building blocks", including nanospheres and nanofibers or nanotubes of a different composition. We first end-functionalized PS-PCEMA-PAA nanotubes by attaching to them water-dispersible PAA-PCEMA nanospheres and spheres bearing surface carboxyl groups that were prepared from emulsion polymerization [54]. Figure 10.19 shows the steps involved in coupling the nanotubes and PCEMA-PAA nanospheres.

To ensure that the PAA core chains were exposed at the ends, we ultrasonicated the pristine nanotubes to shorten them. The carboxyl groups of the nanotubes were then reacted with the amino groups of a triblock PAES-PS-PAES, where PAES denotes poly[4-(2-aminoethyl)styrene], in the presence of catalyst 1-[3-(dimethylamino)propyl]-3-ethylcarbodiimide hydrochloride (EDCI) and co-catalyst 1-hydroxybenzotriazole (HBA). As PAES-PS-PAES was used in excess, it was grafted onto the nanotubes mainly via one end only. We then purified the sample. Nanotubes bearing terminal PAES chains were subsequently reacted with the carboxyl groups on the surfaces of the nanospheres again using EDCI and HBA as the catalysts. Coupling between the nanotubes and emulsion spheres containing surface carboxyl group was achieved in a similar manner.

Figure 10.20 shows the typical products obtained from coupling the PAES-PS-PAES-treated nanotubes with a batch of emulsion nanospheres bearing surface carboxyl groups. The product in Fig. 10.20a resulted from the coupling between one tube and one sphere. As the spherical "head" is water-dispersible and the tube

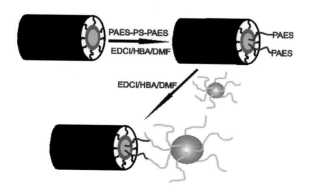

Fig. 10.19 End-functionalization of PS-PCEMA-PAA nanotubes by PAA-PCEMA nanospheres.

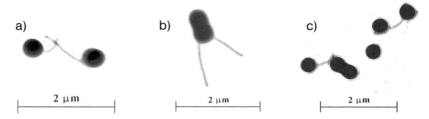

Fig. 10.20 TEM images of nanotube and emulsion nanosphere coupling products.

"tail" is hydrophobic, this structure may be viewed as a macroscopic counterpart of a surfactant molecule or a "super-surfactant". Figure 10.20b shows the attachment of two tubes to one sphere, which had fused with another sphere probably during TEM specimen preparation. "Dumbbell-shaped molecules" were formed from the attachment of one tube to two spheres at the opposite ends, as seen in Fig. 10.20c. The products depicted in Fig. 10.20a–c co-existed regardless of whether we changed the tube to microsphere mass ratio from 20/1 to 1/20. At the high tube to emulsion sphere mass ratio of 20/1, the super-surfactant and dumbbell-shaped species were the major products. At a mass ratio of 1/1 and 1/20, the dumbbell-shaped product dominated. Other than product control by adjusting the stoichiometry, an effective method to eliminate the dumbbell-shaped product was to use nanotubes labeled at only one end by PAES-PS-PAES. These tubes were obtained by using ultrasonication to break up nanotubes that contained end-grafted PAES-PS-PAES chains. For example, the ultrasonication of the PAES-PS-PAES-grafted nanotubes for 8 h reduced the L_w of a nanotube sample from 701 to 252 nm and L_n from 515 to 187 nm. The reaction between the shortened tubes and the nanospheres at a tube to sphere mass ratio of 1/20 yielded almost exclusively the supersurfactant

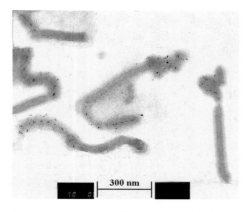

Fig. 10.21 TEM image of nanotube multiblocks.

structure with unreacted nanospheres. The content of the multi-armed structure increased as the nanotube to microsphere mass ratio increased.

More recently, we have used the same chemistry to couple PS-PCEMA-PAA nanotubes with PGMA-PCEMA-PAA nanotubes [55]. Figure 10.21 shows the nanotube multiblocks that we prepared. To facilitate the easy differentiation between the two types of nanotubes, we loaded Pd nanoparticles into the PGMA-PCEMA-PAA nanotubes. Using this method, we succeeded in preparing both nanotube di- and triblocks with structures similar to di- and triblock copolymers. The nanotube multiblocks should self-assemble in a manner similar to the block copolymers.

10.5
Concluding Remarks

Block copolymer nanofibers and nanotubes can now be readily prepared with high yields. Such nanofibers have interesting chemical and physical properties. More pressing challenges in this area of research are to find and to realize the commercial applications for these nanostructures. The latter will be greatly facilitated with more participation from industrial partners.

On the fundamental research side, the construction and study of superstructures prepared from the coupling of different nano- and micro-components are a very interesting and promising area of frontier research. The super-surfactants of Fig. 10.20a may, for example, self-assemble as do surfactant molecules to form supermicelles or artificial cells, which contain structural order to several length scales. The multi-armed structure of Fig. 10.20b and its analogues will be of particular interest if the nanotubes are replaced by PS-PCEMA-PAA/γ-Fe$_2$O$_3$ hybrid nanofibers. Figure 10.22 shows, for example, just such an interesting structure, which can be prepared by attaching three PS-PCEMA-PAA/γ-Fe$_2$O$_3$ hybrid nanofibers to one microsphere.

These fibers will attract one another in a magnetic field due to their magnetization and thus grab a nanoobject. Such a nanoobject can be moved around by focussing a laser beam on the microsphere that is trapped through an optical tweezing mechanism. The nanoobject can then be released at a desired spot by turning off the magnetic field. Thus, such a device can function as an "optical magnetic nanohand". Last but not the least, studies on the concentrated solution properties and nanofiber composite properties should be performed to see if there are any novel desirable applications for these materials.

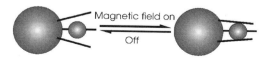

Fig. 10.22 Schematic illustration of the operation of an optical magnetic nanohand.

Acknowledgements

G. L. is very grateful for financial support from the Natural Sciences and Engineering Research Council of Canada, Defense Research and Development Canada, Canada Research Chairs Program, Canada Foundation for Innovations, Ontario Innovation Trust. G. L. would also like to thank Drs. Xiaohu Yan, Zhao Li, Sean Stewart, Jianfu Ding, and Lijie Qiao for carrying out the work reviewed in this chapter.

References

1 X. S. Wang, H. Wang, N. Coombs, M. A. Winnik, I. Manners, *J. Am. Chem. Soc.* **2005**, *127*, 8924–8925.
2 X. S. Wang, A. Arsenault, G. A. Ozin, M. A. Winnik, I. Manners, *J. Am. Chem. Soc.* **2003**, *125*, 12686–12687.
3 J. F. Ding, G. J. Liu, *Macromolecules* **1998**, *31*, 6554–6558.
4 M. F. Zhang, A. H. E. Muller, *J. Polym. Sci. A: Polym. Chem.* **2005**, *43*, 3461–3481.
5 K. Yu, L. F. Zhang, A. Eisenberg, *Langmuir* **1996**, *12*, 5980–5984.
6 J. Raez, I. Manners, M. A. Winnik, *J. Am. Chem. Soc.* **2002**, *124*, 10381–10395.
7 S. A. Jenekhe, X. L. Chen, *Science* **1999**, *283*, 372–375.
8 J. Grumelard, A. Taubert, W. Meier, *Chem. Commun.* **2004**, 1462–1463.
9 S. Stewart, G. Liu, *Angew. Chem., Int. Ed. Engl.* **2000**, *39*, 340–344.
10 X. H. Yan, G. J. Liu, F. T. Liu, B. Z. Tang, H. Peng, A. B. Pakhomov, C. Y. Wong, *Angew. Chem., Int. Ed. Engl.* **2001**, *40*, 3593–3596.
11 G. J. Liu, L. J. Qiao, A. Guo, *Macromolecules* **1996**, *29*, 5508–5510.
12 Y. Kim, P. Dalhaimer, D. A. Christian, D. E. Discher, *Nanotechnology* **2005**, *16*, S484–S491.
13 G. A. Silva, C. Czeisler, K. L. Niece, E. Beniash, D. A. Harrington, J. A. Kessler, S. I. Stupp, *Science* **2004**, *303*, 1352–1355.
14 S. I. Stupp, *MRS Bull.* **2005**, *30*, 546–553.
15 J. A. Massey, M. A. Winnik, I. Manners, V. Z. H. Chan, J. M. Ostermann, R. Enchelmaier, J. P. Spatz, M. Moller, *J. Am. Chem. Soc.* **2001**, *123*, 3147–3148.
16 C. B. W. Garcia, Y. M. Zhang, S. Mahajan, F. DiSalvo, U. Wiesner, *J. Am. Chem. Soc.* **2003**, *125*, 13310–13311.
17 T. Kowalewski, N. V. Tsarevsky, K. Matyjaszewski, *J. Am. Chem. Soc.* **2002**, *124*, 10632–10633.
18 C. B. Tang, A. Tracz, M. Kruk, R. Zhang, D. M. Smilgies, K. Matyjaszewski, T. Kowalewski, *J. Am. Chem. Soc.* **2005**, *127*, 6918–6919.
19 F. S. Bates, G. H. Fredrickson, *Phys. Today* **1999**, *52*, 32–38.
20 N. S. Cameron, M. K. Corbierre, A. Eisenberg, *Can. J. Chem.* **1999**, *77*, 1311–1326.
21 M. Templin, A. Franck, A. DuChesne, H. Leist, Y. M. Zhang, R. Ulrich, V. Schadler, U. Wiesner, *Science* **1997**, *278*, 1795–1798.
22 K. Ishizu, T. Ikemoto, A. Ichimura, *Polymer* **1999**, *40*, 3147–3151.
23 Y. F. Liu, V. Abetz, A. H. E. Muller, *Macromolecules* **2003**, *36*, 7894–7898.
24 X. Yan, G. Liu, H. Li, *Langmuir* **2004**, *20*, 4677–4683.
25 J. Tao, S. Stewart, G. J. Liu, M. L. Yang, *Macromolecules* **1997**, *30*, 2738–2745.
26 Y. Y. Won, H. T. Davis, F. S. Bates, *Science* **1999**, *283*, 960–963.
27 P. Dalhaimer, H. Bermudez, D. E. Discher, *J. Polym. Sci. B: Polym. Phys.* **2004**, *42*, 168–176.
28 Q. G. Ma, E. E. Remsen, C. G. Clark, T. Kowalewski, K. L. Wooley, *Proc. Natl. Acad. Sci. USA* **2002**, *99*, 5058–5063.
29 G. J. Liu, Z. Li, X. H. Yan, *Polymer* **2003**, *44*, 7721–7727.

30 X. H. Yan, G. J. Liu, *Langmuir* **2004**, *20*, 4677–4683.
31 G. J. Liu, J. Y. Zhou, *Macromolecules* **2003**, *36*, 5279–5284.
32 C. Price, *Pure Appl. Chem.* **1983**, *55*, 1563–1572.
33 U. Breiner, U. Krappe, V. Abetz, R. Stadler, *Macromol. Chem. Phys.* **1997**, *198*, 1051–1083.
34 G. J. Liu, X. H. Yan, X. P. Qiu, Z. Li, *Macromolecules* **2002**, *35*, 7742–7747.
35 G. J. Liu, X. H. Yan, S. Duncan, *Macromolecules* **2003**, *36*, 2049–2054.
36 W. R. Moore, *Progr. Polym. Sci.* **1969**, *1*, 3–43.
37 B. H. Zimm, D. M. Crothers, *Proc. Natl. Acad. Sci. USA* **1962**, *48*, 905.
38 H. Yamakawa, M. Fujii, *Macromolecules* **1974**, *7*, 128–135.
39 H. Yamakawa, T. Yoshizaki, *Macromolecules* **1980**, *13*, 633–643.
40 M. Bohdanecky, *Macromolecules* **1983**, *16*, 1483–1492.
41 P. Dalhaimer, F. S. Bates, D. E. Discher, *Macromolecules* **2003**, *36*, 6873–6877.
42 Y. Geng, F. Ahmed, N. Bhasin, D. E. Discher, *J. Phys. Chem. B* **2005**, *109*, 3772–3779.
43 Y. Y. Won, H. T. Davis, F. S. Bates, M. Agamalian, G. D. Wignall, *J. Phys. Chem. B* **2000**, *104*, 9054.
44 Y. Y. Won, K. Paso, H. T. Davis, F. S. Bates, *J. Phys. Chem. B* **2001**, *105*, 8302–8311.
45 G. J. Liu, J. F. Ding, L. J. Qiao, A. Guo, B. P. Dymov, J. T. Gleeson, T. Hashimoto, K. Saijo, *Chem. Eur. J.* **1999**, *5*, 2740–2749.
46 L. Onsager, *Ann. N.Y. Acad. Sci.* **1949**, *51*, 627–659.
47 P. J. Flory, *Proc. R. Soc. London, Ser. A: Math. Phys. Sci.* **1956**, *234*, 73–89.
48 R. J. Hunter, *Foundations of Colloid Science*, Vol. 1, Oxford University Press, Oxford, **1989**.
49 Y. N. Xia, P. D. Yang, Y. G. Sun, Y. Y. Wu, B. Mayers, B. Gates, Y. D. Yin, F. Kim, Y. Q. Yan, *Adv. Mater.* **2003**, *15*, 353–389.
50 C. M. Lieber, *MRS Bult.* **2003**, *28*, 486–491.
51 R. F. Ziolo, E. P. Giannelis, B. A. Weinstein, M. P. Ohoro, B. N. Ganguly, V. Mehrotra, M. W. Russell, D. R. Huffman, *Science* **1992**, *257*, 219–223.
52 Z. Li, G. J. Liu, *Langmuir* **2003**, *19*, 10480–10486.
53 X. H. Yan, G. J. Liu, M. Haeussler, B. Z. Tang, *Chem. Mater.* **2005**, *17*, 6053–6059.
54 G. Liu, X. Yan, Z. Li, J. Zhou, S. Duncan, *J. Am. Chem. Soc.* **2003**, *125*, 14039–14045.
55 X. Yan, G. Liu, Z. Li, *J. Am. Chem. Soc.* **2004**, *126*, 10059–10066.

11
Nanostructured Carbons from Block Copolymers

Michal Kruk, Chuanbing Tang, Bruno Dufour, Krzysztof Matyjaszewski, and Tomasz Kowalewski

11.1
Introduction

Over the last two decades, carbon nanostructures and nanomaterials have attracted a lot of attention. This interest was stimulated by the discovery of fullerenes [1] and carbon nanotubes [2–4], which are discrete carbon molecules and nanoobjects. These nanostructures are typically synthesized via techniques that involve deposition of carbon from a vapor phase. In addition to these emerging carbon nanostructures, there has been a long-standing interest in and significant demand for conventional high surface area carbon materials, such as activated carbons or carbon blacks. These conventional materials are synthesized from organic precursors of natural or synthetic origin, which are subjected to carbonization, that is, to a heat treatment at ≈800–1000 °C under an inert atmosphere to increase the carbon content by decreasing the content of other elements, such as hydrogen, oxygen and nitrogen. Some materials can be additionally activated using steam, carbon dioxide, zinc chloride and other reagents in order to generate readily accessible porosity. The precursors of natural origin typically contain biopolymers, such as cellulose, or are derived from biopolymers. To gain a better control over the formation of carbonaceous porous materials, one can substitute the precursors of natural origin with the synthetic ones, which are usually polymer-based. For instance, activated carbons can be derived from polymers, such as styrene–divinylbenzene random copolymers.

Most of the present uses of porous carbons are related to their high surface areas, which result predominantly from the presence of micropores (pores of diameter below 2 nm [5]). However, there is currently a significant interest in the synthesis of carbons with mesopores [6–11], that is, pores of diameter in the range from 2 to 50 nm [5]. Such carbons are promising as chromatographic packings [6], catalyst supports [12], components of electrochemical double-layer capacitors [8] and electrochemical batteries [13]. The mesoporous carbons are often prepared using mesoporous silicas as hard templates [6–8], which usually involves the prepara-

Block Copolymers in Nanoscience. Massimo Lazzari, Guojun Liu, Sébastien Lecommandoux
Copyright © 2006 WILEY-VCH Verlag GmbH & Co. KGaA, Weinheim
ISBN: 3-527-31309-5

tion of a template, followed by the infiltration of its pores with a carbon precursor, the carbonization of the precursor and the removal of the template. Although this approach provides a variety of opportunities in the fabrication of mesoporous carbons with tailored structures and some degree of tunability of the pore size, it involves multiple steps, including the use of HF or NaOH to dissolve the template. Moreover, it is difficult to obtain these mesoporous carbons in the form of films, which is a serious disadvantage in the development of numerous film-based devices. The synthesis of carbons from readily processable polymeric precursors would constitute a major simplification in the preparation of carbons with tailored pore structures and desired forms.

One of the proposed approaches to achieve this goal involves the use of blends of two immiscible polymers, one of which converts upon heating to carbon, and the other is volatilized, thus creating the pore voids [14, 15]. The feasibility of this approach has been demonstrated, although depending on the composition of the blend, open pores [14, 15] or closed pores [16] can be created. In principle, the use of polymer blends is expected to provide the means for the pore diameter control through the adjustment of the size of the phase-separated domains and for the pore volume control through the selection of an appropriate proportion of the components of the blend. However, such an approach is associated with numerous difficulties, such as: (a) phase separation may occur on the macroscopic rather than on microscopic scale; (b) the shrinkage during carbonization; (c) the possible changes in blend component domain sizes during the heat treatment, and (d) the propensity to the formation of closed pores.

In recent years we have developed an alternative (and potentially more robust) approach to the synthesis of well-defined carbon nanoobjects and nanoporous materials, which relies on the pyrolysis of block copolymers containing an immiscible carbon precursor and sacrificial blocks (Scheme 11.1) [17–22]. Polyacrylonitrile (PAN) has been the carbon precursor block of choice, because it is well known to be an excellent precursor for the synthesis of carbon fibers [23], and at the same time, its block copolymers can be readily synthesized [17, 18, 20, 24, 25] using controlled radical polymerization methods [26]. PAN block copolymers can serve as precursors for the synthesis of isolated carbon nanoobjects, in which case one uses either films of phase-separated block copolymers [17], or shell cross-linked micelles [18]. Phase-separated block copolymers can also be used for the synthesis of nanoporous materials [21]. In this case, porosity of the final carbon product can be enhanced through the introduction of auxiliary, thermally stable components, such as silica introduced at certain stages of the synthesis [20]. This chapter provides an overview of several different strategies that we have developed for the synthesis of well-defined carbon nanoobjects and nanoporous materials using block copolymers as precursors.

It should be noted that as far as the synthesis of inorganic nanostructures is concerned, block copolymers (BCs) can be used as precursors not only for carbon materials, but also for ceramic materials. In particular, films of polyisoprene–poly(pentamethyldisylylstyrene)–polyisoprene (asymmetric) triblock copolymers, which form double gyroid structures, were used to obtain silicon oxycarbide

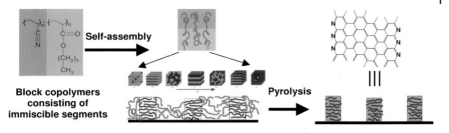

Scheme 11.1 Nanostructured carbon templated from PAN block copolymers.

ceramic nanoporous materials [27, 28]. The silicon-containing block served as a ceramic precursor, while the PI block was decomposed to create the porosity. Copolymers with poly(ferrocenyldimethylsilane) (or similar) blocks (denoted PFS) and with appropriate sacrificial (or auxiliary) blocks have been studied as precursors for ceramic nanoobjects and nanostructured films [29–33]. PFS can be oxidized, for instance via reactive ion etching [29], to decrease the carbon content considerably, with a concomitant increase in oxygen content and appreciable retention of silicon and iron. Under the same conditions, the sacrificial blocks, such as polyisoprene (PI) [29, 32, 33] or polystyrene [31], can essentially be eliminated fully. Thus far, the use of block copolymers with PFS blocks allowed for the preparation of ordered 2D-arrays of nanoscale protrusions [29, 31], nanorods or nanofibers [30, 32, 33], which can be oriented on the surface of the substrate [32.] These nanostructures are promising as resists for nanolithography [29, 30] and as magnetic nanopatterns [32].

11.2 Discussion

11.2.1 Well Defined PAN Polymers and Copolymers

Our studies were focused on the use of polyacrylonitrile (PAN) as a carbon source. PAN is well known as an excellent precursor for the carbon fiber synthesis [23, 34, 35]. An efficient conversion of PAN into carbon typically requires the so-called stabilization, which involves heating at 200–300 °C under an oxygen-containing atmosphere (e. g. air), prior to the carbonization, which is accomplished at $\approx$900 °C under an inert atmosphere (e. g. nitrogen). PAN is typically synthesized using conventional (uncontrolled) free radical polymerization methods. The disadvantage of this approach is that one cannot control the molecular weight, molecular weight distribution and architectures of the obtained polymer. To control the molecular weight and to achieve low polydispersity, PAN can be synthesized using anionic polymerization [36] and controlled radical polymerization, including atom transfer radical polymerization (ATRP) [37, 38] and nitroxide-mediated polymerization (NMP) [39]. More recently, controlled radical polymerization of AN has also been accomplished using reversible addition–fragmentation chain transfer (RAFT) [25], organotellurium-mediated radical polymerization (TERP) [40] and organostibine-mediated radical polymerization [41]. In addition to the controlled synthesis of PAN homopolymers, the formation of block copolymers and other polymer architectures with PAN blocks was achieved. In particular, as shown in Scheme 11.2, poly(n-butyl acrylate)–polyacrylonitrile (PBA-b-PAN) copolymers were synthesized via ATRP, NMP and RAFT [17, 24, 25] and poly(t-butyl acrylate)–polyacrylonitrile (PtBA-b-PAN) copolymer was synthesized via ATRP and subsequently transformed into poly(acrylic acid)–polyacrylonitrile (PAA-PAN) through hydrolysis of t-butyl ester groups [18]. Moreover, poly(ethylene oxide)–polyacrylonitrile copolymers were synthesized by chain-extending commercially available poly(ethylene oxide) blocks (synthesized via anionic polymerization) with PAN blocks formed via ATRP [20]. In addition, 3-, 4- and 6-arm star PAN polymers were synthesized via ATRP [42].

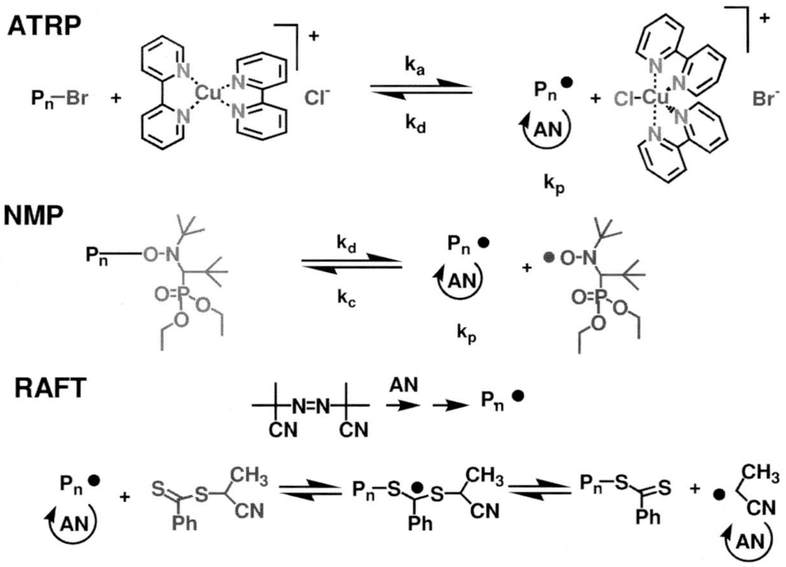

Scheme 11.2 Principles of PAN block copolymer synthesis by controlled radical polymerization (AN = acrylonitrile).

11.2.2
Carbon Films from Phase-separated Block Copolymers

There are a number of ways in which PAN block copolymers can be transformed into nanostructured carbons. A common theme in these methods is the need for the stabilization of PAN via heat treatment at 200–300 °C under air, which in-

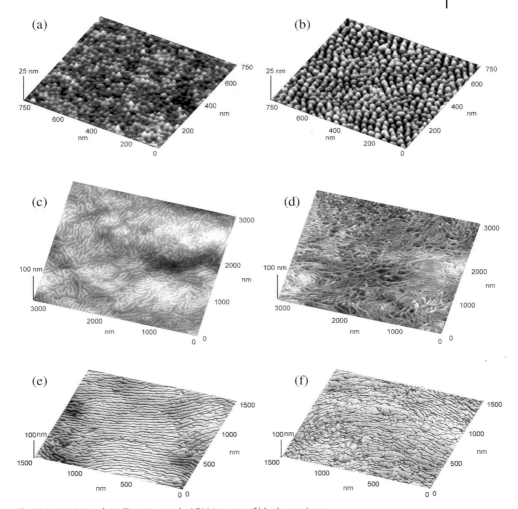

Fig. 11.1 (a), (c) and (e) Tapping mode AFM images of block copolymer films with different morphology of PAN domains: (a) spherical (after 230 °C stabilization); (c) cylindrical; and (e) lamellar.
(b), (d) and (f) Tapping mode AFM images of the corresponding carbon films after pyrolysis: (b) dots; (d) filaments; (f) sheets (reproduced from [17], copyright (2002) with permission from The American Chemical Society, and from [22], copyright (2003) with permission from Springer-Verlag).

creases the carbon yield and facilitates the conversion into carbon, with retention of the original morphology of the PAN domains [17]. Otherwise, the methods are based on a variety of principles. The first approach to be explored, which is perhaps the most elegant and powerful, was based on the well-known phenomenon of nanoscale phase-separation of block copolymers. As shown in Fig. 11.1, the phase separation leads to the formation of PAN domains of distinct morphologies, such as spheres, cylinders, connected cylinders and so forth, in the continuous matrix of domains formed by the other component of the copolymer. Subsequently, PAN is stabilized and the domains composed of the second component of the copolymer are eliminated, for instance via thermal treatment, which also leads to the carbonization of the PAN block. Because of the nature of this process, these thermally unstable blocks of the copolymers will be referred to as "sacrificial blocks".

It can be expected that the ability to convert PAN into carbon with the retention of the morphology and distance between the domains in the structure will depend on the connectivity between the PAN domains. For instance, the structure of isolated PAN spheres or cylinders will collapse after the removal of the sacrificial phase. The same problem is encountered in the synthesis of ordered mesoporous carbons using silica templates [7], wherein the ability to obtain a faithful inverse replica of the template depends on the connectivity between the pores of the template, which are filled with the carbon precursor [43, 44]. In the synthesis of carbon nanostructures from phase-separated PAN block copolymers, it is easy to form thin films of the copolymer precursor, and in such a case, the underlying substrate may support the carbon nanostructure obtained, thus mitigating the structural rearrangement upon removal of the sacrificial phase. In particular, spherical PAN domains can transform into (hemi)spherical protrusions on the surface [17], and edge-on lamellas can retain their periodic distance [19]. However, lifting of such a carbon nanostructure off the surface can modify periodicity of the structure [19].

The use of block copolymers as carbon precursors offers significant benefits, as these precursors are soluble in organic solvents and can readily be obtained in the form of thin films [17, 19], which allows one to fabricate nanostructured carbon films. Moreover, the domains of the copolymer can be aligned [19], for instance using a directional casting method referred to as zone-casting [19, 45], and subsequently converted into an oriented carbon film [19]. The zone-casting involves a deposition of the polymer solution from a syringe maintained at a controlled temperature through a nozzle onto a moving substrate, the temperature of which is also controlled (Fig. 11.2a). The supply of the solution from the syringe and the movement of the support are controlled independently using two motors (stepper motors or dc servo motors), which allows films of thickness from several tens of nanometers to several micrometers to be cast over areas of several or several tens of square centimeters. We have used the zone-casting method to cast oriented thin films of poly(n-butyl acrylate)–polyacrylonitrile (PBA-b-PAN) copolymer of formula BA_{240}-AN_{124} using N,N-dimethylformamide (DMF) as a solvent [19]. The films exhibited the structure of edge-on lamellae with a lamellar spacing of ≈ 36 nm, as determined using grazing incidence small-angle X-ray scattering (GISAXS) (Fig. 11.2c) and atomic force microscopy (AFM) (Fig. 11.2b). The film was oriented over an area of 3 cm by 5 cm, as seen from AFM and GISAXS. Interestingly, the lamellae were oriented perpendicular to the casting direction, which indicates that the formation of the domains was induced along the solvent evaporation front (phase separation zone) [19]. The lamellar BA_{240}-AN_{124} film was heated under air at 280 °C to stabilize the PAN domains, and then heated again under nitrogen at 800 °C,

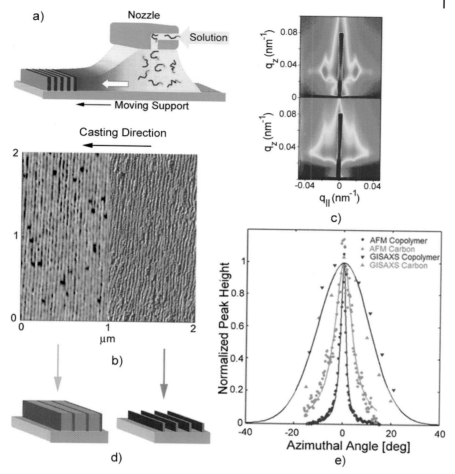

Fig. 11.2 Long-range order in thin films of PBA-*b*-PAN block copolymers prepared by zone-casting (a) and in nanostructured carbons prepared by subsequent pyrolysis: (b) AFM phase images (left: copolymer; right: carbon); (c) GISAXS patterns acquired at 90 degrees to the casting direction (top: copolymer; bottom: carbon); (d) schematic illustration of lamellar order (left: copolymer; right: carbon); (e) azimuthal profiles of maxima in 2D-Fourier transforms of AFM images and maxima in GISAXS patterns corresponding to the lamellar period (reprinted from [19], copyright (2005) with permission from The American Chemical Society).

to convert PAN into carbon and to volatilize the PBA domains. Thus, a lamellar carbon film was obtained, which exhibited essentially the same spacing as that observed in the copolymer film, whereas the height of the carbon domains was smaller than that of the PAN domains (Fig. 11.2d). The orientation of the domains was preserved, although it was somewhat distorted in comparison with the block copolymer precursor film. Hence, the removal of the PBA phase and the loss of volume related to the dehydrogenation and denitrogenation, with the concomitant increase in density upon the conversion of PAN into carbon, did not lead to any appreciable structural changes in the supported film. It should be noted that the lifting of this carbon film from the support resulted in a significant decrease in the spacing between the carbon sheets (from ≈37 nm to 12–16 nm), which suggests that the carbon sheets collapsed on one another during this procedure. However, the use of PBA-b-PAN copolymers with larger PAN content gives access to continuous morphologies of PAN domains, in which case the derived carbon films are expected to undergo lifting off from the substrate without the structural rearrangement.

As mentioned above, the zone-casting allowed us to fabricate PBA-b-PAN films with a long-range lamellar order. In contrast, a more commonly used drop-casting technique afforded PBA-b-PAN films with short-range order [46]. It is clear that the appropriate precursor processing is important in the synthesis of ordered, nanostructured carbon films.

It should be noted here that the use of tapping-mode atomic force microscopy (TMAFM) was crucial in the characterization of both PAN-b-PBA copolymer film precursors and the resulting carbon films. In fact, TMAFM is particularly suitable for the investigation of the nanoscale structure of phase-separated block copolymers. TMAFM allows one to obtain not only the height image, but also a phase image, which provides the map of mechanical differences between domains. The observed phase shift is related to the energy dissipated through the tip–sample interactions [47, 48], and thus reflects viscoelastic properties of the surface. Phase-separated domains of block copolymers often differ significantly in their viscoelastic or mechanical properties, which creates the contrast between the domains, thereby allowing one to map out their boundaries using TMAFM, even in the absence of any height differences. To appreciate the power of this approach further, one should keep in mind that it could be difficult to study the nanostructures of the phase-separated block copolymers by other techniques. For instance, in the case of TEM imaging, the contrast between the domains of particular blocks tends to be weak, and staining of one of the blocks is often needed to improve the contrast. Likewise, in SAXS or XRD measurements, the intensity of the patterns could be weak due to the insufficient contrast. Therefore, TMAFM offers a unique opportunity for the characterization of phase-separated block copolymers.

The conventional stabilization and carbonization of PAN typically takes minutes or hours. However, in the case of thin films of PAN-b-PBA-b-PAN triblock copolymers, the time of preparation to carbon films was greatly reduced by heating the copolymer films in a methane–oxygen flame, in which case the overall pyrolysis time is in the order of 10 s [17]. More specifically, the sample was gradually brought closer to the flame, maintained for several seconds in the ≈200 °C zone of the flame and then gradually moved to the zone with a temperature above 500 °C. Without the stabilization for several seconds in the ≈200 °C zone, the nanostructure was poorly preserved. The above simple and convenient flame pyrolysis method is potentially highly useful for the fast and cost-efficient fabrication of thin carbon films.

The block copolymers can be used not only as carbon precursors, but also as templates that ensure the correct arrangement of the polymeric carbon precursor. In his review on the synthesis of porous carbons published in 2000 [15], Kyotani outlined an ambitious goal of the supramolecular-templated synthesis of ordered mesoporous carbons, along the same lines as in the synthesis of inorganic mesoporous materials templated by surfactant micelles [57, 58]. The feasibility of this approach has been demonstrated recently. Firstly, Liang et al. [59] showed that a thin film of a diblock copolymer can be used to template the formation of resorcinol-formaldehyde (RF) resin, which can subsequently be converted into a carbon film with a 2D-hexagonal structure. More specifically, a complex of resorcinol with polystyrene–poly(4-vinyl pyridine) (PS-P4VP) copolymer was cast as a thin film, which exhibited a 2D-hexagonal arrangement of cylindrical domains of PS aligned perpendicular to the substrate and a continuous phase of P4VP with complexed resorcinol. The film thus obtained was exposed to formaldehyde gas, which led to the formation of the RF resin within the P4VP domains. The heat treatment of this film resulted in the carbonization of the RF resin and the decomposition of the PS-P4VP copolymer template, thus forming a carbon film with a 2D-hexagonal arrangement of uniform cylindrical pores aligned perpendicular to the substrate. More recently, Tanaka et al. [60] demonstrated that one can use a commercially available poly(ethylene oxide)–poly(propylene oxide)–poly(ethylene oxide) ($EO_m PO_n EO_m$) copolymer to structure the RF resin, which is subsequently converted into ordered mesoporous carbon via the heat treatment. Further research efforts in this direction are expected.

11.3
Conclusion

The synthesis of nanostructured carbons from phase-separated block copolymers is a new and very promising direction of research. Using copolymers with polyacrylonitrile blocks, the fabrication of carbon films, including those with nanoscale ordering and large-scale orientation, was successfully accomplished. The orientation and long-range ordering was achieved using a directional casting method, whereas short-range ordering was commonly observed even in the case of drop-cast films. The carbon film structures realized so far are arrays of isolated nanoparticles and lines, in addition to continuous nanostructured frameworks. All of these morphologies can be derived from phase-separated block copolymers, whereas isolated carbon nanoobjects can also be fabricated using shell cross-linked micelles, which are water-soluble carbon precursors.

In addition to the successful fabrication of thin carbon films, bulk porous carbons with fairly high specific surface areas (up to $\approx 450 \text{ m}^2 \text{ g}^{-1}$) were obtained from phase-separated block copolymers, pointing to the usefulness of our approach in applications requiring self-standing or thick carbon films free of binders. The use of silica as an auxiliary component of the sacrificial phase in the phase-separated block copolymer allows one to synthesize nanoporous carbon powders with high specific surface areas (≈ 800 m^2 g^{-1}) and very large pore volume (up to 2.8 cm^3 g^{-1}). Some of these high surface area carbon nanostructures are short-range ordered.

In addition to these advances in the synthesis of carbon nanostructures using copolymers with PAN blocks, it was possible to synthesize high surface area ordered nanoporous carbon from a silica–PEO-b-PPO-b-PEO composite in the presence of sulfuric acid. Also, the block copolymer directed synthesis of ordered nanoporous carbons from homopolymer precursors was demonstrated.

On the basis of all of these recent results, the synthesis of carbon nanostructures from block copolymers promises to be an exciting area for future developments, especially as it is capable of affording thin-film nanostructured carbons with controlled morphologies and arrays of isolated, well defined carbon nanoobjects.

Acknowledgments

The National Science Foundation (T. K. DMR-0304508, DMR-0210247, and K. M. DMR-0090409) and the CRP Consortium at Carnegie Mellon University are acknowledged for funding.

References

1 KROTO, H. W., HEATH, J. R., OBRIEN, S. C., CURL, R. F., SMALLEY, R. E. *Nature (London)* **1985**, *318*, 162–163.
2 IIJIMA, S. *Nature (London)* **1991**, *354*, 56–58.
3 ANDREWS, R., JACQUES, D., QIAN, D., RANTELL, T. *Accounts Chem. Res.* **2002**, *35*, 1008–1017.
4 DAI, H. *Accounts Chem. Res.* **2002**, *35*, 1035–1044.
5 SING, K. S. W., EVERETT, D. H., HAUL, R. A. W., MOSCOU, L., PIEROTTI, R. A., ROUQUEROL, J., SIEMIENIEWSKA, T. *Pure Appl. Chem.* **1985**, *57*, 603–619.
6 KNOX, J. H., KAUR, B., MILLWARD, G. R. *J. Chromatogr.* **1986**, *352*, 3–25.
7 RYOO, R., JOO, S. H., JUN, S. *J. Phys. Chem. B* **1999**, *103*, 7743–7746.
8 LEE, J., YOON, S., HYEON, T., OH, S. M., KIM, K. B. *Chem. Commun.* **1999**, 2177–2178.
9 RYOO, R., JOO, S. H., KRUK, M., JARONIEC, M. *Adv. Mater.* **2001**, *13*, 677–681.
10 LEE, J., HAN, S., HYEON, T. *J. Mater. Chem.* **2004**, *14*, 478–486.
11 YANG, H. F., ZHAO, D. Y. *J. Mater. Chem.* **2005**, *15*, 1217–1231.
12 JOO, S. H., CHOI, S. J., OH, I., KWAK, J., LIU, Z., TERASAKI, O., RYOO, R. *Nature (London)* **2001**, *412*, 169–172.
13 LONG, J. W., DUNN, B., ROLISON, D. R., WHITE, H. S. *Chem. Rev.* **2004**, *104*, 4463–4492.
14 OZAKI, J., ENDO, N., OHIZUMI, W., IGARASHI, K., NAKAHARA, M., OYA, A., YOSHIDA, S., IIZUKA, T. *Carbon* **1997**, *35*, 1031–1033.
15 KYOTANI, T. *Carbon* **2000**, *38*, 269–286.
16 FUKUYAMA, K., KASAHARA, Y., KASAHARA, N., OYA, A., NISHIKAWA, K. *Carbon* **2001**, *39*, 287–290.
17 KOWALEWSKI, T., TSAREVSKY, N. V., MATYJASZEWSKI, K. *J. Am. Chem. Soc.* **2002**, *124*, 10632–10633.
18 TANG, C., QI, K., WOOLEY, K. L., MATYJASZEWSKI, K., KOWALEWSKI, T. *Angew. Chem., Int. Ed. Engl.* **2004**, *43*, 2783–2787.
19 TANG, C. B., TRACZ, A., KRUK, M., ZHANG, R., SMILGIES, D. M., MATYJASZEWSKI, K., KOWALEWSKI, T. *J. Am. Chem. Soc.* **2005**, *127*, 6918–6919.

20 Kruk, M., Dufour, B., Celer, E. B., Kowalewski, T., Jaroniec, M., Matyjaszewski, K. *Chem. Mater.* **2006**, *18*, 1417–1424.

21 Tang, C., Kruk, M., Celer, E. B., Jaroniec, M., Matyjaszewski, K., Kowalewski, T. *J. Am. Chem. Soc.*, to be submitted.

22 Kowalewski, T., McCullough, R. D., Matyjaszewski, K. *Eur. Phys. J. E: Soft Matt.* **2003**, *10*, 5–16.

23 Donnet, J.-B., Wang, T. K., Rebouillat, S., Peng, J. C. M. (eds.) *Carbon Fibers*, 3rd edn, Marcel Dekker, New York, **1998**.

24 Tang, C., Kowalewski, T., Matyjaszewski, K. *Macromolecules* **2003**, *36*, 1465–1473.

25 Tang, C., Kowalewski, T., Matyjaszewski, K. *Macromolecules* **2003**, *36*, 8587–8589.

26 Matyjaszewski, K., Xia, J. *Chem. Rev.* **2001**, *101*, 2921–2990.

27 Avgeropoulos, A., Chan, V. Z. H., Lee, V. Y., Ngo, D., Miller, R. D., Hadjichristidis, N., Thomas, E. L. *Chem. Mater.* **1998**, *10*, 2109–2115.

28 Chan, V. Z. H., Hoffman, J., Lee, V. Y., Latrou, H., Avgeropoulos, A., Hadjichristidis, N., Miller, R. D. *Science* **1999**, *286*, 1716–1719.

29 Lammertink, R. G. H., Hempenius, M. A., van den Enk, J. E., Chan, V. Z. H., Thomas, E. L., Vancso, G. J. *Adv. Mater.* **2000**, *12*, 98–103.

30 Massey, J. A., Winnik, M. A., Manners, I., Chan, V. Z. H., Ostermann, J. M., Enchelmaier, R., Spatz, J. P., Moeller, M. *J. Am. Chem. Soc.* **2001**, *123*, 3147–3148.

31 Temple, K., Kulbaba, K., Power-Billard, K. N., Manners, I., Leach, K. A., Xu, T., Russell, T. P., Hawker, C. J. *Adv. Mater.* **2003**, *15*, 297–300.

32 Cao, L., Massey, J. A., Winnik, M. A., Manners, I., Riethmuller, S., Banhart, F., Spatz, J. P., Moller, M. *Adv. Funct. Mater.* **2003**, *13*, 271–276.

33 Wang, X. S., Arsenault, A., Ozin, G. A., Winnik, M. A., Manners, I. *J. Am. Chem. Soc.* **2003**, *125*, 12686–12687.

34 Chung, D. D. L. *Carbon Fiber Composites*, Butterworth-Heinemann, Newton, **1994**.

35 Peebles, L. H. *Carbon Fibers: Formation, Structure, and Properties*, CRC Press, Boca Raton, **1995**.

36 Sivaram, S., Dhal, P. K., Kashikar, S. P., Khisti, R. S., Shinde, B. M., Baskaran, D. *Macromolecules* **1991**, *24*, 1697–1698.

37 Matyjaszewski, K., Jo, S. M., Paik, H.-J., Gaynor, S. G. *Macromolecules* **1997**, *30*, 6398–6400.

38 Matyjaszewski, K., Jo, S. M., Paik, H.-J., Shipp, D. A. *Macromolecules* **1999**, *32*, 6431–6438.

39 Benoit, D., Chaplinski, V., Braslau, R., Hawker, C. J. *J. Am. Chem. Soc.*, *121*, 3904–3920.

40 Yamago, S., Iida, K., Yoshida, J.-I. *J. Am. Chem. Soc.* **2002**, *124*, 13666–13667.

41 Yamago, S., Ray, B., Iida, K., Yoshida, J., Tada, T., Yoshizawa, K., Kwak, Y., Goto, A., Fukuda, T. *J. Am. Chem. Soc.* **2004**, *126*, 13908–13909.

42 Pitto, V., Voit, B. I., Loontjens, T. J. A., van Benthem, R. A. T. M. *Macromol. Chem. Phys.* **2004**, *205*, 2346–2355.

43 Jun, S., Joo, S. H., Ryoo, R., Kruk, M., Jaroniec, M., Liu, Z., Ohsuna, T., Terasaki, O. *J. Am. Chem. Soc.* **2000**, *122*, 10712–10713.

44 Kruk, M., Jaroniec, M., Ryoo, R., Joo, S. H. *J. Phys. Chem. B* **2000**, *104*, 7960–7968.

45 Burda, L., Tracz, A., Pakula, T., Ulanski, J., Kryszewski, M. *J. Phys. D — Appl. Phys.* **1983**, *16*, 1737–1740.

46 Tang, C., Tracz, A., Kruk, M., Matyjaszewski, K., Kowalewski, T. *Polymer Preprints (Am. Chem. Soc., Div. Polym. Chem.)* **2005**, *46*, 424–425.

47 Cleveland, J. P., Anczykowski, B., Schmid, A. E., Elings, V. B. *Appl. Phys. Lett.* **1998**, *72*, 2613–2615.

48 Tamayo, J., Garcia, R. *Appl. Phys. Lett.* **1998**, *73*, 2926–2928.

49 Thurmond, K. B., Kowalewski, T., Wooley, K. L. *J. Am. Chem. Soc.* **1996**, *118*, 7239–7240.

50 Ma, Q. G., Remsen, E. E., Kowalewski, T., Wooley, K. L. *J. Am. Chem. Soc.* **2001**, *123*, 4627–4628.

51 Bagshaw, S. A., Prouzet, E., Pinnavaia, T. J. *Science* **1995**, *269*, 1242–1244.

52 Kruk, M., Jaroniec, M., Sayari, A. *Langmuir* **1997**, *13*, 6267–6273.

53 Tang, C., Kruk, M., Celer, E. B., Laurent, A., Jaroniec, M., Matyjaszewski, K., Kowalewski, T. *Preprints of Symposia – Am. Chem. Soc., Div. Fuel Chem.* **2005**, *50*, 118–120.

54 Sadezky, A., Muckenhuber, H., Grothe, H., Niessner, R., Poschl, U. *Carbon* **2005**, *43*, 1731–1742.

55 Wang, Y., Alsmeyer, D. C., McCreery, R. L. *Chem. Mater.* **1990**, *2*, 557–563.

56 Kim, J., Lee, J., Hyeon, T. *Carbon* **2004**, *42*, 2711–2719.

57 Kresge, C. T., Leonowicz, M. E., Roth, W. J., Vartuli, J. C., Beck, J. S. *Nature (London)* **1992**, *359*, 710–712.

58 Zhao, D., Feng, J., Huo, Q., Melosh, N., Frederickson, G. H., Chmelka, B. F., Stucky, G. D. *Science* **1998**, *279*, 548–552.

59 Liang, C., Hong, K., Guiochon, G. A., Mays, J. W., Dai, S. *Angew. Chem., Int. Ed. Engl.* **2004**, *43*, 5785–5789.

60 Tanaka, S., Nishiyama, N., Egashira, Y., Ueyama, K. *Chem. Commun.* **2005**, 2125–2127.

12
Block Copolymers at Interfaces

Mark Geoghegan and Richard A. L. Jones

12.1
Introduction

Interfaces perturb immiscible block copolymers and mixtures of homopolymers by breaking the symmetry of ordering processes. Examples include the stratifying effect of surface directed spinodal decomposition in polymer blends, or the imposition of directionality on block copolymer phase behavior. The directionality provided by the interface in the case of block copolymer ordering can be fairly dramatic. The surface can provide direction to a lamellar block copolymer phase some distance into the bulk of a mixture, before the effect of defects in the structure disrupt the direction of the ordering. The reason for the directionality of ordering is because the two different components of a blend have different surface energies. In a thin film, the block copolymer structure is perturbed by the need for thicknesses that are commensurate with the ordering of the copolymer. For a symmetric block copolymer with lamellar spacing L, a thin film would need to have a thickness that is an integer multiple of $2L$ if the surface and substrate favor the same block. However, if different blocks are favored, the preferred film thickness is $2(n + 0.5)L$, where n is an integer. If these conditions are not satisfied there will be regions of varying film thickness, with the different regions satisfying the relevant condition [1]. Various stable thin films are shown in Fig. 12.1. Even if the order–disorder transition (ODT) is crossed into the one-phase region, there will still be some layering very close to the substrate, although this will decay with a correlation length linked to the distance of the one-phase blend from the ODT [2].

Of course, not all block copolymer films form lamellar structures in the ordered phase. Lateral structures may be useful as they can provide the basis for making templates on very small length scales. For example, different plasma etch rates of various copolymer blocks have been demonstrated to be useful for creating negative and positive photoresists [3]. The use of block copolymers as lithographic templates would be particularly appealing, were it not for defects inherent in self-assembled block copolymer films, which limit the importance of such lithographic methods to systems where only short-range order is important. Another important weakness

---

Block Copolymers in Nanoscience. Massimo Lazzari, Guojun Liu, Sébastien Lecommandoux
Copyright © 2006 WILEY-VCH Verlag GmbH & Co. KGaA, Weinheim
ISBN: 3-527-31309-5

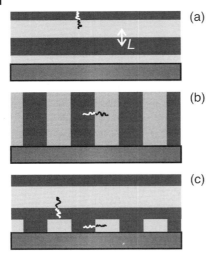

Fig. 12.1 Various thin film morphologies of block copolymers are shown. In (a) the surface favors one block, whilst the substrate the other. The situation where the block copolymer is neutral to either surface (b) is very unusual. It is possible to create a neutral substrate for the copolymer, and it may tolerate high energy interfaces in order to satisfy the likelihood that one of the blocks favors the air interface (c).

of templating with block copolymers is that the final structures are generally very ordered, and generating surfaces with irregular structures is difficult. Experiments to address these problems will be described in this chapter.

Surfaces can invoke ordering in free block copolymer chains, but tethered chains are particularly interesting. Grafted chains allow control of materials against aggregation (colloidal stabilization) and can also compatibilize inorganic with organic materials. The behavior of grafted chains in response to environmental effects is particularly important in nanoscience as it allows switchability; the reversible tailoring of surface properties is important in molecular actuators. Temperature, pH, electric fields and solvent environment are examples of external stimuli that can be useful here.

We will discuss block copolymer films and grafted block copolymer films in turn, highlighting their properties, with particular relevance to nanoscience. We end the chapter with a discussion of how (graft) copolymers can be used in the reconstruction of a polymer film surface. Although this chapter does not specifically concern ordering of block copolymers, the perturbing effect of surfaces and interfaces does provide a means to order films. Chapter 9 of this volume provides a much more comprehensive treatment of ordering processes and complements some of the material contained within the present chapter.

12.2
Block Copolymer Films

Block copolymer films can be easily created by spin casting the polymer from solution. The resultant film is rarely in equilibrium, but the polymer may approach equilibrium through heat treatment, or by using solvent vapor to increase mobility in the film. We mentioned in the Introduction that films will order with the block with the lowest surface energy segregating to the surface, and the block that prefers the substrate will also segregate there. Whilst various microscopic techniques are capable of looking at the lateral structure, the observation of lamellar structure in polymer films is difficult without sectioning, but can be achieved using neutron reflectometry. We show in Fig. 12.2 examples of block copolymer structures obtained using both transmission electron microscopy (TEM) of sectioned samples

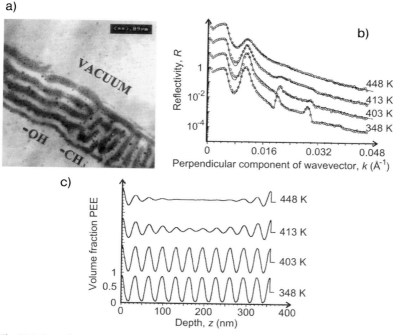

Fig. 12.2 Data illustrating lamellar structure in block copolymer films. (a) TEM image of a polystyrene-*block*-poly(2-vinyl pyridine) showing a lamellar morphology when covering an OH-terminated self-assembled monolayer [4]. The darker regions correspond to the poly(2-vinyl pyridine) block, which was stained with iodine to enhance contrast. Neutron reflectometry data (b) and volume fraction-depth profiles (c) are shown for a block copolymer of poly(ethyl ethylene) and poly(ethylene propylene) [5]. The ODT for the block copolymer described in (b) and (c) were measured as 398 K (data shown are used with permission and are copyright (1997) of The American Chemical Society and (1992) of The American Institute of Physics).

[4] and neutron reflectometry [5]. The reflectometry data are a good illustration of the increase in ordering as the ODT is reached. Surface segregation of the blocks is achieved in the one-phase region with complete ordering reached at around the ODT.

Consideration of lamellar block copolymers on surfaces is relatively straightforward, but can be complicated by the structure present in the surface. The TEM cross-sectional image in Fig. 12.2a is an example of a block copolymer on a heterogeneous surface.

12.3
Block Copolymers on Heterogeneous Surfaces

Lateral patterning changes the way block copolymers behave on surfaces. It is clear that patterning surfaces so that different blocks favor different parts of the surface will have an obvious effect on the final morphology. However, in the image shown in Fig. 12.2a, the methyl- and hydroxy-terminated self-assembled monolayers would both favor the polystyrene block (which is not visible in the image because there is no contrast between it and the embedding resin used for the TEM image). The perpendicular alignment of domains over the methyl-terminated monolayer indicates the complicated effects of subtle changes in surface energy. In this case, the different orientation was attributed to the effect of the edges.

A simpler experiment is one in which the different parts of the surface favor different blocks. Such an experiment has been performed with polystyrene-*block*-poly(methyl methacrylate) (PS-*b*-PMMA) on silicon substrates with a regular pattern of gold and silicon [6]. In this case, the commensurability between the blocks and the surface pattern was the dominant controlling factor. If there were only a slight mismatch ($\approx 10\%$) between the size of the blocks and the surface structure, the copolymers would orient perpendicularly to the substrate. Otherwise the polystyrene adsorbs onto the gold and the PMMA onto the silicon. In fact, if the substrate is very carefully prepared by using an appropriate lithographic technique, block copolymer structures may be formed by epitaxial means [7]. Lithography presents great opportunities for enhancing the structures of surfaces patterned with block copolymers, for example, the possibility of structuring block copolymer films around corners [8]. The reader is invited to refer to Chapter 9 of this volume where these examples, and many more, are discussed in detail.

Heterogeneous surfaces need not simply mean chemical patterning. Topographical effects are also important, and in particular it is worth noting that the roughness of the substrate can play a role in the orientation of the copolymer. For example, thin films of PS-*b*-PMMA were spin cast onto polyimide and indium tin oxide (ITO) substrates and annealed above the glass transition temperature of both blocks. These substrates are particularly important in many applications; ITO is a synthetic anode of great utility in optoelectronic applications due to its transparency in the visible part of the electromagnetic spectrum, whilst polyimides can be used as flexible substrates in applications where exposure to extremes of temperatures is

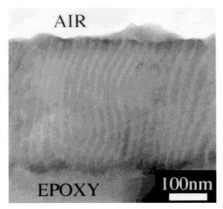

Fig. 12.3 Block copolymer of polystyrene and PMMA after being cast onto ITO and subsequently annealed. For TEM sectioning the film was transferred to an epoxy substrate [9] (image reproduced with permission and is copyright (2002) of The American Chemical Society).

a hazard. If the substrate roughness is above a critical value, the copolymers orient perpendicularly to the surface [9, 10]. In Fig. 12.3 a TEM image of PS-*b*-PMMA is shown aligning perpendicularly after being cast onto an ITO substrate. The transition from lamellar to perpendicular orientation arises because, on a rough surface, perpendicular orientations avoid the chain deformations that are necessary for a lamellar structure [10]. The surface energy cost of not having a lamellar structure is paid for by the chains not having to deform in trying to align with a non-planar surface. We note that it is also possible to obtain perpendicular alignment on flat substrates by, for example, coating the substrate with a random copolymer of the same components as the block copolymer and in the same ratio [11]. This will mean that the surface is neutral and no preferential adsorption can take place. However, the easiest method to align copolymers is to use an external field, a subject that we shall address next.

12.4
Environmental Control of Block Copolymer Films

The means to order copolymers has been described above in terms of the preparation of the appropriate substrates. A much easier method involves the use of external fields as a switch. As an example, it has been shown that PS-*b*-PMMA films can be switched from a lamellar structure to a perpendicular structure by the use of an electric field [12]. An electric field across the film induces dipoles in the polymers due to their polarizability, which in turn causes the copolymers to align as cylinders parallel to the electric field. To achieve this it is not necessary to use large voltages. This is because the films are thin, and a few volts across a sub-micron distance will result in fields in the order of 10^7 V m^{-1}, which is easily sufficient to orient copolymers. In this particular work, the PMMA block was degraded and the polystyrene block fixed by exposure to UV radiation. Electrodeposition was used to deposit cobalt in the spaces previously occupied by PMMA cylinders. In this fashion, a magnetic nanowire array was created. The application of nanowires is certainly very interesting, but the use of an electric field represents a very powerful tool in controlling block copolymer behavior because it is an easily measured field and can be varied in a controlled fashion.

The electric field induced ordering transition has been studied in detail by small angle X-ray scattering (SAXS) [13, 14] and by cross-sectional TEM [14]. Despite the relative simplicity of the experiment, alignment of copolymers in electric fields is a substantial theoretical challenge. For low fields, surface energies will dominate and copolymers may be oriented parallel to the surface, but this should flip over to normal alignment with increasing field strength. There is every possibility that there could be a mixed region, with normal alignment of copolymers in the center of the film and parallel alignment at the surface. Space precludes us from discussing the theory describing such structures in more detail and the interested reader should refer to the original articles [15–19]. Aside from pure theory, it is very likely that simulations will also make important contributions in this area [20]. It should also be noted that cobalt nanoarrays have been created by the use of the copolymer lithographic mask methods that we described earlier [3], to create a positive resist. The copolymer (polystyrene-*block*-polyferrocenyldimethylsilane) is placed on a tungsten and cobalt layer. The polystyrene matrix block etches faster in an oxygen plasma and so after various etching steps, a series of cobalt dots (from the lower lying layer) can be created [21].

We have already discussed how the nature of the substrate affects polymer orientation, but it is also possible to alter the "air" (free) surface by the use of various solvent atmospheres. Different copolymer blocks have different interactions with solvents, and so exposing a film to solvent vapor plasticizes the film in a non-uniform manner. Experiments on a polystyrene-*block*-poly(2-vinyl pyridine)-*block*-poly(*tert*-butyl methacrylate) triblock copolymer (PS-*b*-P2VP-*b*-PtBMA) have shown the effect of short-time *ex situ* exposure to solvent vapor [22] (Fig. 12.4) and also of prolonged annealing in solvent vapor [23]. With such a triblock copolymer, exposure to solvent vapor cannot bring a copolymer to equilibrium because the polymer is not dissolved in the solvent and the glass transition of all of the components is well above room temperature (at which the experiments were performed). The exposure to solvent is interesting because it shows behavior which is perhaps not expected *a priori*. One would imagine that on exposure to solvent vapor, the copolymer would swell most in the blocks where there is the most favorable interaction. In fact, it is the opposite effect that is observed; the polymer block which has the greatest affinity for the solvent is in regions below the other blocks. This is because the polymer is dried before imaging with scanning force microscopy (SFM). At low degrees of solvent exposure, all blocks will swell to some degree, but the block with the most favorable interaction with the solvent will continue to shrink, even after the other phase has had its structure frozen in. An example of some data showing the resultant morphologies is shown in Fig. 12.3. A wider variety of structures have

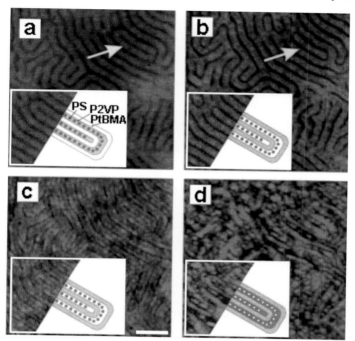

Fig. 12.4 SFM images of PS-*b*-P2VP-*b*-PtBMA triblock copolymer exposed to different solvent vapors for a short period of time [22]. The vapors are: (a) THF, a preferential solvent for the polystyrene block; (b) cyclohexane, preferential solvent for polystyrene and PtBMA; (c) toluene, polystyrene and PtBMA; and (d) methanol (P2VP). The scale bar is 200 nm and the height differences are not large, ~1 nm. The arrows in (a) and (b) indicate inversion in the morphology. The insets represent the regions indicated by the arrows to show the location of the different blocks (reproduced with permission and is copyright (1999) of The American Chemical Society).

been obtained by longer annealing times in different solvents [23]. One of the challenges in such an area of research is to predict the structures obtained. Although it is possible to simulate structures, and although the use of Flory–Huggins interaction parameters within accepted mean-field theory is likely to yield decent results, it will probably be some time before we can completely control block copolymer structures using solvent annealing.

Solvent environments may well be used to control the surface structure of a copolymer film, but they have great utility in increasing the ordering of such films. The suggestion that block copolymers could have applications in, for example, photonics, has been seen as fairly far-fetched on account of the inability of block copolymers to order over more than a micron at best. Controlled solvent evaporation is one means of addressing these limitations, and in one study poly(ethylene oxide)-*block*-polystyrene (PEO-*b*-PS) cylindrical structures (PEO cylinders) were created by casting thin copolymer films onto different substrates and then annealing

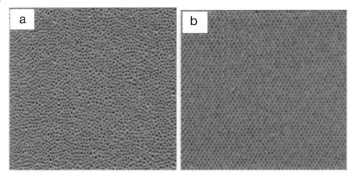

Fig. 12.5 PEO-*b*-PS film before (a) and after (b) exposure to benzene vapor [24]. The vapor increases mobility in the film, which allows ordering to take place. The images are SFM phase images (no topographic information, but the cylinders represent the PEO blocks) of 2 μm × 2 μm (reproduced with permission and is copyright (2004) of Wiley-VCH).

them in an atmosphere of benzene. The evolution of the ordering is independent of substrate type, which suggests that the ordering begins at the surface, where a glassy polystyrene matrix vitrifies first on solvent evaporation, and proceeds deep into the film [24]. An example of the improvement that might be obtained in the ordering in a diblock copolymer film is shown in Fig. 12.5. The same principle has also been applied to PS-*b*-P2VP-*b*-PtBMA films using chloroform vapor to allow the ordering to grow [25]. More control over the structure and length scales can be obtained with the addition of a homopolymer to the block copolymer [26]. The addition of a homopolymer is significant because it allows the length scale of the ordering to be changed over a relatively wide range. This is important, because the molecular weight scaling of domain size in ordered block copolymers is rather weak. Tailoring relatively large length scale structures by molecular weight control is, therefore, problematic, because high molecular weight block copolymers are difficult to synthesize and have prohibitively slow ordering kinetics. To be effective at swelling ordered block copolymer domains, a homopolymer needs to be miscible with one of the components, and not too large. The wrong choice of parameters will push the system to macroscopic phase separation, which is obviously undesirable [27].

12.5
Block Copolymer Brushes

A single layer of polymer molecules attached to a surface by one end is known as a *polymer brush*. The significance of polymer brushes in colloid science has long been known; they are one of the most important ways of conferring stability to colloidal particles, and brushes of water soluble polymers, particularly PEO, provide an effective method of creating surfaces that resist non-specific protein adsorption. The theory of the structure of polymer brushes has a long history, and accurate statistical mechanical models are now well established. The central idea of these theories is that the chains in polymer brushes are often "strongly stretched". The degree of stretching is determined by a competition between the excluded volume interaction between chain segments forced into proximity and the Gaussian stretching energy of the chains. The magnitude and sign of the excluded volume interaction can be altered by changing the quality of the solvent, and this gives rise to the idea of "responsiveness"; if the nature of the interaction between the solvent and the polymer can be altered by a change in environment, then the polymer brush will respond by changing its disposition and its consequent properties. One can distinguish between brushes that are physically anchored to the surface by weak and reversible bonds, and brushes that are chemically anchored by strong covalent bonds. Chemically anchored brushes can themselves be made in two ways. In "grafting to" one prepares a functionalized surface, for example by creating a self-assembled monolayer containing functional groups, to which one grafts end-functionalized chains. "Grafting from" methods also start out with a self-assembled layer, but this is used to initiate polymerization from the surface.

Block copolymers are relevant as polymer brushes in three ways. Firstly, the use of a diblock copolymer in which the various blocks have different affinities to an interface provides a very straightforward method of making a physically anchored brush. Secondly, it may be desirable to have an outer (surface) block that has a different functionality to the inner block (the attached block has the properties required for a film and the outer block is a coating layer). Finally, in a chemically anchored brush, which is itself composed of diblock copolymers, a differential interaction between each block and the solvent can result in the formation of the structure within the brush layer, and a change of this interaction can give very strong surface switching effects.

The idea of using diblock copolymers to create physically adsorbed brushes is now very well established. Although the individual interaction between a segment of the anchoring block and the interface may be rather weak, the essentially additive nature of the interaction means that the total binding energy for a diblock soon becomes rather large. Physically anchored diblock copolymer brushes can be prepared either at the interface between a solid substrate and a solvent [28, 29], at an air–liquid interface [30–32], or at an interface between a substrate and a melt of the homopolymer corresponding to one of the blocks of the copolymer [33–36]. The basic theory for this situation of block copolymer adsorption from a selective solvent has been described by Marques et al. [37], and more detailed calculations have been presented by Szleifer and Carignano [38]. The kinetics of the situation can be important [39]; as chains adsorb, the previously attached chains present a barrier to the adsorption of new ones, hence equilibrium may take a long time to reach. Both equilibrium and kinetic considerations mean that denser brushes are obtained in melt or Θ-solvent conditions [40] than from good solvents.

The theoretical situation in instances in which the soluble block is a polyelectrolyte is much more complex and not yet fully understood; nonetheless such systems have received some experimental attention [41–44], because of their potential in making environmentally responsive surfaces. A block copolymer brush for environmental response requires one block that strongly adsorbs to a surface, and another that has the relevant environmental behavior. Another possibility is to consider *two* (or more) brushes, a mixed brush, in the same film, with two different responsive blocks and the same adsorbing block. This ensures different functionalities depending on the environment, because the exposed surface layer changes depending upon conditions. Such an experiment has been applied to mixed chemically grafted polymers [45]. The work on the triblock PS-*b*-P2VP-*b*-PtBMA [22, 23] shows another route to a similar end, with the P2VP remaining adsorbed on a silicon substrate while either end-block will be preferentially attracted to the surface depending on the solvent vapor environment.

Another way of making a diblock copolymer brush is to use "grafting from" to synthesize a diblock from an initiator grafted at the surface. This has been achieved using the controlled radical polymerization method ATRP (atom transfer radical polymerization) [46, 47]. Briefly, the effectiveness of ATRP is that it includes an activation–deactivation mechanism, so that the growth of the polymer is controlled. The terminal block of the growing polymer is usually an alkyl halide, which can transfer its halide by complexing with a metal salt. The resulting radical on the chain allows propagation, but this is quickly terminated by the reformation of the halide end-block from the complex. Propagation of the chain can continue on further complexation with the metal salt. An important advantage of this and other "grafting from" techniques is that the entropic barrier to chain adsorption does not exist and dense brush layers can be synthesized. ATRP can be a very effective method of making surfaces with a strong response to changing solvent conditions. A strong interaction between the two blocks has been shown theoretically to produce dramatic conformational changes within the brush [48].

Work has also been carried out on diblock copolymer brushes whereby the outer component is used to provide functionality. Recent examples of this include experiments to provide a low surface energy coating by using a fluorinated end-block [49] or to provide a biocompatible surface by using a PEO end-block [50], in both instances with a polystyrene block grafted to a silicon substrate. In both examples a form of controlled living polymerization, not too dissimilar to ATRP (but avoiding the use of a metal catalyst), was used to chemically graft the polymers to the surface. Physically grafting polymers is often unsuitable because large grafting blocks are required for a sufficiently high energy of adsorption, resulting in poor surface coverage by the surface block.

We have emphasized the role of switchability in this chapter, but modern technologies can also benefit from surfaces with well-controlled structure, for example, for high throughput molecular sorting technologies. Examples of this are gradient copolymer brushes, whereby the molecular weight of the brush varies in a controlled fashion from one part of a surface to another [51–53]. The production of such films is relatively easy and the work so far published demonstrates a two-stage procedure, with the first stage being the growth of the first block from the substrate using ATRP. During the growth of the block, the substrate is removed at constant speed from the reaction bath (see Fig. 12.6) and so the part of the sub-

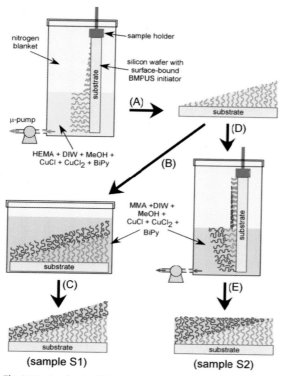

Fig. 12.6 Synthesis of block copolymer films with controlled thickness may be achieved by the careful removal of the substrate during the reaction [52]. PHEMA is grown from the silicon substrate using ATRP, with a varying thickness caused by the withdrawal of the substrate from the solution (A). In the second step, PMMA is grown on to the PHEMA and by control of the retraction (B and D) different variations in thickness can be created (C and E) (reproduced with permission and is copyright (2003) of The Royal Society of Chemistry).

strate remaining in the longest corresponds to a longer block. The end of these blocks can be used as the initiation for the second block. It is possible to add the second block to the film by the same process and to create a range of thickness variations. (Although the ATRP route has more possibilities due to the much greater control in the synthesis of the second block, if the copolymers were first synthesized and then grafted to the surface, the retraction of the substrate would be a means of varying grafting density.) This route has been used to generate chemically attached poly(hydroxyl ethyl methacrylate)-*block*-PMMA (PHEMA-*b*-PMMA), with the PMMA block being at the air surface. The PHEMA block is of some use in biotechnology as it is generally biocompatible; PMMA and PHEMA are also important components in hard tissue replacement, with various biotechnological uses, notably in dentistry.

12.6
Surface Regeneration

A hindrance in many technologies is wear on surface coatings used in hostile environments. A means of completely regenerating a surface affected by exposure to aggressive chemicals, solvents, or other conditions, without the involvement of the user would be very welcome. A possible means to do this has been discussed in terms of protein fouling of surfaces. Polymers readily adsorbing onto surfaces due to the number of contact points of the monomers is not always an advantage, as it is for this reason that proteins readily adsorb onto surfaces. This is a problem in a huge number of coating technologies, from the hulls of ships to biomedical implants. PEO, also known as poly(ethylene glycol), is an important material because of its biocompatibility; it is generally resistant to protein adsorption. A good surface coating for biocompatible purposes would therefore need a PEO coating. If this coating deteriorates with use, it needs to be re-applied. One group has attacked this problem by using the phenomenon of surface segregation to regenerate surfaces [54, 55]. A graft copolymer with a hydrophobic backbone and PEO blocks grafted along the backbone is mixed with a hydrophobic matrix. Because PEO is the only water-soluble component of the mixture, the graft copolymer will segregate to an aqueous surface, giving the required surface properties. However, there is a limit to the amount of segregation possible in a miscible system [56] and so some of the graft copolymer will remain in the bulk. In these first experiments, the graft copolymer could be removed from the surface by acid treatment. Subsequent heat treatment can regenerate this surface; the annealing process is necessary here to give the polymers enough mobility to equilibrate. It has also been demonstrated that environmental responsiveness (switchability) can be built into the system by grafting polymethacrylic acid blocks to the backbone to create a responsive membrane [57].

12.7
Conclusions and Outlook

Block copolymers have properties that relate strongly to nanoscience, particularly because they readily self-assemble into nanoscale structures. These structures can be controlled at surfaces, because different surfaces have different energies, and because in thin films, thickness also plays a determining role in the final morphology. Thin film structures can also be controlled by changing the environment that the film finds itself in, and this can also be achieved reversibly. As well as reversibility, it is also possible to create different structures by altering this environment. Applications of such science are limited, because block copolymers are invariably ordered, and technology also demands tailor-made structures, but we have seen here that such challenges are presently being met with a hybrid top down–bottom up approach to block copolymer lithography. However, block copolymers are not limited to lithographic or patterned nanostructures, but are also useful for functionality.

Here, we have shown that graft-copolymers can be used to create functional membranes for anti-fouling, and we can also create polymer brushes with reversible properties. Such brushes give us the opportunity to create surfaces that respond to the environment, with possibilities in controlled friction, wettability and adhesion.

Finally, the length scales obtainable with ordered block copolymer films make them exceptional candidates for applications in optoelectronics. One of the major areas of investment in polymer science today is in thin film transistors and polymers for optoelectronic applications, including light-emitting diodes (LEDs) and photovoltaic cells. If one imagines a block copolymer with one block that has high electron mobility and the other with high hole mobility, one can envisage applications in optoelectronic devices. Both LEDs and photovoltaics depend on excitons (electron–hole pairs) for operation. In a photovoltaic device for example, an incident photon creates an exciton, which diffuses before decaying (preferably, but not always) by dissociation into its components. If this happens at an interface between materials with high mobility for electrons and holes then a current is generated, thereby demonstrating photovoltaic operation. As the typical diffusion length for an exciton is of the order of several nanometers, block copolymers are ideal materials for their operation, because the interface at which dissociation can take place is inherent in the ordered structure. To date, there has been some success in the synthesis of different semiconducting block copolymers [58–60], but little in the way of working devices. The effort spent in creating the devices means that it is likely that there will be more success in fabricating working devices [61] in the very near future.

References

1 P. F. GREEN (Wetting and dynamics of structured liquid films) *J. Polym. Sci. B: Polym. Phys.* 41, 2219–2235 (**2003**).

2 G. H. FREDRICKSON (Surface ordering phenomena in block copolymer melts) *Macromolecules* 20, 2535–2542 (**1987**).

3 M. PARK, C. HARRISON, P. M. CHAIKIN, R. A. REGISTER, D. H. ADAMSON (Block copolymer lithography: periodic arrays of ~10^{11} holes in 1 square centimeter) *Science* 276, 1401–1404 (**1997**).

4 J. HEIER, E. J. KRAMER, S. WALHEIM, G. KRAUSCH (Thin diblock copolymer films on chemically heterogeneous surfaces) *Macromolecules* 30, 6610–6614 (**1997**).

5 M. D. FOSTER, M. SIKKA, N. SINGH, F. S. BATES, S. K. SATIJA, C. F. MAJKRZAK (Structure of symmetric polyolefin block copolymer thin films) *J. Chem. Phys.* 96, 8605–8615 (**1992**).

6 L. ROCKFORD, S. G. J. MOCHRIE, T. P. RUSSELL (Propagation of nanopatterned substrate templated ordering of block copolymers in thick films) *Macromolecules* 34, 1487–1492 (**2001**).

7 S. O. KIM, H. H. SOLAK, M. P. STOYKOVICH, N. J. FERRIER, J. J. DE PABLO, P. F. HEALEY (Epitaxial self-assembly of block copolymers on lithographically defined nanopatterned substrates) *Nature (London)* 424 411–414 (**2003**); see also: Register, R. A., (On the straight and narrow), *Nature (London)* 424, 378–379 (**2003**).

8 M. P. STOYKOVICH, M. MÜLLER, S. O. KIM, H. H. SOLAK, E. W. EDWARDS, J. J. DE PABLO, P. F. NEALEY (Directed assembly of block copolymer blends into nonregular device-oriented structures) *Science* 308, 1442–1446 (**2005**).

9 E. Sivaniah, Y. Hayashi, M. Iino, T. Hashimoto, K. Fukunaga (Observation of perpendicular orientation in symmetric diblock copolymer thin films on rough substrates) *Macromolecules* 36, 5894–5896 (**2003**).

10 E. Sivaniah, Y. Hayashi, S. Matsubara, S. Kiyono, T. Hashimoto, K. Fukunaga, E. J. Kramer, T. Mates (Symmetric diblock copolymer thin films on rough substrates. Kinetics and structure formation in pure block copolymer thin films) *Macromolecules* 38, 1837–1849 (**2005**).

11 P. Mansky, T. P. Russell, C. J. Hawker, M. Pitsikalis, J. Mays (Ordered diblock copolymer films on random copolymer brushes) *Macromolecules* 30, 6810–6813 (**1997**).

12 T. Thurn-Albrecht, J. Schotter, G. A. Kästle, N. Emley, T. Shibauchi, L. Krusin-Elbaum, K. Guarini, C. T. Black, M. T. Tuominen, T. P. Russell (Ultrahigh-density nanowire arrays grown in self-assembled diblock copolymer templates) *Science* 290, 2126–2129 (**2000**).

13 T. Thurn-Albrecht, J. DeRouchey, T. P. Russell, R. Kolb (Pathways toward electric field induced alignment of block copolymers) *Macromolecules* 35, 8106–8110 (**2002**).

14 T. Xu, C. J. Hawker, T. P. Russell (Interfacial energy effects on the electric field alignment of symmetric diblock copolymers) *Macromolecules* 36, 6178–6182 (**2003**).

15 B. Ashok, M. Muthukumar, T. P. Russell (Confined thin film diblock copolymer in the presence of an electric field) *J. Chem. Phys.* 115, 1559–1564 (**2001**).

16 A. Onuki, J. Fukuda (Electric field effects and form birefringence in diblock copolymers) *Macromolecules* 28, 8788–8795 (**1995**).

17 G. G. Pereira, D. R. M. Williams (Diblock copolymer melts in electric fields: The transition from parallel to perpendicular alignment using a capacitor analogy) *Macromolecules* 32, 8115–8120 (**1999**).

18 Y. Tsori, D. Andelman (Thin film diblock copolymers in electric field: Transition from perpendicular to parallel lamellae) *Macromolecules* 35, 5161–5170 (**2002**).

19 Y. Tsori, F. Tournilhac, D. Andelman, L. Leibler (Structural changes in block copolymers: coupling of electric field and mobile ions) *Phys. Rev. Lett.* 90, 145504 (**2003**).

20 A. V. Kyrylyuk, A. V. Zvelindovsky, G. J. A. Sevink, J. G. E. M. Fraaije (Lamellar alignment of diblock copolymers in an electric field) *Macromolecules* 35, 1473–1476 (**2002**).

21 J. Y. Cheng, C. A. Ross, V. Z.-H. Chan, E. L. Thomas, R. G. H. Lammertink, G. J. Vancso (Formation of a cobalt magnetic dot array via block copolymer lithography) *Adv. Mater.* 13, 1174–1178 (**2001**).

22 H. Elbs, K. Fukunaga, R. Stadler, G. Sauer, R. Magerle, G. Krausch (Microdomain morphology of thin ABC triblock copolymer films) *Macromolecules* 32, 1204–1211 (**1999**).

23 H. Elbs, C. Drummer, V. Abetz, G. Krausch (Thin film morphologies of ABC triblock copolymers prepared from solution) *Macromolecules* 35, 5570–5577 (**2002**).

24 S. H. Kim, M. J. Misner, T. Xu, M. Kimura, T. P. Russell (Highly oriented and ordered arrays from block copolymers via solvent evaporation) *Adv. Mater.* 16, 226–231 (**2004**).

25 S. Ludwigs, A. Böker, A. Voronov, N. Rehse, R. Magerle, G. Krausch (Self-assembly of functional nanostructures from ABC triblock copolymers) *Nat. Mat.* 2, 744–747 (**2003**).

26 S. H. Kim, M. J. Misner, T. P. Russell (Solvent-induced ordering in thin film diblock copolymer/homopolymer mixtures) *Adv. Mater.* 16, 2119–2123 (**2005**).

27 F. S. Bates, W. Maurer, T. P. Lodge, M. F. Schulz, M. W. Matsen, K. Almdal, K. Mortensen (Isotropic Lifshitz behavior in block copolymer-homopolymer blends) *Phys. Rev. Lett.* 75, 4429–4432 (**2005**).

28 T. Cosgrove, J. S. Phipps, R. M. Richardson, M. L. Hair, D. A. Guzonas (Adsorbed block copolymer of poly(2-vinylpyridine) and polystyrene studied by neutron reflectivity and surface force techniques) *Macromolecules* 26, 4363–4367 (**1993**).

29 G. Hadzioannou, S. Patel, S. Granick, M. Tirrell (Forces between surfaces of block copolymers adsorbed on mica) *J. Am. Chem. Soc.* 108, 2869–2876 **(1986)**.

30 J. Bowers, A. Zarbakhsh, J. R. P. Webster, L. R. Hutchings, R. W. Richards (Structure of a spread film of a polybutadiene-poly(ethylene oxide) linear diblock copolymer at the air-water interface as determined by neutron reflectometry) *Langmuir* 17, 131–139 **(2001)**.

31 I. Hopkinson, F. T. Kiff, R. W. Richards, D. G. Bucknall, A. S. Clough (Equilibrium concentration profiles of physically end tethered polystyrene molecules at the air-polymer interface) *Polymer* 38, 87–98 **(1997)**.

32 M. S. Kent, J. Majewski, G. S. Smith, L. T. Lee, S. Satija (Tethered chains in poor solvent conditions: An experimental study involving Langmuir diblock copolymer monolayers) *J. Chem. Phys.* 110, 3553–3565 **(1999)**.

33 A. Budkowski, J. Klein, L. J. Fetters (Brush formation by symmetric, by highly asymmetric diblock copolymers at homopolymer interfaces) *Macromolecules* 28, 8571–8578 **(1995)**.

34 A. C. Costa, R. J. Composto, P. Vlček, M. Geoghegan (Block copolymer adsorption from a homopolymer melt to an amine-terminated surface) *Eur. Phys. J. E* 18, 159–166 **(2005)**.

35 A. C. Costa, M. Geoghegan, P. Vlček, R. J. Composto (Block copolymer adsorption from a homopolymer melt to silicon oxide: Effects of nonadsorbing block length and anchoring block-substrate interaction) *Macromolecules* 36, 9897–9904 **(2003)**.

36 H. Retsos, A. F. Terzis, S. H. Anastasiadis, D. L. Anastassoppoulos, C. Toprakcioglu, D. N. Theodorou, G. S. Smith, A. Menelle, R. E. Gill, G. Hadziioannou, Y. Gallot (Mushrooms and brushes in thin films of diblock copolymer/homopolymer mixtures) *Macromolecules* 35, 1116–1132 **(2002)**.

37 C. Marques, J. F. Joanny, L. Leibler (Adsorption of block copolymers in selective solvents) *Macromolecules* 21, 1051–1059 **(1988)**.

38 I. Szleifer, M. A. Carignano (Tethered polymer layers: phase transitions and reduction of protein adsorption) *Macromol. Rapid Commun.* 21, 423–448 **(2000)**.

39 A. Johner, J. F. Joanny (Block copolymer adsorption in a selective solvent: A kinetic study) *Macromolecules* 23, 5299–5311 **(1990)**.

40 R. Hariharan, W. B. Russel (Enhanced colloidal stabilization via adsorption of diblock copolymer from a nonselective Θ solvent) *Langmuir* 14, 7104–7111 **(1998)**.

41 E. P. K. Currie, A. B. Sieval, G. J. Fleer, M. A. Cohen Stuart (Polyacrylic acid brushes: Surface pressure and salt-induced swelling) *Langmuir* 16, 8324–8333 **(2000)**.

42 D. Julthongpiput, Y.-H. Lin, J. Teng, E. R. Zubarev, V. V. Tsukruk (Y-shaped polymer brushes: Nanoscale switchable surfaces) *Langmuir* 19, 7832–7836 **(2003)**.

43 J. N. Kizhakkedathu, R. Norris-Jones, D. E. Brooks (Synthesis of well-defined environmentally responsive polymer brushes by aqueous ATRP) *Macromolecules* 37, 734–743 **(2004)**.

44 G. Romet-Lemonne, J. Daillant, P. Guenoun, J. Yang, J. W. Mays (Thickness and density profiles of polyelectrolyte brushes: Dependence on grafting density and salt concentration) *Phys. Rev. Lett.* 93, 148301 **(2004)**.

45 A. Sidorenko, S. Minko, K. Schenk-Meuser, H. Duschner, M. Stamm (Switching of polymer brushes) *Langmuir* 15, 8349–8355 **(1999)**.

46 S. G. Boyes, B. Akgun, W. J. Brittain, M. D. Foster (Synthesis, characterization, and properties of polyelectrolyte block copolymer brushes prepared by atom transfer radical polymerization and their use in the synthesis of metal nanoparticles) *Macromolecules* 36, 9539–9548 **(2003)**.

47 A. M. Granville, S. G. Boyes, B. Akgun, M. D. Foster, W. J. Brittain (Synthesis and characterization of stimuli-responsive semifluorinated polymer brushes prepared by atom transfer radical polymerization) *Macromolecules* 37, 2790–2796 **(2004)**.

48 P. G. Ferreira, L. Leibler (Copolymer brushes) *J. Chem. Phys.* 105, 9362–9370 (**1996**).

49 L. Andruzzi, A. Hexemer, X. Li, C. K. Ober, E. J. Kramer, G. Galli, E. Chiellini, D. A. Fischer (Control of surface properties using fluorinated polymer brushes produced by surface-initiated controlled radical polymerization) *Langmuir* 20, 10498–10506 (**2004**).

50 L. Andruzzi, W. Senaratne, A. Hexemer, E. D. Sheets, B. Ilic, E. J. Kramer, B. Baird, C. K. Ober (Oligo(ethylene glycol) containing polymer brushes as bioselective surfaces) *Langmuir* 21, 2495–2504 (**2005**).

51 R. R. Bhat, M. R. Tomlinson, J. Genzer (Orthogonal surface-grafted polymer gradients: A versatile combinatorial platform) *J. Polym. Sci. B: Polym. Phys.* 43, 3384–3394 (**2005**).

52 M. R. Tomlinson, J. Genzer (Formation of surface-grafted copolymer brushes with continuous composition gradients) *Chem. Commun.* 1350–1351 (**2003**).

53 M. R. Tomlinson, J. Genzer (Evolution of surface morphologies in multivariant assemblies of surface-tethered diblock copolymers after selective solvent treatment) *Langmuir* 21, 11552–11555 (**2005**).

54 J. F. Hester, P. Bannerjee, A. M. Mayes (Preparation of protein-resistant surfaces on poly(vinylidene fluoride) membranes via surface segregation) *Macromolecules* 32, 1643–1650 (**1999**).

55 J. F. Hester, A. M. Mayes (Design and performance of foul-resistant poly(vinylidene fluoride) membranes prepared in a single-step by surface segregation) *J. Membrane Sci.* 202, 119–135 (**2002**).

56 M. Geoghegan, G. Krausch (Wetting at polymer surfaces and interfaces) *Prog. Polym. Sci.* 28, 261–302 (**2003**).

57 J. F. Hester, S. C. Olugebefola, A. M. Mayes (Preparation of pH-responsive polymer membranes by self-organization) *J. Membrane Sci.* 208, 375–388 (**2002**).

58 L. Boiteau, M. Moroni, A. Hilberer, M. Werts, B. de Boer, G. Hadziioannou (Synthesis of a diblock copolymer with pendent luminescent and charge transport units through nitroxide-mediated free radical polymerization) *Macromolecules* 35, 1543–1548 (**2002**).

59 B. de Boer, U. Stalmach, P. F. van Hutton, C. Melzer, V. V. Krasnikov, G. Hadziioannou (Supramolecular self-assembly and opto-electronic properties of semiconducting block copolymers) *Polymer* 42, 9097–9109 (**2001**).

60 M. H. van der Veen, B. de Boer, U. Stalmach, K. I. van der Wetering, G. Hadziioannou (Donor-acceptor diblock copolymers based on PPV and C_{60}: Synthesis, thermal properties, and morphology) *Macromolecules* 37, 3673–3684 (**2004**).

61 S. M. Lindner, S. Hüttner, A. Chiche, M. Thelakatt, G. Krausch (Charge separation at self-assembled nanostructure bulk interface in block copolymers) *Ang. Chem. Int. Ed.* 45, 3364–3368 (**2006**).

13
Block Copolymers as Templates for the Generation of Mesostructured Inorganic Materials

Bernd Smarsly and Markus Antonietti

13.1
Introduction

In this chapter, a new facet of block copolymers will be discussed, which has been emerging over the past ten years and has brought an entirely new perspective to this field. While microphase separation of block copolymers has been known for a long time, the use and development of suitable amphiphilic block copolymers (soft matter) to create inorganic replicas (rigid, hard matter) has initiated cross fertilization in both communities, enabling the access to inorganic materials with nanoscaled morphologies that had previously been inaccessible. The idea to convert a self-assembled block copolymer mesostructure into its inorganic counterpart (e.g., providing mesoporous silica, metal oxides and metals) has motivated countless studies, and led to the regular reporting of a rich and amazing diversity of new inorganic materials on the nanometer scale (also termed "mesostructured" materials in the following discussion).

Taking into account the excellent recently published reviews in this field, in the following only a selection of the various aspects will be discussed, addressing features which have not been depicted in other overviews. For instance, the general considerations concerning the self-assembly of block copolymers (BC) and the formation of nanoparticles within BC micelles were described in excellent reviews by Förster and Antonietti [1]. A general overview of ordered mesoporous oxides was given by Ciesla and Schüth [2]. Therefore, this chapter will focus on the generation of mesoporous inorganic structures, placing special emphasis on the role of block copolymers as templates in the templating mechanism and in metal oxide formation. Furthermore, recent progress in preparing mesoporous metal oxides with crystalline nanoparticles to build up the pore walls will be presented, which in turn is also a question of the rational choice of suitable block copolymers. As most of the research has focused on oxydic materials, a separate section is reserved for the interesting emerging field of mesoporous metals.

It should be emphasized that certain aspects of templating towards porous inorganics will not be mentioned, because they are not related to block copolymers.

Block Copolymers in Nanoscience. Massimo Lazzari, Guojun Liu, Sébastien Lecommandoux
Copyright © 2006 WILEY-VCH Verlag GmbH & Co. KGaA, Weinheim
ISBN: 3-527-31309-5

Although templating using surfactants such as cetyltrimethylammonium bromide (CTAB) has contributed significantly to materials science [3], such low molecular weight surfactants follow different templating mechanisms (cooperative assembly). Instead, this overview will be restricted to so-called "true liquid-crystal templating" [4, 5], which is also termed "nanocasting" [6]. In contrast to cooperative assembly mechanisms, only nanocasting involves a direct copy of amphiphilic BC phases in water or other solvents, and can therefore be regarded as "block copolymer templating" in its true sense. In essence, in the present case the templating process itself is at the heart of the overview.

13.2
General Mechanism

The general idea of block copolymer templating ("nanocasting", true liquid-crystal templating) is based on the fact that amphiphilic block copolymers can self-assemble in certain solvents to form regular super-structures, featuring structural motives on the nanometer scale (lyotropic liquid crystalline phases). Typically, the constituting entities (cylindrical or spherical micelles, etc.) are 10–70 nm in size.

This self-assembly is governed by the microphase separation dictated by the mutual incompatibility of the different blocks, one being soluble and the other being insoluble in the solvent to be used ("amphiphilic" polymers). In most cases, aside from classical nonionic surfactants (such as the "Brij" surfactants, which can be regarded as oligoblocks composed of an oligoethylene and oligoethyleneoxide), diblock or triblock copolymers are used featuring polystyrene [5a], polybutadiene [7] and poly(propylene oxide) as the hydrophobic block [8], and poly(ethylene oxide) or poly(vinyl pyridines) [9] as the hydrophilic blocks. Recently, a semi-commercial family of "KLE" block copolymers [KLE = Kraton-liquid-b-poly(ethylene oxide)] was introduced, combining good chemical accessibility, high mesophase robustness with a superb application profile [10]. Kraton liquid is a hydrogenated poly(ethylene-co-butylene) hydrophobic block with low glass transition. In principle, block copolymers (in contrast to low M_w surfactants) self-assemble or microphase separate in a variety of solvents, in particular water, but also certain alcohols, THF and other water soluble solvents can be employed or admixed.

In a typical block copolymer templating procedure (see Fig. 13.1) such regularly ordered mesostructures of an amphiphile in water are converted into their 1:1 replicas (their "negative") by solidifying the hydrophilic domains between the micelles. This is carried out by replacing a majority of the solvent by a metal or metal oxide precursor with similar polarity and condensing this precursor around the aggregates. A particularly suitable precursor for this procedure is, for instance, hydrated silicic acid, which shows a very similar polarity and proton bridging behavior to water. Depending on the final material of choice, this precursor has to be a hydrolysable silica species, such as an alkoxysilane, but can also be a metal salt (metal chloride, etc.), and the final structure can be made from a metal oxide or elemental metal. As for the oxides, the intended inorganic moiety is created by

Fig. 13.1 Illustration of the typical block copolymer templating procedure ("nanocasting"), which also holds true for the preparation of thin films by evaporation-induced self-assembly.

condensation, induced by appropriate pH values and heat treatment. Hence, the organic BC can be regarded as a place holder to prevent the inorganic reaction from occurring, which becomes a void space after removal of the BC, either by extraction or thermal decomposition.

One main advantage of block copolymers over low molecular weight surfactants is the significantly larger molecular weight and its variability, enabling a rich diversity of structures (hexagonal, cubic, etc.) and much larger pore sizes, currently up to up to 80 nm [7]. Initial work was pioneered by Göltner and coworkers using PS-b-PEO [5] or PS-b-PVP block copolymers [9]. Later, presumably independently, this work was extended by Stucky and coworkers towards Pluronic block copolymers [8]. Figures 13.2 and 13.3 show examples of the mesostructures that can be obtained by variation of the polymers [7b].

The assumption that the final silica mesostructure represents a 1:1 copy of the lyotropic phase was later proven by comparing the replicas with species obtained by cross-linking of lyotropic phases using γ-radiation [11].

Figure 13.4 shows a recent example of the surprising precision of the nanocasting procedure, using KLE polymers as templates. Using transmission electron microscopy (TEM) images, it can be seen that the mesoporous silica contains spherical mesopores ca. 14 nm in diameter, which are located on the sites of a closed-packed

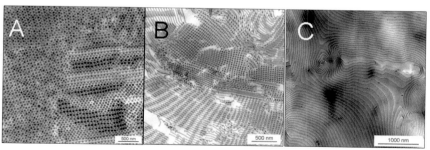

Fig. 13.2 Thin cuts of lyotropic phases of block copolymers, as they are used for nanocasting: (A) BCC phase, (B) hexagonal phase, (C) lamellar phase. The images are similar to those in [7b, c].

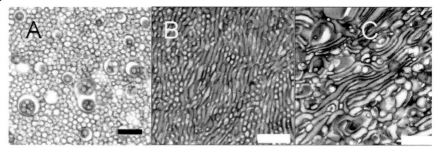

Fig. 13.3 TEM of various mesoporous silicas obtained from PB-*b*-PEO templates [7], graphically illustrating the wide range of accessible structures, presenting a micellar phase at the borderline to a lamellar phase (A: containing some vesicles as precursor structures, scale bar 100 nm), a sponge-like L_3-phase (B: scale bar 500 nm) and a branched lamellar phase (C: scale bar 500 nm) not usually found in classical phase diagrams.

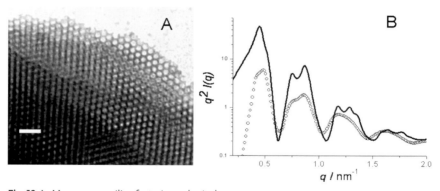

Fig. 13.4 Mesoporous silica featuring spherical mesopores (size ca. 13 nm) obtained from the KLE block copolymer [poly(ethylene-*co*-butylene)-*b*-poly(ethylene oxide)] [10a]. The TEM image (A) reveals a highly ordered structure of spherical mesopores, arranged on an fcc lattice. The scale bar is 80 nm. (B) Small-angle X-ray scattering curve of the mesoporous silica (circles) and a fitting (solid line) based on a model of spherical mesopores with a certain polydispersity and lattice distortions [10a].

cubic lattice, here a face-centered cubic arrangement. Evidently, not only are the single mesopores extremely uniform in shape and size, but also the 3D-arrangement is virtually undisturbed over large areas up to the micron range. This example illustrates the amazing order and control of material structures, which can be achieved by the nanocasting procedure, in spite of the mechanical stresses during the condensation of the matrix. While TEM images can only illustrate the mesostructure on a local scale of the specimen, small-angle X-ray scattering (SAXS) is usually applied to provide an overall, averaged characterization of mesoporous materials.

In this case (Fig. 13.4b), the presence of distinct SAXS patterns with pronounced maxima and minima confirms the high regularity of the mesostructure and provides a further tool to determine the mesopore size and the lattice type. It has to be emphasized again that one of the most important advantages of block copolymers is that the mesophase structure can be conveniently addressed by the relative composition or lengths of both blocks, according to the general phase diagram of block copolymers [12].

A good example for the perfect way in which the nanocasting process converts a soft, lyotropic structure into its hard replica is seen in Fig. 13.5. The two TEM images show the original lyotropic phase (Fig. 13.5A) and the templated, porous silica (Fig. 13.5B), and it is readily seen that the one is the "negative" of the other, although it is almost impossible to differentiate without *a priori* knowledge.

In the case of metal oxides, the ultimate composition is readily obtained by heat-treatment in air or oxygen to yield the amorphous or crystalline oxide form, and mesoporous metals are obtained by an *in situ* or final reduction step, e. g., electrochemically or chemically.

These general concepts do not only hold true for the generation of bulk materials (monoliths or powders), but also for thin films, which can be simplified by the introduction of the technique of "evaporation-induced self-assembly" (EISA) [13]. EISA proceeds along a different methodology, but the underlying templating mechanism is identical to nanocasting. The only difference lies in the fact that to start with a significantly larger amount of volatile solvent is required, being added initially to provide homogeneous solutions with low viscosity. Coatings of excellent macroscopic homogeneity can typically be generated by EISA. After evaporation of the volatile solvent, the situation of nanocasting, that is the presence of a high concentration template phase, is regenerated, although with a controlled outer geometry. Neglecting the peculiarities of the film formation, the final structure ideally represents a 1:1 copy of the lyotropic block copolymer phase formed after evapora-

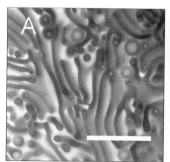 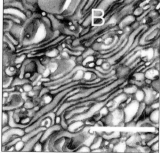

Fig. 13.5 Original or copy? (A) The lyotropic phase (TEM obtained thin cuts) leading to the corresponding silica (B) shown in Fig. 13.3C. This example is a good illustration that the final porous structures are an almost perfect 1:1 "negative" of the original block copolymer mesophase. The scale bars correspond to 1 μm.

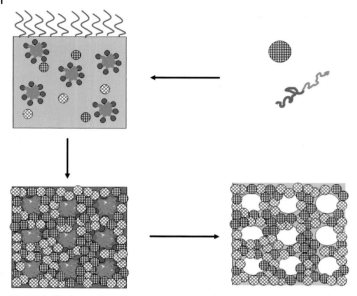

Fig. 13.6 Block copolymer templating starting from inorganic nanoparticles, using the EISA process.

tion of the solvent. For EISA in thin films, the strong interaction with the substrate usually leads to a high degree of preferred orientation of the mesostructure with respect to the surface, much higher than when compared with the bulk materials [14]. An orientation is usually established that provides a maximum interaction with the substrate via the densest lattice plane [15].

While the procedure discussed above begins with molecular, inorganic precursors or monomers, recent papers have shown that similar mesoporous materials can also be achieved using preformed inorganic nanoparticles, with sizes of only several nanometers, in the EISA process for structure build-up, e. g., TiO_2 [16] or CeO_2 [17], which is illustrated in Fig. 13.6.

It is evident that this approach faces different obstacles, for instance, the colloidal stability of the initial sol, and only recently a novel procedure for the synthesis of nanoparticles provided particles which could be dissolved in common solvents such as water and THF [18]. On the other hand, the material is already crystallized, and further temperature treatment is less critical and also does not lead to pronounced shrinkage [17].

While the overall principle of BC templating seems to be surprisingly straightforward, it is worthwhile stressing the peculiarities of this strategy from a block copolymer point of view.

- The whole concept relies on the selective swelling of one part of a microphase separated block copolymer phase with inorganic salts or other precursors. It is evident that the homogeneous, molecular distribution of the precursor within the block copolymer mesostructure requires a suitable solvent or solvent mixture.

The final mesostructure is, however, sensitive towards changes in the solvent environment. PEO is known to form different conformations, such as helices, depending on the solution conditions and temperature [19]. Hence, nanocasting is eventually fairly dependent on the external humidity or vapor pressure.

- All components have to be compatibilized by the block copolymers, lowering all involved interfacial energies throughout the process. In particular, it has to be emphasized that the chemical nature of the blocks has to be well chosen to avoid macrophase separation upon addition of the water, the solvent and the molecular precursors. The intrinsic property of amphiphilic block copolymers to lower the surface energy between any two materials of choice can be regarded as the basic principle of the templating process.

- As the formation of microphase-separated block copolymer phases is a thermodynamic phenomenon, templating is governed to a great extent by numerous physical parameters, and together with solvent effects, the combination makes it difficult to predict pore structure *a priori*. Hence, in many cases optimum templating conditions were found by trial-and-error experiments. In addition to these thermodynamic parameters, templating is further aggravated by strong kinetic effects. For instance, Gibaud et al. showed that the rate of evaporation of the volatile solvent, with all the other parameters being kept the same, can dramatically change the final mesophase structure [15]. This is why "robust templates", such as KLE block copolymers, where the mesophase structure is less sensitive to minor changes in the conditions, are recommended ahead of "sensitive templates", such as Pluronics.

- A general problem of true liquid-crystalline sol–gel templating is the simultaneous condensation of inorganic species and the self-assembly process. Usually, experimental procedures are needed to delay the condensation significantly after micelle formation and to allow for a controlled co-assembly of the inorganic and organic species. As the self-assembly of amphiphilic copolymers into micelles occurs at low concentrations, and because of the higher segregation strength (i.e., the incompatibility between the hydrophilic and hydrophobic block), BC

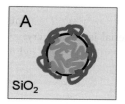 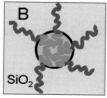 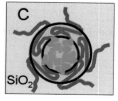

Fig. 13.7 Various scenarios for the distribution of the hydrophilic block [in most instances poly(ethylene oxide), PEO] in block copolymer templating. (A) Collapse of the PEO chains on the hydrophobic core. (B) "Two-phase model" with the PEO chains being entirely embedded in the surrounding inorganic matrix. (C) "Three-phase model", which assumes only partial penetration of the PEO block in the matrix.

self-assembly is less affected by condensation phenomena than low molecular weight surfactants. On the other hand, the use of block copolymers is limited by the increasingly reduced mobility in the case of very high block lengths. Hence, the upper limit for block copolymer templating is, under practical conditions, currently ca. 80 nm in structural size [7].

13.3
Details of the BC Templating Mechanism

As already indicated, the apparently simple templating mechanism depicted in Fig. 13.1 bears certain delicate features, which had been neglected in the initial research on BC templating. We will focus here on one of the major issues, the behavior of the polymer chains on being exposed to the solvent, i. e., in most cases the hydrophilic poly(ethylene oxide) blocks. The solvation of these blocks drastically affects both the self-assembly and the final pore structure through the conformation and degree of swelling. As nanocasting can be expected to create a 1:1 replica of the initial self-assembled structure, it can also be used to depict the fine details of the templating process based on the final porous structure.

In principle, three scenarios were envisaged for the distribution of the PEO chains around the BC micelles [20–22]: in the "two-phase" model (see Fig. 13.7B) the hybrid material would consist of the hydrophobic core and the metal oxide (silica) phase, also containing all of the PEO chains. The other extreme situation would be the collapse of the PEO chains onto the micellar core in the course of templating (Fig. 13.7A). In an intermediate scenario, a part of the PEO chain may form an additional surface layer ("third phase") between the hydrophilic block and the silica phase (Fig. 13.7C), and this was termed the "three-phase" model [21]. As an important consequence, "two-phase" or "three-phase" systems would result in additional porosity in the pore walls after template removal due to the PEO chains embedded in the matrix. Indeed, various studies reported an extra microporosity (pore size below 2 nm) located in the matrix separating the ordered mesopores [23]. More importantly, systematic studies even proved the existence of varying degrees of the penetration of PEO chains into the silica walls, that is the "three-phase model" is the most realistic one [21]. While physisorption techniques could not give any unambiguous conclusions, detailed small-angle X-ray and neutron scattering (SAXS, SANS) analyses of PS-b-PEO templated silicas provided quantification of the microporosity [22, 24–26] and proved that the micropores actually contributed significantly to the overall surface area. More importantly, *in situ* SANS-nitrogen sorption experiments provided evidence that the micropores are indeed located between the mesopores, as required for the aforementioned models (the solvating polymer chains form their own "molecular" pores) [26].

These *in situ* sorption–SANS studies also suggested that the two-phase situation occurs for relatively large ratios between the hydrophobic and hydrophilic block lengths, while smaller ratios lead to the three-phase case. In essence, it was concluded that long PEO blocks cannot be sufficiently solubilized and therefore "phase" separate into a fraction collapsing onto the micellar core and a fraction penetrating the silica matrix (scenario (C) in Fig. 13.7). These and further studies showed that the solvent environment plays a crucial role in the conformation of the PEO chains and thus for the microporosity. Indeed, it was observed that the microporosity can be influenced by the temperature and the salt addition during the templating process [27]. With increasing temperature the microporosity decreases, which can be attributed to the sensitive phase behavior of PEO in an aqueous environment: PEO has an upper critical solution temperature and therefore becomes more and more water insoluble at elevated temperatures, thus resulting in the collapse of PEO chains on the micellar core and in the consequent absence of microporosity.

For the interested reader, further studies, which provide more details on this topic of nonionic, small surfactants, are recommended [21].

13.4
Crystalline, Mesoporous Metal Oxides

One of the major promising areas, coupled with current scientific trends in BC templating, is the creation of mesoporous metal oxides with the pore walls being entirely nanocrystalline. Such materials are suggested to show high promise for a number of vital problems in materials chemistry, because various metal oxides possess interesting inherent physical functionality, which is improved or becomes accessible by the introduction of mesopores. For instance, mesoporous, crystalline TiO_2 could be beneficial for various applications such as solid-state solar cells ("Grätzel cell") [28] or photocatalysis [29]. In all these applications, porosity would be helpful to improve the contact between the oxide and other species (e. g., sensitizers, in the case of solar cells).

For numerous catalytic purposes, it would be desirable to generate mesoporous metal oxides such as CeO_2, ZrO_2 and V_2O_5. While BC templating offers an elegant method of creating such species with mesostructure, the necessity for a high crystallinity imposes a severe dilemma: it is evident that the soft self-assembled structures of ABCs are not very compatible with the high thermal and mechanical load required for and created by the crystallization processes. The use of preformed, soluble metal oxide nanoparticles, which would be partitioned around the BC micelles in the course of the self-assembly process (Fig. 13.6) evidently has certain elegance because of the absence of a separate nucleation step for the oxide, which is usually achieved by vigorous heat-treatment. Indeed, well-ordered mesoporous SnO_2 and CeO_2 powders were recently realized by Niederberger and coworkers who used this approach [17], and similarly, a pre-hydrolyzed solution of titanium isopropoxide, containing nanometer-sized particles, was used to create mesoporous TiO_2 films [16]. However, the small number of systems prepared in this way also indicates some difficulties. Firstly, the metal oxide under consideration has to be synthesized as nanometer-sized, highly-crystalline particles, which has not been achieved for too many oxides. Secondly, and more seriously, dispersing the nanoparticles in solvents such as water, ethanol or THF without aggregation is a major difficulty, and is possible for only a few systems. Thirdly, the physical properties (conductivity, etc.) may be severely aggravated by insufficient connectivity between the nanoparticles. Hence, the most frequently employed strategy starts from molecular metal precursors, such as metal chlorides or alkoxides (e. g., $TiCl_4$, $CeCl_3.7H_2O$), and is regarded as being general. Many metal oxide precursors are indeed available, being amenable to sol–gel templating and soluble in common solvents. This approach also faces severe difficulties, in particular due to the nucleation and crystallization step, which is inevitable when generating a highly crystalline metal oxide. Evidently, the crystallization and crystal growth can severely disrupt the fragile, soft ABC mesostructure, which would lead to pore collapse, loss of mesostructural order and uncontrolled crystal growth. Therefore, fairly delicate and tedious heat-treatment procedures usually have to be used to avoid mesostructural breakdown and to balance condensation, crystallization and calcination of the template.

Previous experiments with classical templates gave highly-crystalline mesoporous oxides in only a few cases, in particular for TiO_2 [30–32]. In the majority of these studies the materials showed either moderate crystallinity or significant damage to the mesostructural regularity (pore shape and porosity). Grosso et al. developed an *in situ* experiment to study the morphological changes upon heat-treatment by simultaneous small- and wide-angle X-ray scattering (SAXS–WAXS), leading to an improved understanding of the crystallization and changes in the mesostructure [33].

Again, progress in the rational design of novel amphiphilic block copolymer templates helped to partially resolve these difficulties and to achieve a more precise understanding of the templating of crystalline oxides, because KLE templates with

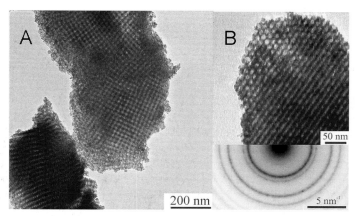

Fig. 13.8 Crystalline, mesoporous CeO_2 films prepared from molecular precursors (A) [35c], and mesoporous powder obtained from preformed CeO_2 nanoparticles (B) [17a]. The selected area electron diffraction pattern (bottom part of B) proves the crystallinity of the mesopore walls.

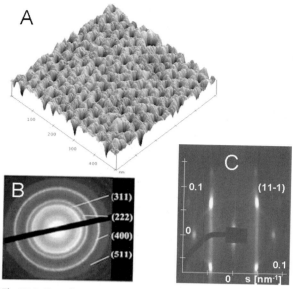

Fig. 13.9 Crystalline, mesoporous films of γ-Al$_2$O$_3$, heat-treated at 900 °C [36a]. (A) Atomic force microscopy image; (B) selected area electron diffraction; and (C) 2D-SAXS pattern obtained at a low angle of incidence with respect to the substrate.

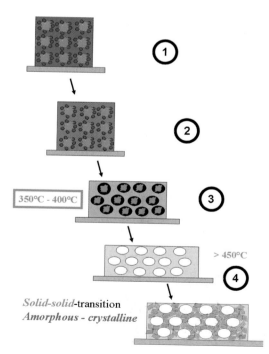

Solid-solid-transition
Amorphous - crystalline

Fig. 13.10 Formation mechanism of crystalline metal oxide films, as found for HfO$_2$ [36b] using KLE block copolymers.

low polydispersity represented an ideal model case to perform nanocasting and structuration operations, and for the crystalline oxide derivatives.

These simple amphiphilic diblock copolymers [poly(ethylene-co-butylene)-b-poly (ethylene oxide)] feature more robustness and more hydrophobic "contrast" than the commonly applied polymers of the Pluronics family. In addition, owing to their larger and chemically flexible molecular weight, the pore structure can be adapted to match the nanocrystal size. In addition to KLE, polybutadiene-b-poly(ethylene oxide) (PB-b-PEO) [7] and polyisobutylene-b-poly(ethylene oxide) (PIB-b-PEO) [34] were established as ABC with improved templating properties.

These polymers allowed for the generation of various metal oxides (TiO_2, CeO_2, ZrO_2) as highly crystalline coatings (Fig. 13.8) with high structural regularity of the mesopores (spherical, size larger than 12 nm) [35]. The preparation of crystalline films with very high crystallization temperatures (HfO_2, Al_2O_3, Fig. 13.9) was even possible with these polymers [36]. Furthermore, complex materials such as Perovskites (e. g., $SrTiO_3$) can now be obtained as mesoporous films, which is particularly interesting, because such materials are more difficult to accomplish in the desired stoichiometric composition by sol–gel methods [37].

A detailed investigation of these systems by appropriate, temperature-dependent *in situ* SAXS–WAXS experiments provided clarification of the parameters governing the crystallization and the role of the block copolymer template. Smarsly and coworkers revealed three major aspects of the BC template for the successful generation of mesoporous, crystalline metal oxides [36b], which come into play at different stages during the self-assembly, condensation and crystallization (Fig. 13.10):

- A strong driving force for block copolymer self-assembly is inevitably needed, which manifests itself in a greater hydrophobic–hydrophilic contrast between the two different types of blocks (step 1 in Fig. 13.10). This is advantageous, because micellization/self-assembly is less sensitive to parameters such as the type of solvent and the presence of inorganic species and their hydrolysis. This issue is of particular significance for transition metal salts such as zirconium species, which have a strong tendency towards hydrolysis. Furthermore, a highly hydrophobic block prevents water and inorganic salts from entering the micellar core, which could lead to disruption of the micelles upon heat-treatment (step 2 in Fig. 13.10).
- The temperature stability of the amphiphilic BCs was found to be a further crucial factor. Case studies on Al_2O_3 [36a], and particularly HfO_2 [36b], demonstrated that the transformation of the initial self-assembled composite, consisting of the block copolymer and weakly condensed metal oxy-hydrates, into the final mesoporous, crystalline metal oxide proceeds through various steps. As HfO_2 and Al_2O_3 have fairly high crystallization temperatures, well above the decomposition of the polymers, these substances allowed independent observation of the separated condensation, template removal and crystallization steps. *In situ* WAXS–SAXS experiments performed during the heat-treatment of these oxides revealed that the crystallization had to be adjusted to be a solid–solid transforma-

tion from an almost completely dry, water-free amorphous metal oxide into its crystalline counterpart (step 3 in Fig. 13.10). It is essential that the BC template is sufficiently stable to allow for the stabilization of the mesostructure up to a temperature (350–400 °C in an atmosphere of nitrogen) where the anhydrous metal oxide species is formed. KLE templates show a significantly higher decomposition temperature than the Pluronics templates, which therefore helped to facilitate and improve the generation of crystalline metal oxides. If the BC templates decompose too early during the heat-treatment, the rearrangement of the matrix due to capillary forces and enforced condensation (e.g., in the form of Me–OH groups) drastically disturbs the mesostructrure.

- The crystallization of metal oxides from sol–gel precursors usually leads to a minimum initial crystal size of ca. 2–5 nm, depending on the respective oxide. Evidently the pore walls separating the BC micelles have to be sufficiently thick to keep the mesostructure intact while forming these crystals, i.e., ca. 8–10 nm (step 4 in Fig. 13.10). Based on the minimum nanocrystal size, mechanical stability can only be achieved if the micelles/mesopores are sufficiently larger than the crystals, that is ca. 10–15 nm at minimum. These prerequisites impose the necessity of large block-lengths of the amphiphilic BCs, and all successfully prepared mesoporous metal oxides have been obtained from larger BCs.

Very recently, some new approaches were developed to overcome some of the aforementioned problems. For instance, in the approach by Holmes and coworkers and Watkins and coworkers, the films are exposed to solutions of metal oxide precursors in supercritical CO_2 [38]. This procedure has the advantage that capillary forces are absent and that the films have higher temperature stability, because of the supercritical conditions. In a similar method, Chmelka and coworkers first cross-linked the mesophases of block copolymers featuring double bonds [39], and then the inorganic precursors were infiltrated. This approach reduces the disruption of the BC mesophase upon condensation of the metal salts.

A further interesting approach was introduced by Domen and coworkers [40]. Firstly, an amorphous mesoporous oxide is prepared, which is then followed by carbon deposition inside the pores, acting as a temperature-stable scaffold. The treatment at elevated temperature leads to crystallization, with the carbon preventing mesostructural collapse. Finally, carbon is oxidized by calcination in oxygen.

It is our personal view that these new materials will have a significant impact on nanoscience and nanotechnology, especially for diverse electrochemical applications where complex metal oxides are required with high surface areas. It seems justifiable to predict that various fields, such as electrochromism, batteries and electrode materials, will dramatically benefit from progress in the preparation of mesoporous oxides.

13.5
Mesoporous Metals

Aside from mesoporous metal oxides, mesoporous elemental metals, e.g., Pt, Pd and Ni, were obtained by block copolymer templating. In principle, mesoporous metals are also accessible through the schematic principle depicted in Fig. 13.1. Similar to mesoporous oxides, in the first step, reducible metal salts ($PdCl_2$, H_2PtCl_6, etc.) are distributed around the block copolymer domains. The preparation of the corresponding metals is more delicate because of the additional reduction step, which can be performed by electrodeposition (for instance on gold electrodes) or chemically, e.g., by metallic zinc. A second problem of these strategies is the high densification of matter throughout the metal formation.

The group working with Attard reported the electrochemical preparation of mesoporous metal electrodes (Fig. 13.11A) [41], which could be interesting for various applications such as fuel cells (Ni) or electrochemical hydrogenation of organic species (Pd, Pt). Yamauchi et al. [42] described the generation of mesostructured particles of Ni (Fig. 13.5) and Ni–Co alloys, partly by electrodeless methods, in particular by using organic reducing agents, such as dimethylamineborane or sodium borohydride. Furthermore, Kucernak demonstrated that it was possible to obtain mesostructured, otherwise unaccessible mesoporous "alloys" such as Pt–Ru [43]. Such two-phasic, mesostructured metal blends show high promise for catalytic applications, e.g., in fuel cells.

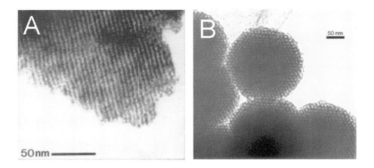

Fig. 13.11 Mesoporous Pt (A) [41a] and mesoporous Ni particles (B) [42a].

13.6
Conclusion and Outlook

Block copolymer templating, particularly nanocasting, is now commonly regarded as a standard, well established approach in materials science, well documented by the broad variety of new materials described above. However, in spite of the richness of new structures it seems worthwhile to take a step back again and reconsider this methodology in the light of the "classical" classification of materials science. Surprisingly, the use of amphiphilic block copolymers and their lyotropic mesophases, which represent "soft", fragile matter in its truest sense, allows for the generation of mesostructured, "hard" matter, even such durable and mechanically stable species as mesoporous, crystalline Al_2O_3, i. e., a material with entirely opposite properties to those of polymers. This consideration shows that block copolymers, as a relatively new class of polymers, have revolutionized materials by the concept that soft matter organization can be converted into hard matter through this "hardcopy process". Amphiphilic block copolymers therefore have added a further importance to polymer sciences in general, complementing the thermoplastic properties already well known for BCs. Thus, amphiphilic BCs have established a bridge between "hard" and "soft" matter, which were considered as totally separate domains of research. Through the hardcopy approach, BCs will contribute to the further development of completely different areas of materials science, e. g., sensing, catalysis, electrochromism, solar cells. In other words, one could even postulate the theorem that advances in these areas will significantly depend on progress in the field of BCs.

As we have seen, nanocasting could provide a strategy to create completely novel materials with mesostructure, such as novel types of carbon, complex oxides and semiconductors. Moreover, BCs in combination with nanocasting can be expected to create further diversity to the mesostructure. For instance, by using templates of different sizes one could even generate hierarchical pore structures through the nanocasting approach [44, 45], which is almost impossible by any other method.

In conclusion, BC templating has become a mature area of research in material science, and further important progress can be expected by the appropriate design of polymers. It could be expected that in the near future important technological improvements will be achieved with inorganic materials obtained through block copolymer templating.

References

1 a) S. FORSTER, M. ANTONIETTI, *Adv. Mater.* **1998**, *10*, 195; b) S. FORSTER, T. PLANTENBERG, *Angew. Chem., Int. Ed. Engl.* **2002**, *41*, 689–714.

2 U. CIESLA, F. SCHUTH, *Micropor. Mesopor. Mater.* **1999**, *27*, 131–149.

3 C. T. KRESGE, M. LEONOWICZ, W. J. ROTH, J. C. VARTULI, J. S. BECK, *Nature (London)* **1992**, *359*, 710–712.

4 a) G. S. ATTARD, J. G. GLYDE, C. G. GÖLTNER, *Nature (London)* **1995**, *378*, 366–368; b) G. S. ATTARD, C. G. GÖLTNER, J. M. CORKER, S. HENKE, R. H. TEMPLER, *Angew. Chem., Int. Ed. Engl.* **1997**, *109*, 1372–1374.

5 a) C. G. GÖLTNER, S. HENKE, M. C. WEISSENBERGER, M. ANTONIETTI, *Angew. Chem., Int. Ed. Engl.* **1998**, *110*, 633–636; b) C. G. GÖLTNER, M. ANTONIETTI, *Adv. Mater.* **1997**, *9*, 431; c) M. C. WEISSENBERGER, C. G. GOLTNER, M. ANTONIETTI, *Ber. Bunsenges. Phys. Chem.* **1997**, *101*, 1679–1682.

6 S. POLARZ, M. ANTONIETTI, *Chem. Commun.* **2002**, 2593–2604.

7 a) C. G. GÖLTNER, B. BERTON, E. KRÄMER, M. ANTONIETTI, *Adv. Mater.* **1999**, *11*, 395–398; b) S. FORSTER, B. BERTON, H. P. HENTZE, E. KRAMER, M. ANTONIETTI, P. LINDNER, *Macromolecules* **2001**, *34*, 4610–4623; c) H. P. HENTZE, E. KRAMER,

B. Berton, S. Forster, M. Antonietti, M. Dreja, *Macromolecules* **1999**, *32*, 5803–5809.

8 a) D. Zhao, Q. Huo, J. Feng, B. F. Chmelka, G. D. Stucky, *J. Am. Chem. Soc.* **1998**, *120*, 6024–6036; b) D. Zhao, J. Feng, Q. Huo, N. Melosh, G. H. Fredrickson, B. F. Chmelka, G. D. Stucky, *Science* **1998**, *279*, 548–552.

9 E. Krämer, S. Förster, C. Göltner, M. Antonietti, *Langmuir* **1998**, *14*, 2027–2031.

10 a) A. Thomas, H. Schlaad, B. Smarsly, M. Antonietti, *Langmuir* **2003**, *19*, 4455–4459; b) M. Groenewolt, M. Antonietti, S. Polarz, *Langmuir* **2004**, *20*, 7811–7819; c) C. G. Göltner, B. Berton, E. Krämer, M. Antonietti, *Chem. Commun.* **1998**, 2287–2288.

11 H. P. Hentze, E. Krämer, B. Berton, S. Förster, M. Antonietti, *Macromolecules* **1999**, *32*, 5803–5809.

12 a) F. S. Bates, G. H. Fredrickson, *Annu. Rev. Phys. Chem.* **1990**, *41*, 525–557; b) I. W. Hamley, *Philos. Trans. R. Soc. London A — Math. Phys. Eng. Sci.* **2001**, *359*, 1017–1044.

13 C. J. Brinker, Y. F. Lu, A. Sellinger, H. Y. Fan, *Adv. Mater.* **1999**, *11*, 579–585.

14 a) W. Ruland, B. Smarsly, *J. Appl. Crystallogr.* **2004**, *37*, 575–684; b) W. Ruland, B. Smarsly, *J. Appl. Crystallogr.* **2005**, *38*, 78–86.

15 A. Gibaud, D. Grosso, B. Smarsly, A. Baptiste, J. F. Bardeau, F. Babonneau, D. A. Doshi, Z. Chen, C. J. Brinker, C. Sanchez, *J. Phys. Chem. B* **2003**, *107*, 6114–6118.

16 F. Bosc, A. Ayral, P. A. Albouy, C. Guizard *Chem. Mater.* **2003**, *15*, 2463–2468.

17 a) A. S. Desphande, N. Pinna, B. Smarsly, M. Antonietti, M. Niederberger, *Small* **2005**, *1*, 313–316; b) J. Ba, J. Polleux, M. Antonietti, M. Niederberger, *Adv. Mater.* **2005**, *17*, 2509–2512.

18 a) N. Pinna, G. Garnweitner, M. Antonietti, M. Niederberger, *J. Am. Chem. Soc.* **2005**, *127*, 5608–5612; b) N. Pinna, G. Garnweitner, M. Antonietti, M. Niederberger, *Adv. Mater.* **2004**, *16*, 2196; c) M. Niederberger, M. H. Bard, G. D. Stucky, *J. Am. Chem. Soc.* **2002**, *124*, 13642–13643.

19 M. Björling, P. Linse, G. Karlström, *J. Phys. Chem.* **1990**, *94*, 471–481.

20 S. M. De Paul, J. W. Zwanziger, R. Ulrich, U. Wiesner, H. W. Spiess, *J. Am. Chem. Soc.* **1999**, *121*, 5727.

21 B. Smarsly, S. Polarz, M. Antonietti, *J. Phys. Chem. B* **2001**, *105*, 10473–10483.

22 C. G. Göltner, B. Smarsly, B. Berton, M. Antonietti, *Chem. Mater.* **2001**, *13*, 1617–1624.

23 a) M. Jaroniec, M. Kruk, C. H. Ko, R. Ryoo, *Chem. Mater.* **2000**, *12*, 1961–1968; b) R. Ryoo, C. H. Ko, M. Kruk, V. Antochshuk, M. Jaroniec *J. Phys. Chem. B* **2000**, *104*, 11465.

24 B. Smarsly, M. Thommes, P. I. Ravikovitch, A. V. Neimark, *Adsorption* **2005**, *11*, 653–655.

25 M. Impéror-Clerc, P. Davidson, A. Davidson, *J. Am. Chem. Soc.* **2000**, *122*, 11925–11933.

26 B. Smarsly, C. Göltner, M. Antonietti, W. Ruland, E. Hoinkis, *J. Phys. Chem. B* **2001**, *105*, 831–840.

27 a) A. Galarneau, N. Cambon, F. Di Renzo, R. Ryoo, M. Choi, F. Fajula, *New J. Chem.* **2003**, *27*, 73–79; b) B. L. Newalkar, S. Komarneni, *Chem. Mater.* **2001**, *13*, 4573–4579; c) K. Miyazawa, S. Inagaki, *Chem. Commun.* **2000**, 2121–2122.

28 M. Grätzel, *Nature (London)* **2001**, *414*, 338–344.

29 I. Mora-Seró, *J. Phys. Chem. B* **2005**, *109*, 3371–3380.

30 D. M. Antonelli, J. Y. Ying, *Curr. Opin. Coll. Interf. Sci.* **1996**, *1*(4), 523–529.

31 P. D. Yang, T. Deng, D. Y. Zhao, P. Y. Feng, D. Pine, B. F. Chmelka, G. M. Whitesides, G. D. Stucky, *Science* **1998**, *282*, 2244.

32 a) O. Dag, I. Soten, O. Celik, S. Polarz, N. Coombs, G. A. Ozin, *Adv. Func. Mater.* **2003**, *13*, 30–36; b) S. Y. Choi, M. Mamak, N. Coombs, N. Chopra, G. A. Ozin, *Adv. Func. Mater.* **2004**, *14*, 335–344.

33 a) D. Grosso, G. J. D. A. Soler-Illia, E. L. Crepaldi, F. Cagnol, C. Sinturel, A. Bourgeois, A. Brunet-Bruneau, H. Amenitsch, P. A. Albouy, C. Sanchez, *Chem. Mater.* **2003**, *15*, 4562–4570; b) E. L. Crepaldi et al., *J. Am. Chem. Soc.* **2003**, *125*, 9770–9786.

34 M. Groenewolt, T. Brezesinski, H. Schlaad, M. Antonietti, P. W. Groh, B. Ivan. *Adv. Mater.* **2005**, *17*, 1158.

35 a) B. Smarsly, D. Grosso, T. Brezesinski, N. Pinna, C. Boissière, M. Antonietti, C. Sanchez, *Chem. Mater.* **2004**, *16*, 2948–2952; b) T. Brezesinski, M. Groenewolt, N. Pinna, M. Antonietti, B. Smarsly, *New J. Chem.* **2005**, *29*, 237–242; c) T. Brezesinski, C. Erpen, K. I. Iimura, B. Smarsly, *Chem. Mater.* **2005**, *17*, 1683–1690.

36 a) M. Kuemmel, D. Grosso, C. Boissière, B. Smarsly, T. Brezesinski, P. A. Albouy, H. Amenitsch, C. Sanchez, *Angew. Chem., Int. Ed. Engl.* **2005**, *44*, 4589–4592; b) T. Brezesinski, B. Smarsly, K. I. Iimura, D. Grosso, C. Boissière, H. Amenitsch, M. Antonietti, C. Sanchez, *Small* **2005**, *1*, 889–898.

37 D. Grosso, C. Boissière, B. Smarsly, T. Brezesinski, N. Pinna, P. A. Albouy, H. Amenitsch, M. Antonietti, C. Sanchez, *Nature Mater.* **2004**, *3*, 787–792.

38 a) K. X. Wang, B. D. Yao, M. A. Morris, J. D. Holmes, *Chem. Mater.* **2005**, *17*(19), 4825–4831; b) R. A. Pai, R. Humayun, M. T. Schulberg, A. Sengupta, J. N. Sun, J. J. Watkins, *Science* **2004**, *303*, 507–510.

39 R. C. Hayward, B. F. Chmelka, E. J. Kramer, *Macromolecules* **2005**, *38*, 7768–7783.

40 J. N. Konodo, T. Katou, D. Lu, M. Hara, K. Domen, *Stud. Surf. Sci. Catal.* **2004**, *154*, 951–957.

41 a) G. S. Attard, P. N. Bartlett, N. R. B. Coleman, J. M. Elliott, J. R. Owen, J. H. Wang, *Science* **1997**, *278*, 838-840; b) P. A. Nelson, J. M. Elliott, G. S. Attard, J. R. Owen, *Chem. Mater.* **2002**, *14*, 524-529; c) J. M. Elliott, G. S. Attard, P. N. Bartlett, N. R. B. Coleman, D. A. S. Merckel, J. R. Owen, *Chem. Mater.* **1999**, *11*, 3602–3609.

42 a) Y. Yamauchi, T. Momma, T. Yokoshima, K. Kuroda, T. Osaka, *J. Mater. Chem.* **2005**, *15*, 1987–1994; b) Y. Yamauchi, T. Yokoshima, T. Momma, T. Osaka, K. Kuroda, *J. Mater. Chem.* **2004**, *14*, 2935–2940; c) Y. Yamauchi, T. Yokoshima, T. Momma, T. Osaka, K. Kuroda, *Chem. Lett.* **2004**, *33*, 1576–1579; d) Y. Yamauchi T. Yokoshima, H. Mukaibo, M. Tezuka, T. Shigeno, T. Momma, T. Osaka, K. Kuroda, *Chem. Lett.* **2004**, *33*, 542–543.

43 a) J. Jiang, A. Kucernak, *Chem. Mater.* **2004**, *16*, 1362–1367; b) J. Jiang, A. Kucernak, *J. Electroanal. Chem.* **2003**, *543*, 187–199.

44 T. Sen, G. J. T. Tiddy, J. L. Casci, M. W. Anderson, *Angew. Chem., Int. Ed. Engl.* **2003**, *42*, 4649–4653.

45 D. B. Kuang, T. Brezesinski, B. Smarsly, *J. Am. Chem. Soc.* **2004**, *126*, 10534–10535.

14
Mesostructured Polymer–Inorganic Hybrid Materials from Blocked Macromolecular Architectures and Nanoparticles

Marleen Kamperman and Ulrich Wiesner

14.1
Introduction

Today, both existing and emerging technologies require materials with structure control over multiple length scales, including nanoscale functional features [1]. The availability of such materials would lead to important advances in many industries, including microelectronics and energy conversion, and in technology-enabling processes such as electroluminescence, molecular separation and catalysis. For decades, scientists and engineers have been intrigued by bulk block copolymeric macromolecular architectures leading to distinct property profiles associated with phase-separated morphologies with nanometer spatial control [2]. AB diblock architectures and ABC triblock copolymers have been studied extensively, with the third block leading to an enormous range of structural variability [3–5]. While fascinating bulk morphologies, such as the double gyroid and the "knitting pattern", or solution structures, such as multicompartment micelles, have been revealed for these materials [6, 7], using their intrinsic nanostructure to create nanoscale functionality remains a challenge [8, 9]. Recent advances in studies of small-molecule synthetic peptide-amphiphiles suggest [10], however, that higher order blocked macromolecular architectures could combine functionality imparted by distinct blocks with multiscale order phenomena. Combined with the ability of block copolymers to sequester inorganic species and nanoparticles into specific domains with nanoscale precision [11–15], this offers tremendous opportunities for the generation of materials with complex property profiles. Emerging from these considerations is a vision that, in analogy to biology, the monomer (block) sequence information of synthetic macromolecules can be used to encode information about near-molecular level structure and functionality of co-assemblies with dissimilar nanosized materials, such as ceramics, or even metals. This may lead to the design of entirely new classes of materials with properties that have no analogue in the natural world.

Achieving multiscale order and multiple nanoscale functional features will have a profound impact in a broad range of areas, including power generation and stor-

---

Block Copolymers in Nanoscience. Massimo Lazzari, Guojun Liu, Sébastien Lecommandoux
Copyright © 2006 WILEY-VCH Verlag GmbH & Co. KGaA, Weinheim
ISBN: 3-527-31309-5

age, advanced microelectronics and photonics. The following example elucidates the need for these materials and demonstrates the challenges in their design. In electrocatalysis, multifunctionality is obligatory. For high performance, combinations of surface reactivity, electron conductivity, ionic conductivity, separation of electron–hole pairs, or facile mass transport of molecules through porous media must be provided to enhance the molecular conversion. Electrocatalyzed reactions for energy conversion in low-temperature proton-exchange membrane fuel cells dominate the current literature on electrocatalysis because of the societal interest in generating clean energy. In current fuel cell electrodes, platinum group nanoparticle catalysts are supported on electronically conductive carbon microparticles that are immersed in a proton-conducting polymer. Additionally, some degree of porosity must remain to allow transport of fuel and reaction products. For the cost-effective fabrication of more efficient fuel cells, a self-assembly approach to such multiscale structures with multiple nanoscale functional features is highly desirable.

In this chapter recent progress in the field of mesostructured polymer–inorganic hybrid materials directed from blocked macromolecular architectures is described. Rather than attempting a general overview of this rapidly growing and exciting research area, general features, challenges and future trends will be highlighted through examples of work by the Wiesner research group at Cornell University. The chapter is structured into four parts. In the first part, work on a particular block copolymer–silica-type inorganic hybrid material is highlighted introducing general concepts and co-assembly mechanisms. In the second part, the work is generalized to include other blocked macromolecular amphiphiles as structure-directing agents. The third part generalizes the concepts to other inorganic materials systems, and the final section describes the transition from bulk mesostructured hybrids to mesostructured thin films.

14.2
AB Diblock Copolymers as Structure-directing Agents for Aluminosilicate Mesostructures

To the best of our knowledge it was in the second half of the 1990s that block copolymers were used for the first time as structure-directing agents for silica-type ceramic materials [16–19]. In the Wiesner group, research focused on a particular hybrid system employing poly(isoprene-*block*-ethylene oxide) diblock copolymers (PI-*b*-PEO) as structure-directing agents for organically modified ceramic (ormocer) precursors, (3-glycidyloxypropyl) trimethoxysilane (GLYMO) and aluminum-*sec*-butoxide [Al(O^{s}Bu)$_3$] [17]. After epoxy ring opening triggered through the necessary presence of aluminum (Lewis acid), the 3-glycidyloxypropyl-ligand of the silane precursor makes the inorganic mixture compatible with the PEO block of the block copolymer at any mole fraction. This compatibility allows for the selective swelling of the PEO block when a mixture of pre-hydrolyzed GLYMO–Al(O^{s}Bu)$_3$ is added to PI-*b*-PEO effectively increasing the PEO volume fraction with respect

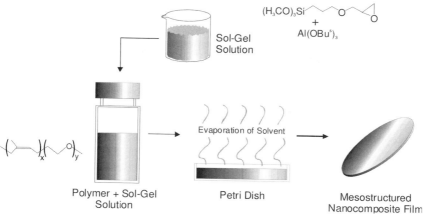

Fig. 14.1 Synthesis schematic of the procedure for preparing aluminosilicate mesostructures from AB diblock copolymer amphiphiles as structure-directing agents. A solution of the pre-hydrolyzed sol–gel precursors is added to the polymer solution from which a film is cast in a Petri dish.

to the PI block. This results in a rational design of hybrid morphology based on current understanding of the phase behavior of block copolymers and copolymer–homopolymer mixtures [2, 20–24]. Figure 14.1 shows a schematic of the synthesis procedure.

In a typical preparation, in a 20-mL vial, 0.5 g of PI-*b*-PEO is dissolved in a 1:1 weight ratio of chloroform and THF (10 g in total). The polymer solution is stirred rapidly for 1 h. Al(O^{s}Bu)$_3$ (1.4 g), GLYMO (5.3 g) (20:80 mol-%) and an excess of KCl (40 mg) are stirred rapidly for 1–2 min at 0 °C in a beaker before 0.27 g of 0.01 M HCl is added dropwise. After 15 min of stirring the sol at 0 °C followed by stirring for 15 min at room temperature, an additional 1.7 g of 0.01 M HCl are added dropwise and stirred for another 20 min. The clear sol is filtered through a 0.2-µm PTFE filter and added dropwise to the polymer solution. The organic–inorganic solution is stirred for 1 h before being transferred into a Petri dish and heated at 50 °C on a heating mantel for 1 h to evaporate off the solvents. The condensation products are removed from the film upon additional heating for 1 h at 130 °C in a vacuum oven giving a film of 0.5–1 mm in thickness. For a description of the specific chemistry taking place under these materials preparation conditions, the interested reader is referred to an earlier review [24] and references cited therein.

Figure 14.2 shows transmission electron microscopy (TEM) images of thin sections of eight hybrid demonstrating (a) the quality of the structure control and (b) that practically the entire phase space of AB diblock copolymers is accessible for mesostructure formation. The morphologies identified in the composites are the sphere (Sph, PEO-inorganic spheres in a matrix of PI), the hexagonal cylinder (Cyl, PEO–inorganic cylinders in a matrix of PI), a cubic bicontinuous morphology (3D-cubic bicontinuous PEO–inorganic channels in a matrix of PI), the lamellar (Lam,

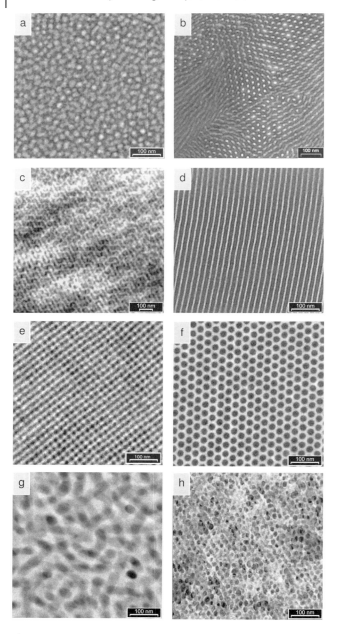

Fig. 14.2 Representative TEM micrographs of eight morphologies observed in mesostructured hybrids from PI-b-PEO–Glymo–Al(O^{s}Bu)$_3$ mixtures: the sphere (a), the hexagonal cylinder (b), a cubic bicontinuous morphology (c), the lamellar (d), the inverse cubic bicontinuous morphology (e), the inverse hexagonal cylinder (f), the wormlike micelle (g) and inverse sphere (h) structures.

PEO–inorganic layers alternating with PI layers), the inverse cubic bicontinuous morphology (3D-cubic bicontinuous PI channels in a matrix of PEO–inorganic), the inverse hexagonal cylinder (invCyl, PI cylinders in a matrix of PEO–inorganic), the wormlike micelle, (WM, PI worms in a matrix of PEO–inorganic) and inverse sphere (invSph, PI spheres in a matrix of PEO–inorganic) structures. The images were taken using energy filtering TEM (EFTEM) and adjusted for structure sensitive contrast ($\Delta E = 250$ eV) making the PEO–inorganic domains appear light and the PI domains appear dark.

Small angle X-ray scattering (SAXS) data from samples with the cubic bicontinuous morphology (Fig. 14.2 c and e) are consistent with the Plumber's Nightmare morphology and its inverse [25, 26, 48]. This is remarkable because in diblock copolymer self-assembly usually the double-gyroid structure is found [2]. The reasons for this switch of morphology are not fully understood and are still under investigation. Overall, the degree of order in these composites is surprisingly high. As pointed out earlier, the block copolymer has two important features that enable this degree of mesostructure control [17]. Firstly, the hydrolysis products of the metal alkoxides preferentially swell the hydrophilic PEO block as a result, e.g., of hydrogen bonding. Secondly, the low glass transition temperature, T_g, of the blocks introduces high mobility at ambient temperatures allowing rapid structure formation even in the bulk.

14.2.1
Formation Mechanisms

In order to understand the mesostructure formation mechanism it is instructive to take a closer look at what is known about the sol–gel process of silica precursors and then apply this knowledge to the current PI-*b*-PEO plus GLYMO–Al(O^{s}Bu)$_3$ system. Two-step acid catalyzed sol–gel polymerization of traditional silica sources under the reaction conditions employed here has been shown to result in nanoparticles, which are mass fractals with a fractal dimension D_f around 2.0 [27–29]. It can thus be concluded, and was experimentally verified, that at the stage in the synthe-

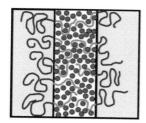
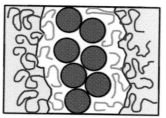

Fig. 14.3 AB diblock copolymer–nanoparticle co-assemblies as a function of nanoparticle size. The left-hand side shows nanoparticles smaller than the radius of gyration of the block copolymer leading to a particle distribution throughout the B domain. The right-hand scenario shows much larger nanoparticles segregating out from the block copolymer.

sis procedure where solvents are evaporated off, the PI-*b*-PEO block copolymer is mixing with nanoscopic silica-type particles [15]. The resulting morphologies then depend on the amount of sol added to the block copolymer, through mixing of the organically modified inorganic particles (referred to as the "inorganic") with the PEO block of the block copolymer [17, 24].

Based on these considerations, the Wiesner group recently developed a two-step model for the block copolymer–inorganic hybrid formation [15]. In the first step, nanoscopic (<5 nm) silica-type particles are formed through hydrolysis and condensation of the GLYMO and $Al(O^sBu)_3$ precursors. After mixing with the PI-*b*-PEO AB diblock copolymer solution, through evaporation of the volatile organic solvents they start to selectively swell the PEO block B. When nanoparticles are mixed into the B domain, the resulting structural changes depend strongly on the particle size. As analyzed in detail, e. g., for low particle volume fractions, in recent work by Thompson et al. [30], when the particles are small in comparison with the radius of gyration of the polymer chain, they can swell the polymer domain without significantly perturbing the polymer chain conformations. In this instance, the translational entropy of the particles dominates the behavior leading to a dispersion of the nanoparticles within the B (PEO) domain, as depicted on the left of Fig. 14.3. In contrast, when the particle size is of the order of the radius of gyration of the chains or larger, incorporation into the B domain (PEO) significantly perturbs the chain conformations, as B chains have to stretch to get around the spheres resulting in a loss in chain conformational entropy. Larger particles then segregate into a particle rich core, which prevents the chain entropy penalty (at the expense of translational entropy of the particles), as shown on the right of Fig. 14.3.

From early NMR results we know that a segregation of a silica domain from a PEO domain does not occur in the PI-*b*-PEO–GLYMO–$Al(O^sBu)_3$ system at sufficiently high inorganic loading [31]. This suggests that the scenario depicted on the left of Fig. 14.3 closely describes the hybrid formation process. The small nanoparticles only weakly perturb the PEO chain conformations and their subsequent crosslinking through further condensation reactions leads to the final condensed state of the hybrid after additional thermal annealing for 1 h at 130 °C in a vacuum oven, leaving the PEO chains completely immersed in the organic–inorganic framework.

At this point it is interesting to revisit solid-state ^{29}Si NMR results that indicated that the local inorganic environments are not significantly perturbed by the presence of the PEO chains [31]. Irrespective of the PI-*b*-PEO–inorganic mixing ratio, ^{29}Si spectra showed about 30–35 % T^3 groups, 49–54 % T^2 groups and 12–18 % T^1 groups, with no obvious trends as a function of composition [32]. In contrast, intuitively for a homogeneous mixing process one would rather expect that the incorporation of PEO chains into the inorganic network would lead to a significant suppression of the most highly condensed T^3 groups and populate the less condensed T^1 and T^2 groups. This effect should be strongest for small amounts of inorganic (e. g., micelles of inorganic–PEO in PI matrix morphology) and systematically weaken as the inorganic loading is increased (e. g., wormlike PI micelles in inorganic–PEO matrix morphology).

A condensation mechanism that is consistent with these earlier NMR results and knowledge of the sol–gel process is the following. During pre-hydrolysis, mass fractals in the form of silica-type nanoparticles are first formed. The 2-step (hydrolysis and condensation) acid catalysis produces a distribution of T^1–T^3 species as expected for classic polycondensation of multifunctional monomers [33]. After mixing with the block copolymer solution as the solvents evaporate and hydrolysis and condensation proceed, the nanoparticles start to impinge on each other. The open network structure generated in this way and filled with the PEO chains is likely to be qualitatively similar to what is typically observed during the formation of a conventional alcogel. After supercritical evaporation of the solvents the pore structure of the resulting aerogel was imaged in previous work by, for example, Ruben et al. [34]. Furthermore, at this point the number of additional bonds that need to be formed in order to generate a full three-dimensional network of nanoparticles is small compared with the number of bonds already present. This may explain why the relative populations of the different silicon T-sites in hybrids do not vary much as a function of inorganic content in the block copolymer and do not differ significantly from those in the pure hydrolyzed GLYMO–aluminum-*sec*-butoxide mixture [31].

This suggested condensation mechanism is not only consistent with the nanoparticle picture and NMR results, but also with heterogeneities as observed in SFM phase images of the resulting material after dissolution into small nanoobjects (data not shown) [15]. The sol–gel process is not expected to lead to a homogenous distribution of nanoparticles but rather to an open network structure with free volume that can be occupied by the PEO chains. The nanoparticle-rich domains will be much harder than the PEO-rich domains. It can thus be assumed that it is these heterogeneities in the network density that were phase imaged by SFM. Finally, the suggested mechanism and the resulting "open" structure is consistent with the observation of microporosity in the walls of mesoporous materials in the case where PEO containing block copolymers rather than, for example, ionic surfactants are used as the silica structure-directing agent [35]. The microporosity in the wall is a direct result of the heterogeneity described above.

It is interesting to consider the broader implications of the suggested hybrid formation mechanism. Whereas, most theoretical [30, 36–41] and experimental [11–14, 42–47] studies of block copolymers with nanoparticles up to the present date describe the low nanoparticle density regime, the present work strongly suggests that working in the high nanoparticle density regime might lead to some exciting new possibilities. As an example, it should be possible to generalize the hybrid formation mechanism beyond the present silica-based system. In principle, any inorganic material should be applicable that can: (a) be generated in small nanoparticle form, i.e., with particle dimensions significantly smaller than the radius of gyration of the polymer chains and (b) be made compatible (e.g., through the appropriate surface modifications) with one block of the block copolymer or any other amphiphile to provide nanostructurability through co-assembly. If an appropriate surface chemistry allows further cross-linking of the inorganic, the nanostructure can additionally be fixed permanently. Thus, not only will amorphous materials be nanostructured through the help of block copolymers, but also crystalline inorganic materials (see below) or even metals. This may open an interesting playground for the design of novel nanostructured, functional hybrids.

14.2.2
Flow-induced Alignment of Mesostructured Block Copolymer–Sol Nanoparticle Co-assemblies

Co-assembly of block copolymers with inorganic materials leads to hierarchical order with phase-separated regions forming grains with sizes of the order of a few microns. Cylindrical or lamellar mesophases are directional, which results in local mechanical anisotropy in the grains of these materials. However, the grains are relatively small in comparison with the dimensions of samples needed for many applications, and often randomly distributed, which puts severe limitations on their utility. Apart from the large grain forming bicontinuous cubic hybrids [25, 48], which are challenging to obtain experimentally because of their restricted window of stability, the more common phases require development of simple approaches for the generation of macroscopically aligned bulk samples for realization of the full application potential of these hybrid materials.

Flow fields have been successfully used to change the structure and long-range order in pure block copolymers [2]. Orientation of the microstructure both in the melt [49–51] and in solution [52], either parallel or perpendicular to the direction of flow can be achieved by carefully manipulating the flow fields and conditions. However, co-assembly in polymer sol–gel nanoparticle hybrids is accompanied by inorganic nanoparticle gelation, preventing straightforward extension of these approaches to obtain macroscopically aligned nanostructured bulk hybrids. Alignment in such systems has therefore been obtained using surface effects, extensional flow or considerable modification *during* the hybrid synthesis procedures [17, 53–56]. As an example, dip coating has been used to achieve high degrees of orientation in hybrid thin (<1 µm) films obtained through evaporation induced self assembly (EISA) [53, 56]. In our first publication on block copolymer directed aluminosilicate mesostructures, we reported that simple film casting leads to substantial macroscopic orientation of the lamellae parallel to the substrate [17]. While these techniques clearly demonstrate the ability to achieve orientation effects in nanostructured silica-type hybrids, they are usually applicable only to thin films and/or not adaptable to large-scale synthesis of bulk materials. We therefore recently presented a straightforward scheme for the macroscopic alignment of bulk polymer–nanoparticle hybrid materials from solutions subjected to hydrodynamic flow based on a simplified version of the "roll casting" method used by Albalak and Thomas [57] to produce oriented pure block copolymer samples. Initially the technique requires relatively low viscosity starting solutions. This is advantageous for solutions, such as the examples discussed in this chapter, which are not viscous enough to work well with roll-casting. Also, this method may be extended in a straightforward way to a larger scale commercial process.

14.2 AB Diblock Copolymers as Structure-directing Agents for Aluminosilicate Mesostructures | 317

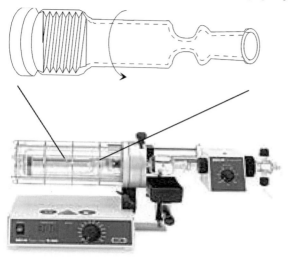

Fig. 14.4 Schematic of the custom-built flow device with a Teflon screw cap operated inside the Büchi evaporator. The arrow indicates the direction of flow.

Alignment experiments were performed as follows. After mixing the polymer with sol nanoparticles from the pre-hydrolyzed precursors in organic solvents (tetrahydrofuran–chloroform in a ratio of 1:1 by weight), the solution was stirred in a vial for 1 h. Oriented films were formed by transferring the solution into a custom-built flow device (Fig. 14.4) placed horizontally in a Büchi evaporator at 50 °C under dynamic vacuum and the solution was rotated for 1–2 h until all the solvents evaporated off. The rotation speed of the cylinder is an important parameter in the set-up for effective alignment of the nanostructure. It was chosen such that a stable flow profile was ensured during the course of film formation. A high rotation speed is likely to induce a larger orientation effect but at the same time leads to instabilities in the flow. A trade-off was therefore needed to maximize the orientation. Based on these considerations, for an inner cylinder diameter of 25 mm of the flow device, a rotation speed of ≈15 rpm was used, which resulted in optimum orientation effects while ensuring a uniform film thickness. Further cross-linking of the gel and removal of the condensation products from both control samples as well as flow-aligned samples were achieved by additional heating at 130 °C in vacuum resulting in 0.5–1 mm thick nanostructured films.

Bulk hybrid nanostructured films with different morphologies were cast from organic solvents under flow and were found to be macroscopically aligned. SAXS was used to characterize the oriented hybrids, see Fig. 14.5. Hexagonal samples with low inorganic content showed quasi-single crystalline SAXS patterns when the X-ray beam was parallel to the flow direction. They exhibited considerable angular spread along the other two directions (Fig. 14.5a). Maximum orientation effects were found for the lamellar samples with order parameters for the lamellar par-

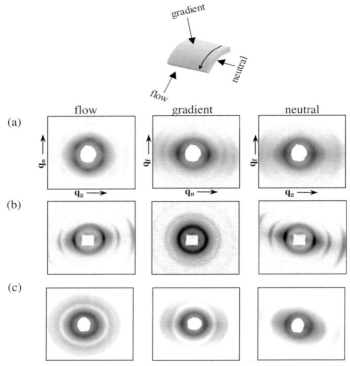

Fig. 14.5 2D-SAXS data for three orthogonal incident X-ray beam directions as indicated in the top sketch for: (a) a flow cast hexagonal hybrid, (b) a flow cast lamellar hybrid, and (c) a flow cast inverse hexagonal hybrid.

allel to the substrate as high as 0.77 (Fig. 14.5b). After flow alignment, hexagonal hybrids with an inorganic matrix did not show quasi-single crystalline patterns but exhibited similar order parameters around 0.4 for the orientation distribution of the cylinders along the flow direction, as in the regular hexagonal samples (Fig. 14.5c).

The extension of the roll casting technique [57] thus holds promise for considerably improving the macroscopic order of sol–gel derived bulk polymer–inorganic hybrids. Alignment of these technologically relevant materials is vital for realizing their true potential, and development of simple techniques such as the one described here may hold the key for successful implementation of these functional materials in real applications. Furthermore, the orientation technique described herein is not limited to sol–gel derived hybrids but rather should be applicable to any block copolymer–nanoparticle assemblies formed under evaporation of the solvent(s), which exclude conventional post-synthesis alignment methods.

14.3
Generalization to Other Blocked Macromolecular Amphiphiles as Structure-directing Agents for Mesostructured Materials

More recently, the range of organic structure directing agents has been extended to other molecular architectures in order to realize advanced functionalities. For example, Aida and co-workers used charge transfer complexes based on an amphiphilic triphenylene electron donor as structure-directing agents [58]. They tuned the color of organic–inorganic nanohybrids depending on the electron acceptor. In another example, Kimura et al. used a disc-shaped phthalocyanine based amphiphile [59]. They prepared a hexagonally arrayed cylindrical nanohybrid material through sol–gel polymerization of inorganic precursors. In addition, they employed a polymeric phthalocyanine, which organized into potentially conductive one-dimensional molecular cables [60].

Besides using various low molecular weight surfactant structures, it is particularly interesting to explore the wide variety of polymer architectures to design novel structure-directing agents. In particular, to the best of our knowledge, dendrimer-type macromolecules had not been used before to direct silica-type mesostructured materials, despite their fascinating structural features [61]. To this end the Wiesner group recently described the synthesis and characterization of a novel family of blocked extended amphiphilic dendrons, see Fig. 14.6 [62, 63]. From a structural point of view, the hydrophilic fraction consists of linear PEO attached to a PEO-like dendritic core, while the hydrophobic fraction is composed of eight docosyl branches [64]. The interface of a phase segregated bulk phase of this compound runs through the middle of the dendritic part of the molecule, and is expected to have an inherent curvature, which is unique compared with the aforementioned conventional linear amphiphiles. Indeed, we recently prepared extended amphiphilic dendrons in the shape of macromolecular dumbbells with identical hydrophilic volume fractions but with branched second- and third-generation dendritic architectures, respectively [65]. Structural data for the second-generation based species were consistent with a 2D-hexagonal columnar mesophase, while those of the third-generation based compounds were consistent with a $Pm\bar{3}n$ micellar cubic mesophase. These results demonstrated that compared with linear block copolymers, self-assembly can be fine-tuned in more detail, because, in addition to volume fraction, generation dependent molecular shape is an independent parameter for tailoring bulk supramolecular architecture.

These novel amphiphiles are very fascinating materials in their own right. For example, we have demonstrated that the present extended amphiphilic dendrons self-assemble into an unexpected sequence of crystalline lamellar, $Pm\bar{3}n$ micellar cubic, hexagonal columnar, $Ia\bar{3}d$ continuous cubic and lamellar mesophases as a function of volume fraction and temperature [66]. Furthermore, local core-shell topologies were revealed by monitoring the ion-conductivity of lithium ion-doped samples, which is strongly correlated to mesophase behavior and mechanical properties. We were thus able to study, for the first time, charge transport within a nanostructured material in which the conducting medium is confined to either micelles

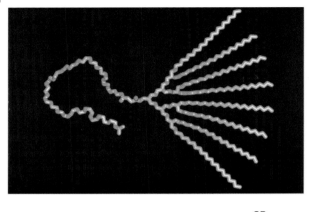

Fig. 14.6 Molecular architecture and model of two extended amphiphilic dendrons **1** and **2** with different PEO chain length. Oxygen containing segments are represented in red.

(zero-dimensional), cylinders (one-dimensional), lamellae (two-dimensional), or is a continuous (three-dimensional) network throughout the entire macroscopic sample (Fig. 14.7). These results may have significant implications in areas where charge transport in nanostructured materials and devices is becoming increasingly, important such as ion conductors, photovoltaics or electroluminescence for which the present extended amphiphilic dendrons may provide an advanced molecular design concept.

Sol–gel polymerizations of GLYMO and $Al(O^sBu)_3$ in the presence of extended amphiphilic dendrons as structure-directing agents were carried out using procedures similar to the one described in the previous section [63]. Volatiles were evaporated at 63 °C in order for the sol–gel condensation to take place in the melt phase and to suppress the docosyl chain crystallization. Hybrids with lamellar and cylindrical morphologies were obtained upon addition of different amounts of sol, confirming that the underlying mechanism of mesostructure formation is a co-assembly process, as in the case of linear macromolecular amphiphiles. Mesostructures were identified by employing a combination of SAXS, TEM and atomic force microscopy (AFM), see Fig. 14.8. Based on these successful first ef-

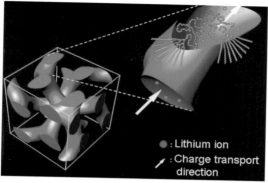

Fig. 14.7 Lithium ion charge transport in a nanostructured double-gyroid morphology from novel blocked extended amphiphilic dendrons.

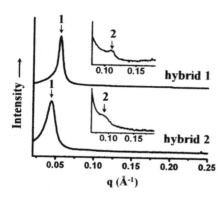

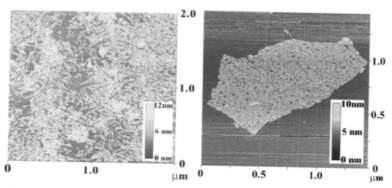

Fig. 14.8 Top: SAXS traces of lamellar (hybrid 1) and hexagonal (hybrid 2) silica mesostructures from extended amphiphilic dendrons as structure-directing agents. Bottom: AFM images of individual cylindrical (left) and lamellar (right) nanoparticles obtained through dissolution of the bulk materials.

forts, the versatility of the dendron architecture now allows various functionalities to be introduced into the mesostructured hybrids. Thus, the present extended amphiphilic dendron architecture may provide an exciting novel materials platform for formation of multifunctional mesostructured hybrid materials.

14.4
Generalization to Other Inorganic Materials Systems

This section will be divided into two parts. Firstly, developments of the approaches presented so far towards functional mesostructured silica hybrid materials will be discussed. In particular, we will concentrate on magnetic properties. The second part will describe initial efforts to move away from oxide materials altogether towards mesostructured high temperature, non-oxide materials.

14.4.1
Mesoporous Aluminosilicate Materials with Superparamagnetic γ-Fe$_2$O$_3$ Particles Embedded in the Walls

Expanding the functionality of silica-based mesostructured and mesoporous materials by the incorporation of functional organic compounds [67–69], substitution or addition of other inorganic materials [70] or templating into carbon-based materials [71] has sparked much interest among researchers. In this section we will focus on recent work of the Wiesner group on magnetic mesoporous materials [72, 73]. Although research had been performed on surfactant-based templating and sonochemical approaches to layered iron oxide–oxyhydroxide mesophases [74, 75], polymer derived magnetic bulk ceramics [76] and bulk iron oxide silicates prepared through sol–gel techniques [77–80], research on mesostructured iron containing silicates was scarce and limited to surfactant based systems where the iron compound was added in a post-synthesis loading step [81, 82]. Backfilling the pores is a common technique to functionalize mesoporous materials [83], but it requires more synthesis and characterization steps and, more importantly, risks clogging the pore structure. In contrast, we presented a simple block copolymer based "one-pot" self-assembly approach to multifunctional γ-iron oxide–aluminosilicates that are mesoporous and exhibit superparamagnetic behavior [72]. Nanoscopic iron oxide particles are incorporated in the walls of the aluminosilicate matrix; therefore, blocking of the pores, as observed in earlier studies, on backfilled materials was overcome even for high iron loadings [82]. This simple and versatile block copolymer directed approach facilitating large pores may lead to new techniques for the separation of magnetically labeled biological macromolecules that combine size exclusion and magnetic interactions. Also, the robust matrix with thick walls (> 10 nm) makes the material stable to temperatures as high as 800 °C, allowing for catalytic applications at elevated temperatures.

The synthesis is unique in allowing for precise control over the structure and composition of the final materials. Although solely the inverse hexagonal cylinder morphology was described (cylindrical pores in an aluminosilicate matrix), several morphologies were observed in the Wiesner laboratories similar to those seen for diblock copolymers and their mixtures with homopolymers [2, 24]. These include the hexagonal cylinder (inorganic cylinders in a polymer matrix) and the lamellar phases. The approach can also be extended to other transition metal oxide systems. Iron oxide was used in the study for its potential magnetic properties, but also serves as an example of what should be possible from a wide range of commercially available metal alkoxides. The actual composition of the resulting materials can be tailored according to the application, which becomes particularly important in catalyst technology.

Polymer–inorganic composites with 25 mol-% Fe on a cation basis were prepared using a sol of GLYMO and Al(O^{s}Bu)$_3$ mixed with a solution of the structure directing PI-b-PEO diblock copolymer and iron(III) ethoxide powder. For the production of a crystalline magnetic iron oxide phase, the as-made composites were calcined at elevated temperatures in air. Results on model materials suggested that in a bulk aluminosilicate matrix, γ-Fe$_2$O$_3$ is the only detected crystalline phase after heat treatment at elevated temperatures (550 °C). Comparable results, i.e., the formation of only γ-Fe$_2$O$_3$ in iron containing silicate materials prepared by similar means, were also obtained by other groups [77–80, 84]. Although pure γ-Fe$_2$O$_3$ usually decomposes into non-magnetic α-Fe$_2$O$_3$ at ca. 350 °C, the remaining amorphous aluminosilicate matrix apparently hinders the transformation; thus, stabilizing the γ-phase even when heated to 750 °C [80]. This is fortunate as one can now exploit the magnetic behavior of γ-Fe$_2$O$_3$. TEM was performed on as-made hybrid samples and materials calcined to 750 °C. The TEM micrographs in Fig. 14.9 demonstrate that upon calcination the inverse hexagonal morphology of the as-made material is preserved. The higher resolution image in Fig. 14.9d shows homogeneously distributed dark spots within the matrix. A histogram of the size distribution of these precipitates revealed an average particle diameter of approximately 5 nm. Considering that TEM is a local technique, this number was in excellent agreement with a value of 5.6 nm calculated for the γ-Fe$_2$O$_3$ particle size from SQUID data (see below), which represented measurements integrated over macroscopic sample dimensions. This remarkable result indicated that the particle size distribution reflected from the TEM micrograph in Fig. 14.9d was representative of the entire sample suggesting the amazing structural control of the present approach.

The magnetic properties of samples before and after calcination at different temperatures were investigated using a SQUID (super-conducting quantum interference device) magnetometer (Fig. 14.10). As expected, the magnetization of the as-made composite exhibits paramagnetic behavior. However, after calcination at 550 °C the onset of an inflection point and the absence of residual magnetization at zero applied field in the field (H) versus magnetization (M) plot indicated superparamagnetic behavior (data not shown). As the calcination temperatures were increased above 550 °C, the measured magnetizations also increased. At the same time, the sigmoidal nature of the superparamagnetic response became more apparent being most developed for the composite calcined at 750 °C. The magnetic

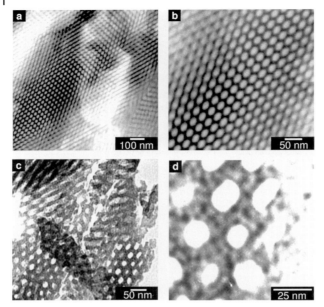

Fig. 14.9 Bright-field TEM images of (top row) an as-made hybrid sample at two different magnifications revealing the inverse hexagonal morphology (a, b). The same material (bottom row) after calcination to 750 °C at two different magnifications (c, d). The image in (c) demonstrates that the hexagonal order is preserved after calcination. At higher magnification (d), the iron oxide precipitates show up as dark spots within the matrix of the inverse hexagonal morphology.

response of the ceramic material suggested the nucleation and growth of small γ-Fe_2O_3 particles. To determine the iron oxide precipitate size quantitatively, zero-field cooling (ZFC) and field cooling (FC) experiments were performed on samples calcined to 750 °C. The data in the inset of Fig. 14.10 indicate a blocking temperature of 25 K and a collapse temperature (the temperature where the two curves deviate from each other) of 35 K. Both the narrowness of the ZFC curve around the maximum and the low collapse temperature were strong indications of a narrow crystallite size distribution of particles within the matrix [85]. Application of superparamagnetic theory resulted in a calculated average γ-Fe_2O_3 particle size of 5.6 nm [86].

14.4.2
Ordered Mesoporous Ceramics Stable up to 1500 °C from Diblock Copolymer Mesophases

Non-oxide ceramics have been investigated extensively because of their excellent mechanical and thermal properties. To produce non-oxide ceramics polymer-derived ceramics (PDCs) are often used, because they can be processed at

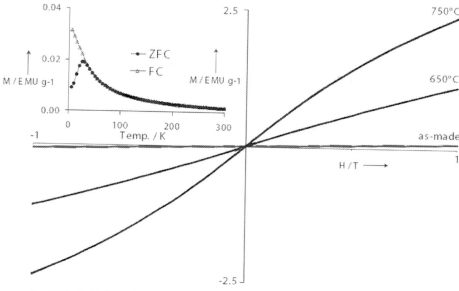

Fig. 14.10 Field dependence of the magnetic properties of an as-made sample and the same sample calcined to different temperatures, as indicated in the figure. Inset: zero-field cooling (ZFC) and field cooling (FC) curves for a sample calcined to 750 °C. Data were collected at 300 K for the H versus M curves and a 200-Oe field was used for the ZFC–FC experiments.

temperatures lower than those required in conventional fabrication methods [87]. The demand for uniform porosity and a large surface area, required for applications such as catalyst supports or filters operating in harsh environments, makes the preparation of mesoporous non-oxide ceramics an attractive challenge. In 2003 the Wiesner group showed that PI-b-PEO can serve as a structure-directing agent for a poly(ureamethylvinyl)silazane pre-ceramic polymer (PUMVS) commercially known as Ceraset® [88]. While working with PI-b-PEO prooved difficult, switching to a new block copolymer, poly(isoprene-*block*-dimethylamino ethyl methacrylate) (PI-b-PDMAEMA), as a structure-directing agent for PUMVS enabled, to the best of our knowledge, the first successful synthesis to be made of mesoporous ceramic materials stable up to 1500 °C [89]. It is interesting to note that the work in this section relates to block copolymer–homopolymer blends rather than block copolymer–nanoparticle mixtures. It should be emphasized, however, that the physical principles guiding the structure formation in both cases are very similar. For example, segregation phenomena in block copolymer–homopolymer mixtures are also governed by homopolymer size relative to the copolymer block size [90–93].

Both the molecular structure of the block copolymer and the pre-ceramic polymer, and details of the heat treatment to convert the liquid polysilazane precursor

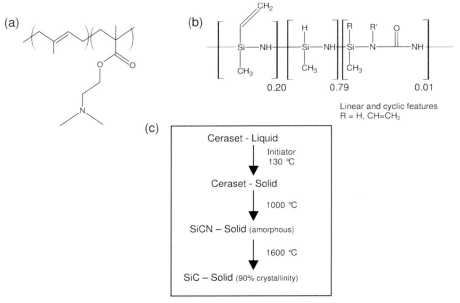

Fig. 14.11 Chemical structure of (a) poly(isoprene-*block*-dimethylamino ethyl methacrylate) and (b) poly(ureamethylvinyl)silazane (Ceraset, Kion Corp.). No information is available about the R'. (c) Details of the heat treatment of PUMVS resulting in amorphous SiCN or crystalline SiC.

into a ceramic, are shown in Fig. 14.11. Deviation from PEO as the hydrophilic block is motivated by the limited PEO–PUMVS miscibility and PEO crystallization in polymer–PUMVS composites. In contrast, both blocks in PI-*b*-PDMAEMA are amorphous polymers.

Blending PUMVS with the block copolymer is expected to lead to selective swelling of the PDMAEMA microdomains with PUMVS primarily due to the polar nature of the molecule (see Fig. 14.11). This increases the effective volume fraction of the PDMAEMA domains. Different mesophases similar to those in the PI-*b*-PEO–aluminosilicate system are observed by increasing the polysilazane to block copolymer weight fraction [2, 24]. The structure is permanently set by cross-linking the silazane oligomer with a radical initiator. The nanostructured hybrid materials can be converted into high temperature SiCN-type ceramic materials upon calcination. This treatment up to 1500 °C under Ar–H_2 results in a mesoporous ceramic with open and accessible pores.

As an example, composites with a polymer to PUMVS weight ratio of 0.5 were prepared by dissolving all the components in toluene and casting a film. The SAXS pattern of the composite after cross-linking at 130 °C (Fig. 14.12a) and the TEM results (Fig. 14.12c) are consistent with hexagonally packed domains. The contrast in the micrograph arises from the density difference between PI and PDMAEMA–PUMVS domains, the latter appearing darker. The SAXS diffractogram of the

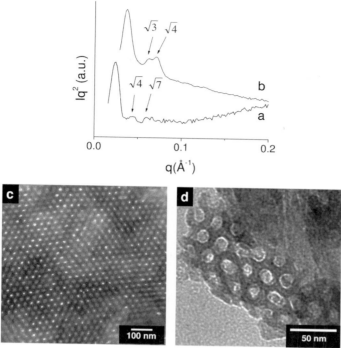

Fig. 14.12 SAXS traces of the as-made composite (a) and ceramic after calcination to 1500 °C (b). Bright-field TEM images of the as-made composite (c) and ceramic calcined to 1500 °C (d) demonstrating that the hexagonal structure is preserved during heat treatment.

composite after heat treatment to 1500 °C together with TEM images (Fig. 14.12b, d) clearly show the preservation of the hexagonal structure. There is a significant shift of the X-ray spectrum to higher q-values compared with the composite, indicating sample shrinkage upon high temperature treatment. Separate microprobe studies of the composition of these materials treated up to 1400–1500 °C under N_2–CO and Ar–H_2 typically showed the following number percentages: 29 % C, 9 % O, 30 % N and 32 % Si. Nitrogen adsorption–desorption isotherms were measured to ensure the pores are open and accessible (data not shown) [89].

An interesting aspect of this system is that the PUMVS–block copolymer hybrids can be annealed at elevated temperatures before cross-linking of the PUMVS is initiated through radical polymerization. This has the potential to overcome severe limitations of sol–gel processes to produce long-range order in thin films as the network already forms during hybrid synthesis, thereby limiting molecular rearrangements. Both blocks of the copolymer exhibit low glass transition temperatures and the polysilazane is a liquid at room temperature, which provides the hybrid system with high molecular mobility for defects to heal out. This suggests that nanostructured composites, combining properties of block copolymers with those of solid-state materials, may be obtained in a cheap co-assembly process that allows macroscopic order, one of the most important requirements for practical applications.

14.5
Generalization from Bulk Mesostructured Hybrids to Mesostructured Thin Films

In this last section the transition from bulk mesostructured hybrids to mesostructured thin films (sub-100 nm) is discussed. The routine formation of nm-size surface structures remains a challenge that limits advances in many fields of nanotechnology. Increasingly "bottom-up" self-assembly approaches for the nanometer-scale patterning of surfaces are competing with traditional "top-down" lithographic processes, such as scanned probe lithography or high-resolution e-beam lithography. Block copolymer thin films (<100 nm) are among the more promising materials being examined as they offer ease of processing combined with phase separation induced structure formation on the nanometer-scale. Nanostructured block copolymer thin films [94, 95] have been investigated for the production of high-density magnetic media [96], quantum dot arrays [97–100] and photonic crystals [101–103]. While all-organic approaches to nanostructuring have considerable real-world disadvantages, such as low etch resistivity, low thermal stability, the generalization of block copolymer–inorganic nanoparticle hybrid systems from the bulk to the thin film regime has the potential to improve performance significantly. The first results of an investigation of such organic–inorganic hybrid thin films by the Wiesner group reported on samples with a majority inorganic component to enhance adhesion to the substrate, provide greater mechanical stability and to help facilitate characterization [104]. Different parameters, such as solution composition and concentration, spin velocity, acceleration and duration were varied in the spin-coating process and the subsequent films were characterized to elucidate film morphology.

After the calcination step, the films were probed with tapping mode AFM to gain topographical information. Figure 14.13 shows in the upper left a representative AFM image of these films captured over a 2 µm × 2 µm area. The image reveals a porous structure. Note the uniformity in pore size and medium-range order exhibited. Power spectral density analysis revealed a strong peak for each film (32 nm for the film in Fig. 14.13). The depth profiles of the films were obtained by cleaving the wafers and viewing the newly exposed surfaces with field-emission SEM. In the upper right of Fig. 14.13 is a representative cross-sectional image. It reveals a multilayer structure where the pore cross-sections are not spherical, but rather ellipsoidal. The deviation from expected spheres to ellipsoids is an effect of treatment after the spin-casting process. The sizes of the structures (pore size and center-to-center distance) observed in cross-section agree with the values obtained with AFM. The same figure also shows characterization of the monolayer films, which was reported in a subsequent publication indicating the degree of control over film formation [105].

14.5 Generalization from Bulk Mesostructured Hybrids to Mesostructured Thin Films | 329

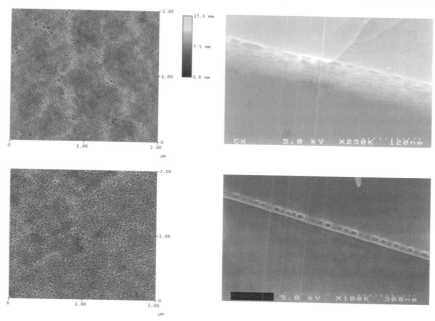

Fig. 14.13 Left: AFM images taken in tapping mode of film surfaces created after calcinations from mixtures of PI-b-PEO diblock copolymers with Glymo–Al(O^{s}Bu)$_3$. The imaged film area is 2 μm × 2 μm in both cases. Right: Cross-sectional SEMs of the film shown on the left revealing multilayer and monolayer formation.

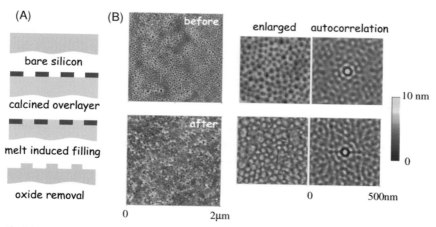

Fig. 14.14 (A) Schematic of the 3-step nanopillars fabrication process. (B) AFM images from as-calcined nanopore film (before) and fully formed nanopillars film (after). Sections of each film are shown enlarged, with the autocorrelation image showing equivalent hexagonal symmetry.

In order to test the thin film structures on macroscopic length scales and to demonstrate that the AFM images of Fig. 14.13 are representative, grazing incidence small angle X-ray scattering (GISAXS) was employed on the monolayer films (data not shown) and quantitatively analyzed [105]. Pore-structure dimensions obtained from all simulations were in good agreement with experimental results from AFM and SEM.

Nanostructuring of silicon-type surfaces employing organic molecule self-assembly is limited to low-temperature processing. In contrast, ordered nanostructures that can tolerate high temperature require conventional — and expensive — lithographic patterning methods. In the final example we will see that the monolayer thin films can serve as templates for a laser-induced capillarity-driven filling process to form Si nanopillars. This leads to an extremely efficient method for generating large arrays of nanostructured surfaces. The process allows control of the surface structure geometry and feature sizes down to the near-molecular level. As this process can be extended to materials other than Si and to gallery-type structures, it offers enormous scientific and technological promise in a wide range of fields from nanofluidics, to biosensors and nanoelectronics.

Silicon nanopillars were created using a nanosecond laser-induced transient melt and solidification process. Capillary forces were exploited to fill the pores with liquid Si; the time-scale of laser-induced melts is sufficiently short (<50 ns) to avoid sintering collapse of the pores. Samples were irradiated with a 30-ns pulse from a XeCl excimer laser ($\lambda = 308$ nm) at a fluence sufficient to melt the underlying Si substrate. Fluences just above the single-crystal melt threshold (125 % of E_{melt}) for several (5–15) pulses were required to melt the underlying crystal Si, break through the native oxide and fill the nanopores. The duration of the melt was determined *in situ* to be 20 ns, corresponding to a "filling" velocity of order 0.5–1 m s^{-1}. Despite the low viscosity of liquid Si (0.34 centistokes), these velocities still correspond to laminar flow through the pores. With adequate control, it should be possible to ensure that these pillars grow epitaxial from the substrate,

From AFM, in Fig. 14.14 the nanopore lattice spacing of the monolayer template is 40 nm with a physical thickness after calcination of 15 nm, as determined by a combination of scanning electron microscopy and X-ray diffraction (data not shown). The same figure shows an AFM image of the sample after irradiation with five laser pulses and removal of the skeletal silica using a 48 % HF solution. As compared with the sample before laser treatment, the regular array of dots changes from dark to bright in color, confirming the conversion of the nanopores to nanopillars. Auger spectroscopic measurements confirmed complete removal of the oxides; the remaining structures are clearly Si. The hexagonal symmetry of the nanopores is replicated in the nanopillars, as demonstrated by corresponding FFT autocorrelations shown in the same figure.

These preliminary results give clear evidence that inorganic nanopores can be transformed into nanopillars of various materials — currently demonstrated for Si, but likely to be extensible to any material with a reasonable melting temperature and liquid viscosity. Large areas of such nanostructures can thus be fabricated in an extremely inexpensive, rapid and flexible manner by the general community, without the use of any traditional nanofabrication methods such as photo- or e-beam lithography. In addition, use of the polymer precursors offers tremendous variability in structural parameters, such as pore size, pore-to-pore correlation, pore wall thickness, etc. The symmetry of the phase is not restricted to hexagonal ordering as various structures are known to exist in the phase diagram of diblock copolymers, including simple cubic, lamellar and cubic bicontinuous (see Section 14.2).

14.6
Conclusions

The development and application of sophisticated mesostructured nanoparticles–block copolymer hybrid systems is an area of tremendous scientific and technological promise. Since the start in the 1990s of the use of block copolymers to structure direct silica-type materials, we have gained a great deal of fundamental understanding about these systems. Moreover, in the meantime the concept has been successfully generalized towards other blocked macromolecular amphiphiles, other inorganic materials systems and towards thin films.

Our understanding of the of co-assembly mechanisms of mesostructured block copolymer-silica type hybrids improved significantly by mapping the problem onto the behavior of nanoparticles in specific block copolymer domains. There is a critical size of the particles where the assembly behavior of polymers and particles fundamentally changes. This critical size is the result of a competition between chain conformational entropy loss of the polymers and translational entropy gain of the particles. Below the critical size, which turns out to be close to the radius of gyration of the polymer, the nanoparticles can be treated as small molecular weight additives or solvents that are selectively added to one of the blocks, thereby swelling it, resulting in a rational design of hybrid mesostructuring.

The prospects for the design of such block copolymer–nanoparticle based mesostructured materials, with multiple nanoscale functional features for applications in nanotechnology, is tremendous. As pointed out in the Introduction, there starts to nucleate a vision that, in analogy to biology, near-molecular level structure and functionality control of co-assemblies of soft and hard materials can be encoded (and therefore predicted) by the monomer (block) sequence of synthetic macromolecules. Experiments described in this chapter and work of many others in the field so far probably only scratch the surface of things to come, considering, e. g., that in the future, work on multi-block copolymers with multiple nanoparticles will be conducted. In the long term, such powerful bottom-up strategies will develop into alternatives to, or converge with, expensive top-down approaches, thereby revolutionizing the way materials will be fabricated.

References

1 O. Ikkala, G. ten Brinke, *Science* **2002**, *295*, 2407.
2 I. W. Hamley, *The Physics of Block Copolymers*, Oxford University Press, Oxford, **1998**.
3 Y. Matsushita, H. Choshi, T. Fujimoto, M. Nagasawa, *Macromolecules* **1980**, *13*, 1053.
4 C. Auschra, R. Stadler, *Macromolecules* **1993**, *26*, 2171.
5 F. S. Bates, G. H. Fredrickson, *Phys. Today* **1999**, *52*, 32.
6 U. Breiner, U. Krappe, E. L. Thomas, R. Stadler, *Macromolecules* **1998**, *31*, 135.
7 Z. B. Li, E. Kesselman, Y. Talmon, M. A. Hillmyer, T. P. Lodge, *Science* **2004**, *306*, 98.
8 T. Goldacker, V. Abetz, R. Stadler, I. Erukhimovich, L. Leibler, *Nature (London)* **1998**, *398*, 137.
9 S. Stewart, G. Liu, *Angew. Chem., Int. Ed. Engl.* **2000**, *39*, 340.
10 J. D. Hartgerink, E. Beniash, S. I. Stupp, *Science* **2001**, *294*, 1684.
11 V. Sankaran, C. C. Cummins, R. R. Schrock, R. E. Cohen, R. J. Silbey, *J. Am. Chem. Soc.* **1990**, *112*, 6858.
12 M. Moffitt, L. McMahon, V. Pessel, A. Eisenberg, *Chem. Mater.* **1995**, *7*, 1185.
13 J. P. Spatz, A. Roescher, S. Sheiko, G. Krausch, M. Möller, *Adv. Mater.* **1995**, *7*, 731.
14 M. Antonietti, E. Wenz, L. Bronstein, M. Seregina, *Adv. Mater.* **1995**, *7*, 1000.
15 A. Jain, U. Wiesner, *Macromolecules* **2004**, *37*, 5665.
16 C. G. Göltner, M. Antonietti, *Adv. Mater.* **1997**, *9*, 431.
17 M. Templin, A. Franck, A. Du Chesne, H. Leist, Y. Zhang, R. Ulrich, V. Schädler, U. Wiesner, *Science* **1997**, *278*, 1795.
18 D. Zhao, J. Feng, Q. Huo, N. Melosh, G. H. Fredrickson, B. F. Chmelka, G. D. Stucky, *Science* **1998**, *279*, 548.
19 C. G. Göltner, S. Henke, M. C. Weissenberger, M. Antonietti, *Angew. Chem. Int. Ed. Engl.* **1998**, *37*, 613.
20 L. Leibler, *Macromolecules* **1980**, *13*, 1602.
21 F. S. Bates, G. H. Fredrickson, *Annu. Rev. Phys. Chem.* **1990**, *41*, 525.
22 M. W. Matsen, F. S. Bates, *Macromolecules* **1996**, *29*, 1091.
23 G. Floudas, B. Vazaiou, F. Schipper, R. Ulrich, U. Wiesner, H. Iatrou, N. Hadjichristidis, *Macromolecules* **2001**, *34*, 2947.
24 P. F. W. Simon, R. Ulrich, H. W. Spiess, U. Wiesner, *Chem. Mater.* **2001**, *13*, 3464.
25 A. C. Finnefrock, R. Ulrich, G. E. S. Toombes, S. M. Gruner, U. Wiesner, *J. Am. Chem. Soc.* **2003**, *125*, 13084.
26 A. C. Finnefrock, R. Ulrich, A. Du Chesne, C. C. Honeker, K. Schumacher, K. K. Unger, S. M. Gruner, U. Wiesner, *Angew. Chem. Int. Ed. Engl.* **2001**, *40*, 1208.
27 P. Meakin, *Annu. Rev. Phys. Chem.* **1988**, *39*, 237.
28 R. K. Iler, *The Chemistry of Silica*, John Wiley & Sons, New York, **1979**.
29 H. E. Bergna, *The Colloid Chemistry of Silica, Advances in Chemistry Series*, American Chemical Society, Washington DC, **1994**.
30 R. B. Thompson, V. V. Ginzburg, M. W. Matsen, A. C. Balazs, *Science* **2001**, *292*, 2469.
31 S. M. De Paul, J. W. Zwanziger, R. Ulrich, U. Wiesner, H. W. Spiess, *J. Am. Chem. Soc.* **1999**, *121*, 5727.
32 For organosilicate species with a direct Si–C bond, the coordination of the silicon atom is described by the notation T_n ($n = 0, 1, 2$ or 3) where n is the number of bridging (Si–O–Si or Si–O–Al) oxygens. Different T_n species resonate at different ppm values in a ^{29}Si NMR spectrum.
33 C. J. Brinker, G. W. Scherer, *Sol-Gel Science*, Academic Press, San Diego, **1990**.
34 G. C. Ruben, L. W. Hrubesh, T. M. Tillotson, *J. Non-cryst. Solids* **1995**, *186*, 209.
35 C. G. Göltner, *Curr. Opin. Colloid Inter. Sci.* **2002**, *7*, 173.
36 A. I. Chervanyov, A. C. Balazs, *J. Chem. Phys.* **2003**, *119*, 3529.

37 J. Y. Lee, Z. Shou, A. C. Balazs, *Macromolecules* **2003**, *36*, 7730.
38 G. J. A. Sevink, A. V. Zvelindovsky, B. A. C. Vlimmeren, N. M. Maurits, J. G. E. M. Fraaije, *J. Chem. Phys.* **1999**, *110*, 2250.
39 J. Huh, V. V. Ginzburg, A. C. Balazs, *Macromolecules* **2000**, *33*, 8085.
40 J. Y. Lee, A. R. C. Baljon, D. Y. Sogah, R. F. Loring, *J. Chem. Phys.* **1987**, *112*, 9112.
41 J. Groenewold, G. H. Fredrickson, *Eur. Phys. J. E: Soft Matt.* **2001**, *5*, 171.
42 Y. N. C. Chan, R. R. Schrock, R. E. Cohen, *Chem. Mater.* **1992**, *4*, 24.
43 R. T. Clay, R. E. Cohen, *Supramol. Sci.* **1995**, *2*, 183.
44 R. S. Kane, R. E. Cohen, R. Silbey, *Chem. Mater.* **1996**, *8*, 1919.
45 M. R. Bockstaller, Y. Lapetnikov, S. Margel, E. L. Thomas, *J. Am. Chem. Soc.* **2003**, *125*, 5276.
46 R. Ulrich, J. W. Zwanziger, S. M. De Paul, R. Richert, U. Wiesner, H. W. Spiess, *Polym. Mater. Sci. Eng.* **1999**, *80*, 610.
47 R. Ulrich, J. W. Zwanziger, S. M. De Paul, A. Reiche, H. Leuninger, H. W. Spiess, U. Wiesner, *Adv. Mater.* **2002**, *14*, 1134.
48 A. Jain, G. E. S. Toombes, L. M. Hall, S. Mahajan, C. B. W. Garcia, W. Probst, S. M. Gruner, U. Wiesner, *Angew. Chem., Int. Ed. Engl.* **2005**, *44*, 1226.
49 K. A. Koppi, M. Tirrell, F. S. Bates, K. Almdal, R. H. Colby, *J. Phys. II* **1992**, *2*, 1941.
50 H. Leist, D. Maring, T. Thurn-Albrecht, U. Wiesner, *J. Chem. Phys.* **1999**, *110*, 8225.
51 Z.-R. Chen, J. A. Kornfield, S. D. Smith, T. J. Grothaus, M. M. Satkowski, *Science* **1997**, *277*, 1248.
52 J. L. Zryd, W. R. Burghardt, *Macromolecules* **1998**, *31*, 3656.
53 Y. Lu, R. Ganguli, C. A. Drewien, M. T. Anderson, C. J. Brinker, W. Gong, Y. Guo, H. Soyez, B. Dunn, M. H. Huang, J. I. Zink, *Nature (London)* **1997**, *389*, 364.
54 N. A. Melosh, P. Davidson, P. Feng, D. J. Pine, B. F. Chmelka, *J. Am. Chem. Soc.* **2001**, *123*, 1240.
55 P. Yang, D. Zhao, B. F. Chmelka, G. D. Stucky, *Chem. Mater.* **1998**, *10*, 2033.
56 C. J. Brinker, Y. Lu, A. Sellinger, H. Fan, *Adv. Mater.* **1999**, *11*, 579.
57 R. J. Albalak, E. L. Thomas, *J. Polym. Sci.: Part B: Polym. Phys.* **1993**, *31*, 37.
58 A. Okabe, T. Fukushima, K. Ariga, T. Aida, *Angew. Chem., Int. Ed. Engl.* **2002**, *41*, 3414.
59 M. Kimura, K. Wada, K. Ohta, K. Hanabusa, H. Shirai, N. Kobayashi, *J. Am. Chem. Soc.* **2001**, *123*, 2438.
60 M. Kimura, K. Wada, Y. Iwashima, K. Ohta, K. Hanabusa, H. Shirai, N. Kobayashi, *Chem. Commun.* **2003**, 2504.
61 a) D. A. Tomalia, H. Baker, J. Dewald, M. Hall, G. Kallos, S. Martin, J. Roeck, J. Ryder, P. Smith, *Polym. J.* **1985**, *17*, 117; b) F. W. Zeng, S. C. Zimmenerman, *Chem. Rev.* **1997**, *97*, 1681; c) A. W. Bosman, H. M. Janssen, E. W. Meijer, *Chem. Rev.* **1999**, *99*, 1665; d) G. R. Newkome, E. He, C. N. Moorefield, *Chem. Rev.* **1999**, *99*, 1689; e) S. M. Grayson, J. M. J. Fréchet, *Chem. Rev.* **2001**, *101*, 3819; f) S. D. Hudson, H.-T. Jung, V. Percec, W.-D. Cho, G. Johansson, G. Ungar, V. S. K. Balagurusamy, *Science* **1997**, *278*, 449; g) J. Iyer, K. Fleming, P. T. Hammond, *Macromolecules* **1998**, *31*, 8757; h) E. D. Sone, E. R. Zubarev, S. I. Stupp, *Angew. Chem., Int. Ed. Engl.* **2002**, *41*, 1706; i) L. Li, E. Beniash, E. R. Zubarev, W. Xiang, B. M. Rabatic, G. Zhang, S. I. Stupp, *Nat. Mater.* **2003**, *2*, 689.
62 B.-K. Cho, A. Jain, J. Nieberle, S. Mahajan, U. Wiesner, S. M. Gruner, S. Tuerk, H. J. Raeder, *Macromolecules* **2004**, *37*, 4227.
63 B.-K. Cho, A. Jain, S. Mahajan, H. Ow, S. M. Gruner, U. Wiesner, *J. Am. Chem. Soc.* **2004**, *126*, 4070.
64 M. Jayaraman, J. M. J. Fréchet, *J. Am. Chem. Soc.* **1998**, *120*, 12996.
65 B.-K. Cho, A. Jain, S. M. Gruner, U. Wiesner, *Chem. Commun.* **2005**, *16*, 2143.
66 B.-K. Cho, A. Jain, S. M. Gruner, U. Wiesner, *Science* **2004**, *305*, 1598.
67 Y. Lu, Y. Yang, A. Sellinger, M. Lu, J. Huang, H. Fan, R. Haddad,

G. Lopez, A. R. Burns, D. Y. Sasaki, J. Shelnutt, C. J. Brinker, *Nature (London)* **2001**, *410*, 913.

68 T. Asefa, M. J. MacLachlan, N. Coombs, G. A. Ozin, *Nature (London)* **1999**, *402*, 867.

69 S. Inagaki, S. Guan, T. Ohsuna, O. Terasaki, *Nature (London)* **2002**, *416*, 304.

70 P. Yang, D. Zhao, D. I. Margolese, B. F. Chmelka, G. D. Stucky, *Nature (London)* **1998**, *396*, 152.

71 S. H. Joo, S. J. Choi, I. Oh, J. Kwak, Z. Liu, O. Terasaki, R. Ryoo, *Nature (London)* **2001**, *412*, 169.

72 C. B. W. Garcia, Y. Zhang, F. DiSalvo, U. Wiesner, *Angew. Chem., Int. Ed. Engl.* **2003**, *42*, 1526.

73 C. B. W. Garcia, Y. Zhang, S. Mahajan, F. DiSalvo, U. Wiesner, *J. Am. Chem. Soc.* **2003**, *125*, 13310.

74 G. Wirnsberger, K. Gatterer, H. P. Fritzer, W. Grogger, P. Behrens, M. F. Hansen, C. B. Koch, *Chem. Mater.* **2001**, *13*, 1153.

75 Y. Wang, L. Yin, A. Gedanken, *Ultrason. Sonochem.* **2002**, *9*, 285.

76 M. J. MacLachlan, G. Madlen, N. Coombs, T. W. Coyle, N. P. Raju, J. E. Greedan, G. A. Ozin, I. Manners, *Science* **2000**, *287*, 1460.

77 F. del Monte, M. P. Morales, D. Levy, A. Fernandez, M. Ocana, A. Roig, E. Molins, K. O'Grady, C. J. Serna, *Langmuir* **1997**, *13*, 3627.

78 G. Ennas, A. Musinu, G. Piccaluga, D. Zedda, D. Gatteschi, C. Sangregorio, J. L. Stanger, G. Concas, G. Spano, *Chem. Mater.* **1998**, *10*, 495.

79 C. Cannas, M. F. Casula, G. Concas, A. Corrias, D. Gatteschi, A. Falqui, A. Musinu, C. Sangregorioc, G. Spanob, *J. Mater. Chem.* **2001**, *11*, 3180.

80 M. F. Casula, A. Corrias, G. Paschina, *J. Non-cryst. Solids* **2001**, *293–295*, 25.

81 M. Fröba, R. Köhn, G. Bouffaud, O. Richard, G. van Tendeloo, *Chem. Mater.* **1999**, *11*, 2858.

82 L. Zhang, G. C. Papaefthymiou, J. Y. Ying, *J. Phys. Chem. B* **2001**, *105*, 7414.

83 X. He, D. Antonelli, *Angew. Chem., Int. Ed. Engl.* **2002**, *41*, 214.

84 D. Niznansky, N. Viart, J. L. Rehspringer, *J. Sol-Gel Science Technol.* **1997**, *8*, 615.

85 B. H. Sohn, R. E. Cohen, G. C. Papaefthymiou, *J. Magn. Magn. Mater.* **1998**, *182*, 216.

86 A. Aharoni, *Relaxation Processes in Small Particles*, North Holland, Amsterdam, 1992.

87 R. Raj, R. Riedel, G. D. Soraru, *J. Am. Ceram. Soc.* **2001**, *84*, 2158.

88 C. B. W. Garcia, C. Lovell, C. Curry, M. Faught, Y. Zhang, U. Wiesner, *J. Polym. Sci. B: Polym. Phys.* **2003**, *41*, 3346.

89 M. Kamperman, C. B. W. Garcia, P. Du, H. Ow, U. Wiesner, *J. Am. Chem. Soc.* **2004**, *126*, 14708.

90 T. Hashimoto, H. Tanaka, H. Hasegawa, *Macromolecules* **1990**, *23*, 4378.

91 H. Tanaka, H. Hasegawa, T. Hashimoto, *Macromolecules* **1991**, *24*, 240.

92 K. R. Shull, K. I. Winey, *Macromolecules* **1992**, *25*, 2637.

93 M. Matsen, *Macromolecules* **1995**, *28*, 5765.

94 M. Park, C. Harrison, P. M. Chaikin, R. A. Register, D. H. Adamson, *Science* **1997**, *276*, 1401.

95 V. Z.-H Chan, J. Hoffman, V. Y. Lee, H. Iatrou, A. Avgeropoulos, N. Hadjichristidis, R. D. Miller, E. L. Thomas, *Science* **1999**, *286*, 1716.

96 T. Thurn-Albretcht, J. Schotter, G. A. Kastle, N. Emley, T. Shibauchi, L. Krusin-Elbaum, K. Guarini, C. T. Black, M. Tuominen, T. P. Russell, *Science* **2000**, *290*, 2126.

97 S. Forster, *Ber. Bunsenges.-Phys. Chem. Chem. Phys.* **1997**, *101*, 1671.

98 D. E. Fogg, L. H. Radzilowski, R. Blanski, R. R. Schrock, E. L. Thomas, *Macromolecules* **1997**, *30*, 417.

99 P. Leclere, V. Parente, J. L. Bredas, B. Francois, R. Lazzaroni, *Chem. Mater.* **1998**, *10*, 4010.

100 R. R. Li, P. D. Dapkus, M. E. Thomson, W. G. Jeong, C. Harrison, P. M. Chaikin, R. A. Register, D. H. Adamson, *Appl. Phys. Lett.* **2000**, *76*, 1689.

101 A. Urbas, Y. Fink, E. L. Thomas, *Macromolecules* **1999**, *32*, 4748.
102 A. Urbas, R. Sharp, Y. Fink, E. L. Thomas, M. Xenidou, L. J. Fetters, *Adv. Mater.* **2000**, *12*, 812.
103 Y. Fink, A. M. Urbas, M. G. Bawendi, J. D. Joannoupoulos, E. L. Thomas, *J. Lightwave Technol.* **1999**, *62*, 1852.
104 P. Du, J. S. Gutmann, P. F. W. Simon, C. B. W. Garcia, K. Guarini, C. T. Black, U. Wiesner, *Polym. Prepr. (Am. Chem. Soc. Dev. Polym. Chem.)* **2002**, *43*, 438.
105 P. Du, M. Li, K. Douki, X. Li, C. B. W. Garcia, A. Jain, D.-M. Smilgies, L. J. Fetters, Sol M. Gruner, U. Wiesner, C. K. Ober, *Adv. Mater.* **2004**, *16*, 953.

15
Block Ionomers for Fuel Cell Application

Olivier Diat and Gérard Gebel

15.1
Introduction

In electrochemical converters, polymer conducting materials are an alternative to pure inorganic assemblies as they can be chemically "tuned". They are easier to process particularly for making large, thin and flexible films at relatively low cost. Nevertheless, they also have drawbacks, such as their low charge mobility and poor chemical aging, which are fundamental problems for fuel cells based on proton exchange membranes (PEMFC) or other applications, such as the photovoltaic organic cells. Research is being undertaken to understand these problems and their origins in order to develop new materials that are more competitive.

The principle of the H_2-cell is shown in Fig. 15.1 [1], and is as follows.

1. H_2 (the fuel) undergoes catalytic oxidation on the anode side after diffusion through channels grooved in polar plates and through carbon tissues, termed the gas diffusion layer (GDL).
2. On the cathode side, the catalytic reduction of oxygen requires protons and electrons to produce water, and heating occurs.
3. Once both electrodes are connected and a potential difference is obtained, electrons and protons are transferred through the external electrical circuit and the electrolyte, respectively.

The membrane-electrode assembly (MEA) is a key component of the cell, and for PEMFCs it is made of a thin polymeric film (a few tens of microns thick) coated on both sides with an active layer. These layers are obtained by the deposition of an active polymer solution in which a carbon catalyst is dispersed.

Under operating conditions, a polarization curve, voltage U versus current density J, can be recorded, its shape characterizes the operation of the cell. At low current densities, the voltage drop is mainly due to the oxygen reduction overpotential (kinetics losses). As J increases, ohmic (ionic and electrical) losses are at the origin of the gradual decrease of the overall cell potential and at higher current densities, mass-transport limitations become predominant [2].

Block Copolymers in Nanoscience. Massimo Lazzari, Guojun Liu, Sébastien Lecommandoux
Copyright © 2006 WILEY-VCH Verlag GmbH & Co. KGaA, Weinheim
ISBN: 3-527-31309-5

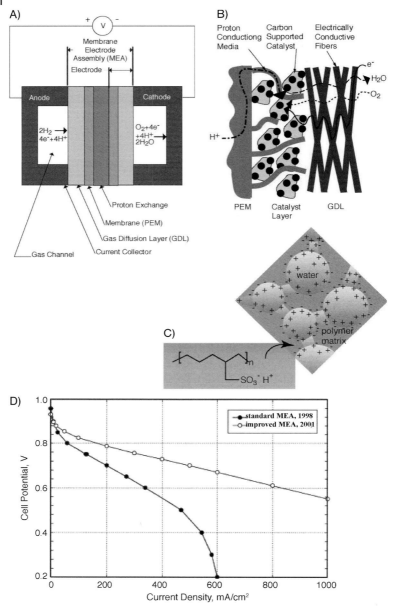

Fig. 15.1 (A) Schematic view of a fuel cell with an electrode–membrane assembly sandwiched between two polar plates (anode and cathode) for the gas distribution and the current collection [1]. (B) Transport of gas, proton, electron and water in PEM electrodes [1]. (C) Representation of the swollen ionomer membrane structure, showing the separation between hydrophobic and hydrophilic domains. (D) Graph of the potential versus current density for cells with standard and improved electrode–membrane assemblies [49] (Reprinted from [1] (A, B) and [49] (D) with permission of Elsevier).

PEMFCs are considered as fairly promising systems for automotive, stationary and portable electrical power generation because they are quiet, zero-emission power units and use an energy vector of the future, the di-hydrogen. Other hydrocarbon fuels can be used either directly, such as methanol (in DMFC) or ethanol (DEFC), or indirectly using an H_2 reformer system placed upstream to the cell. For a brief history of the milestones of fuel cell developments and the economic challenges for the future, the reader can refer to two articles by Stones and Morrison [3] and Costamagna and Srinivasan [4].

The polymer electrolyte membrane is one of the most important active components of the fuel cell. The PEM has to meet many stringent properties, not all of which are completely compatible. Firstly, the membrane must have a high ionic intrinsic conductivity ($5 \times 10^{-2} - 10^{-1}$ S cm^{-1}) but a low ion content to avoid excessive water swelling. The polymer can be either a neutral polymer doped with ionic species or a functionalized polymer with chemically bonded acid (or basic) functions. It is important to point out that this chapter will mainly focus on cationic (acid) polymers, even though promising research on anionic (basic) systems has recently started again, in particular, for the direct methanol fuel cell (DMFC) application [5]. Secondly, a low permeability to (gas or liquid) reactants is required. The membrane should be able to swell slightly with water to allow the hydrated (proton) charge to diffuse through the membrane. Thirdly, it is important that the membrane has excellent physical properties, which combine flexibility, ductibility and swelling capacity without breaking. Fourthly, a good chemical and electrochemical stability is essential, as during operation the environment is fairly corrosive (an acid and oxidizing medium). Moreover, it is preferable that the operation temperature exceeds 100 °C in order to increase the charge mobility, the catalyst efficiency [1] and the heat-exchange capacity. However, this has a direct consequence on the MEA lifetime due to an increase in the oxidative degradation and the hydrolytic etching, which are thermally activated reactions. This aging is revealed by hole and crack formation in the film resulting in a decrease in the performances of the FCs. As a general rule, proton conducting membranes require a minimum water content to ensure good ionic conductivity and this water content has to be maintained under fuel cell operation, whatever the temperature and current density conditions. The conductivity can be enhanced by an increase in the ionic group content but to the detriment of the mechanical resistance, as a result of excessive water swelling. Finally, the polymer should not be electron-dispersive, which means it cannot be a conjugated polymer. These ion-containing polymers are called ionomers and contain less than 20 mol-% of acid (or basic) functions, contrary to polyelectrolytes, which are not soluble in water even at high temperature.

Synthetic materials can fulfill all these requirements independently, but the challenge arises in obtaining all of these properties simultaneously. It is thus clear that some compromises have to be made depending on the specifications of the fuel cell application.

For scientists just starting out in this field, they will quickly find out that the benchmark proton exchange membrane (PEM) is Nafion® from DuPont. This perfluorosulfonated polymer was developed in the 1960s for electrochemical separators and more specifically for chlor-alkali electrolyzers [6]. It is a comb-like copolymer (see Scheme 15.1) where one monomer is functionalized with a sulfonic acid at the end of a fluoro-vinylic ether pendant chain. Although other similar polymers have been synthesized (by Dow Chemicals, Asahi-Glass, 3M, Solvay, etc.), Nafion remains the ionomer of reference for cationic membranes for PEM fuel cells in terms of performance and stability [7], at least in a stationary regime. Its performances are related, firstly, to the presence of the sulfonic functions for which the degree of dissociation is known to be high in a fluorinated structure (a strong acid function). Secondly, the hydrated Nafion® membrane structure is characterized by a sharp nanoscale separation between hydrophobic and hydrophilic domains, which seems to favor good proton diffusion within the membrane. The characteristics of alternative polymer membranes are always compared with those of Nafion® or its derivatives. Nafion® can be dispersed in an aqueous solution, a very important advantage in the preparation of active-layer solutions that are compatible with the electrolyte membrane [8, 9]. Thus, Nafion is used in nearly all of the industrial prototypes, despite its poor resistance to thermal or swelling cycles and its excessive cost (due to complex chemistry and costly recycling).

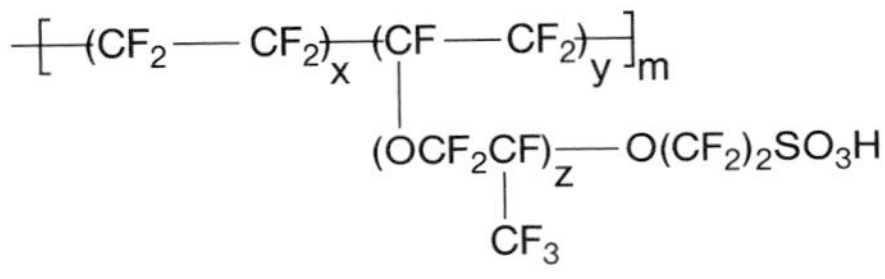

$$[-(CF_2-CF_2)_x-(CF-CF_2)_y-]_m$$
$$|$$
$$(OCF_2CF)_z-O(CF_2)_2SO_3H$$
$$|$$
$$CF_3$$

Scheme 15.1 The chemical structure of Nafion.

So why not design PEMs with a polymer that has an equivalent molecular structure, at a lower cost? Partially or fully hydrocarbonated polymer materials are less chemically stable than the perfluorinated systems, once they are functionalized with acid groups. Moreover, for a similar charge content, the conductivity of hydrocarbonated polymer membranes is almost one order of magnitude lower. The ideal membrane structure for obtaining the best performances in fuel cells is not known; indeed, the molecular organization inside a PEM is complex with a multiscale architecture that leads to non-affine physical properties. For example, the water sorption and transport mechanism of a Nafion® membrane is completely different when the membrane is partially or fully hydrated. The molecular structure and the correlation between structure and ionic transport for Nafion® are still being debated, after more than 30 years and thousands of publications. Finally, different routes are followed to synthesize the polymer and different processes are used to make films (extrusion or solution cast) but none of them can really be controlled. For the same membrane, different treatments have different effects on the structure and lead to protonic conductivity values that differ by more than one order of magnitude. Next, the question is how to design suitable macromolecules and membranes for fuel cells? This question can really be tackled once we have a much greater understanding of the chemical and physical properties of the system.

Ionomers with a low charge content (<5 % mol) have found applications in packaging, as elastomers, adhesives, polymer additives and rheology modifiers, and in the 1980s numerous investigations were performed and patents were awarded [10, 11]. The ionomers were derivatives from commercial polymers with partially ionized carboxylated functions. To improve the ion selectivity properties, ionomers containing sulfonic acids were thus studied, and for 15 years the fuel cell activity has boosted the development of new ionomer membranes. Recent studies have shown that the morphology of the films and their transport properties are strongly correlated. This research has also shown that the problem of the chemical and mechanical aging of these membranes has to be overcome and must be studied in relation to the structure and the ionic transport. However, this topic is beyond the scope of the present chapter and will not be dealt with in detail, even if it is a determining factor in the choice of the development of alternative ionic conducting membranes.

When the ion content is between 5 and 15 mol-%, the structure of the hydrated membrane is characterized by two interpenetrating and continuous networks with an interface which is more or less well defined: one region is hydrophilic and proton conducting, the other is hydrophobic and forms a dense matrix (see Fig. 15.1c) that ensures mechanical resistance and prevents the permeation of the reactants. The characteristic size of this separation falls between the nanometer and the micron scale, depending on the system. The first approach to understanding the interactions between the protons and the co-anions, the latter being covalently bonded to the polymer backbone, is to characterize the topology of the hydrophilic region.

The state-of-the-art in ionomer membranes has already been described in other recent review articles [12–16, 17]. They have detailed the chemical and physical characteristics relative to their use for FC application. In this article, our objective is to present one strategy that involves controlling the phase separation between ionic and non-ionic domains in this type of polymeric material and the effects on the performance of the membrane observed qualitatively or quantitatively. This strategy consists of the copolymerization of functionalized and non-functionalized segments in order to control the charge distribution along the polymer backbone. This technique should help to obtain a self-segregation with a defined topology between ionic and hydrophobic domains while having the required physical and chemical characteristics [18]. Moreover, one of the objectives of this review is to highlight, when possible, the differences compared with the post-functionalization of a non-ionic polymer and the blend process, which is the "classical" way of making ion-conducting membranes.

In the first section of this chapter, we briefly summarize the various techniques used to characterize the structure and the transport properties of an ionomer membrane by considering the studies carried out on Nafion® as examples.

The second part describes the advantages and drawbacks of the post-sulfonation methods, the blend and grafting approaches. Then we will show some examples of long block copolymers and the differences between random and sequenced arrangements as well as the di- and triblock ionomers used for their self-assembly properties.

15.2
Definitions and Investigations

To understand some of the graphs and parameters used to describe the structural and transport properties in these materials, a few definitions and investigation techniques have to be introduced.

Firstly, the main characteristic of this type of polymer is the amount of ionic groups attached to the polymer or in other words, the ionic exchange capacity (IEC, or the equivalent weight, EW = 1000/IEC) and is expressed in millimoles of sulfonic groups per gram of dry polymer. The IEC can be determined by titration or by NMR [19] and sometimes, if the polymer does not contain any other sulfur atoms than those in the acid functions, by elemental analysis or using scanning electron microscopy (SEM), analyzing the fluorescent signal at the sulfur absorption edge. For example, Nafion 112 (EW = 1100 g mol^{-1}, thickness 2 × 25 μm) corresponds to a polymer with 0.91 mmol of charge per gram of dry Nafion®. However, it is important to point out that this quantity does not take into account the polymer density, which can be very different from one polymer to another and should not be neglected when comparing ionomer properties as a function of their IEC. Another critical point is that this value is not directly related to the charge carrier concentration, the degree of dissociation depending on the chemical bonds around the acid groups and on the water content.

Another characteristic of these ionic polymers is the water uptake expressed as a mass fraction, λ, and corresponds to the number of water molecules absorbed per acid function; λ is of course dependent on the water vapor partial pressure. For Nafion 117, this value can vary from 1 in a dry state up to 20–22 in 100 % relative humidity or when immersed in water [20]. Isotherm measurement using a sorption balance is a suitable technique to determine this quantity. However, the equilibrium time is often difficult to achieve and this value can be underestimated. This value corresponds to an averaged value that does not take into account large-scale structural and ionic inhomogeneities that have been evidenced in these polymer membranes.

Small-angle scattering and microscopy [transmission electron (TEM) or atomic force microscopy (AFM)] studies have shown that these volume heterogeneities are in the order of thousands of angstroms [21–30]. Both techniques are suitable for mapping this ionic distribution over large length scales. For example, the Nafion® structure is usually depicted as an array of nanometric ionic clusters made up of ionic groups that are chemically bonded to the polymer backbone and that are swollen by water to form a continuous and ion-conducting network (see Fig. 15.1c) [31]. The water swelling described in terms of percolation enables us to account semi-quantitatively for scattering and transport data [31–34]. However, it develops shortcomings when used to describe membranes over a wide range of swelling states and kinetics. Powerful maximum-entropy methods for the inversion of small angle X-ray diffraction patterns [35, 36] or 2D-chord analysis [37] can be used to obtain the structural information, but the limited q-range in the scattering data leads to some uncertainties on the long-range correlation from intra- and/or inter-scattered particles. Also, care has to be taken performing AFM measurements in the tapping mode on soft materials because many artifacts can be recorded [38]. In addition, TEM pictures on microtomed films have to be interpreted with care because they can represent non-equilibrium morphologies due to the special sample preparation [39]. The structure of ionomer membranes being usually "multiscale", their properties have to be analyzed over the largest possible length- and time-scale. Through performing high-resolution scattering techniques, spectroscopy and microscopy and correlating the collected data, the Nafion® structure was recently characterized from nanometers up to microns [40].

The conductivity, which is proportional to the concentration of the charge carrier (the protons) and to their mobility, has to be measured. It obviously depends on the IEC, on the water content (or the relative humidity) and on the temperature. The conductivity is usually determined by impedance spectroscopy. However, although it seems that this technique is well controlled, the dispersion of the data published in numerous articles shows that it is, in fact, fairly difficult to obtain an absolute value of the intrinsic proton conductivity of a given membrane, some problems being recurrent (the problem of electrode–membrane contact, cell geometry with various cell constants, difficulties in measuring simultaneously the degree of hydration, the temperature and the variation of the membrane dimensions [41–43]).

The NMR techniques, such as the fast-field cycling relaxometry or the pulsed-field gradient (PFG) NMR, are suitable techniques for determining the dynamic dispersion and more especially the hydration-specific proton self-diffusion through the membrane [44]. Another characteristic of ion transport that is important for the FC application is the drag electro-osmotic parameter. The drag mechanism has been investigated using permeation techniques or electrophoretic NMR techniques. By determining the permeation and the electro-osmotic drag coefficient as a function of lambda, recent analysis in Nafion® and a sulfonated PEEK system demonstrated that the charge transport is really of a collective nature. However, the relaxation times associated with the proton dynamics in these complex systems vary from a few picoseconds up to a few seconds. Therefore, to study the transport properties within the ionic membranes, several complementary techniques such as quasi-elastic neutron scattering [45, 46] or field cycling NMR relaxometry [47, 48] also have to be used if we really want to obtain the maximum amount of information on the interactions between the ions and the interfaces. Care has to be made on the pre-treatment (synthesis, storage, condition of acidification, temperature and so on), which has a great influence on the measurements, making the comparison of data for the same membrane and between membranes of various natures and structures very difficult.

Recording of the polarization curve $U(J)$ is also often used for comparison of the membrane performance in FC working conditions; unfortunately too many external parameters, related to the MEA, come into play to be able to make reasonable comparisons. Until a standard method of testing these membranes in a fuel cell is defined and accepted by all those working in this field, it will be hard to really evaluate the evolutions and advantages of new methods of membrane synthesis, preparation and processing [49].

The very important problem of chemical and electrochemical aging depends on the chemical and physical structure of the ionomer film and finally, the notion of molecular weight (MW) in these systems is extremely important. The total MW is rarely considered, even for Nafion. It is assumed to be a critical parameter defining mechanical and electrical durability [50] because the scission of the chains due, for example, to chemical etching, leads to water soluble oligomers. However, its determination is not an easy task; the chemical heterogeneity with polar and non-polar groups does not help in finding a good solvent to obtain a soluble polymer. Moreover, the dipolar interactions are often at the origin of large aggregates that prevent the correct determination of the MW. In polar solvents, the system has to be exchanged with salt to screen the charges, the optimum salt concentration depending on the chemical structure [51, 52].

15.3
Polymer Modification

To obtain a proton conducting membrane, thermostable, polymers are usually modified using various physical or chemical routes: the most commonly used methods are either direct sulfonation of the polymer backbone (modification by post-sulfonation) or the doping of a neutral or an amphoteric polymer with some acidic (or basic) species. Other procedures used involve the chemical or physical grafting of a functional sequence, which is post-acidified (or quaternized), or the blending of a fully charged system and a hydrophobic component [15]. More recently, the synthesis of functionalized and non-functionalized monomers in a random or controlled way was tested; however more chemical development is required. In all the cases, the aim is to obtain a dense film composed of two interpenetrating networks, one which ensures the mechanical properties and the other which allows an optimal ionic conduction.

15.3.1
Post-sulfonation

A post-sulfonation method [53] is frequently used to functionalize a polymer that presents good mechanical and thermal stabilities. Aromatic polymers are good candidates and are easily sulfonated. The pre-formed film is immersed in either concentrated sulfuric acid, chlorosulfonic acid or sulfur trioxide complexes. However, this post-functionalization is often fairly drastic; the corresponding degree of sulfonation and the acid group distribution depend strongly on the reaction time. This method usually induces some heterogeneous reactions [54] and some defects, the concentration and distribution of which depend on the chemical (electron substituents) and structural morphology of the polymers (orientation, crystallinity, free volume). As a general rule, a long reaction time results in a homogeneous sulfonation, but with a degree so high that the polymer becomes water-soluble at room temperature. If the sulfonic agent is a good solvent for the non-sulfonated polymer then the swelling sometimes results in a homogeneous distribution of the acid groups, especially at low IEC. Soft methods have been developed using, for example, trimethylsilyl chlorosulfonate sulfonating agent [54] but the resulting film has a low degree of sulfonation and consequently a low ionic conductivity. It is difficult to control the location of the functionalization even if the active site is known; generally, the electron-rich aromatic site is the most active but the least stable, contrary to the electron-deficient site, which is more chemically stable [55]. Moreover, the functionalization of one site can inhibit another neighboring site. Polyaromatic polymers are often crystalline and these crystalline domains prevent a homogeneous distribution of the sulfonation. Even though much progress has been achieved and certain results are convincing [56], it is clear that this method does not permit the "tailoring" of the phase separation between the hydrophobic and the hydrophilic domains.

Another drawback of the post-sulfonation method is the cost of its exploitation on a large scale, due to the production of waste. The agents and solvents of sulfonation are polluted after the chemical treatment, thus increasing the actual cost of the transformation.

15.3.2
Grafting

Another post-functionalization method is the introduction of an appropriate chemical function, chemically grafted after irradiation of the base-polymer material. This function can then be acidified. It is a relatively simple method of modifying an existing polymer. For example, perfluorinated (PTFE) or partially fluorinated (PVDF, FEP or ETFE) pre-formed membranes, commercially available, can firstly be treated by introducing polystyrene grafts, whose lengths can be controlled by the styrene fraction and the grafting temperature, into irradiated membranes (see Fig. 15.2) [57, 58a]. The sulfonation of styrene monomers has been reported extensively in the literature and the polymerization chemistry is well controlled. For example, using chlorosulfonic acid in a chlorinated solvent, the sulfonation can be performed as a function of time, to obtain one sulfonic acid group per aromatic ring. However, as for the previous method, this treatment can be non-homogeneous; these fluorinated materials are often semi-crystalline and only the amorphous parts contain the styrene grafts and thus the acid groups. Moreover, the thermal treatments and the orientations of the polymer chains due to the pre-formage have some unpredictable influences on the final products. If the irradiation and grafting conditions are taken into account, the number of parameters is too high to be able to monitor precisely the final membrane structure (except with swift heavy ions). The addition of a cross-linker agent would possibly improve the mechanical properties but reduce the water uptake and hence the conductivity. Increasing

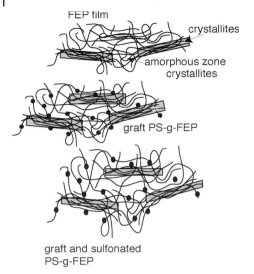

Fig. 15.2 Schematic representation of the structural change after grafting and sulfonation of fluorinated film [58b].

the degree of grafting would improve the conductivity but increase the reactant permeation and thus speed up membrane degradation.

Nevertheless, this radiation grafting method does indeed result in a defined interface between the hydrophobic matrix and the styrene based parts, which correspond to large domains, themselves constituted of nanometric ionic domains [59].

The grafting and the sulfonation conditions give some flexibility to this relatively low cost method but the ideal degrees of grafting and functionalization are still being debated [57, 60].

It is still possible to find new routes to sulfonate [61] or acidify thermostable polymers without degrading the polymer, but the functionalization is not fully guaranteed and its location not well controlled [16].

15.3.3
Blends

Another route in the PEMFC development, which will not be extensively developed here, concerns the blend systems. If we want to have a dense matrix that forms the mechanical architecture of the film into which is embedded a highly efficient ionic conducting network, one appropriate way is to blend two polymeric systems that match these properties. Yet, if we blend two polymers that fulfill these specifications, it is clear that due to their incompatibilities the system will tend to separate. If the mixing is forced mechanically, an irregular and more or less coarse structure is formed. On the other hand, this phase separation can be subtly quenched by choosing the right thermodynamic conditions (cooling below the crystallization or glass transition temperature of one component) or by cross-linking the polymer chains together. This cross-linking procedure can be either physical, using the Van der Waals, ionic, H-bond interactions between some suitable molecules, or chemical (covalent bonds). In the former case, these interactions are not stable at high temperature and in the latter case they effectively reduce the swelling rate. The reader can refer to a recent review article by Kerres [62] that presents the advantages and the drawbacks of this approach.

As for the two previous methods, the degree of freedom for this procedure is so large that we can always expect to get an appropriate material. For the moment, none of these methods leads to the ideal PEM with specific mechanical and chemical properties for fuel cell application.

15.4
Copolymerization of Functionalized Monomers

As mentioned previously, one recent strategy for making two interpenetrating domains is to functionalize the polymer during synthesis so that two molecular architectures can be built. One looks the same as the Nafion®, i. e., with a molecular structure comprising a hydrophobic backbone on which functionalized side chains would be grafted, the nature, location and the length of the grafts being of course adjusted (a comb-like structure). The other structure corresponds to the sequential construction of the backbone with a random or sequenced distribution of ionic and neutral blocks. Then by "tuning" the thermodynamic parameters, the self-organized properties of the polymer can be used in order to control the molecular organization and separation between these two components, which are chemically bonded.

McGrath and coworkers, pioneers in the study of the correlation between the structure and transport properties, introduced this concept of block ionomers [63, 64] using hydrocarbon- and heterocyclic-based polymers. In the last five years other studies have been undertaken and have dealt mainly with the synthesis and the measurement of the required properties (e. g., swelling, conductivity, permeation) of these block ionomers. Very little work exists on the analysis of the influence of this tailored distribution. Nevertheless, in the following sections, we will detail some examples concerning what we will call the long co-ionomer blocks, separating the "main chains" and the "side chains" cases. Then, some examples will be given concerning "short copolymers", i. e., the di- and triblock ionomer systems.

15.4.1
Main-chain Type Co-ionomer

15.4.1.1 Sulfonated Polyimides
The sulfonated polyimides (sPI) was the first copolymer system for which the synthesis was controlled so as to produce random or block (segmented) sulfonated copolymers [65]. They were prepared in two steps by polycondensation in m-cresol using various aromatic diamines, some of them (the non-functionalized) contain-

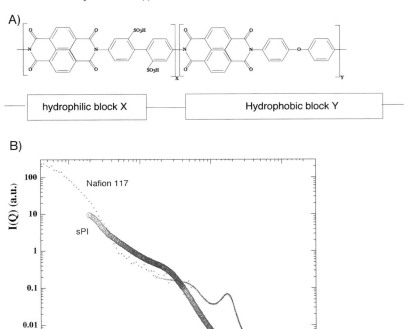

Fig. 15.3 (A) The chemical structure of NTDA–ODA sulfonated polyimides.
(B) Typical small angle scattering curves of Nafion and sPI samples.

ing phenylether bonds and/or bulky side groups. The first step involves transforming the sulfonated sequences in an ammonium form to ensure solubility. The chain length or mole number X of the sulfonated oligomers can be modified by varying the molar ratio of the starting diamines and dianhydrides. The second stage is the polycondensation reaction, which consists of adding the adequate amount of non-sulfonated diamines and dianhydrides. Knowing the non-sulfonated diamine number, Y, the ion-exchange capacity (IEC) of the final sulfonated block copolyimide can be adjusted using the molar ratio (X/Y) [66] (see Fig. 15.3A). Ideally, the length of the sulfonated block influences the size of the ionic domains, and the Y/X ratio determines the number of ionic domains per unit volume or the distance between the ionic domains. The first generation based on oxydiphthalic dianhydride (ODPA), called the phthalic system, is commercially available. However this system was not stable enough under fuel-cell conditions [67] and was quickly abandoned (lifetime <100 h). The second generation, based on 1,4,5,8-naphthalene tetracarboxylic dianhydride (NTDA) instead of ODPA, was more stable and more studies were carried out on its structure and transport properties [68, 69].

The membranes, which were prepared by a solution-casting method, were characterized by the sulfonated block distribution and length by determining the water uptake and swelling, and the proton conductivity. These properties were correlated with the morphology of the polymer, which was studied using small angle scattering and microscopy techniques [70]. The data obtained from the scattering and microscopy studies do not reveal a well-defined structure but do show a strong anisotropy between the planar and the normal directions of the membrane, which depends on the size of the blocks [71, 72]. A large bump at small scattering angles was identified as a characteristic ionomer peak and analyzed as the distance between the ionic domains (see Fig. 15.3B). In the case of sPI, it was concluded that large ionic domains exist in comparison with the Nafion® structure. They are a function of the ionic block length X and not really of the charge amount, X/Y [73]. A glassy and highly oriented foliated structure was proposed [71, 74, 75] with a nanoscopic porosity that can be filled with water without modifying the structure. The water molecules simply cause a swelling of the membrane, which is mainly related to the mass, with a weak geometrical variation along the membrane thickness. Lambda does not depend on the IEC value as it does for the other systems, a specific characteristic of the sPI (see Fig. 15.4A). IR spectroscopy reveals a strong interaction of the water molecules with the sulfonated groups and also with the COO groups present in the non-functionalized sequences [76]. The structural anisotropy has, of course, an influence on the ionic transport [75, 77] and some recent PFG NMR studies [78] revealed that the preparation method and the casting process have a strong influence on the molecular orientation and thus the transport properties.

An interesting point, which is not often established for the other systems, is a comparison between the random and the sequenced block copolymers. Figure 15.4B demonstrates the interest in the copolymerization of functionalized blocks to control partially the local organization of the ionic groups by blocks, permitting an increase in the conductivity without a significant increase in λ. However, when the distance between these ionic blocks is too large, the increase in conductivity due to the percolation between the ionic domains is lost.

One important problem with these materials is chemical degradation, the imide ring is very likely to hydrolyze in the presence of the water molecules around the sulfonated groups and close to the diamines [79–81]. However, owing to specific properties, such as their weak swelling and their low permeation coefficient, research into designing new structures was pursued. The insertion of bulky groups [82, 83] or of flexible diamines to increase solubility was tested [66]. To improve the stability and mainly the degree of dissociation of the acid groups, fluorinated functions were added to the hydrophobic blocks. However, the results are not very convincing and another approach is to modify the functionalized blocks and to move the acid function out of the imide rings. The insertion of stable groups between the naphthalenic block and the sulfonated diamine has given some fairly encouraging results. Okamoto and coworkers demonstrated that polyimide membranes based on 4,4'-diaminodiphenyl-ether-2,2'-disulfonic acid (ODADS) have a more flexible structure than the corresponding BDSA-based ones and are much

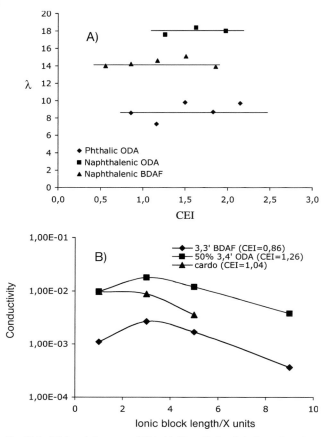

Fig. 15.4 (A) Lambda versus CEI (with $X = 5$) for phthalic and naphthalenic sulfonated polyimide. (B) Conductivity versus ionic block length (X) for three different naphthalenic sulfonated polyimides ($X = 1$ corresponds to the statistic copolymerization).

more stable in the presence of water [84]. They argued that the structural conformation permits an easy relaxation of the chain and a reversible recombination of the opening and closing of the ring due to the hydrolysis. They have also reported that polyimides, with high basicity of the sulfonated diamine moieties slow down the hydrolysis of the imide ring due to the electron donating effect. These recent synthesis developments of this type of ionic block copolymer have led to improvements in solubility and chemical stability, so that these membranes are promising materials for use in PEMFCs.

15.4.1.2 Polyarylene Systems

Polyaryl membranes appear to be good candidates for PEMFC. Polymer such as polysulfones (PSU), poly(ether ketones) (PEK), polybenzimidazole (PBI) already present excellent thermal and mechanical stabilities. They are known to be chemically stable in severe hydrolytic and oxidative conditions, an important property for the fuel cell application. Their chemistry is well controlled, commercial products are available and many different chemical structures exist, opening up a large variety of membranes, blends and composite systems [13]. More often than not the functionalization of these polymers to obtain a protonic conducting material is done by post-sulfonation, with all the disadvantages that this entails. Very little research has been carried out to date on the direct copolymerization of sulfonated and non-sulfonated monomers to obtain either random (statistical) or block copolymers. This is mainly because new syntheses have to be developed to obtain soluble polymers with relatively high molecular weights.

The Sulfonated Poly(ether sulfone) There are two commercially available references for the poly(arylene ether sulfone) based solid polymer (PSU), Udel® P-1700 and Vitrex® PES 5200P. The sulfonation procedure can be performed for both of these materials either in solution [85] or using a slurry procedure [86]. As before, the conductivity increases with the degree of sulfonation until the mechanical properties degrade. Above about 30 % sulfonation the polymer becomes, in general, soluble. One option is to use this type of polymer with a lower degree of sulfonation and to associate it with an inorganic acid [13, 54]. The second option is copolymerization. A good example of this is the copolymerization of biphenol, disulfonated activated aromatic halide monomers with the precursor-activated aromatic halide monomer (BPSH-*xx*). BP stands for 4,4′-biphenol, S for 4,4′-dichlorodiphenyl sulfone and H for the acid form; *xx* is the mole-% of disulfonated units relative to non-sulfonated activated halide monomers S. It has been demonstrated that for an equivalent IEC, the proton conductivity is higher and the mechanical/chemical stability better than the reference PSU system. Some TM-AFM images are presented in [55], showing a clear segregation between domains for which the phase response is different at the nanometer scale. This difference is interpreted as being due to water molecules present in ionic domains and forming a continuous cluster within the hydrophobic matrix.

The water uptake behavior of these materials is very similar to that of the post-sulfonated PSU membranes. An excessive swelling is measured when more than 30 mol-% of disulfonated blocks are incorporated (IEC > 1.5 for reasonable conductivity values). Similar observations were performed by Scherer and coworkers, who showed some degradation of the mechanical properties (brittle samples) by decreasing the number of sulfonated blocks in order to achieve the correct water uptake [87]. No thorough study has yet been published presenting the relationship between the structural morphology and the different ratios of sulfonated and non-sulfonated blocks, or the effect of the incorporation of various chemical moieties that bring other specific properties to these membranes (affinity with fluorinated system or inorganic host particles).

More recently work dedicated to PEEK polymers was carried out and the sulfonated hydrophilic and the hydrophobic blocks were synthesized successfully by a two-step process [88, 89]. To combine both types of blocks, the same polycondensation procedure was used as for the sPI samples. Consequently, the total molar mass is not so high and is in general limited to 150–200 kg mol^{-1} depending on the X/Y ratio, but nevertheless remains sufficient to make thin films. Although the second monomers are added at the desired stochiometry to adjust the length between the ionic blocks, some discrepancies still exits between the desired IEC and the theoretical value, which means the chemistry can still be improved.

One advantage of this technique is the selective solubility of the sequenced co-ionomer. It is usually difficult to find a common and good solvent for the different blocks, which also makes the MW determination complicated. However, by "tuning" the chemistry and by adjusting the ratio of blocks, some improvement in solubility can be achieved. On the other hand, it is possible to choose a specific solvent that has a better affinity for one segment than the other, leading to the quenching of certain heterogeneities that exist in the solution. This results in a membrane morphology that can be easily reproduced. For example, for a 30/70 system a cylindrical morphology would be expected for a di-block composition, however a spherical organization was observed (see Fig. 15.5) [88]. These workers argued that this structure is reinforced by the DMF solvent that swells the hydrophobic phase more than the hydrophilic one, leading to collapsed ionic spheres. This particular structure is not favorable for ionic transport as the ionic domains are no longer connected together; however, perhaps by using another solvent the inverse structure could be achieved.

Another point is the possibility to multi-functionalize the hydrophilic block, this means many (2, 3 or 4) acid groups can be created on one block in order to potentially increase the number of charge carriers without increasing the charge dispersion along the chain or within the matrix.

As for sPI, it is clear that this approach is very encouraging with a large variety of precursors possible (especially with different fluorinated or alkyl biphenol units) [89]. The correct distribution of blocks would lead to a polymer that can be

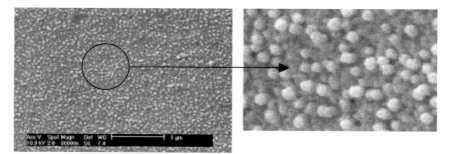

Fig. 15.5 Electron microscopy images of the block copolymer membranes. SEM sulfonyl chloride forms of copolymerized polyether ketone as cast, after drying [88] (Reprinted from [88] with permission of Elsevier).

15.4 Copolymerization of Functionalized Monomers

dispersed in a solution, a state which is very useful for film processing and for the fabrication of polymer compatible electrodes – this is not possible with post-functionalized systems. Further development of new synthetic procedures would also lead to an increase in the total molecular weight of the polymer.

Sulfonated Polybenzimidazole As for the polyarylene polymers, the choice of the polybenzimidazole (PBI) for use in fuel cells is related to its commercial availability and its highly hydrothermal stability. Moreover, it is a basic polymer that can form complexes with organic or inorganic acids. Therefore, the functionalization by acid doping [90–92] is one way to obtain a protonic conductor material. The main difference from the previous systems is the charge carrier, which is not solvated by the water but by the phosphoric acid itself. This is a big advantage for high temperature operation (>100 °C), the protonic conductivity being thermally enhanced as opposed to water mediated ionic transport. Nevertheless, the conductivity remains a few orders of magnitude lower than for the swollen sulfonated systems. Also, the elution of phosphoric acid, when the produced water diffuses into the electrolyte, becomes a problem.

Chemical grafting of protogenic groups or direct sulfonation by electrophilic substitution on the polymer backbone leads to a significant improvement in their proton conduction ability, increasing the number of mobile charges. Both hopping and diffusion transport mechanisms are combined, but often at the expense of the mechanical properties due to an excess of water and therefore an excess of swelling.

Good performances can be obtained in a fuel cell but only at high temperatures and under stationary conditions.

More recently, some innovative solutions were proposed by Mercier and collaborators [93]. The synthesis of high molecular weight sulfonated PBI from a sulfonated monomer was investigated. A comparison between random, sequenced block copolymer and blend membranes demonstrated that for an equivalent IEC, the intrinsic proton conductivity is higher for the block copolymer membrane than for the statistic or the blend system, and increases with the sulfonated sequence length. Nevertheless, these values remain rather low ($< 5 \times 10^{-4}$ S cm^{-1}) but, by combining the copolymerization and the doping routes, promising results could be obtained.

15.4.2
"Side Chain" Co-ionomers

Nafion® is a good example of a side chain or a comb-like chemical structure. However, we have already described the morphology of the Nafion® membrane so in this part other examples will be given. We can consider that the polystyrene sulfonic acid polymer, used in the PEMFC for the auxiliary power unit of NASA's Gemini flight, is the first co-ionomer with a side-group structure. This polymer was rapidly given up for this application because of the strong oxidative etching of the C–H bounds (due to the high potential in FC). It has been suggested that trifluorostyrene-based ionomers (sPTFS) should overcome these difficulties [94]. Ballard Advanced Materials Corporation have developed substituted sPTFS membranes (BAM® Ionomers). The polymer is a random molecular system both in the chemical composition along the backbone and in the sulfonic groups distribution. Processable materials were obtained and they satisfy the requirements for fuel-cell operation in terms of lifetime and performance [95].

SAXS and SANS studies of water swollen BAM® membranes were performed [96, 97] and as is often the case for these types of systems, data analysis differed from one group to another. The first group pointed out the presence of an ionomer peak and they analyzed the SAXS spectra using a model of spherical ionic domains randomly distributed in the polymer matrix. The data analysis suggested that these domains are very small ($R < 15$ Å) and that only a small percentage of the ionic groups are located in these spherical domains. However, the measured water volume fractions are extremely large for a proton conducting membrane (between 80 and 90 % for most of the membranes). Therefore, the majority of the hydrophobic part of the polymer (fluorinated backbone and phenyl rings) and water should be intimately mixed to avoid any contribution to the SAXS intensity even at relatively large angles. It is somewhat questionable that ionic domains, containing ions and water, could exist while dispersed in a matrix containing more than 80 % water.

The second group analyzed BAM membranes with low equivalent weights of 417 and 477 g mol^{-1}, respectively. Using small-angle neutron scattering (SANS) and over an extended angular range, they studied the swelling behavior, analyzing the ionomer peak position as a function of the water volume fraction (see Fig. 15.6). These results were interpreted as being indicative of a connected network of relatively small rod-like particles whose connectivity decreases as the water content increases. From the data obtained using a contrast agent (protonated and deuter-

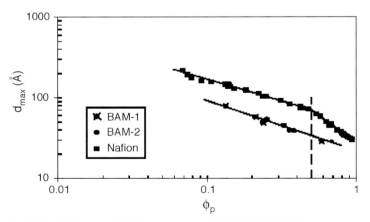

Fig. 15.6 Dilution law of BAM membranes with IEC of 2.4 (BAM-1) and 2.1 (BAM-2) compared with the Nafion swelling behavior. d_{max} corresponds to the characteristic size determined from the position of the ionomer peak in the scattering curves [97].

ated ammonium ions), they deduced that as in the Nafion® system, a counterion condensation occurs at the interface between the hydrophilic and the hydrophobic domains. Moreover the data analysis revealed a network of connected rod-like agregates, composed of a perfluorinated core ($R = 4$ Å) and a phenyl ring shell ($R = 2.5$ Å). The density of this network is a function of the water swelling. However, these data could not really be compared with those of Nafion® 1100, firstly, because the IEC was about twice as high for the BAM system if we consider that the polymer densities are equivalent and, secondly, the structural change is too important when the swelling is excessive.

Even if these styrene based materials show poor oxidative stability that prevents their use in high temperature fuel cells, research should be pursued as they do show some interesting water swelling properties and good conductivity.

Indeed, one advantage of styrene is that its chemistry is well known. A radical polymerization can be used to control the sequence lengths.

A more recent ionomer developed by Holdcroft's group is a good example of controlled graft copolymers [98]. The scheme of the copolymer is presented in Fig. 15.7A, it is synthesized in two steps: firstly, the preparation of the grafts in an Na-form using a stable free radical polymerization and then the copolymerization with styrene to form PS-g-*mac*PSSNa, which can be acidified. As for the sequenced block copolymer (sPI or sPEEK), the length of the graft chains affects the size of the ionic domains, whereas the graft density controls the ionic domain density. IEC is related to X/Y. Comparison with the random equivalent system shows that although, for a given ion content, the water uptake of the random system is higher than for the sequenced one, the conductivity is much lower (see Fig. 15.7B and C). This is a very interesting result, which shows that the conductivity, or even more so the charge mobility, is not just related to the amount of water; it is clearly dependent on a parameter that characterizes the structural organization. The question is what is the pertinent scale? Holdcroft and collaborators have shown, using TEM techniques, that the charge distribution along the PS chains affects the ionic aggregation morphology, and they observed some bi-continuous structure with mesoscopic characteristic lengths (see Fig. 15.7D).

This approach of grafting pendant chains to a hydrophobic backbone is different in the case of sPI. This is mainly due to the problem of chemical degradation that forces chemists to synthesize copolymers, which are much more stable in the presence of water than those derived from the widely used sulfonated diamine 2,2c-benzidinedisulfonic acid (BDSA). Aliphatic groups were incorporated into a sulfonated polyimide ionomer both in the main chain and in the side chains between the acidic function and the backbone in order to improve the hydrolytic stability. However, different conclusions were proposed by different research groups. Asano et al. showed that their new polyimide ionomer was very stable to hydrolysis and oxidation [99]. The proton conductivity was comparable to or even better than that of Nafion® 112 at high temperatures and under high relative humidity conditions. Okamoto and coworkers concluded that NTDA–DAPPS membranes showed poorer oxidative stability than ODADS and BAPFDS-based polyimide membranes, probably due to its alkoxy linkage in the side chains [100].

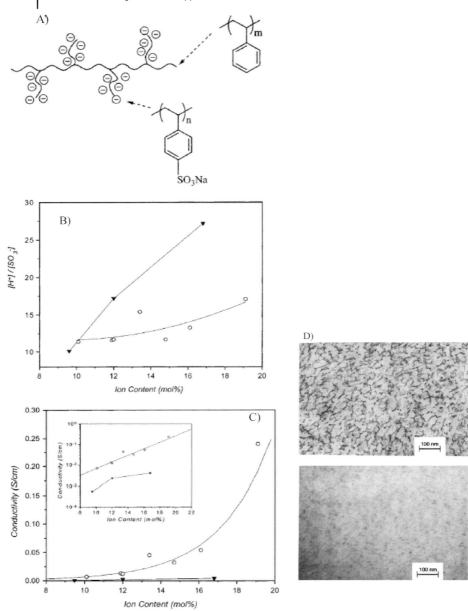

Fig. 15.7 (A) Schematic view of the PS-g-macPSSA graft polymer from [98]. (B) Lambda versus ion content (in mol-%) for a statistic (PS-r-PSSA, upper curve) and a sequenced (PS-g-PSSA, lower curve) copolymer [98]. (C) Proton conductivity versus ion content (in mol-%) for a statistic (PS-r-PSSA, lower curve) and a sequenced (PS-g-PSSA, upper) copolymer [98]. (D) TEM pictures for a statistic (PS-r-PSSA, lower) and a sequenced (PS-g-PSSA, upper graph) copolymer [98] (Reprinted with permission from [98], Copyright (2002) American Chemical Society).

This last example shows that a comparison of the different methods for improving these polymers is still extremely difficult for many reasons, such as, the various processing methods, the lack of reproducibility, the different conditions for measurements and thus the fast development of PEMs is not easy.

15.5
Di- and Triblock Ionomers

As we detailed previously, the statistical or block copolymers synthesized for PEMFC have a fairly high polydispersity in length. Depending on the equivalent weight, the structure and connectivity of the swollen ionic network through the low dielectric polymer matrix is quite difficult to characterize, this leads to a non-realistic permeation and ionic transport model.

Self-assembled morphologies in di- or tri-block co-ionomers present some ordered microstructures including cylinders, lamellae and other ordered or disordered structures [101]. They are good candidates for model membranes for studying the structure and ionic transport relationship.

Indeed, regular morphologies with different symmetries can be created as a function of polymer type, length and composition. The thermodynamic control of the morphologies (competition between the de-mixing entropy and the chain elasticity to minimize the interfacial energy) and being able to orient them macroscopically allow us to identify clearly the characteristic features of the ionic transport, such as ionic domain size, connectivity, interaction between blocks and so on. However, for ionomers, the electrostatic interaction among ion pairs has to be taken into account. It becomes a limiting factor for the self-assembly property of the blocks and can allow the formation of new phases.

In the 1980s and 1990s, several studies were performed to analyze the effects of the ionic interaction. An example is the case of the SEBS, a commercial styrene (S) based triblock product where the mid-block after hydrogenation of the butadiene is essentially a random copolymer of ethylene and butylene (EB). Weiss and coworkers showed that SAXS profiles and peaks of samples with 25 mol-% sulfonation of the styrene block can be analyzed by considering a three-phase structure with a Bragg spacing of approximately 3–4 nm for the ionic domains and 20–30 nm for the polystyrene domains [102, 103] embedded in the EB matrix. They demonstrated that the long-range order of the PS–EB separation decreases with increasing degree of sulfonation. Moreover, changing the ionic strength by exchanging the counterions from Na^+ to Zn^{2+} and H^+ led to the same effect [104, 105]. The mechanical properties are also influenced with the development of a new plateau region above the T_g of the polystyrene-rich phase [106].

In order to analyze the structural organization in real space, observations were made using the TEM technique, but as indicated in the publications, the sample preparation had a strong effect on the results. Indeed, TEM pictures representing non-equilibrium morphologies and meaningful TEM images are sometimes not available [39].

SEPS triblock systems with ethylene-*alt*-propylene instead of EB were also examined, and transitions in morphologies were observed upon sulfonation [107]. Either the χ parameter was influenced due to an increase in the thermodynamic repulsion between the styrene and the EP blocks or the ionic interaction behaving as physical cross-linkers, led to an apparent increase in the degree of polymerization N.

However, these studies were not performed with the aim of quantifying the ionic and more especially the protonic conductivity in such materials. In the last 5 years, the research on di- and triblock ionomers has mainly been concerned with the possibility of decreasing methanol permeability through the membranes, a major problem with most of the long block copolymers (for direct methanol fuel cell, DMFC). The objective is to create a suitable and well-organized structure so that the macroscopic orientation hinders the alcohol crossover without penalizing the protonic conductivity.

As for the multiblock co-ionomers, we found some studies on systems for which the functionalization was performed either *a posteriori* [108] or during the synthesis; the films obtained from copolymers can be molded, cast, laminated or extruded before or after the acidification.

The hydrophobic blocks are mainly hydrocarbon subunits such as styrene, ethylene/butylene, isobutylene [33, 109] isoprene, butadiene [110, 111], propylene [107, 112], bipyridine [104] and more recently fluorinated blocks [39] have been synthesized. The hydrophilic blocks are either cationic or anionic. The former includes either some weak acid function with an acrylic segment [108] or strong acid groups obtained by the sulfonation of polystyrene blocks. The latter can be obtained by the quaternization of vinyl pyridine groups. One of the advantages of using an OH^- conducting polymer for DMFC application is that the methanol crossover is significantly reduced. This crossover corresponds to the diffusion of the fuel (alcohol and water molecules) from the anode to the cathode through the membrane. When an anionic polyelectrolyte is used, the hydroxyl ions diffuse (with water molecules) from the cathode to the anode, a transport mechanism that slows down the crossover of the alcohol molecules [5].

Similar synthesis difficulties exist for the block copolymer systems as for the previous systems. An ABA- or BAB-type of triblock sytem depends on the order of monomer addition. When synthesizing polymers by living anionic polymerization, it is known that the sequence of monomer addition must be in the following order: dienes, styrene, acrylates, methacrylates, vinyl pyridines, ethylene oxide [113]. The choice of solvent is also not trivial, as it must be compatible with both blocks. The control of the degree of acidification without resulting in mechanical degradation is delicate, especially because the MW is lower than for the long block copolymer, thus the effect on the mechanical properties is more drastic.

Nevertheless, several structural [114], mechanical [115] and transport studies [32, 101, 114] have already been carried out [116].

The structural dependence on the relative sequence lengths, the degree of functionalization with acidic groups, the counterions and the swelling have been investigated. In general, increasing the degree of sulfonation results in a loss of the long-range order existing in the neutral systems and generally modifies the phase diagram of the system due to some structural rearrangement during ionization in the swollen state [117]. As for the block copolymer systems reviewed previously, an increase in the glass transition temperature (T_g) is observed with an increase in the modulus and the tensile strength, but with a decrease in elongation at break. The water sorption affects the morphology and the long-range order slightly.

However, the problem of excessive swelling for a relative amount of charge is also observed for example in sSEBS [118] and thus the cross-linking method becomes an alternative for reducing the solvent sorption and permeation [114].

Recently, progress has been made [33, 109] in preserving, at least partially, the long-range organization of the copolymer while maintaining good protonic conductivity (and also methanol permeability) for H_2 (see Fig. 15.8A and B) or methanol fuel cell application (see Fig. 15.8C). Indeed, sulfonated SIBS triblock systems are interesting for analyzing the dependence of the mesoscopic order and the structure orientation on the transport properties.

More recently, new model di-block copolymers consisting of sulfonated poly([vinylidene difluoride-co-hexafluoropropylene]-b-styrene) block copolymers (P[VDF-co-HFP]-b-PS) were synthesized by Shi and Holdcroft. It is known that although the chemistry is more complex and costly, fluorinated blocks are more chemically and electrochemically stable. Moreover, a new method of synthesis of the polymers to obtain low polydispersity of the block length was developed and well-defined structures, due to the strong incompatibility between the blocks, were studied [39]. The ionic exchange capacity (IEC) was controlled by adjusting either the length or degree of sulfonation of the polystyrene chains. Cast membranes from THF solutions were investigated by TEM and scattering techniques [119] and the results clearly show differences due to the degree of sulfonation.

The T_g of the P(VDF-co-HFP) segment is about –40 °C, whereas the T_g of the sulfonated polystyrene is higher than 90 °C (for molecular weights of about 12 000 g mol^{-1}) [39]. This difference in mechanical properties between both blocks, and thus both domains, is important as the resulting di-block copolymer is a soft film but with the desired transport properties. One question arises: is it not preferable to have a rubber phase that is an ionic conductor embedded in a glassy matrix?

Let us consider a triblock system with a sulfonated block as the middle one [120] flanked by two hydrophobic blocks [121]. The length ratio should be chosen in order to obtain either a hexagonal or a lamellar structure in which the hydrophilic domain can swell without causing dilution of the phase. The choice of the hydrophilic block as the middle one, embedded in a glassy matrix, should prevent excessive swelling of the ionic domains. It would also be very interesting to choose a hydrophobic polymer block that is impermeable to methanol when it is in its glassy state.

Finally, an electrical field (the dielectric constant will be very different in both domains) or a shear technique can be used to orient macroscopically the system in a specific direction to optimize the transport properties, depending on the electrochemical applications [33, 109]. An interesting study still related to polymer materials and based on self-organization is the work done by Ikkala and coworkers [122, 123]. The complexation between supramolecules and amphiphile molecules

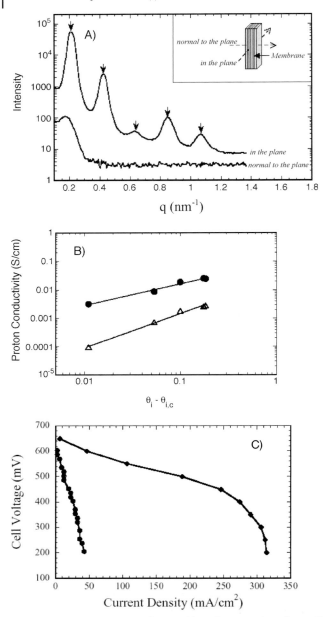

Fig. 15.8 (A) Scattering curve of a typical lamellar mesophase from a film made of a sulfonated SIBS triblock copolymer [109], showing the excellent orientation of the phase. (B) In-plane (upper curve) and out-of-plane (lower curve) proton conductivity versus the water concentration [109]. (C) Polarization curves obtained from a DMFC: the upper curve corresponds to a cell using a Nafion membrane and the lower curve corresponds to the sSIBS film test [109] (Reprinted from [109] with permission of Elsevier).

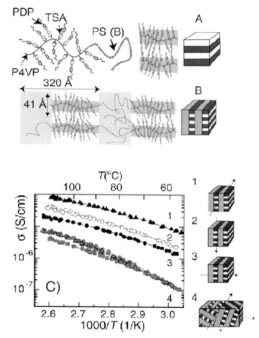

Fig. 15.9 (A) and (B) Schematic views of the hierarchical nanostructures of the proton conductive P4VP(TSA)PDP and PS-P4VP(TSA)PDP macromolecules, respectively [124]. (C) Proton conductivity versus temperature for various orientations of the structure into the film [124]. These orientations are induced by an overall alignment of the structures using an oscillating shear flow [125].

via, e.g., hydrogen bonding to form comb-shaped supramolecules, is very interesting. The self-organized structure is formed due to repulsive forces between the polymer backbone and the tails of the amphiphiles. Moreover, this structure can be combined with a block copolymer structure to obtain a large hierarchical self-organized structure. The larger structure between PS and P4VP(MSA)1.0PDP1.0 has a periodicity of 200–1000 Å and the smaller structure 20–50 Å inside the P4VP(MSA)1.0PDP1.0. TEM shows that the smaller structure tends to orient perpendicularly to the larger structure (see Fig. 15.9). However, the hierarchical order also makes the conductivity behavior more complex! The conductivity is low as the transport is characterized by a hopping mechanism, which is strongly dependent on the orientation of the blocks as is shown in Fig. 15.9B. Clearly the proton conductivity is greater when the large-scale orientation of the blocks is perfectly ordered in 3D [124]. This ideal orientation of the structure in 3D is obviously a requirement that is not compatible with an industrial-scale application. This example shows perhaps that a bi-continuous structure, which is isotropic, would be more suitable for a fuel-cell membrane.

15.6
Conclusion

Within the framework of the proton exchange membranes for fuel-cell applications, sulfonated block-copolymer materials as substitutes for the current sulfonated-perfluoro polymers show promising results. However, many stringent properties are required of these materials. To summarize, these membranes act as solid electrolytes and separators, separating two gas (or liquid) compartments. They need to be thin, dense films which should be good proton conductors and resistant to various mechanical and chemical aggressive constraints. In this chapter, a few examples of studies concerning ionomer membranes are described in order to highlight the two different routes, namely pre- and post-functionalization with acid groups and their advantages and drawbacks. It is shown that molecular structures and ionic transport properties are strongly related. It is clear that by controlling the structure it is possible to improve the chemical and physical properties of these membranes. Unfortunately, the post-functionalization of polymer membranes does not allow this control, making it impossible to define a membrane "building" strategy that takes into account the different interactions between the polymer chains and between the solvent and the polymer.

Of course, this latter method would be easier to develop as compared with imagining new routes of synthesis.

The chemical and physical ageing of the membranes and of the MEA interfaces is still a problem in the development of PEMFC. An organic structure, such as these polymer structures, always has some weak bonds which will undergo various chemical and physical etchings. One solution to prevent or slow down these reactions is to modify the chemical functions locally. The copolymerization of different functionalized blocks could help to solve this problem. A fully fluorinated ionomer system is certainly more inert and more appropriate for FC application than the partially hydrocarbonated one, but it entails costly recycling processes, a point that has to be taken into account for industrial scale applications.

The second critical point is the aging of the MEA interfaces between the pure solid electrolytes and the active layers. Active layers are usually made from a Nafion® solution in which the carbon and catalyst particles are dispersed. Both parts undergo similar (electro-) chemical and physical stresses but due to the different structures their responses are very different and create some resistance to the proton diffusion through the interfaces. If we want to increase the compatibility between these layers, the chemical structures of both should be similar; one alternative is to use the same polymer solution to make the membrane and the electrodes. Thus, the development of a strategy of tailoring the complete MEA set-up means that a study concerning the physical chemistry of the ionomer solution is also necessary. The behavior of Nafion® aggregates in aqueous solution [22, 126, 127] has been reported, and similar studies should be carried out for the other ionomer systems. In this chapter, we have shown that by using the pre-functionalization method the solubility of the block-ionomers is improved, and further research should be carried out to understand the formation of aggregates as a function of the solvent concentration. Indeed, the conformation of the ionomer chains that can form, for example, some loops [128] or aggregates [129] greatly influences the viscosity and hence all the film formation mechanisms.

Finally, model systems using short co-ionomer blocks will perhaps not be suitable materials for making membranes for fuel cells but do remain good research support for correlating the structural and transport properties and analyzing the different degradation mechanisms. We will end this chapter by citing a remark by McGrath: "Issues of molecular weight, mechanical properties, and chemical/physical degradation mechanisms need to be addressed with much more rigor than they have been in the past". Indeed much more fundamental research on the structure and property relationships of PEMs has to be carried out if any innovative breakthroughs are to be made in the development of PEM fuel cells as power sources for the future.

References

1. S. LITSTER, G. MCLEAN, *J. Power Sources* 130, 61–76 (**2004**).
2. C.-Y. WANG, *Chem. Rev.* 104, 4727–4766 (**2004**).
3. C. STONES, A. E. MORRISON, *Solid State Ionics* 152–153, 1–13 (**2002**).
4. P. COSTAMAGNA, S. SRINIVASAN, *J. Power Sources* 102, 242–252 (**2001**).
5. J. R. VARCOE, R. C. T. SLADE, *Fuel Cells* 5, 188–200 (**2005**).
6. C. GAVACH, G. POURCELLY, in *Proton Conductors Solids, Membranes and Gels-Material and Devices*, ed. P. Colomban, Cambridge University Press, Cambridge, **1992**, pp. 294–310, 487–498.
7. C. HEITNER-WIRGUIN, *J. Membrane Sci.*,120, 1 (**1996**).
8. W. G. GROT, C. CHADDS, *Eur. Patent* 0 066 369(**1982**).
9. B. LOPPINET, *Thesis in Physics*; Université Joseph Fourier, Grenoble, **1994**.
10. K. A. MAURITZ, C. J. HORA, A. J. HOPFINGER, in *Ions in Polymers*, ed. A. A. Eisenberg, *ACS Advances in Chemistry Ser.* No. 187, American Chemical Society, Washington, DC, **1980**, pp. 124–154.
11. S. SCHLICK, (ed.) *Ionomers: Characterization, Theory and Applications*, CRC Press, Boca Raton, **1996**.
12. O. SAVADOGO, *J. New Mater. Electrochem. Syst.* 1, 47–66 (**1998**).
13. O. SAVADOGO, *J. Power Sources* 127, 135–161 (**2004**).
14. J. R. JURADO, M. T. COLOMER, *Chem. Ind.* 56, 264–272 (**2002**).
15. J. ROZIÈRE, J. DEBORAH, *Annu. Rev. Mater. Res.* 33, 503–555 (**2003**).
16. M. HICKNER, H. GHASSEMI, Y. S. KIM, B. R. EINSLA, J. E. MCGRATH, *Chem. Rev.* 104, 4587–4612 (**2004**).
17. K. A. MAURITZ, R. B. MOORE, *Chem. Rev.* 104, 4535–4585 (**2004**).
18. Y. YANG, S. HOLDCROFT, *Fuel Cells* 5, 171–186 (**2005**).
19. A. NOSHAY, L. M. ROBESON, *J. Appl. Polym. Sci.* 20 (7), 1885–1903 (**1976**).
20. K. D. KREUER, *Solid State Ionics* 97, 1 (**1997**).
21. B. CHU, J. WANG, Y. LI, D. G. PEIFFER, *Macromolecules* 25, 4229 (**1992**).
22. G. GEBEL, J. LAMBARD, *Macromolecules* 30, 7914–7920 (**1997**).
23. Y. LI, D. G. PEIFFER, B. CHU, *Macromolecules* 26, 4006-4012 (**1993**).
24. M. CHOMAKOVA-HAEFKE, R. NYFFENEGGER, E. SCHMIDT, *Appl. Phys. A* 59, 151–153 (**1994**).
25. P. J. JAMES, J. A. ELLIOTT, T. J. MCMASTER, J. M. NEWTON, A. M. ELLIOTT, S. HANNA, M. J. MILES, *J. Mater. Sci.* 35, 5111 (**2000**).
26. P. J. JAMES, T. J. MCMASTER, J. M. NEWTON, M. J. MILES, *Polymer* 41, 4223 (**2000**).

27 A. Lehmani, S. Durand, P. Turq, *J. Appl. Polym. Sci.* 68, 503 (1998).
28 R. S. McLean, M. Doyle, B. B. Sauer, *Macromolecules* 33, 6541 (2000).
29 Z. Porat, J. R. Fryer, M. Huxham, I. Rubinstein, *J. Phys. Chem.* 99, 4667–4671 (1995).
30 T. D. Gierke, G. E. Munn, F. C. Wilson, *J. Polym. Sci., Polym. Phys. Ed.* 19, 1687–1704 (1981).
31 K. D. Kreuer, *J. Membrane Sci.* 185, 29–39 (2001).
32 C. A. Edmondson, J. J. Fontanella, *Solid State Ionics* 152–153, 355–361 (2002).
33 Y. A. Elabd, E. Napadensky, J. M. Sloan, D. M. Crawford, C. W. Walker, *J. Membrane Sci.* 217, 227–242 (2003).
34 R. Wodski, A. Narebska, W. K. Nioch, *J. Appl. Polym. Sci.* 30, 769–780 (1985).
35 J. A. Elliott, S. Hanna, *J. Appl. Crystallogr.* 32, 1069–1083 (1999).
36 J. A. Elliott, S. Hanna, A. M. S. Elliott, G. E. Cooley, *Macromolecules* 33, 4161–4171 (2000).
37 V. Barbi, S. Funari, R. Gehrke, N. Scharnagl, N. Stribeck, *Polymer* 44, 4853–4861 (2003).
38 P. J. James, M. Antognozzi, J. Tamayo, T. J. McMaster, J. M. Newton, M. J. Miles, *Langmuir* 17, 349 (2001).
39 Z. Shi, S. Holdcroft, *Macromolecules* 37, 2084–2089 (2004).
40 L. Rubatat, G. Gebel, O. Diat, *Macromolecules* 37, 7772–7783 (2004).
41 J. Fontanella, M. McLin, M. C. Wintersgill, J. Calame, S. G. Greenbaum, *Solid State Ionics* 66, 1 (1993).
42 C. L. Gardner, A. V. Anantaraman, *J. Electroanal. Chem.* 395, 67 (1995).
43 T. A. Zawodzinski, M. Neeman, L. O. Sillerud, S. Gottesfeld, *J. Phys. Chem.* 95, 6040 (1991).
44 K. D. Kreuer, S. J. Paddison, E. Spohr, M. Schuster, *Chem. Rev.* 104, 4637–4678 (2004).
45 A. A. Pivovar, B. S. Pivovar, *J. Phys. Chem. B* 109, 785–793 (2005).
46 a) F. Volino, M. Pineri, A. J. Dianoux, A. de Geyer, *J. Polym. Sci.: Polym. Phys. Ed. B* 109, 785–793 (1982); b) F. Volino, J.-C. Perrin, S. Lyonnard, *J. Phys.*

Chem. B 110, 11217 (2006); c) F. Volino, M. Pineri, A. J. Dianoux, A. de Geyer, *J. Polym. Sci.: Polym. Phys. Ed.* 20, 481–496 (1982).
47 P. Levitz, J.-P. Korb, D. Petit, *Eur. Phys. J. E* 12, 29–33 (2003).
48 J.-C. Perrin, S. Lyonnard, A. Guillermo, P. Levitz, *J. Phys. Chem. B* 110, 5439 (2006).
49 S. Gamburzev, A. J. Appleby, *J. Power Sources* 107, 5–12 (2002).
50 F. Wang, J. E. McGrath, *Polym. Prepr.* 43, 492 (2002).
51 F. Wang, M. Hickner, Y. S. Kim, T. A. Zawodzinski, J. E. McGrath, *J. Membrane Sci.* 197, 231–242 (2002).
52 N. Volk, D. Vollmer, M. Schmidt, W. Oppermann, K. Huber, in *Polyelectrolytes with Defined Molecular Architecture II*, ed. M. Schmidt, Springer, Berlin, 2004.
53 A. Noshay, L. M. Robeson, *J. Appl. Sci.* 20, 1885 (1976).
54 P. Genova-Dimitrova, B. Baradie, D. Foscallo, C. Poinsignon, J.-Y. Sanchez, *J. Membrane Sci.* 185, 59–71 (2001).
55 F. Wang, M. Hickner, Y. S. Kim, T. A. Zawodzinski, J. E. McGrath, *J. Membrane Sci.* 197, 231, 242 (2002).
56 S. Kaliaguine, S. D. Mikhailenko, K. P. Wang, P. Xing, G. Robertson, M. Guiver, *Catal. Today* 82, 213–222 (2003).
57 N. Walsby, F. Sundholm, T. Kallio, G. Sundholm, *J. Polym. Sci.; Part A, Polym. Chem.* 39, 3008–3017 (2001).
58 (a) B. Gupta, F. N. Büchi, G. G. Scherer, A. Chapiro, *J. Membrane Sci.* 118, 231–238 (1996); (b) B. Gupta, G. G. Scherer, *Chimia* 48, 127 (1994).
59 G. Gebel, O. Diat, *Fuel Cells* 5, 261–276 (2005).
60 N. Walsby, S. Hietala, S. Maunu, F. Sundholm, T. Kallio, G. Sundholm, *J. Appl. Polym. Sci.* 86, 33–42 (2002).
61 J. Kerres, W. Cui, S. Richie, *J. Polym. Sci.* 34, 2421 (1996).
62 J. A. Kerres, *Fuel Cells* 5, 230–247 (2005).
63 R. D. Allen, T. L. Huang, D. K. Mohanty, S. S. Huang, H. D. Qin, J. E. McGrath, *Am. Chem. Soc. Div. Polym. Chem. Polym. Prepr.* 24, 41 (1983).

64. R. D. Allen, I. Yilgor, J. E. McGrath, *Coulombic Interactions in Macromolecular Systems*, Eds. A. Eisenberg, F. E. Bailey, ACS Symp. Series, No. 302, Chap. 6, Washington, D. C. **1986**.
65. M. Sauviat, R. Salle, B. Sillion, French Patent 2 050 251, **1969**, p. 7214.
66. C. Genies, R. Mercier, B. Sillion, N. Cornet, G. Gebel, M. Pineri, *Polymer* 42, 359–373 **(2001)**.
67. C. Genies, R. Mercier, B. Sillion, R. Petiaud, N. Cornet, G. Gebel, M. Pineri, *Polymer* 42, 5097–5105 **(2001)**.
68. C. Perrot, G. Meyer, L. Gonon, G. Gebel, *Fuel Cell* 6(1), 10–15 **(2006)**.
69. S. Faure, R. Mercier, P. Aldebert, M. Pineri, B. Sillion, French Patent 96 05707, **1996**.
70. N. Cornet, *Thesis in Physical Chemistry*, Université Joseph Fourier, Grenoble, **1999**.
71. J. F. Blachot, O. Diat, A.-L. Rollet, L. Rubatat, C. Valois, M. Müller, G. Gebel, *J. Membrane Sci.* 214, 31–42 **(2003)**.
72. O. Diat, J. F. Blachot, A.-L. Rollet, L. Rubatat, C. Valois, M. Müller, G. Gebel, *J. Phys. IV: Proceedings* 12, 63–71 **(2002)**.
73. F. Piroux, E. Espuche, M. Escoubes, R. Mercier, M. Pineri, *Macromol. Symp.* 188, 61–71 **(2002)**.
74. A.-L. Rollet, P. Porion, A. Delville, O. Diat, G. Gebel, *Magn. Reson. Imaging* 23, 367–368 **(2005)**.
75. A. L. Rollet, J. F. Blachot, A. Delville, O. Diat, A. Guillermo, P. Porion, L. Rubatat, G. Gebel, *Eur. Phys. J. E* 12, 131–134 **(2003)**.
76. D. Jamroz, Y. Marechal, *J. Mol. Struct.* 693, 35–48 **(2004)**.
77. A. L. Rollet, O. Diat, G. Gebel, *J. Phys. Chem. B* 108, 1130–1136 **(2004)**.
78. A. N. Galatanu, A.-L. Rollet, P. Porion, O. Diat, G. Gebel, *J. Phys. Chem. B* 109, 11332–11339 **(2005)**.
79. G. Meyer, C. Perrot, L. Gonon, G. Gebel, J.-L. Gardette, *Preprints of Symposia–ACS, Division of Fuel Chemistry* 49(2), 608–609 **(2004)**.
80. G. Meyer, C. Perrot, G. Gebel, L. Gonon, S. Morlat, J.-L. Gardette, *Polymer*, in press.
81. G. Meyer, G. Gebel, L. Gonon, P. Capron, D. Marsacq, C. Marestin, *J. Power Sources* 157, 293–301 **(2006)**.
82. Y. Zhang, M. H. Litt, R. F. Savinell, J. S. Wainright, *Polym. Prepr.* 40, 480 **(1999)**.
83. Y. Zhang, M. H. Litt, R. F. Savinell, J. S. Wainright, J. Vendramint, *Polym. Prepr.* 41, 1561, **(2000)**.
84. J. Fang, X. Guo, S. Harada, T. Watari, K. Tanaka, H. Kita, K. Okamoto, *Macromolecules* 35, 9022–9028 **(2002)**.
85. J.-T. Wainright, R. F. Savinell, M. H. Litt, *Second International Symposium on New Materials for Fuel Cells and Modern Battery Systems*, Montréal, Canada, July 6–10, **1997**, pp. 808.
86. M. J. Coplan, G. Gootz. US Patent 4,413,106, **1983**.
87. C. K. Shin, G. Maier, G. G. Scherer, *J. Membrane Sci.* 245, 163–173 **(2004)**.
88. C. K. Shin, G. Maier, B. Andreaus, G. G. Scherer, *J. Membrane Sci.* 245, 147–161 **(2004)**.
89. C. Marestin, V. Martin, R. Brunel, E. Chauveau, R. Mercier, unpublished results **(2005)**.
90. B. B. X. Glipa, B. Mula, D. Jones, J. Rozière, *J. Mater. Chem.* 9, 3045–3049 **(1999)**.
91. R. H. Q. Li, J. O. Jensen, N. J. Bjerrum, *Fuel Cells* 4, 147–159 **(2004)**.
92. J. S. Ma, M. H. Litt, R. F. Savinell, *J. Electrochem. Soc.* 151, A8–A16 **(2004)**.
93. J. Jouanneau, R. Mercier, L. Gonon, G. Gebel, *Polym. J.* submitted for publication **(2005)**.
94. R. B. Hodgdon Jr., *J. Polym. Sci. Part 1: Polym. Chem.* 6, 171–191 **(1968)**.
95. A. Steck, C. Stones, *Second International Symposium on New Materials for Fuel Cells and Modern Battery Systems*, Montreal, Canada, July 6–10, **1997**, pp. 792.
96. P. D. Beattie, F. P. Orfino, V. I. Basura, K. Zychowska, J. Ding, C. Chuy, J. Schmeisser, S. Holdcroft, *J. Electroanal. Chem.* 503, 45–56 **(2001)**.
97. G. Gebel, O. Diat, C. Stones, *J. New. Mater. Electrochem. Syst.* 6, 17–23 **(2003)**.
98. J. Ding, C. Chuy, S. Holdcroft, *Macromolecules* 35 1348–1355 **(2002)**.
99. N. Asano, K. Miyatake, M. Watanabe, *Chem. Mater.* 16, 2841–2843 **(2004)**.

100 Y. Yin, J. Fang, Y. Cui, K. Tanaka, H. Kita, K. Okamoto, *Polymer* 44, 4509–4518 (**2003**).

101 J. M. Serpico, S. G. Ehrenberg, J. J. Fontanella, X. Jiao, D. Perhia, K. A. McGrady, E. H. Sanders, G. E. Kellogg, G. E. Wnek, *Macromolecules* 35, 5916–5921 (**2002**).

102 X. Lu, W. P. Steckel, R. A. Weiss, *Macromolecules* 26, 6525-6530 (**1993**).

103 R. A. Weiss, A. Sen, C. L. Willis, L. A. Pottick, *Polymer* 32, 1867 (**1991**).

104 J.-P. Gouin, C. E. Williams, A. Eisenberg, *Macromolecules* 22, 4573–4578 (**1989**).

105 X. Lu, W. P. Steckle, R. A. Weiss, *Macromolecules* 26, 5876–5884 (**1993**).

106 R. A. Weiss, A. Sen, C. L. Willis, L. A. Pottick, *Polymer* 32, 2787–2792 (**1991**).

107 S. Mani, R. A. Weiss, C. E. Williams, S. F. Hahn, *Macromolecules* 32, 3663–3670 (**1999**).

108 L. N. Venkateshwaran, G. A. York, C. D. Deporter, J. E. McGrath, G. I. Wilkes, *Polymer* 33, 2277 (**1992**).

109 Y. A. Elabd, C. W. Walker, F. L. Beyer, *J. Membrane Sci.* 231, 181–188 (**2004**).

110 A. Mokrini, J. L. Acosta, *Polymer* 42, 9 (**2001**).

111 A. Mokrini, C. D. Rio, J. L. Acosta, *Solid State Ionics* 166, 375 (**2004**).

112 G. Zhang, L. Liu, H. Wang, M. Jiang, *Eur. Polym. J.* 36, 61 (**2000**).

113 H. L. Hsieh, R. P. Quirk, *Anionic Polymerization: Principles and Practical Applications*, Marcel Dekker, New York, **1996**.

114 J. Won, H. H. Park, Y. J. Kim, S. W. Choi, H. Y. Ha, H. S. Oh, Y. S. Kang, K. J. Ihn, *Macromolecules* 36, 3228–3234 (**2003**).

115 S. K. Ghosh, D. Khastgir, P. P. De, S. K. De, *J. Appl. Polym. Sci.* 77, 816–825 (**2000**).

116 M. Rikukawa, K. Sanui, *Prog. Polym. Sci.* 25, 1463–1502 (**2000**).

117 D. Crawford, *Polym. Mater. Sci. Eng.* 83, 473–474 (**2000**).

118 J. Kim, B. Kim, B. Jung, *J. Membrane Sci.* 207, 129–137 (**2002**).

119 L. Rubatat, Z. Shi, O. Diat, S. Holdcroft, B. Frisken, *Macromolecules* 39(2), 720–730 (**2006**).

120 J. C. Yang, J. W. Mays, *Macromolecules* 35, 3433–3438 (**2002**).

121 K. Busse, J. Kressler, D. van Eck, S. Höring, *Macromolecules* 35, 178–184 (**2002**).

122 J. Ruokolainen, R. Mäkinen, M. Torkkeli, R. Serimaa, T. Mäkelä, G. ten Brinke, O. T. Ikkala, *Science* 280, 557–560 (**1998**).

123 O. T. Ikkala, J. Ruokolainen, R. Mäkinen, M. Torkkeli, R. Serimaa, T. Mäkelä, G. ten Brinke, *Synthetic Met.* 102, 1498–1501 (**1999**).

124 R. Mäki-Ontto, K. de Moel, E. Polushkin, G. Alberda van Ekenstein, G. ten Brinke, O. Ikkala, *Adv. Mater.* 14, 357–361 (**2002**).

125 K. de Moel, R. Mäki-Ontto, M. Stamm, G. ten Brinke, O. T. Ikkala, *Macromolecules* 34, 2892–2900 (**2001**).

126 G. Gebel, P. Aldebert, M. Pineri, *Macromolecules* 20, 1425–1428 (**1987**).

127 B. Lopinnet, G. Gebel, *Langmuir* 14, 1977–1983 (**1998**).

128 Y. Xiaohu, Y. Xuehai, C. Rongshi, *J. Polym. Sci. B: Polym. Phys.* 36, 2677–2681 (**1998**).

129 C. Wu, K. Woo, M. Jiang, *Macromolecules* 29, 5361–5367 (**1996**).

16
Structure, Properties and Applications of ABA and ABC Triblock Copolymers with Hydrogenated Polybutadiene Blocks

Vittoria Balsamo, Arnaldo Tomás Lorenzo, Alejandro J. Müller, Sergio Corona-Galván, Luisa M. Fraga Trillo, and Valentín Ruiz Santa Quiteria

16.1
Introduction

Over time, material requirements have become more demanding and diverse, and have been increasingly difficult to satisfy with a single polymer. In many cases, differentiation and development of unique materials has been solved by the introduction of block copolymers such as polystyrene-*b*-poly(ethylene-*co*-butylene)-*b*-polystyrene (SEBS). These materials behave as elastomers, and provide a wide range of properties that are particularly useful in applications such as adhesives, bitumen modification, compounding and plastic modification.

Initially SEBS, developed by Shell, was introduced in 1972, starting with a few thousand tons per year from a technological "push", with very little market demand "pull", and became the dominant product for 20 years. During the 1990s, first Asahi Kasei and next Dynasol, independently, started the commercial production of SEBS block copolymers. Recently, in the early 2000s, Taiwan Synthetic Rubber Corp (TSRC) started production of SEBS, and Polimeri Europe (former Enichem) has announced commercialization to begin in the near future.

The chemical modification of synthetic elastomers is a useful method for altering and optimizing the physical and mechanical properties of these polymers. It opens a unique route to new materials development, which have resistance to degradation. Thus, hydrogenation of synthetic elastomers of styrene and butadiene block copolymers is an excellent example of a chemical modification process that circumvents the inability to synthesize controlled styrene and olefin block copolymers. Initially these materials were developed utilizing a high-pressure hydrogenation step (Shell), with a nickel catalyst, of anionically synthesized styrene and butadiene block copolymers. The latest commercial production developments of SEBS block copolymers (Dynasol) is a low-pressure route (less than 20 bar) based on an efficient metallocene hydrogenation catalyst. The selectivity of hydrogenation to double bonds is controlled using relatively low temperatures, generally in the

Block Copolymers in Nanoscience. Massimo Lazzari, Guojun Liu, Sébastien Lecommandoux
Copyright © 2006 WILEY-VCH Verlag GmbH & Co. KGaA, Weinheim
ISBN: 3-527-31309-5

range 35–150 °C. Under optimum conditions, selectivity is such that more than 99 % of diene units are saturated.

When one polymerizes butadiene anionically to form the center soft segment (elastomeric) of the triblock copolymers, one can control the 1,2-segment configuration by adjusting the appropriate amount of polar additive in the medium. Thus, one can control the ratio of ethylene (E) and butylene (B) units after hydrogenation. This, in turn, affects the glass transition temperature (T_g) and crystallinity of the segment. The butylene concentration of the commercial products is generally in the range of 30–60 %, depending on the application. Owing to the lack of unsaturations, SEBS has excellent weather resistance degradation to oxygen, ozone and ultraviolet (UV) light, heat-induced ageing resistance, chemical resistance and it is also highly compatible with polyolefin resins (polypropylene and polyethylene), and at the same time it imparts excellent elastomeric properties.

Increasing demand for SEBS over the last decade, due to the unique properties of these polymers, has meant a world market for hydrogenated styrene elastomers of over a hundred thousand tons per year and has led to a new class of blended compounds and materials. This is a consequence of the ability of the block copolymers to separate in nanophases.

As mentioned above, the key feature of SEBS block copolymers is that they are phase-separated systems, where each phase retains many of the properties of the respective homopolymers. Both ends of each polydiene chain are terminated by PS, which forms domains that act as multifunctional junction points to give an elastomeric network. However, in this case, the cross-links are formed by a physical rather than a chemical process. Thus, at room temperature, the material behaves as a conventional vulcanized elastomer, but when it is heated, the domains soften, the network loses its strength and eventually the block copolymer can flow; the changes experienced by the material upon heating are completely reversible [1].

In addition to the conventional SEBS, other ABA polymers have been developed, which may or may not behave as elastomers. They include chemically modified SBS triblock copolymers and copolymers based on PS and PB that have a different structure. Styroflex is a commercial styrenic polymer, produced by BASF, which combines the advantages of SBS elastomers with the properties of transparent, impact modified SBS polymers such as Styrolux (another BASF product with good processability, thermal stability and high transparency). Such a combination makes this polymer attractive for food packaging [2]. As with conventional SBS block copolymers, Styroflex is obtained by sequential anionic polymerization of styrene and butadiene, but a different block sequence is introduced in the middle B block, where polybutadiene (PB) and polystyrene are found. Owing to its composition, Styroflex forms nanometric PS spheres in an SB rubber as the matrix. Nevertheless, the borders of the spheres are diffuse, indicating that the system is close to the order–disorder transition (ODT). Structures such as tapered block copolymers of styrene and butadiene can also be found. They have been used in particular for asphalt modification and in plastic modification for HIPS and ABS productions. They are characterized by high modulus and hardness, low shrinkage, good extrusion, high resistance to abrasion, low brittle point, transparent and glossy appearance and thermoplastic behavior [3].

From a molecular point of view, when ABA thermoplastic elastomers (TPEs) are studied, the way the polymer chains are linked throughout the phases should be considered. Figure 16.1 shows that in ABA copolymers, the connecting rubbery chains of phase B can form loops or bridges [4, 5]. The bridges are considered fundamental to the rubber elastic behavior of the materials. The loops do not contribute to the mechanical strength of the material and tend to decrease the elastic modulus. The fraction of chains that can form bridges in ABA triblock copolymers has been estimated both theoretically and experimentally, and values ranging from 0.4 to 0.6 have been obtained [6–10]. It is generally agreed that the population of bridges decreases with increases in the molecular weight of the block copolymer.

As it is also desirable to increase the number of bridges to maximize the elastic properties of TPEs, one possible answer is to switch from ABA to ABC architectures (see Fig. 16.1) where A and C are immiscible. In order to keep the end-blocks well segregated, $\chi_{AC} N_{AC}$ has to be large (where χ_{AC} is the segmental interaction parameter between the end-blocks and N_{AC} the degree of polymerization of the A and the C blocks). Unfortunately, a high molecular weight and/or high incompatibility leads to a very high melt viscosity and therefore poor processing. One way to produce phase segregation without having to increase the molecular weight too much is to employ crystallizable blocks, as crystallization can be a strong driving force for microphase separation, as will be shown in Section 16.3. Furthermore, the A and C blocks could form one phase in the melt in a wide composition range if they have a low χ_{AC} parameter at high temperatures, a fact that could lead to a comparatively low melt viscosity and easy processing characteristics [5].

Besides TPEs, the morphological richness of ABC triblock copolymers diversifies the type of material that can be obtained. Thus, Balsamo et al. obtained superductile materials with polystyrene-b-polybutadiene-b-poly(ε-caprolactone) (SBC) triblock copolymers that, even with PS contents higher than 50%, exhibit elongations up to 700% [11].

The examples given above are only a small sample of the variety of materials that have been reported. Because of space limitations, we will focus our attention on ABA and ABC triblock copolymers that contain at least one hydrogenated PB

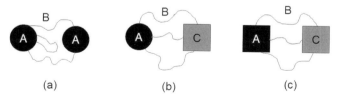

Fig. 16.1 Schematic representation of the middle block chains in triblock copolymers. (a) Amorphous ABA triblock copolymers, loops and bridges; ABC triblock copolymers with one (b) or two (c) semicrystalline end-blocks, only bridges (taken from [5] with permission of the author).

Table 16.1 Trade names, grades and main applications of some commercial SEBS.

Kraton TM [a], manufactured by Kraton Polymers

Property	G1650	G1651	G1652	G1654	G1657	G1726	FG1901	FG1924
Styrene content/%	30	33	30	31	13	30	30	13
Diblock (%)	0	0	0	0	30	70	—	—
Tensile strength (psi)	5000	5500	4500	3500	3400	350	5000	3400
300% modulus (psi)	800	900	700	900	350	—	—	—
Elongation (%)	500	—	500	700	750	200	500	750
Hardness (°Shore A, 10 s)	72	61	69	63	47	70	71	49
Brookfield viscosity[b] (cps)	8000	>50 000	1800	>50 000	4200	200	5000	19 000
Melt index[c] (g per 10 min)	<1	<1	5	>1	22	65	22	40
Comments							1.7% BF[d]	1.0% BF[d]
Main applications[e]	A, S, C, AM	Comp, PH	A, S, C	Comp, PH	A, S, C, Comp, P, AM	A, S, C, AM	A, S, C, Comp, P	A, S, C, Comp, P

Tuftec TM [a], manufactured by Asahi Kasei

Property	H1031	H1041	H1043	H1051	H1052	H1053	H1062	H1141
Styrene content (%)	30	30	67	42	20	29	18	30
Tensile strength (MPa)	12.7	21.6	10.3	32.3	11.8	24.6	15	2.7
300% tensile stress (MPa)	3.2	3.4	—	8.3	2.5	4.8	4.3	2.8
Elongation (%)	650	650	20	600	700	550	670	520
Hardness (°Shore A)	82	84	72	96	67	79	67	84
Main applications[f]	A, S, PM	A, S, PM	PM, Compz	PM, PM	A, S, PM, C	A, S, PM, AM	PM, Compz	A, S

(HPB) block. As mentioned above, the hydrogenation of block copolymers with low or high 1,4-PB contents lead to amorphous and semicrystalline materials, respectively. As amorphous triblock copolymers have been thoroughly studied, we will focus our attention on some of the most recent applications of SEBS triblock copolymers (Section 16.2), whereas in Section 16.3 we will review some aspects of semicrystalline ABA and ABC triblock copolymers with one or more HPB blocks. In the latter, emphasis will given to the study of confinement of semicrystalline blocks in nanoscopic domains, a phenomenon that determines the final properties of the copolymers.

16.2
Applications of SEBS Triblock Copolymers

As with most conventional vulcanized rubbers, SEBS block copolymers are rarely used commercially as pure materials. To achieve the particular properties for each application, SEBS are compounded with other polymers, processing oils, resins, fillers, stabilizers, etc. Trade names and various grades of some commercial products, together with their characteristics and applications, are listed in Table 16.1.

Table 16.1 (continued)

Calprene $^{TM\ a)}$, manufactured by Dynasol Elastomers				
Property	CH-6110	CH-6120	CH-6140	CH-6170
Styrene content (%)	30	30	31	33
Hardness (°Shore A)	75	75	75	76
Brookfield viscosity (cps)	400$^{f)}$	1700$^{f)}$	400$^{g)}$	2000$^{g)}$
Main applications$^{e)}$	A, PM, OG, AM	A, PM, AM, Comp	PM, Comp	PM, Comp

Septon $^{TM\ a)}$, manufactured by Kuraray Co., Ltd.				
Property	8007		8076	8104
Styrene content (%)	30		30	60
Hardness (°Shore A)	77		72	98
100% modulus (MPa)	3.5		1.1	12.9
Elongation (%)	550		530	500
Solution viscosity (mPa s)	20$^{g)}$		21$^{h)}$	80$^{h)}$
Applications$^{e)}$	A, S, C, PM, WCC, Compz, Comp		A, S, C	PM, Compz

a) Information obtained from the web pages of Kraton Polymers, Dynasol Elastomers, Kuraray Co., Ltd, and Asahi Kasei.
b) 25% wt in toluene solution, at 77 °F.
c) 230 °C, 5 kg.
d) BF, bond functionality.
e) A, adhesives; S, sealants; C, coatings; AM, asphalt modification; Comp, compounding; PH, personal hygiene; P, packaging; PM, plastic modification; OG, oil gels; WCC, wire and cable coating; Compz, compatibilizer.
f) 20 % solids content in toluene.
g) 10 % solids content in toluene.
h) 15 % solids content in toluene.

With the exception of Kraton FG1924 and Kraton FG1901, which are SEBS containing grafted maleic anhydride, all the others are non-functionalized SEBS. Apart from Kraton, Asahi Kasei, has several grades of functionalized reactive SEBS not listed in Table 16.1. Although in many respects SEPS, i.e., thermoplastic elastomers obtained from hydrogenation of polystyrene-b-polyisoprene-b-polystyrene (SIS), are similar to SEBS, they are beyond the scope of this review.

According to the information gathered in Table 16.1, the main applications in which SEBS are used include compounding, plastics and asphalt modification, adhesives, coatings and sealants. In the following, the most recent advances on the use of SEBS in the majority of these applications are discussed.

16.2.1
Adhesives, Sealants and Coatings

Styrenic block copolymers (SBCops), both hydrogenated and non-hydrogenated, are widely used in adhesives, sealants and coatings where they bring the advantage of processing as high solid content solutions or as solvent-free hot melts [12]. SBCops have practically no inherent adhesive character; therefore, in order to properly balance the adhesive properties it is necessary to add combinations of resins, stabilizers, plasticizers or fillers to the formulation. Whereas SEBS triblock copolymers are preferred for use in weather-resistant adhesives and applications such as sealants [13], the non-hydrogenated ones cannot be used in high-temperature applications or where resistance to solvents or plasticizers is required.

As the temperature at which the product is used approaches the T_g of the PS end-blocks, the hard nanometric domains soften and the product loses cohesive strength. Also, as the plasticizer migrates into the hard PS domains, the T_g of the domains is reduced, leading to a loss of cohesive strength. To be useful at temperatures higher than the T_g of the PS, it is common to chemically cross-link SBCops. This approach is affordable in SBCops having unsaturated PB or PI blocks, but not in SBCops that have hydrogenated middle blocks, as for example SEBS. In this instance, SEBS polymers containing grafted maleic anhydride have been cross-linked with aluminum acetylacetonate (AlAcAc), obtaining pressure-sensitive adhesives (PSA) with increased service temperature [14]. As shown in Table 16.2, typical general purpose PSA formulations, containing a maleic anhydride grafted SEBS (Kraton FG1924) and AlAcAc exhibit an increase in shear adhesion failure temperature (SAFT) and an improvement in solvent resistance with increasing concentration of AlAcAc. Other important properties were not significantly affected. These workers claimed that a ligand exchange took place, in which the 2,4-pentanedione from the AlAcAc leaves the aluminum and it is replaced by the acid–anhydride functional groups on the polymer to generate cross-linking reactions.

Similar results to above were obtained when formulations for hot-melt PSAs were prepared. In addition, this approach was also useful in the preparation of PSAs containing the plasticizer dioctyl phthalate (DOP), restoring the SAFT back to the level of the PS with no DOP when the polymers were cross-linked with

Table 16.2 Cross-linked solvent-based PSA [14].

Composition, % wt of solids	1	2	3	4
Kraton FG1924	30.0	30.0	30.0	30.0
Resin R1090	54.4	54.4	54.4	54.4
Oil D34	15.5	15.5	15.5	15.5
Aluminum acetylacetonate	—	0.2	0.5	1.0
PSA solution, time to gel (h)	—	18	3.5	2.0
Properties[a]				
Rolling ball tack (mm)	42	24	24	31
Polyken probe tack (kg)	0.9	1.1	1.0	1.1
Loop tack (N m^{-1})	820	800	670	800
180° peel (N m^{-1})	420	420	370	400
Shear (h)	>60	>60	>60	>60
SAFT (°C)	64	79	83	94
Gel soaked in toluene	No	Yes	Yes	Yes

[a] Adhesives were cast from 40 % wt solids solution in 75 + 25 toluene + isopropyl acetate onto 25-μ polyester at 35-μ dry adhesive thickness, baked 10 min at 180 °C and stored at 25 °C for 1 week before testing.

AlAcAc. This technology has also been demonstrated to be useful in the preparation of hot-melt sealants. At temperatures around 70 °C, sealants made with SBCops usually soften and flow out of the joint. Cross-linking the sealant with AlAcAc almost eliminated the slump at 130 °C.

In another report [15] selectively hydrogenated SEBS were functionalized with hydroxy, acid or amine moieties by a melt reaction. The hydrogenated block copolymers were cured with isocyanate, obtaining cross-linked products that overcome the creep and solvent resistance, while maintaining the advantages of conventional block copolymers in adhesives, coatings and sealants due to the separated structure of the nanophase.

Hot-melt adhesives are usually prepared at high temperature, it is therefore necessary that the SBCops used in the formulation resist those temperatures. If an SEBS is used, as seen in Fig. 16.2, thermally stable hot-melt adhesives are obtained for longer times than those formulated with a linear SB, as seen by the lack of viscosity decrease with time for the SEBS formulation.

Recent developments in the area have been devoted to obtaining PSAs that combine low melt viscosity with elevated service temperature. This has been obtained by developing SEBS with medium molecular weight and medium polystyrene content [17].

16.2.2
Bitumen Modification

Since the late 1960s, diene polymer rubbers such as SBS have been used to improve the thermal and mechanical properties of asphalts. The practical application of asphalt modification requires that the blended product retains improved properties

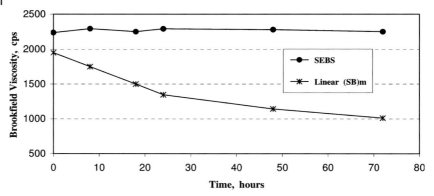

Fig. 16.2 Stability test at 350 °F of a hot melt adhesive prepared with SEBS and comparison against a formulation made with a linear (SB)m [16].

and homogeneity during transportation, storage and final processing. Long-term performance of elastomer-modified asphalts also depends on the ability of the blend to maintain thermal and chemical stability. The two main applications of polymer-modified asphalts are in paving or as roofing materials, each one of them having specific requirements. For instance, for paving applications, the polymer must be fully mixable in asphalt and remain like that in subsequent processing, the mixtures must have appropriate rheological properties to prevent permanent deformation (caused by continuous traffic loads) and the mixture must be able to recover rather than just resist deformation, in addition to having good low-temperature properties in order to resist or avoid cracking. In the case of roofing materials, the mixtures must have enough resistance to flow at high temperature and sufficient flexibility at low temperatures (both of these being possible due to nanophase separation), must be workable according to conventional roofing technique methods, must exhibit adequate hardness (to prevent deformation during walking on the roof) and hot storage stability, and, if it is to be used as an adhesive, it must exhibit sufficient adhesion.

The addition of SBCops to asphalt, depending on the amount added, can almost double the temperature range of the asphalt application, which also increases the cohesive strength and the elasticity of the mixtures. However, the upper service temperature is usually no higher than 90 °C. As in the case of adhesives, the addition of an SEBS grafted with maleic anhydride in the presence of AlAcAc gives cross-linked polymers that increase the softening point of the mixtures above 175° C [14].

It has been reported that the addition of SEBS and SEBS grafted with maleic anhydride to asphalt, produced compatible and stable blends, when compared with blends prepared with SBS of similar styrene content. The improvement in compatibility and stability in the case of maleated SEBS containing blends was attributed to the possibility of chemical reactions between the functional groups of the components of the polymer–asphalt blends. However, these possible reactions would not explain the substantial improvement in the stability of the SEBS-modified asphalt [18].

Blends of asphalt with SEBS grafted with maleic anhydride or SEBS functionalized with vinyl triethoxysilane have been reported to improve adhesion to polar substrates. For joint sealant applications, improved adhesion between sealant and substrate were observed, as demonstrated by the higher elongations observed with the asphalt-functionalized SEBS compared with a blend prepared with non-functionalized SEBS. For a hot mix asphalt concrete, less stripping was obtained with blends containing the silane and maleic anhydride functionalized polymers, by comparison with blends made with conventional SEBS, meaning that there was an increase in the adhesion and a decrease in detachment between the asphaltic binder and the aggregate caused by the action of water [19].

16.2.3
Compounding and Plastic Modification

According to a recent report, SEBS compounds participate in a broad range of market sectors as shown by the estimated demand in North America and Europe listed in Table 16.3 [20]. Although their property profile and market-sector distributions are somewhat different, in most of these markets, SEBS compounds compete directly with thermoplastic vulcanizates (TPV). The automotive sector accounts for 40–50 % of the total demand for SEBS compounds and TPVs in North America and Europe, and is a key motivator for new TPE technologies, which are often adapted to other sectors. Examples are 3D-blow molding, foam extrusion, coated fabrics and two-shot molding of large parts.

The very diverse area of molded and extruded goods is supplied primarily either by thermoplastic vulcanizates or by SBCops compounds, among which SEBS-based compounds are becoming more important due to their higher use temperature

Table 16.3 Markets for SEBS compounds in Europe and North America for 2003 [20].

Market sector	Comments
Auto	—
Fluid handling	Growth in foods
Building/construction	Strong in European non-auto glazing
Consumer/houseware	Strength in consumer soft touch
Medical	Compression set, heat resistance required
Food/pharma packaging	E.g., cap seals, wine
Appliance/tool	Soft touch
Wire/cable	—
Sports/leisure	—
Personal care	—
Footwear	SBS compounds dominate
Consumer electronics	Strong growth in Asia
Coated fabrics	Could be growth sector
Total (kT)	150

and stability. In this application, products are usually manufactured by equipment originally developed to process conventional thermoplastics [4]. A wide variety of products can be obtained, ranging from extremely soft to semi-rigid, which are capable of operating over a temperature range of −50 or −60 to 140–150 °C. Grades capable of being co-molded or co-extruded with a wide variety of engineering thermoplastics have been developed. The list of products include compounds with high-temperature resistance, low compression set grades, compounds for soft-touch applications, among others.

The automotive market is a very important niche for SEBS compounds. They are mainly used in acoustic barriers, airbag doors, body plugs, body seals, damper mounts, glazing seals and wire and cable applications. Glazing seals and body seal penetration by thermoplastic vulcanizates (TPVs) and SEBS is enhanced via improved foaming methods, providing weight savings and better dynamic performance [21]. Foaming increases the value of thermoplastic elastomer (TPE) compounds as a result of increased softness, energy absorption (depending on thickness and foam structure), acoustic properties and the potential for cost savings when they are integrated with the skin or other surface material (e. g., textiles). The combination of foaming with two-shot molding and sheet extrusion offers the potential for both cost savings and the addition of value to the constructions in which they are used. Thus far, the potential added value of TPE foams has been underexploited but is being developed for TPVs, SEBS and TPOs. An additional TPE benefit for SEBS is low-gloss interiors. Several compounds made with SEBS and an ultra-high molecular weight poly(dimethylsiloxane) have been used to produce auto parts. The compounds, known as Multiflex^{TM}, are co-molded with polypropylene (PP) to fabricate exterior pillars, rear light seals and safety belt boxes [22].

Recently, Teknor Apex have developed a high-performance TPV using PP as the matrix, but substituting a "special" hydrogenated styrenic block copolymer for EPDM in the vulcanized phase [23]. The resultant styrenic TPV, Uniprene XL, exhibits improved oil resistance, wet coefficient friction and good long-term compression set. Reportedly, the superior properties are related to a secondary network of nano-scale (30-nm) hard PS domains in the rubber phase, which intensifies cross-linking and boosts elasticity of the TPVs. This compound is aimed to be used in blends with either styrenic or olefinic thermoplastic elastomers and to provide substantial performance improvement over styrenic TPEs and intermediate performance between olefinic TPVs and super TPVs. They may also operate in the competitive range targeted by reactive or cross-linked SEBS compounds [21]. Uniprene XL is aimed at use in molded seals, grommets, plugs, extruded automotive tubing and bellows, glazing strips, weather seals and electrical insulation.

One of the fastest growing sectors for TPEs is soft touch applications, i. e., for parts that have a soft material, either to improve grip or to improve the tactile feel. Usually these parts are fabricated by overmolding a soft styrene block copolymer compound, such as SEBS onto a hard, rigid substrate such as polycarbonate, ABS, polycarbonate–ABS alloys, acrylics and nylon. Newer technologies allow co-injection molding of a very soft SEBS onto a more rigid, but still flexible, SEBS substrate, without the deformation and adhesion difficulties of previous attempts at "soft-on-soft" overmolding. This allows designers to tailor levels of softness in molded parts precisely in ways that are not possible with the old "soft-on-hard" molding. Some examples include toothbrushes, razors, kettles and screwdriver handles, the bodies of cameras and mobile phones. For many of these applications, the preferred rigid part is polypropylene, while styrenic TPEs are one of the preferred polymers for the soft, rubbery part, especially for very soft to medium hardness. Among TPEs, SEBS are the polymers of choice due to their better ageing performance, as they retain their surface grip for a longer time [24]. Ticona and Kraiburg Corp. have developed what they claim is the first system for overmolding thermoplastic elastomer grips, gaskets and other components onto rigid acetal copolymer substrates. In this approach, SEBS compounds bond to acetal compounds either by insert molding or two-shot injection molding [25].

Wire and cable covering is also an important application for SEBS, particularly when compounded with other polymers in the presence of fire retardants. PVC resins are conventionally used in this application, however, the heat resistance is usually low (about 60 °C), and halogen-containing resins are increasingly undesirable [26]. Further improvements in the area should look toward improving chemical resistance, fire retardancy and heat resistance, while maintaining or improving the mechanical properties of the compounds.

16.2.4
Miscellaneous Applications

16.2.4.1 Gels and Nanocomposites

The highest volume applications for SEBS were mentioned above. However, because of their stability and ability to phase separate in nanoscaled domains, SEBS are also used in the fabrication of thermoplastic elastomer gels (TPEGs) and nanocomposites. TPEGs are molecular networks composed of a microphase-separated multiblock copolymer swollen to a large extent by a low-volatility midblock-selective solvent (usually an oil compatible with the EB block). Products can be formulated that range from strong elastomers to weak gels. Oil gels can be used for sealants, greases, strippable coatings, corrosion protection, binders and toys [27].

Polymer nanocomposites constitute a new class of materials that are hybrid structures between an organic polymer and an inorganic phase. The inorganic layer, silicates for instance, are dispersed at a nanometer-level within the polymer matrix and unique physical and chemical properties are observed. Since the pioneering work of Toyota scientists on polymer–clay nanocomposites [28–30], a large amount of research has been performed in this field. The enhanced mechanical properties, dimensional stability, thermal stability, chemical stability and resistance to solvent swelling of nanocomposites make them useful for a plethora of current and potential applications.

Most of the research in the field of nanocomposites has been done by combining nanofillers (layered silicates, nanotubes or spherical particles) with polymers such as polyethylene (PE), poly(vinyl chloride) (PVC), poly(ethylene terephthalate) (PET) or certain thermoplastic elastomers based on PP, PBT, nylon-6 or nylon-6,6 with ethylene α-olefins. The use of SEBS, although limited, can be divided into two lines of research, as a toughening added ingredient, or as the main component in the composite. Following the first approach, PP–vermiculite (VMT) nanocomposites toughened with maleated SEBS were prepared [31]. According to the report, SEBS additions improved the tensile ductility of the ternary nanocomposites. Additionally, SEBS-g-MA addition led to a significant improvement in the impact strength. Tjong and Bao also reported the preparation of nylon-6–montmorillonite nanocomposites toughened with SEBS-g-MA [32], observing that the fracture toughness of the nanocomposite was significantly increased. Likewise, in a recent Solvay patent, the optional addition of SEBS to an ionomeric nanocomposite made of a propylene-based polyolefin–metal salt component and a nanostructured material is described [33]. Thus, it appears that SEBS polymers act effectively as nanocomposite tougheners.

There have been some reports on the use of SEBS as the main ingredient in nanocomposite fabrication. In a recent study [34], SEBS and SEBS grafted with maleic anhydride were melt blended with both a nanoclay (cloisite 10A) and multi-walled carbon nanotubes using conventional polymer processing techniques. Reportedly, addition of 4 wt-% nanoclay resulted in an enhancement in tensile modulus, dynamic modulus, elongation at break and flexural strength, without affecting the hardness of the SEBS. In a similar report [35], SEBS–clay hybrids were prepared by melt blending SEBS and organoclay, using a maleic anhydride modified SEBS as a compatibilizer. These hybrids exhibited enhanced mechanical properties and thermal degradation compared with pristine SEBS. By using a combined melt intercalation method and *in situ* intercalative polymerization, SEBS–clay nanocomposites with greater elongation and tensile strength than unfilled SEBS were obtained [36].

In a recent report [37], the effect of several network-forming nanoscale modifiers on the property development of a TPEG prepared from a microphase-ordered SEBS imbibed with an EB-compatible aliphatic mineral oil was investigated. The resultant nanocomposite TPEGs (NCTPEGs) exhibits an increase in the linear viscoelastic threshold, the dynamic elastic modulus, G', and the flow onset temperature with respect to the parent TPEG. Complementary studies revealed that the NCTPEGs underwent substantial relaxation irrespective of added modifier at temperatures above ≈60 °C, and that the nanoclay particles were swollen with copolymer, which indicates that the nanoparticles are intercalated. All of the above indicates that the continued development of the intercalative strategy can greatly improve the compatibility and properties of a wide range of composite materials.

16.2.4.2 Medical Applications

SEBS polymers are finding applications in a range of medical devices and increasingly as replacements for poly(vinyl chloride). Several publications have provided an overview of the properties of these materials and discussed their potential in medical applications [38, 39].

To name a few examples, SEBS can be used, for instance, as a pure component in the fabrication of tubing for medical devices [40], or in the fabrication of multilayered films to obtain PVC-free intravenous bags [41]. SEBS compounded with polyolefins are non-allergenic alternatives to latex rubber [42], while a similar SEBS-based compound was designed to replace latex rubber and plasticized PVC in medical tubing, in order to avoid allergic effects in some healthcare workers [43]. Further SEBS applications in the medical area include the manufacture of special adhesives that are useful for improving the flux through skin of the anticoagulant warfarin [43], and in the preparation of hydrogel wound dressings and inserts for cavities created by surgery or cavities that are common and natural in mammalian anatomy [44].

Because of their intrinsic properties, SEBS will find increasing use in medical products, offering durability, design flexibility and favorable performance to cost ratios. TPEs based on SEBS could provide medical designers with a broad spectrum of soft-feel, hygienic materials that can readily fulfill accepted medical industry standards.

16.2.5
Future Trends

According to a recent news release, the world market for hydrogenated styrene elastomers in 2004 was about 110 000 tons [45]. SEBS block copolymers are finding continuous use in many applications as a substitute for vulcanized rubber and soft PVC. For these applications, a 10 % annual rate rise in SEBS use is estimated in Europe, America and Asia. Owing to the growing demand, Dynasol Elastómeros, Kuraray and Kraton Polymers have very recently increased their production.

Products with higher heat resistance, abrasion resistance and oil resistance have been demanded by the market. Research oriented to fulfill this demand will continue, along with research oriented towards new applications, such as nanocomposites and biomedical applications.

16.3
Semicrystalline Triblock Copolymers with One or More HPB Blocks

As some of the most recent applications of modified and unmodified SEBS triblock copolymers have been presented in the previous section, a review of the state-of-the-art of the much less studied "semicrystalline" ABA and ABC triblock copolymers (incorporating HPB) will now be presented. As we mentioned in the Introduction, semicrystalline copolymers are obtained when hydrogenation is performed in block copolymers whose polybutadiene (PB) block has a high 1,4-microstructural units content. In this case, the PB block is converted into a poly(ethylene-*co*-butylene) random copolymer. Although from an industrial point of view these materials have not become widespread, there has been intensive research in order to understand their complex behavior.

The difficulty in understanding the behavior of ABA and ABC triblock copolymers with semicrystalline polyethylene (PE) blocks is due to the fact that the PE block is itself a copolymer of ethylene and butylene, a fact that is additional to the interplay of the "microphase separation–crystallization". Therefore, several parameters ought to be considered: composition, morphology, location of the block (at the end or in the center), T_g, crystallizability of the adjacent blocks and level of segregation. A combination of all these parameters affects the crystallization and melting behavior of the PE block markedly, as compared with similar polyethylene homopolymers.

When semicrystalline blocks are involved, previous works and reviews have focused mostly on AB diblock copolymers. For this reason it is our objective to summarize in this short chapter some interesting results that have been obtained in ABA and ABC semicrystalline triblock copolymers that include a polyethylene block (HPB). We will focus on mechanical and thermal properties, with particular attention on the crystallization and structure of the PE block.

16.3.1
Semicrystalline ABA Triblock Copolymers

Semicrystalline ABA copolymers have been explored and compared with amorphous conventional TPEs, such as SEBS [4, 46–50]. PE based systems such as poly(ethylene-b-isoprene-b-ethylene) and poly(ethylene-b-ethylene-stat-butylene-b-ethylene) triblock copolymers were investigated by Morton [50]. This last system is miscible in the melt and leads to good flow properties besides enhanced chemical resistance. However, compared with ABA materials with PS end-blocks, these semicrystalline triblock copolymers exhibited higher plastic deformations that increased as the PE content in the copolymers increased. This behavior was attributed to the less rigid and more deformable semicrystalline domains as compared with PS domains. Because these systems have identical end-blocks, loops are still present and may limit the elastic properties that could be obtained.

Besides the mechanical properties, the thermal behavior of ABA triblock copolymers containing semicrystalline polyethylene is another aspect that has been investigated. Weimann et al. [51] studied the crystallization phenomena of tethered PE in confined geometries of diblock and triblock copolymers of polyethylene and poly(vinyl cyclohexane) (PVCH). It should be mentioned at this point that Müller et al. [52] recently published an extensive review on the nucleation, crystallization and morphology of diblock and triblock copolymers with one or two crystallizable components. The issues of the different types of nucleation processes and their relationship to the crystallization kinetics of the components were addressed thoroughly in that work, therefore the reader is referred to the review for a detailed explanation on fractionated crystallization or homogeneous nucleation phenomena that are commonly present when the crystallizable component is confined within isolated phases.

The results of Weimann et al. [51] showed that the strong segregation of the PVCH and PE components coupled with the high glass transition temperature

(T_g) of the PVCH component forced the PE chains to crystallize in well-defined geometries determined by the mesophase structure of the block copolymer. These workers evaluated the effect of chain tethering on crystallization by comparing single-tethered PE chains in PE-b-PVCH (EV) diblocks and double-tethered PE in PVCH-b-PE-b-PVCH (VEV) triblock copolymers. The melting temperature (T_m) data reported for the VEV triblocks showed that for identical PE volume fractions, the depression of the T_m was greater in VEV than in EV copolymers, indicating that the crystallization of the PE block was more restricted when it is double-tethered due to topological constraints. These workers argued that the restriction of the crystal growth due to the confinement in the VEV triblock was the primary reason for the large supercooling required to induce crystallization in the dimensionally confined lamellar and cylindrical PE MDs.

Another type of semicrystalline ABA triblock copolymer that has been investigated is poly(cyclohexylethylene)-b-polyethylene-b-poly(cyclohexylethylene) (PCHE-b-PE-b-PCHE) [53]. Ruokilainen et al. evaluated the effect of thermal history and ordered microdomain (MD) orientation on deformation and fracture properties of triblock copolymers containing cylindrical PE domains. Morphological studies showed that after annealing at 190 °C, the PE cylinders clearly oriented normal to the film surface, and that they packed in a hexagonal lattice. In addition they detected a large abundance of small PE crystals that they attributed to the fact that the T_m of the PE block was lower than the glass transition temperature of PCHE, therefore, the crystallization of PE took place within confined nanosized cylindrical domains.

Park et al. [54] studied thin films of semicrystalline polystyrene-b-poly(ethylene-alt-propylene)-b-polyethylene (PS-b-PEP-b-PE) copolymers. They could obtain patterned surfaces using benzoic acid (BA) and anthracene (AN) as crystallizable solvents, indicating that the microstructure of semicrystalline block copolymers is controlled by a combination of the directional crystallization of crystallizable organic solvents and the following epitaxial crystallization of the crystalline block onto the organic crystalline substrate. Thus, they concluded that the method could be successfully used to form periodically ordered microdomain (MD) patterns and, furthermore, to design new microstructures which can be used in various applications for nanostructure fabrication.

16.3.2
Semicrystalline ABC Triblock Copolymers

In order to study the relevance of bridges versus loops, which was mentioned in the Introduction, Schmalz and coworkers [5, 55–58] prepared a series of ABC triblock copolymers with either one or two crystallizable blocks where no loops should be present (see Fig. 16.1). For comparison purposes, different types of triblock copolymers were prepared initially by living anionic polymerization and later by hydrogenation: (a) PB-b-PI-b-PEO and PE-b-PEP-b-PEO; (b) PS-b-PI-b-PS and PS-b-PEP-b-PS; and (c) PS-b-PI-b-PB and PS-b-PEP-b-PE. From these triblock copolymers, the block copolymers in series (b) are amorphous and can be considered conventional TPEs, while only PE-*b*-PEP-*b*-PEO and PS-*b*-PEP-*b*-PE represent TPE alternatives with one or two crystallizable blocks.

In the case of PE-*b*-PEP-*b*-PEO the PE and the PEP blocks are miscible in the melt leading to good flow properties. However, when PEO is present in contents of 20% or less in the copolymer, it does not crystallize when cooled from the melt down to room temperature. The reason is that the material is dispersed into a large number of isolated droplets, a number much larger than the usual number of active heterogeneities that normally nucleate PEO in the bulk. As a result, fractionated crystallization develops and PEO can only crystallize in these triblock copolymers at temperatures below −20 °C [55, 56, 58, 59]. The mechanical properties of these PE-*b*-PEP-*b*-PEO triblock copolymers were poor, they exhibited elongations at break below 100% [5], a fact that may be linked to the rubbery nature of the PEO block, which did not crystallize upon cooling to room temperature. However, when samples were annealed overnight at −30 °C to allow PEO crystallization, no improvements were encountered in elongation at break while the elastic modulus increased somewhat.

When PEO contents were higher, PEO was able to crystallize above room temperature but even so, mechanical properties for $E_{19}EP_{40}EO_{41}$[138] were still not satisfactory and a low elongation at break was obtained. As crystallization occurred from a mixed melt, the PE phase was a continuous semicrystalline domain consisting of interconnected PE crystallites [56]. Such a continuous PE matrix may also contribute to the poor mechanical properties exhibited by the samples in this case. It can be concluded that at least in this type of PE-*b*-PEP-*b*-PEO triblock copolymers, the two crystalline end-blocks and their peculiar nucleation and/or crystallization characteristics do not lead to mechanical properties that are comparable to conventional TPEs. A different result was encountered when PS-*b*-PEP-*b*-PE triblock copolymers were tested and compared with amorphous TPEs [57, 58].

PS-*b*-PEP-*b*-PE triblock copolymers exhibited a two-phase melt as the PS block is strongly segregated from the homogeneous mixture of PEP–PE that forms at high temperatures ($\chi_{PEP/PE} = 0.007$ at 120 °C). Upon cooling, the crystallization of PE is expected to occur from a mixed PEP–PE melt (that is separated from PS) just below the vitrification of the PS phase. The morphology of these copolymers was studied with a combination of scanning electron microscopy (SEM), transmission electron microscopy (TEM) and scanning force microscopy (SFM) [57, 58]. Samples were prepared by solution casting and by compression molding. In these copolymers it is difficult to stain and later differentiate the PE and the PEP blocks when viewed under TEM. However, as there are large differences in stiffness between amorphous and crystalline MDs, SFM can be employed.

Figure 16.3 shows SFM phase contrast images taken from the work of Schmalz et al. [57]. The first image (Fig. 16.3A) is from a thin film of $S_{14}EP_{64}E_{22}$[122] that was spin-coated from a toluene solution; three phases can be clearly distinguished. The darkest phase corresponds to PEP MDs that are rubbery at room temperature. The bright structures that resemble dots and worms correspond to PS glassy cylinders appearing white because of their vitreous nature at room temperature. The structures that appear gray are an interconnected network of percolated PE crystallites.

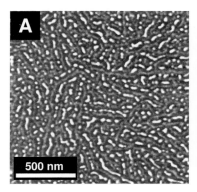 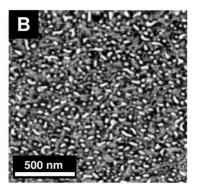

Fig. 16.3 SFM phase contrast images: (A) thin film of $S_{14}EP_{64}E_{22}{}^{122}$ prepared by spin-coating (z range = 50°); (B) $S_{13}EP_{57}E_{30}{}^{112}$ prepared by compression-molding, measurements were performed on a smooth cut surface (z range = 15°) (reprinted from [57], copyright 2001, with permission from The American Chemical Society).

SEM observations and selective swelling experiments of PS MDs have confirmed the above microphase assignment [5, 56, 57]. The PS cylinders appeared distorted because the sample was cast from toluene and the PE probably crystallized before PS vitrification; therefore, the network of PE crystallites was already formed by the time the PS vitrified and had to be accommodated in the remaining spaces. Therefore, depending on composition, these triblock copolymers could be used in applications where the material is subjected to moderate strains.

On the other hand, Fig. 16.3B shows an SFM image of a compression molded sample of $S_{13}EP_{57}E_{30}{}^{112}$. In the compression molded sample, the PS microphase vitrifies from the melt at 100 °C during cooling before the PE phase can crystallize (at approx. 60 °C). As a result, the PE crystallites are smaller in length than in the solution-cast sample even though a higher proportion of PE is present in the compression molded sample and are also more distorted. Yielding was found in some solution-cast samples, while it was absent in the compression molded identical material and it was speculated that differences in morphology are responsible for this behavior [57].

In order to test the elastic recovery of PS-b-PEP-b-PE, Schmalz et al. [58] performed mechanical hysteresis tests and the results were compared with those obtained with PS-b-PEP-b-PS samples (ABC with bridges only versus ABA with loops and bridges). Figure 16.4 (left-hand side) presents the results where the amount of residual plastic deformation is plotted as a function of the specific levels of strain that were applied to a series of ABC and ABA samples with comparable compositions. When the applied strain is 100 or 200%, the performance of the PS-b-PEP-b-PE triblock copolymers is better than the PS-b-PEP-b-PS analogues in the sense that they have better elastic properties as they recover more strain, and therefore their residual plastic deformation is lower. This seems to confirm the hypothesis that the absence of loops is beneficial to the elastic recovery. However,

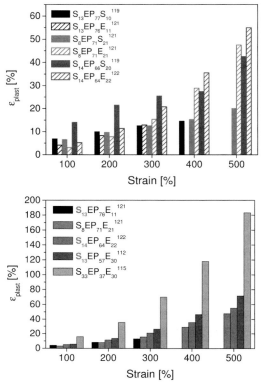

Fig. 16.4 Comparison of plastic deformations (ε_{plast}) obtained from hysteresis measurements on PS-b-PEP-b-PS and PS-b-PEP-b-PE triblock copolymers with comparable composition (reprinted from [58], copyright 2003, with permission from Elsevier).

it must be stressed that the morphology of the samples is not exactly the same, so the comparison may not be completely fair. When the applied strain is 300 % or higher, the results are reversed and the performance of the ABA samples is better than the ABC ones. This is probably a consequence of the higher mechanical resistance to deformation of the PS glassy MDs as compared with semicrystalline PE MDs at high elongations. When a series of PS-b-PEP-b-PE triblock copolymers of identical PS content and several PE compositions are compared, the elastic recovery performance improves as PE content decreases (Fig. 16.4, right-hand side). A similar result was obtained when the PE content was fixed and the PS amount was varied [58].

Different types of ABC systems with a PE block that have been thoroughly investigated are polystyrene-b-polyethylene-b-poly(ε-caprolactone) (SEC) and polyethylene-b-polystyrene-b-poly(ε-caprolactone) (ESC) triblock copolymers [60–63]. In these instances the emphasis was placed on the interplay crystallization–microphase separation. When the crystallization of the PE block within SEC and ESC copolymers was investigated, two parameters had to be considered: the topological constraints due to the number of free ends of the PE block and the fractionated crystallization that can occur as a consequence of confinement. Balsamo and coworkers showed that the thermal transitions (crystallization and melting) and degree of crystallinity within the SEC copolymers decreased as the degree of confinement increased. The magnitude of such a depression depended, in agreement with Weimann et al. [51], on chain tethering as a lower melting point depression was observed when the PE block was only tethered on one end, as in the ESC triblock copolymer [60–63]. However, the influence of the number of free ends on crystallization depended in its turn on composition. When PE formed the matrix or was highly interconnected, the effect was negligible.

With respect to confinement, unlike block copolymers with semicrystalline blocks based on poly(ε-caprolactone) (PCL) and/or poly(ethylene oxide) (PEO), the occurrence of fractionated crystallization has been more difficult to analyze [59, 63]. This is due to the influence of the short branch content, and because the shift in crystallization temperatures to lower temperatures for PE is lower than for PCL or PEO. Therefore, confusion may arise in assigning the crystallization temperature depression to the branch content/molecular weight or to fractionated crystallization, which in turn can be induced by surface or homogeneous nucleation. To determine whether an exotherm is due to homogeneous or surface nucleation, Müller and coworkers have reported that the disappearance of domain II in self-nucleation experiments occurs. This is a general feature of confined isolated MDs, which, in view of their vast number, have serious problems for nucleation and within the limit of confined compositions tend to nucleate either homogeneously or at the surface of the MDs [52, 62–66].

More peculiar is the behavior of SEC and ESC block copolymers upon isothermal crystallization. Studies have been performed employing the "isothermal step crystallization procedure" [63]. The method has the advantage that it is possible to obtain information about the phenomena that take place at the different stages of the crystallization at a specific crystallization temperature (T_c) and that it is possible to widen the temperature range in which the isothermal crystallization is performed when the content of the crystallizable component is low. The following unusual features were observed in SEC copolymers upon melting, with a crystallization time t_c. The melting process of the PE block takes place in a bimodal endotherm at any t_c and the low melting endotherm becomes less important at high T_c [63]. The double melting phenomenon has been reported previously in hydrogenated polybutadiene (HPB) homopolymers and was also observed by us when step isothermal crystallizations were performed in HPB. This behavior has been interpreted as a result of the melting of two lamellar populations arising from the intrinsic short chain branching distribution within the hydrogenated PB [64–66]. Nevertheless, what is different in the block copolymers is the variation of the melting temperature of the peaks within the double endotherm with crystallization time. T_{m1} (the low melting temperature) increased with t_c while, surprisingly, T_{m2} decreased or remained constant and did not follow the expected trend with T_c

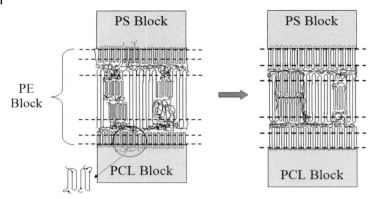

Fig. 16.5 Schematic representation of morphological evolution with t_c during isothermal crystallization of the PE block. The close up in the left-hand figure presents one of the possible structures of the crystalline lamellae that melts at lower temperatures (T_{m1}), where the branches are located in the amorphous regions. The dotted lines schematically represent two lamellar populations of different thickness, where the inner ones increase in size at the expense of the thicker ones.

(i.e., an increase of T_m with T_c). The anomalous behavior was more pronounced as the PE content was lowered. Therefore, although the double melting can be attributed to the short chain distribution within the PE block, the variations with t_c and T_c can not. Thus, the confinement of the PE block introduces a unique feature whose mechanism can be explained by a speculative model proposed by Balsamo et al. [63], in which the morphological evolution process for the two lamellar populations was schematically represented (see Fig. 16.5). In the model the thinner crystals thicken at the expense of amorphous material and also at the expense of the crystalline material present in the fraction that melts at higher temperatures.

Another interesting aspect that should be mentioned is the failure of the Hoffman–Weeks method for the determination of the equilibrium melting temperature (T_m^o) of the isothermally crystallized PE block in SEC copolymers for T_{m2}. Figure 16.6 shows a different temperature dependence of T_{m2} between SEC copolymers and HPB that can be attributed to confinement effects [63].

The crystallization kinetics of the PE block was also strongly affected because as the degree of PE confinement increased, the Avrami index decreased. When PE confinement was maximum and homogenous (or surface) nucleation was observed, Avrami indexes as low as 0.5 or lower were obtained, while HPB exhibited values ranging from 2.7 to 3.3. Such extremely low values in the copolymers can be explained by the nature of the homogenous (or surface) nucleation process that is in between sporadic and instantaneous when the growth is unimportant, as it is so fast that nucleation dominates the kinetics of the overall transformation. This aspect has been treated thoroughly in a previous review [52].

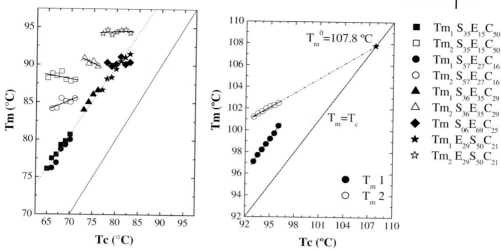

Fig. 16.6 Left: Hoffman–Weeks plots for the PE block of the different SEC and ESC triblock copolymers for samples that were crystallized at a constant time of 200 min (reprinted from [63], copyright 2004, with permission from Elsevier). Right: Hydrogenated polybutadiene homopolymer, crystallized until saturation [67].

16.4 Conclusions

The development of copolymers through chemical modification of well-known poly(styrene-b-butadiene-b-styrene) (SBS) triblock copolymers allows the preparation of new materials that can be used in a wide variety of applications that span a range from conventional to sophisticated ones, such as gels and nanocomposites. This has been possible due to the ability of the modified copolymers to separate in nanoscopic domains that would not necessarily be the same as the original ones.

Additionally, a short review of more complex block copolymer systems that include one and/or two semicrystalline HPB blocks has been presented. These types of materials have played a key role in understanding the influence of loops and bridges on the mechanical performance of triblock copolymers. Copolymers with semicrystalline blocks may exhibit good mechanical properties at moderate strains, depending on composition, as a result of the morphology–crystallization interplay.

Composition and location of the crystallizable blocks are parameters that determine the crystallization behavior, making the triblock copolymers ideal systems (due to their nanometric nature) for the study of interesting phenomena, such as fractionated and topologically constrained crystallization. These parameters should be considered when studying potential applications of these materials as compatibilizing agents in blends, for example, or in the fabrication of nanopatterned devices.

Acknowledgements

The USB team acknowledges financial support from Fonacit through grant G97-000594. We are also indebted to Dr. Holger Schmalz (University of Bayreuth) for providing us with figures from his Ph. D. Thesis.

References

1. G. HOLDEN, N. R. LEGGE, R. QUIRK, H. E. SCHROEDER, *Thermoplastic Elastomers*, 2nd edn, Hanser, Munich (**1996**).
2. K. KNOLL, N. NIESSNER, in *Applications of Anionic Polymerization Research*, ed. R. P. Quirk, ACS Symposium Series 696, ACS, Washington, Chap. 9 (**1996**).
3. S. A. MOCTEZUMA, E. N. MARTÍNEZ, R. FLORES, E. FERNÁNDEZ-FASSNACHT, in *Applications of Anionic Polymerization Research*, ed. R. P. Quirk, ACS Symposium Series 696, ACS, Washington, Chap. 10 (**1996**).
4. G. HOLDEN, H. KRICHELDORF, R. QUIRK, *Thermoplastic Elastomers*, 3rd edn, Hanser, Ohio, USA (**2004**).
5. H. SCHMALZ, *Development of Thermoplastic Elastomers with Improved Elastic Properties Based on Semicrystalline Block Copolymers*, Ph.D. Thesis, Bayreuth University, Bayreuth, Germany (**2003**).
6. M. W. MATSEN, *J. Chem. Phys.* **1995**, 102, 3884.
7. M. W. MATSEN, M. SCHICK, *Macromolecules* **1994**, 27, 187.
8. R. L. JONES, L. KANE, R. SPONTAK, *J. Chem. Eng. Sci.* **1996**, 51, 1365.
9. H. WATANABE, *Macromolecules* **1995**, 28, 5006.
10. K. KARATASOS, S. H. ANASTASIADIS, T. PAKULA, H. WATANABE, *Macromolecules* **2000**, 33, 523.
11. V. BALSAMO, F. VON GYLDENFELDT, R. STADLER, *Macromolecules* **1999**, 32, 1226.
12. I. SKESIT (ed.), *Handbook of Adhesives*, 2nd edn, Hanser, New York (**1996**).
13. H. L. HSIEH, R. P. QUIRK, *Anionic Polymerization: Principles and Practical Applications*, Marcel Dekker, New York (**1996**).
14. D. J. ST. CLAIR, *Adhesives Age* **2001**, 44, 31.
15. D. J. ST. CLAIR, *Eur. Patent 522658 A1*, assigned to Shell (**1993**).
16. Dynasol Elastomers, Technical Service Report (**2002**).
17. N. DE KEYZER, X. MUYLDERMANS, (Hydrogenated Styrenic Block Copolymers Offers benefits for PSAs), *Adhesives & Sealants Industry*, March **2003**.
18. Y. BECKER, A. J. MÜLLER, Y. RODRIGUEZ, *J. Appl. Polym. Sci.* **2003**, 90, 1772.
19. R. GELLES, *US Patent 5,130,354*, assigned to Shell Oil, **1992**.
20. R. ELLER, *Overview of TPE Technology, Markets, Industry Structure and Value Added Growth Opportunities*, TPE Asia, Bangkok, March **2004**.
21. R. ELLER, *Rapra TPE 2004*, paper 10, September **2004**.
22. D. PUCKETT, *Rapra TPE 2004*, paper 11, September **2004**.
23. R. LEAVERSUCH, *Plast. Technol.* **2004**, 50, 56.
24. D. FRASER, *Rapra TPE 98*, paper 4, Sept. **1998**.
25. Web News Release: "Novel System Overmolds to Acetal", *Omnexus, Plastics and Elastomers*, Aug 17, **2001**.
26. S. SATO, H. KUBO, *US Patent. Applic.* 20040102551 A1, **2004**.
27. *Styrenic Block Copolymers in Oil Gels*, Fact sheet K0026, Kraton Polymers, Inc.
28. A. USUKI, M. KAWASUMI, Y. KOJIMA, A. OKADA, T. KURAUCHI, O. KAMIGAITO, *J. Mater. Res.* **1993**, 8, 1174.
29. A. USUKI, Y. KOJIMA, M. KAWASUMI, A. OKADA, Y. FUKUSHIMA, T. KURAUCHI, O. KAMIGAITO, *J. Mater. Res.* **1993**, 8, 1179.
30. Y. KOJIMA, A. USUKI, M. KAWASUMI, A. OKADA, Y. FUKUSHIMA, T. KURAUCHI, O. KAMIGAITO, *J. Mater. Res.* **1993**, 8, 1185.
31. S. C. TJONG, Y. Z. MENG, *J. Polym. Sci., B: Polym. Phys.* **2003**, 41, 2332.

32. S. C. TJONG, S. P. BAO, *J. Polym. Sci., B: Polym. Phys.* **2005**, 43, 585.
33. R. D. DING, N. CLINT, *WO Patent 2004108815*, assigned to Solvay, **2004**.
34. T. MCNALLY, M. LEWIS, *Rapra TPE 2003*, paper 19, **2003**.
35. Y. W. CHANG, J. Y. SHIN, S. H. RYU, *Polym. Int.* **2004**, 53, 1047.
36. M. ÜNAL, *Preparation and Characterization of SEBS-Clay Nanocomposites*, MSc Thesis, Sabanci University (**2004**).
37. G. J. VAN MAANEN, *Morphological and Property Analyses of Multicomponent Block Copolymer Nanocomposites Gels*, MSc Thesis, North Carolina State University (**2004**).
38. K. SIPKENS, *Med. Device Technol.* **2000**, 11, 8–10, 12–13.
39. H. POL, *Med. Device Technol.* **1995**, 6, 18.
40. R. SNELL, S. MARTIN, M. CABLE, E. KING, *WO Patent 200411287*, **2004**.
41. *The Use of Di-2-Ethyhexyl Phthalate in PVC Medical Devices: Exposure, Toxicity and Alternatives*, report published by The Lowell Center for Sustainable Production, March **2005**.
42. W. LEVENTON, Finding Tube Materials that Make the Grade, *Medical Device & Diagnostic Industry*, January **2005**, p. 88.
43. Online article: TPEs Offer Cure for Medical-Tubing Problems, *Plast. Technol.* August **2001**.
44. C. W. PAUL, M. L. SHARAK, L. A. RYAN, M. XENIDOU, M. G. HARWELL, Q. HE, *US Patent Applic. 20040077240 A1*, assigned to National Starch, **2004**.
45. Kuraray Co., Ltd. *News release 10*, April 18, **2005**.
46. R. SÉGUÉLA, J. PRUD'HOMME, *Polymer* **1994**, 30, 1446.
47. J. C. FALK, R. J. SCHLOTT, *Macromolecules*, **1971**, 4, 152.
48. J. C. FALK, R. J. SCHLOTT, *Angew. Makromol. Chem.* **1972**, 21, 17.
49. Y. MOHAJER, G. L. WILKES, I. C. WANG, J. E. MCGRATH, *Polymer* **1982**, 23, 1523.
50. M. MORTON, *Rubber Chem. Technol.* **1983**, 56, 1096.
51. P. A. WEIMANN, D. A. HAJDUK, C. CHU, K. A. CHAFFIN, J. C. BRODIL, F. S. BATES, *J. Polym. Sci. Polym. Phys.* **1999**, 37, 2053.
52. A. J. MÜLLER, V. BALSAMO, M. L. ARNAL, *Adv. Polym. Sci.* **2006**, 190, 1.
53. J. RUOKOLAINEN, G. H. FREDRICKSON, E. J. KRAMER, *Macromolecules* **2002**, 35, 9391.
54. C. PARK, C. DE ROSA, B. LOTZ, L. J. FETTERS, E. L. THOMAS, *Macromol. Chem. Phys.* **2003**, 204, 1314.
55. H. SCHMALZ, A. KNOLL, A. J. MÜLLER, V. ABETZ, *Macromolecules* **2002**, 35, 10004.
56. H. SCHMALZ, A. J. MÜLLER, V. ABETZ, *Macromol. Chem. Phys.* **2003**, 204, 111.
57. H. SCHMALZ, A. BÖKER, R. LANGE, G. KRAUSCH, V. ABETZ, *Macromolecules* **2001**, 34, 8720.
58. H. SCHMALZ, V. ABETZ, R. LANGE, *Comp. Sci. Tech.* **2003**, 63, 1179.
59. A. J. MÜLLER, V. BALSAMO, M. L. ARNAL, T. JAKOB, H. SCHMALZ, V. ABETZ, *Macromolecules* **2002**, 35, 3048.
60. V. BALSAMO, A. J. MÜLLER, F. VON GYLDENFELDT, R. STADLER, *Macromol. Chem. Phys.* **1998**, 199, 1063.
61. V. BALSAMO, A. J. MÜLLER, *Macromolecules* **1998**, 31(22), 7756.
62. V. BALSAMO, Y. PAOLINI, G. RONCA, A. J. MÜLLER, *Macromol. Chem. Phys.* **2000**, 201, 2711.
63. V. BALSAMO, N. URDANETA, L. PÉREZ, P. CARRIZALES, V. ABETZ, A. J. MÜLLER, *Eur. Polym. J.* **2004**, 40, 1033.
64. A. ALIZADEH, L. RICHARDSON, J. XU, S. MCCARTNEY, H. MARAND, Y. W. CHEUNG, *Macromolecules* **1999**, 32, 6221.
65. B. CRIST, E. S. CLAUDIO, *Macromolecules* **1999**, 32, 8945.
66. B. CRIST, D. N. WILLIAMS, *J. Macromol. Sci. Phys.* **2000**, B39(1), 1.
67. A. T. LORENZO, M. L. ARNAL, A. J. MÜLLER, A. BOSCHETTI DE FIERRO, V. ABETZ, *Macromol. Chem. Phys.* **2006**, 207, 39.

17
Basic Understanding of Phase Behavior and Structure of Silicone Block Copolymers and Surfactant–Block Copolymer Mixtures

Carlos Rodríguez, Arturo López-Quintela, Md. Hemayet Uddin, Kenji Aramaki, and Hironobu Kunieda[†]

This work is dedicated to Professor Hironobu Kunieda, who passed away unexpectedly in November 2005.

17.1
Introduction

Amphiphilic block copolymers are macromolecules composed of lipophilic and hydrophilic blocks (see Fig. 17.1). Polymerization of two different monomer types (A and B) produces the so called A-B block copolymers. Among them, siloxane amphiphilic copolymers consist of a methylated siloxane hydrophobe attached to one or more polar chains such as poly(oxyalkylene), as shown in Fig. 17.2. Siloxane amphiphilic copolymers can be used in areas where other types of surfactants are not efficient [1–3]. They are surface active both in aqueous and non-aqueous media, and can reduce the surface tension up to 20 mN m^{-1}, as the poly(dimethylsiloxane) chains have low cohesive energy. Moreover, as Si–O–Si bonds are very flexible, siloxane amphiphilic copolymers are fluid even at very high molecular weights [4] and they do not generally show a Krafft point or a gel point in aqueous media.

Siloxane amphiphiles are used in a wide variety of applications, including foam stabilization in plastic (polyurethane) foams, cosmetics, wetting, emulsification, lubrication and antistatic agents, drug delivery and so forth [5]. They also have potential uses in solid-state batteries [6] and in the preparation of mesoporous materials [7, 8].

Although siloxane copolymers were introduced to the market in the middle of the 20th century [9], the volume of related literature is small when compared with other types of amphiphilic block copolymers, such as PEO-PPE. One reason is that commercially available silicone copolymers are generally very polydisperse, which makes a systematic study difficult. However, there are some reports on the phase behavior of rake (comb or graft) type siloxane copolymers [3, 10–14], trisiloxane surfactants [15–19] and PEO-PDMS diblock or triblock copolymers [15, 20, 21] in different solvents.

Block Copolymers in Nanoscience. Massimo Lazzari, Guojun Liu, Sébastien Lecommandoux
Copyright © 2006 WILEY-VCH Verlag GmbH & Co. KGaA, Weinheim
ISBN: 3-527-31309-5

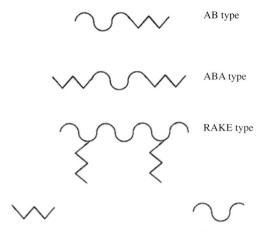

Fig. 17.1 Schematic structure of siloxane–ethylene oxide amphiphilic molecules.

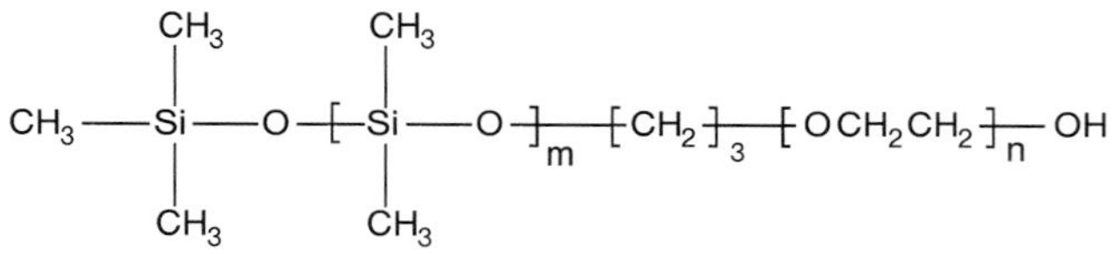

Fig. 17.2 Molecular structure of a typical AB-type siloxane copolymer (PEO-PDMS).

This chapter deals with the phase behavior and self-organized microstructures of a series of AB-type silicone copolymers in the melted and dispersed state both in polar and non-polar solvents, which is not only relevant for the use of these copolymers as a templates or structure-directing agents in the synthesis of nanomaterials, but is also important to understand the mechanism of such a synthesis when it is based on the self-organization of amphiphilic systems. We concentrate on a series of poly(oxyethylene)-poly(dimethylsiloxane) diblock copolymers (PEO-PDMS) with the general formula shown in Fig. 17.2, abbreviated as $Si_m C_3 EO_n$, where m is the total number of dimethylsiloxane groups and n is the average number of ethylene oxide (EO) units. The behavior of mixtures of these block copolymers with non-ionic surfactants will also be addressed.

17.2
General Aspects of Phase Behavior and Liquid Crystal Phases

Above the melting temperature of the solid (S), amphiphilic block copolymers can form a variety of self-organizing structures in selective solvents. The lyotropic phases (namely, those formed upon addition of solvent) found in these systems can be divided into isotropic solution phases and liquid crystalline phases. The isotropic micellar solution phases show a relatively low degree of correlation between aggregates. They are formed by direct micelles with a hydrophobic core (W_m) or reverse micelles with a hydrophilic core (O_m). Some phases, on the other hand, show long-range order and therefore are called liquid crystalline phases. A long hydrophilic chain surfactant or copolymer forms a direct (oil in water) type aggregate with a positive curvature, that is, convex towards water, and is given the subscript 1, while long a hydrophobic chain copolymer forms a reverse (water in oil) type with negative curvature and is given the subscript 2. The following liquid crystalline phases are commonly observed in surfactant and copolymer systems (the nomenclature indicated below will be used in the rest of this chapter).

1. Lamellar phases (L_α): consist of stacked non-polar and polar layers. The non-polar layers are formed by hydrophobic chains in a disordered, liquid-like state. The samples are usually fluid and birefringent (Fig. 17.3).
2. Hexagonal phases (H_1, H_2): consist of long parallel cylindrical aggregates arranged in a two-dimensional hexagonal lattice in the plane perpendicular to the cylinder axis. Hexagonal phases are stiff and optically anisotropic. In the direct hexagonal phase (H_1), the cylinders have a non-polar interior, while the hydrophilic chains face the polar solvent. In the reverse hexagonal phase (H_2), the core of the cylinders consists of the hydrophilic chains with the hydrophobic chains directed towards a non-polar solvent (Fig. 17.4).

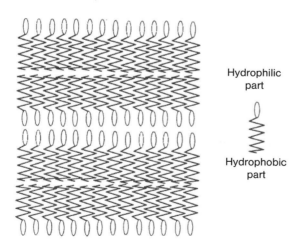

Fig. 17.3 Lamellar (L_α) phase.

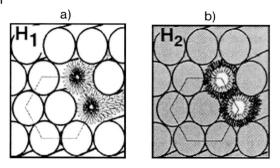

Fig. 17.4 Hexagonal phases: (a) direct hexagonal and (b) reverse hexagonal.

3. Cubic phases: they are stiff (extremely viscous), optically isotropic and ordered in three dimensions. According to their underlying structure, the cubic phases can be divided into two families: micellar cubic phases and bicontinuous cubic phases. The micellar (or reverse micellar) cubic phase consists of discrete direct micelles (or reverse micelles) packed in cubic symmetry. The direct (oil-in-water) micellar cubic phase (I_1) is built up with discontinuous non-polar regions embedded in continuous aqueous media and generally found between the aqueous micellar solution phase (W_m) and the direct hexagonal phase (H_1). The reverse (water-in-oil) micellar cubic phase (I_2) consists of discontinuous polar regions embedded in a continuous non-polar medium and is generally found between the reverse hexagonal phase (H_2) and the reverse micellar solution phase (O_m) in the phase diagram (Fig. 17.5). The bicontinuous cubic phases (V_1 and V_2) have two polar domains separated by a non-polar film (bilayer) or *vice versa*, where the bilayer mid-plane can be modeled as a minimal surface. The V_1 phase is normally found between the H_1 phase and the lamellar phase (L_α) and the V_2 phase is between the L_α phase and the H_2 phase in the phase diagram (Fig. 17.6).

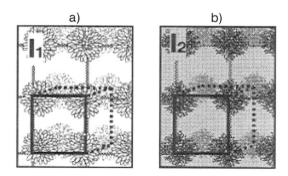

Fig. 17.5 Micellar cubic phases (primitive unit cell): (a) direct and (b) reverse.

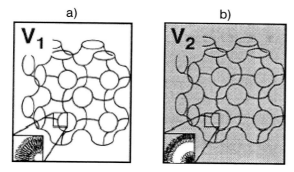

Fig. 17.6 Bicontinuous cubic phases: (a) direct and (b) reverse.

17.3
Phase Behavior and Microstructure of $Si_mC_3EO_n$ Melts

Amphiphilic block copolymer melts form different microstructures induced by block segregation. In the self-consistent mean-field theory [22–25], two parameters, χN and f, are used to represent the copolymer melt phase behavior, where χ is the Flory–Huggins interaction parameter between A and B blocks, N is the overall degree of polymerization and f is the volume fraction of one block in the copolymer. The product χN measures the degree of segregation and usually changes as M_w/T, because N is proportional to the copolymer molecular weight (M_w) and χ is roughly proportional to $1/T$. Block segregation (and hence the stability of ordered morphologies) is favored by a decrease in temperature or an increase in the copolymer size. For small values of χ, ordered structures are formed only in copolymers with high molecular weight. The copolymer compositional parameter f determines the interfacial curvature and morphology.

The thermotropic phase behavior of poly(oxyethylene)–poly(dimethylsiloxane) copolymer ($Si_{25}C_3EO_n$) as a function of PEO chain length (n = 3.2–51.6 for f = 0.06–0.50) is shown in Fig. 17.7. At room temperature, $Si_{25}C_3EO_n$ copolymers are in a liquid (disordered) state up to $n \approx 10$, whereas for $n < 10$ the copolymers are either ordered phases or solids. As poly(dimethylsiloxane) chains are fluid up to high molecular weights [27], the melting temperature of the solid phase (T_m) mainly depends on the chain length of the semicrystalline polyoxyethylene. Hence, with the increase of PEO chain length, T_m increases. The increase in T_m with molecular mass, for low molecular mass polymers, is a well-known phenomenon [28] and it is attributed to the decrease in the number of chain ends (and therefore defects) inside a crystalline domain. Above the melting temperature of the solid phase, $Si_{25}C_3EO_n$ forms either a liquid phase (O_m) containing copolymer micelles (with a PEO core and PDMS corona), or different ordered phases such as micellar cubic (I_2), hexagonal (H_2) and lamellar (L_α) phases, depending on the value of f. The thermal stability of the ordered phases increases with increasing PEO chain length and hence with the increase of the copolymer molecular weight. The two-phase co-

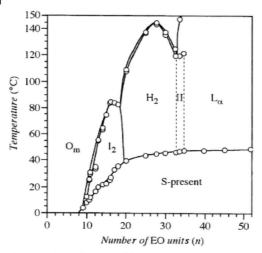

Fig. 17.7 Phase diagram of the $Si_{25}C_3EO_n$ system as a function of the poly(oxyethylene) chain length, n [26].

existence (corresponding to region II in Fig. 17.7) is only observed in the blend of two block copolymers ($Si_{25}C_3EO_{15.8}$ and $Si_{25}C_3EO_{51.6}$) with very different relative block sizes and thus is not a violation of the Gibbs phase rule.

The phase behavior of dry $Si_mC_3EO_{51.6}$ as a function of Si chain length (m = 5.8–52 for f = 0.80–0.33) is presented in Fig. 17.8. The solid copolymers melt to either anisotropic phases or fluid isotropic phases at ≈ 50 °C. I and II indicate transparent and turbid isotropic liquid phases, respectively. There is a large miscibility gap between the two copolymers $Si_{5.8}C_3EO_{51.6}$ and $Si_{14}C_3EO_{51.6}$ (region II). Beyond region II, the two copolymers $Si_{5.8}C_3EO_{51.6}$ and $Si_{14}C_3EO_{51.6}$ mix well with each other, and an isotropic transparent liquid phase (I), probably a direct micellar phase, is observed. Ordered anisotropic phases, L_α and H_2, are observed for $m > 12$ in the phase diagram. In both regions, the samples are completely transparent and strongly birefringent. The H_2 and L_α phases are thermally stable up to at least 150 °C, except in the vicinity of the phase boundaries, where the L_α–I and the H_2–L_α phase transitions take place.

$Si_{25}C_3EO_n$ with $n \geq 7.8$ forms reverse micelles, as can be inferred from the broad correlation peaks represented in Fig. 17.9. A flat diffraction pattern is observed for $Si_{25}C_3EO_{3.2}$, indicating that no aggregates are formed for such a short EO chain copolymer because of the weak segregation tendency between the two blocks. $Si_{25}C_3EO_{3.2}$ does not even aggregate in solvents such as octamethylcyclotetrasiloxane (D_4) or decane. However, $Si_{25}C_3EO_{7.8}$, $Si_{25}C_3EO_{12.2}$ and $Si_{25}C_3EO_{15.8}$ form micelles in these solvents [29]. $Si_{25}C_3EO_{7.8}$ shows a very broad peak at q_{max} = 0.22 nm^{-1}, whereas $Si_{25}C_3EO_9$ shows a comparatively strong correlation peak at q_{max} = 0.35 nm^{-1}. The peak position q_{max} depends on both, the micellar volume fraction, ϕ, and the hard-sphere interaction distance, r_{HS}, if micelles are considered as hard spheres. The scattering intensity can be estimated, assuming a form

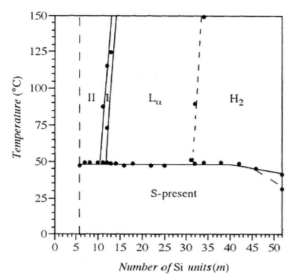

Fig. 17.8 Phase diagram of the $Si_mC_3EO_{51.6}$ system as a function of the poly(dimethylsiloxane) chain length, m. (I) represents a micellar or copolymer liquid phase; (II) represents immiscible copolymer liquids; the dotted line indicates the sample $Si_{5.8}C_3EO_{51.6}$; experiments have not been performed for $m < 5.8$ [26].

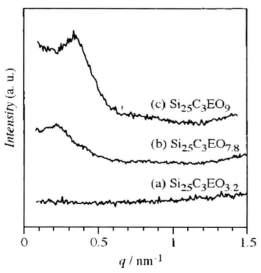

Fig. 17.9 SAXS diffraction patterns obtained from $Si_{25}C_3EO_n$ at 25 °C. (a) $Si_{25}C_3EO_{3.2}$ gives a flat pattern indicating no aggregate formation; (b) $Si_{25}C_3EO_{7.8}$ and (c) $Si_{25}C_3EO_{15.8}$ show wide correlation peaks indicating formation of aggregates [26].

factor of a dense spherical object with a sharp interface of radius r_c and a structure factor determined by the Ornstein–Zernike and Perkus–Yevick approximations [30] using a hard-sphere interaction potential. From the fitting of the diffraction patterns, the following values are obtained: for $Si_{25}C_3EO_{7.8}$, $r_c = 5.9$ nm, $r_{HS} = 1.3r_c$ and $\phi = 0.08$, whereas for $Si_{25}C_3EO_9$, $r_c = 6.2$ nm, $r_{HS} = 1.05r_c$ and $\phi = 0.20$. The values obtained for r_c are close to those calculated from the extrapolated data of the micellar core radius obtained in the cubic phase (see subsequent sections) and the effective length of the siloxane chain ($l_{eff} \approx 2.7$ nm): r_c ($Si_{25}C_3EO_{7.8}$) = 5.5 nm; r_c ($Si_{25}C_3EO_9$) = 5.8 nm. It is interesting to note that repulsive interactions increase with PEO chain length and lead to higher micellar volume fractions ($\phi = 0, 0.08, 0.20, 1$ for $n = 3.2, 7.8, 9$ and 10, see below, respectively).

For longer PEO blocks, the micelles crystallize into a cubic lattice. The I_2 phase is found between the O_m and the H_2 phases from $n = 10$–20 (Fig. 17.7). The samples in the I_2 region are transparent, isotropic and very stiff. SAXS results [26] indicated that this cubic phase most probably belongs to an $Fd\bar{3}m$ space group, with a structure proposed by Luzatti [31] in which the unit cell is filled with 24 micelles of equal core size. This structure has been found in some binary [32] and ternary [33] surfactant or copolymer systems.

For further discussion we have constructed a "phase diagram" as a function of M_w/T versus f for the PEO-PDMS system using order–disorder transition temperatures (T_{ODT}) shown in Fig. 17.7, and have represented this phase diagram in Fig. 17.10 in a way comparable to the χN versus f phase diagram used in copolymer melts. The ordered morphologies range from $f = 0.17$ to 0.69, in which

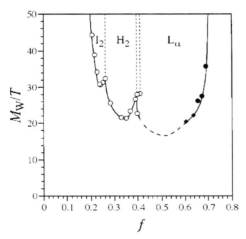

Fig. 17.10 M_w/T ($\approx \chi N$) versus f phase diagram for $Si_mC_3EO_n$ copolymers. Open and filled circles correspond to the transition temperatures obtained from Figs. 17.7 and 17.8, respectively. Filled diamonds correspond to the lamellar-disorder transition temperatures 149 and 155 °C for $Si_{14}C_3EO_{45}$ and $Si_{14}C_3EO_{51.6}$, respectively [26].

$0.17 \leq f \leq 0.27$ corresponds to the I_2 phase, $0.27 \leq f \leq 0.39$ to the H_2 phase, and $0.4 \leq f \leq 0.69$ to the L_α phase.

Comparing with other copolymer melt systems, $Si_mC_3EO_n$ forms ordered morphologies at relatively low molecular weights. For example, in the PI-PS melt system, the sample with the lowest $M_w = 7000$, IS-54, forms lamellar liquid crystals below 124 °C [34]; $Si_{14}C_3EO_{12}$ ($M_w \approx 1600$, $N = 26$) and $Si_{25}C_3EO_{12.2}$ ($M_w \approx 2400$, $N = 37$) melts form H_2 and I_2 phases, respectively, although the molecular weights of these silicone copolymers are about 10 times lower than that of the PI-PS melt. For many monomer pairs the interaction parameter χ is ≈ 0.1, hence the microphase separation should not occur for chains $N < 100$, as for $\chi N < 10$ only a disordered phase is stable [24]. A study of the PDMS-PEO-PDMS triblock system reveals that χ for PEO-PDMS is ≈ 0.4–1.1, with the higher limit corresponding to high PEO volume fractions in the copolymer [35]. Large values of the interaction parameter have also been reported recently for fluorinated diblock copolymers [36]. Even if we assume $\chi = 0.4$, the present copolymer system will undergo microphase separation at about $N = 5$, a value which is consistent with the observed results.

The phase diagram depicted in Fig. 17.10 is asymmetric and shows differences with other theoretically predicted or experimentally studied phase diagrams. This might be attributed to the unusual incompatibility of PDMS and PEO chains even in the low molecular weight range. The system already forms PEO-core micelles in the vicinity of the ordered cubic structure. The observed I_2 phase belongs to an $Fd\bar{3}m$ space group, which is uncommon in the copolymer melt systems and has not been predicted yet by the theoretical models that only account for the existence of the most commonly found bcc phase, although, a normal fcc lattice (close-packed, thus Q^{225} and alternatively hcp) has already been assumed by Matsen [37]. The intermediate bicontinuous cubic and HPL (perforated layers) phases were not observed in the present PDMS-PEO system, probably because of the rigidity of the PEO-PDMS interface (strong segregation). Mixing the copolymers with low molecular weight surfactants increases the flexibility of the polar–non-polar interface, and hence, the mixtures form the V_2 phase in water (see subsequent sections). $Si_mC_3EO_n$ melts form the L_α phase in the range $0.4 \leq f \leq 0.69$, and no direct (PDMS core and PEO corona) morphologies are observed for $f > 0.69$, whereas the PI-PS system forms the H_1 phase at this f value. However, direct hexagonal (H_1) and cubic (I_1) phases are observed in the $Si_{5.8}C_3EO_{51.6}$–water binary system (see the following sections).

Figure 17.10 can be interpreted in terms of a conformational asymmetry parameter, $\varepsilon = \beta_A^2/\beta_B^2$, where $\beta^2 = R_g^2/V = a^2/6v$, with notations similar to those used in [38]. ε accounts for differences in the conformational and space-filling characteristics of each block and $\varepsilon \neq 1$ produces an asymmetric phase diagram around $f = 1/2$. For a compositional symmetric copolymer ($f = 1/2$), such as $Si_{25}C_3EO_{51.6}$, $\varepsilon = \beta_{PEO}^2/\beta_{PDMS}^2 \approx 2.8$, indicating that the space-filling capability of PDMS is larger than that of PEO.

In the $Si_mC_3EO_n$ melt system, when the PDMS chain is very short, the segregation tendency is obviously not strong enough to induce microphase separation.

Packing the more flexible PDMS block in the core of a copolymer micelle requires strong swelling of the polar side of the interface, either by cosurfactants or by water [39]. Therefore, the lamellar phase is observed in a wide range of f values in the melt. The observed asymmetric phase behavior of the PEO-PDMS copolymer melt can be ascribed to the fact that the PDMS block, which is more voluminous at the interface (due to its higher cross-section), has the higher flexibility and lower cohesive energy.

17.4
Phase Behavior and Microstructure $Si_mC_3EO_n$ in Water

17.4.1
Phase Diagrams of Water–$Si_mC_3EO_n$ Systems as a Function of Temperature

The phase diagrams of a series of silicone copolymers, $Si_{5.8}C_3EO_{36.6}$, $Si_{14}C_3EO_n$ (n = 3.2, 7.8, 12, 15.8, 33.1 and 51.6), $Si_{25}C_3EO_n$ (n = 3.2, 7.8, 12.2, 15.8 and 51.6), $Si_{5.8}C_3EO_{51.6}$, and $Si_{5.8}C_3EO_{51.6}$ in water are shown in Figs. 17.11 and 17.12.

$Si_{5.8}C_3EO_{36.6}$ and $Si_{5.8}C_3EO_{51.6}$ form an aqueous micellar solution phase (W_m), a discontinuous direct micellar cubic phase (I_1) and a direct hexagonal phase (H_1) in water (Fig. 17.11). The I_1 phase region is wider for $Si_{5.8}C_3EO_{51.6}$ as it is more hydrophilic. Owing to the long PEO chain of the copolymers, the cloud point curve, namely, the lower boundary of the two-phase region at low concentrations, appears at very high temperatures (>100 °C).

In water–$Si_{14}C_3EO_n$ systems, the morphology of aggregates changes from copolymer liquid or reverse micellar solution (O_m) to lamellar phase (L_α) via discon-

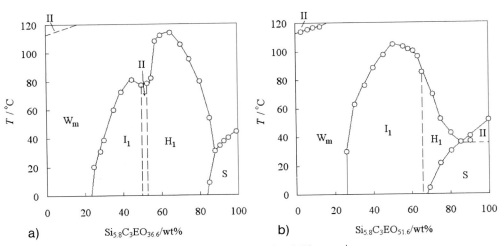

Fig. 17.11 Phase diagrams of binary (a) water–$Si_{5.8}C_3EO_{36.6}$ and (b) water–$Si_{5.8}C_3EO_{51.6}$ systems as a function of temperature [27, 40].

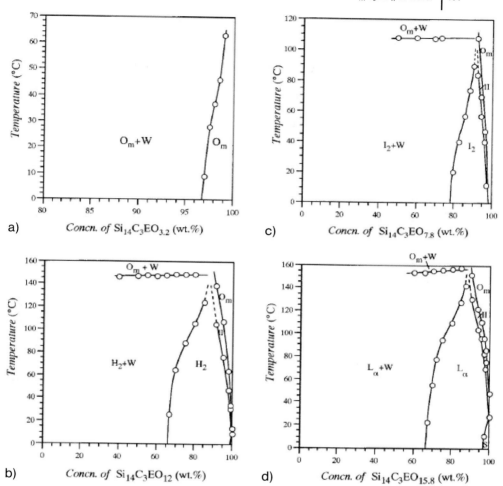

Fig. 17.12 Phase diagrams of binary water–$Si_{14}C_3EO_n$ systems as a function of temperature. (a) $Si_{14}C_3EO_{3.2}$, (b) $Si_{14}C_3EO_{7.8}$, (c) $Si_{14}C_3EO_{12}$ and (d) $Si_{14}C_3EO_{15.8}$ [27].

tinuous reverse micellar cubic (I_2) and reverse hexagonal (H_2) phases when the EO chain length increases from 3.2 to 15.8 as shown in Fig. 17.12. Only one type of morphology is observed in each phase diagram. At high temperatures, the liquid crystals melt and a two-phase region consisting of O_m and excess water phase (W) appears. The liquid crystal regions show a maximum melting temperature similar to an azeotropic point. This point should be adjacent to the phase boundary of the O_m + W region according to the phase rule, but there is a gap probably due to the presence of a small amount of impurity (unreacted polyethylene glycol). Nevertheless, as the transition from liquid crystal + W to O_m + W occurs almost at constant temperature, the influence of the impurity on the phase behavior is not very large.

In $Si_{25}C_3EO_n$ systems, with a longer hydrophobic chain, copolymer liquid or reverse micellar solution (O_m) (n = 3.2), optically isotropic I_2 (n = 7.8, 12.2 and 15.8) and birefringent L_α (n = 51.6) phases are found [27]. In general, the maximum melting temperature of the different liquid crystal phases increases with the increase in both m and n, namely, with the increase of molar mass of the copolymer.

The present phase diagrams clearly show that the copolymer layer curvature tends to be negative as the hydrophobic chain length of the copolymer increases and reverse-type self-organized structures appear. On the other hand, the copolymer layer curvature tends to be positive when the PEO chain length increases and direct-type self-organized structures are formed in a long PEO chain copolymer system.

17.4.2
Phase Diagrams of Water–$Si_mC_3EO_n$ Systemsas a Function of PEO Chain Length

The phase diagram of water–$Si_{25}C_3EO_n$ systems as a function of the PEO chain length, n, at 25 °C is shown in Fig. 17.13. The volume ratio of the hydrophilic moiety to copolymer, f or nV_{EO}/V_P, is plotted vertically. The corresponding EO number (n) is also plotted on the right-hand axis.

O_m, I_2, H_2 and L_α phases are successively formed with the increase of the PEO-chain length of the copolymer and the resulting positive change in the curvature of aggregates.

The solubilization of water in the aggregates tends to increase with PEO chain length, as can be inferred from the shifting of the left-side phase boundary, especially in water–$Si_{25}C_3EO_n$ systems. The homogeneous liquid crystalline phases are

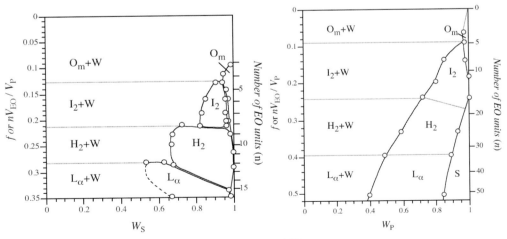

Fig. 17.13 Phase diagrams of (a) water–$Si_{14}C_3EO_n$ and (b) water–$Si_{25}C_3EO_n$ systems as a function of the EO chain length at 25 °C. f or nV_{EO}/V_P is the volume fraction of the PEO chain in the copolymer [27].

in equilibrium with excess water beyond their solubilization limits. In the $H_2 + W$ region, a dispersion of an optically anisotropic liquid crystal is observed by optical microscope, whereas vesicles were found in the $L_\alpha + W$ region [27].

The same type of phase diagrams have been constructed for other poly(oxyethylene) dodecyl [41] or oleyl ethers [42, 43] and poly(oxyethylene) trisiloxane surfactants [17].

17.4.3
Phase Diagram of Water–$Si_mC_3EO_{51.6}$ System as a Function of PDMS Chain Length

The phase diagram of the water–$Si_mC_3EO_{51.6}$ system constructed at 25 °C as a function of the volume fraction of the hydrophilic moiety to the copolymer, f (vertical axis) and the weight fraction of the copolymer in the system, W_p (horizontal axis) is presented in Fig. 17.14. The corresponding number of siloxane units, m, is also shown on the right-hand vertical axis. A large solid-present region (S) is observed at high concentrations. This solid phase consists of semicrystalline PEO and amorphous PDMS chains. Addition of water dissolves the PEO chain and increases the segregation between water and PDMS, driving the formation of different microstructures to a lower molecular weight when compared with the dry copolymer melt.

Short PDMS chain copolymers, $Si_{5.8-10}C_3EO_{51.6}$ ($f = 0.8 \approx 0.7$) form aqueous micellar solution (W_m), discontinuous micellar cubic (I_1) and direct hexagonal (H_1) phases with increasing surfactant concentration. A large lamellar region is observed in the middle of the phase diagram.

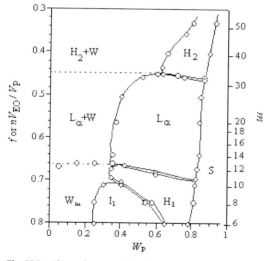

Fig. 17.14 Phase diagram of the binary water–$Si_mC_3EO_{51.6}$ system as a function of the PDMS chain length (m) at 25 °C [44].

At $m \approx 13$–33, vesicles are present in the L_α + W region. Long PDMS chain copolymers $Si_{32-52}C_3EO_{51.6}$ form reverse hexagonal liquid crystals (H_2). The H_2 phase solubilizes less water when compared with the L_α phase. Although the PEO chain length of the copolymer is the same ($EO_{51.6}$), the solubilization of water decreases with the increase of m in the H_2 phase.

17.4.4
Effect of PEO and PDMS Chain Lengths on the Effective Cross-sectional Area per Copolymer Molecule, a_P

The microstructural parameters in different phases observed for the $Si_{25}C_3EO_n$ and $Si_mC_3EO_{51.6}$ systems can be derived from small angle X-ray scattering (SAXS) measurements. To calculate the interfacial area per amphiphilic molecule, a_P, in different phases we assume that the microstructures consist of two components, namely polar (PEO) and non-polar [PDMS + $(CH_2)_3$] domains, separated by a sharp interface. The volume fraction of the poly(oxyethylene) part of the copolymers, $f = nV_{EO}/V_P$, was calculated from the molar volumes of poly(oxyethylene) (V_{EO}) and copolymer (V_P). The volume fraction of the non-polar part $f_{non-polar}$ is $(1-f)$.

The following equations hold for the interlayer spacing, d, and the volume fraction of the lipophilic part of copolymer, ϕ_L for each liquid crystal.

$$\text{For } I_2 \text{ phase, } d = (36\pi)^{1/3} \frac{n_m^{1/3}}{\sqrt{h^2 + k^2 + l^2}} \frac{v_L}{a_P} \frac{(1-\phi_L)^{2/3}}{\phi_L} \quad (17.1)$$

$$\text{For } H_2 \text{ phase, } d = (2\sqrt{3}\pi)^{1/2} \frac{v_L}{a_P} \frac{(1-\phi_L)^{1/2}}{\phi_L} \quad (17.2)$$

$$\text{For } L_\alpha \text{ phase, } d = \frac{2v_L}{a_P} \frac{1}{\phi_L} \quad (17.3)$$

$$\text{For } H_1 \text{ phase, } d = (2\sqrt{3}\pi)^{1/2} \frac{v_L}{a_P} \left(\frac{1}{\phi_L}\right)^{1/2} \quad (17.4)$$

$$\text{For } I_1 \text{ phase, } d = (36\pi)^{1/3} \frac{n_m^{1/3}}{\sqrt{h^2 + k^2 + l^2}} \frac{v_L}{a_P} \left(\frac{1}{\phi_L}\right)^{1/3} \quad (17.5)$$

where a_P and v_L are the effective cross-sectional area and the volume of lipophilic part per copolymer, respectively; n_m is the number of (reverse) micelles in a unit cell; h, k and l are the Miller indices and $h^2 + k^2 + l^2$ is 3 for a face-centered, 2 for a body-centered and 1 for a simple (primitive) cubic array.

The interlayer spacings (d) for each representative liquid crystal are plotted as a function of $(1-\phi_L)^p/\phi_L$ in Fig. 17.15a or $(1/\phi_L)^p$ in Fig. 17.15b, where p is a constant with a value assigned for each liquid crystal according to Eqs. (1) to (5). The data fit straight lines passing through the origin, which is an indication that a_P is practically constant (according to Eqs. 1–5). The values of a_P obtained by extrapolation to 100 % copolymer are plotted in Fig. 17.16a and b as a function f (or nV_{EO}/V_P). The values of a_P for the L_α phase of trisiloxane surfactant [17] are also plotted in Fig. 17.16.

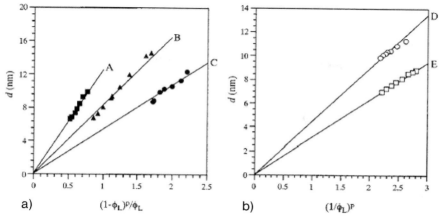

Fig. 17.15 The interlayer spacing, d, of the different types of liquid crystals. (a) In water–$Si_{14}C_3EO_n$ systems; p is a constant with a value assigned according to Eqs. (1–5): line A: $p = 2/3$, I_2 phase in water–$Si_{14}C_3EO_{7.8}$ system; line B: $p = 1/2$, H_2 phase in water–$Si_{14}C_3EO_{12}$ system; line C: $p = 0$, L_α phase in water–$Si_{14}C_3EO_{15.8}$ system. (b) In water–$Si_{5.8}C_3EO_n$ systems: line D: $p = 1/2$, H_1 phase in water–$Si_{5.8}C_3EO_{36.6}$ system; line E: $p = 1/3$, I_1 phase in water–$Si_{5.8}C_3EO_{51.6}$ system [27].

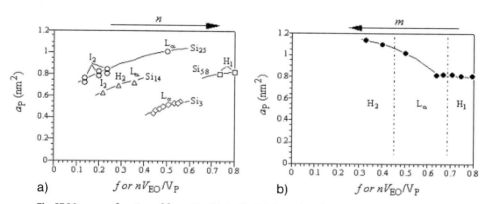

Fig. 17.16 a_P as a function of f or nV_{EO}/V_P in $Si_mC_3EO_n$ and trisiloxane systems. (a) The solid lines labeled with Si_{25}, Si_{14}, $Si_{5.8}$ and Si_3 represent the water + $Si_{25}C_3EO_n$, water + $Si_{14}C_3EO_n$, water + $Si_{5.8}C_3EO_n$ and water + trisiloxane systems, respectively. (b) Water + $Si_mC_3EO_{51.6}$ system [27].

The value of a_P increases on increasing the EO numbers in all the copolymer systems because of the increase in copolymer head group repulsions. Hence, the interfacial curvature tends to be positive with the increasing PEO chain length. It is observed that a_P also increases with PDMS chain length at a fixed PEO chain length. The long PDMS chain adopts a coiled conformation due to its high flexibility and low cohesive energy, therefore a_P expands and consequently the copolymer layer curvature changes from positive to negative. Consequently, I_1–H_1–L_α–H_2 phase transitions take place with the increase in the PDMS chain length.

When the molecular size of the surfactant or copolymer (hydrophilic and/or hydrophobic parts) increases, a_P always increases. Both hydrophilic and hydrophobic parts of the surfactant contribute to the increase in a_P while the interfacial tension tends to reduce a_P.

It should be pointed out that commercial amphiphilic copolymers with a very broad distribution of blocks also form reverse lyotropic mesophases [45].

17.5
Phase Behavior of $Si_mC_3EO_n$ in Non-polar Oil

The phase behavior of $Si_mC_3EO_n$–octamethylcyclotetrasiloxane (D_4) systems was investigated as a function of temperature and the representative phase diagrams are shown in Fig. 17.17. The melting temperature of the copolymer solid mainly depends on the hydrophilic (EO) chain lengths. As the EO chain is insoluble in silicone oil, D_4 does not greatly influence the melting temperature of the copolymer solid. $Si_{14}C_3EO_{33.1}$ ($f = 0.53$), $Si_{14}C_3EO_{51.6}$ ($f = 0.64$) and $Si_{25}C_3EO_{51.6}$ ($f = 0.5$) form the lamellar phase (L_α) and $Si_{52}C_3EO_{51.6}$ ($f = 0.33$) forms the reverse hexagonal phase (H_2) in the melt. Upon addition of oil L_α–H_2–O_m (Fig. 17.17a and b), L_α–H_2–I_2–O_m (Fig. 17.17c) and H_2–I_2–O_m (Fig. 17.17d) phase transitions take place, an indication that the copolymer layer curvature becomes more negative, namely, reverse structures are formed. The thermal stability of the liquid crystalline phases depends on both the hydrophilic and hydrophobic chain lengths, namely, the molecular weight of the copolymer.

The melting temperature of liquid crystals gradually decreases upon addition of oil, which reduces the degree of segregation between the blocks. However, there is an increase in the effective volume fraction of the lipophilic part in the presence of oil, and therefore phase transitions occur. As can be observed in Fig. 17.17, the liquid crystalline phases are more stable for copolymers with a balance between hydrophilic and hydrophobic moieties ($Si_{14}C_3EO_{33.1}$ or $Si_{25}C_3EO_{51.6}, f \approx 0.5$).

The microstructural parameters for the L_α phase of the $Si_{14}C_3EO_{51.6} + D_4$ system (Fig. 17.17b) at 58 °C are shown in Fig. 17.18. The half thicknesses of the non-polar (d_L) and polar (d_H) layers and effective cross-sectional area per copolymer molecule (a_P) were calculated using the following equations:

$$d_H = fd/2 \tag{17.6}$$

$$a_P = \frac{v_P f}{d_H} = \frac{2v_P}{d} \tag{17.7}$$

$$d_L = (1-f)d/2 \tag{17.8}$$

where v_P and f are the molar volume and the volume fraction of the polar core, respectively.

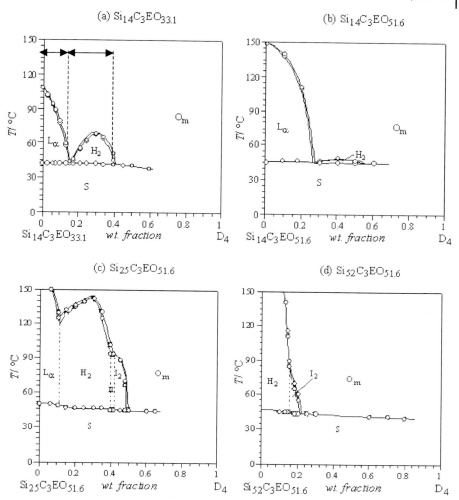

Fig. 17.17 Phase diagrams of a binary $Si_mC_3EO_n$–D_4 system as a function of temperature. (a) $Si_{14}C_3EO_{33.1}$–D_4, (b) $Si_{14}C_3EO_{51.6}$–D_4, (c) $Si_{25}C_3EO_{51.6}$–D_4, and (d) $Si_{52}C_3EO_{51.6}$–D_4 systems [28].

The values for d and d_L increase as the non-polar layer swells with oil; a_P increases with the increasing amount of oil solubilization into the L_α phase, because D_4 molecules penetrate into the poly(dimethylsiloxane) palisade layer and increase the effective cross-sectional area. Hence, the copolymer layer curvature changes to more negative and the L_α–O_m phase transition takes place. The addition of D_4 increases the total lipophilic volume in the system and the penetration of D_4 molecules between the copolymer siloxane chains promotes a negative change in the curvature of aggregates.

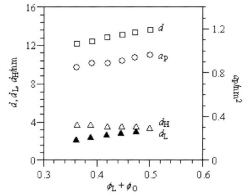

Fig. 17.18 Change in interlayer spacing, d, the half thickness of the non-polar layer, d_L, half thickness of the polar layer, d_H, and the effective cross-sectional area, a_P, in the lamellar phase of $Si_{14}C_3EO_{51.6}$–D_4 systems as a function of the total lipophilic volume fraction (ϕ_O is the volume fraction of oil) [28].

The phase polymorphism observed in Fig. 17.17 contrasts with the behavior of conventional polyoxyethylene-type non-ionic surfactant, which rarely forms aggregates in pure non-polar solvents [46].

When D_4 is replaced with silicone oils having a higher molecular weight, the reverse hexagonal phase in Fig. 17.17a disappears [28]. Moreover, the reverse micellar region (O_m) is changed to a zone in which the lamellar phase coexists with excess oil. Hence, the segregation of surfactant aggregates increases with the molecular weight of the solvent.

17.6
Phase Behavior of $Si_mC_3EO_n$ in Non-aqueous Polar Solvents

Figure 17.19 represents the phase diagrams of $Si_{14}C_3EO_{33.1}$ (f = 0.53) in water and poly(ethylene glycol) (PEG) with two different molecular weights. The bulk $Si_{14}C_3EO_{33.1}$ forms an L_α phase above the solid-melting temperature (43 °C), which melts to an isotropic liquid (O_m) at around 110 °C. Upon addition of water, the solid melting temperature decreases due to the hydration of the PEO chain, whereas the L_α–O_m phase transition temperature increases because the segregation between the two blocks of the copolymer is enhanced by dissolving water in the PEO chain. With further increase in water content, an excess water phase (W) separates from the L_α phase and a two-phase region of L_α + W is formed.

Low molecular weight PEG is more compatible with the PEO chain than water. In conventional surfactant systems, polyols inhibit the formation of liquid crystals [47, 48]. However, silicone copolymer forms self-organized structures in these solvents due to the long lipophilic chain and the strong segregation between lipophilic

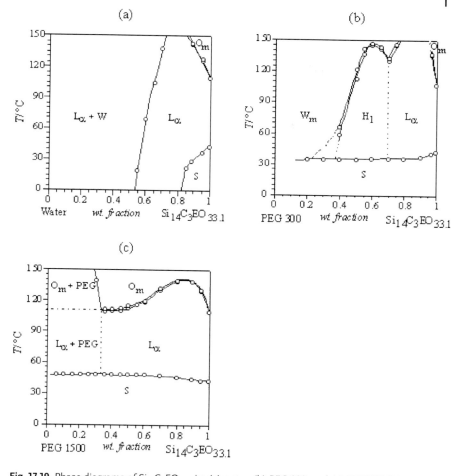

Fig. 17.19 Phase diagrams of $Si_{14}C_3EO_{33.1}$ in: (a) water; (b) PEG 300; and (c) PEG1500 [49].

and hydrophilic chains. The melting temperature of the copolymer solid is not particularly changed upon addition of PEG because it is not soluble in the copolymer PEO chain in a solid state. As shown in Fig. 17.19b, L_α–H_1–W_m transitions occur in a PEG 300 system. Hence, PEG 300 makes the copolymer layer curvature more positive.

Excess PEG phase coexists with the L_α phase in the longer-chain PEG 1500 system as shown in Fig. 17.19c, whereas an isotropic liquid phase coexists with excess liquid PEG phase at high temperatures, as it also occurs for shorter PEO-PDMS copolymers [50]. This suggests that the compatibility of solvent with the silicone copolymer decreases with the increasing molecular weight of the solvent, because there is no micellar solution phase in the dilute region.

Lin and Alexandridis [51] found that in some siloxane copolymers the addition of glycerol led to an increase of the micelle association number. A transition from spherical to ellipsoidal micelles was observed upon an increase in the temperature: this is attributed to the desolvation of the polyether chains.

17.7
Mixing of Poly(oxyethylene)–Poly(dimethylsiloxane) Copolymer and Non-ionic Surfactant in Water

Figure 17.20 represents the ternary phase diagram of the water–$Si_{25}C_3EO_{51.6}$–$C_{12}EO_5$ system at 25 °C [52]. Various liquid crystalline phases are successively formed with the increase in amphiphile concentration. The phase behavior obtained in the mixed amphiphile system is richer than that found in each binary system. Moreover, copolymer and surfactant do not mix. Hence, a large solid-present region (S) is observed along the copolymer-surfactant axis.

As described later in the text, the hydrodynamic radius of the surfactant micelles decreases upon addition of silicone copolymer, due to a rod to sphere transition. On the other hand, vesicles (hydrodynamic radii ≈100 nm) are observed in the dilute region of the water–$Si_{25}C_3EO_{51.6}$ system. A vesicle to micelle transition takes place with the increase of surfactant mixing fraction. As a result, two types of

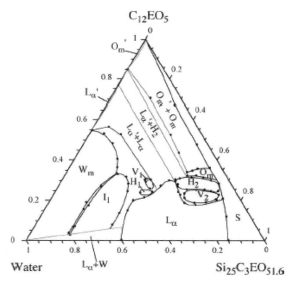

Fig. 17.20 Ternary phase diagram of water–$Si_{25}C_3EO_{51.6}$–$C_{12}EO_5$ system at 25 °C. L_α or L'_α, copolymer-rich or surfactant-rich lamellar; O_m or O'_m, copolymer-rich or surfactant-rich reverse micellar solution or surfactant liquid [52].

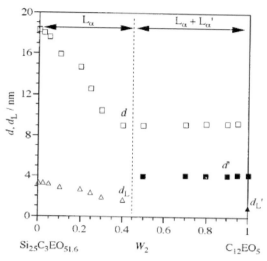

Fig. 17.21 Change in interlayer spacing (d or d') and half-thickness of the hydrophobic part (d_L or d'_L) of two L_α phases as a function of W_2 ($W_S = 0.75$) at 25 °C. The symbols without and with a prime represent the copolymer-rich L_α phase and the surfactant-rich L'_α phase, respectively [39].

aggregates, namely composite micelles and vesicles, coexist at the copolymer-rich region.

Two coexisting phases with surfactant-rich and copolymer-rich aggregates: (L'_α + L_α), (L'_α + H_2) and (O'_m + O_m) were observed at high surfactant to copolymer mixing ratios (see Fig. 17.20). Copolymer molecules are expected to be practically insoluble in the surfactant rich lamellar (L'_α) and in the surfactant liquid or reverse micellar phase (O'_m) because of the large size difference between the copolymer and surfactant molecules. The interlayer spacings of both lamellar phases (d and d') are plotted against the mixing fraction of $C_{12}EO_5$ (W_2) in Fig. 17.21. It can be seen that d and the half-thickness of the hydrophobic part, d_L decrease sharply by replacing the silicone copolymer with surfactant. This means that the silicone chains are compressed in a direction perpendicular to the surface because a_S is not expanded by mixing. However, the poly(dimethylsiloxane) chain cannot be compressed beyond $d_L = 1.7$ nm and, hence, surfactant cannot be incorporated in the L_α phase of $Si_{25}C_3EO_{51.6}$ beyond $W_2 \approx 0.45$. $Si_{25}C_3EO_{51.6}$ is practically insoluble in the $C_{12}EO_5$ L'_α phase, because the half-thickness is too small ($d'_L = 0.82$ nm at $W_2 = 1$) to allow the poly(dimethylsiloxane) chain in the surfactant bilayer. In the two-phase region, both interlayer spacings, d and d', are constant and d' is exactly the same as that in the pure $C_{12}EO_5$ system.

For ideal mixing, the average $\bar{a}_S$ should change linearly according to:

$$\bar{a}_S = x_1 a_{S(1)} + x_2 a_{S(2)} \tag{17.9}$$

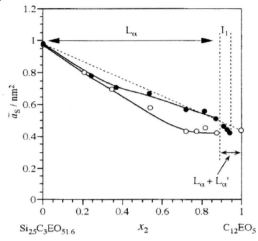

Fig. 17.22 Change in a_S as a function of mole fraction of $C_{12}EO_5$ in $C_{12}EO_5 + Si_{25}C_3EO_{51.6}(x_2)$ at $W_S = 0.75$ (○) and $W_S = 0.55$ (●). The dotted line is calculated from Eq. (17.9) [39].

where $a_{S(1)}$ and $a_{S(2)}$ are the effective cross-sectional areas for $Si_{25}C_3EO_{51.6}$ and $C_{12}EO_5$ in each aqueous binary (non-mixed) system, respectively. However, as shown in Fig. 17.22, the values of $\bar{a}_S$ are lower than those corresponding to ideal mixing (dotted line). As the segregation tendency between the surfactant and the poly(dimethylsiloxane) chain of the copolymer is enhanced in the presence of water, the monomeric solubility of the surfactant in the poly(dimethylsiloxane) moiety must be very low. Hence, the tendency of $\bar{a}_S$ implies that $a_{S(1)}$ and/or $a_{S(2)}$ decrease by mixing.

In non-ionic surfactant systems, at the same PEO chain length, a_S values are similar for both hydrocarbon and poly(dimethylsiloxane) surfactants having similar molar volume. Therefore, the reduction of $\bar{a}_S$ is mainly attributed to the reduction in repulsion of copolymer chains on the both sides of the A–B interface, due to the mixing of amphiphiles with different chain lengths. As shown schematically in Fig. 17.23, when amphiphiles with different chain lengths are mixed, the spatial constraints for the long chain are reduced. As a result, there is a reduction in entropy loss and in repulsion forces on both sides of the A–B interface.

The effect of added $Si_mC_3EO_{3.2}$ ($m = 25, 14, 5.8$) on the cloud temperature of $C_{12}EO_5$ aqueous solution is shown in Fig. 17.24. The cloud temperature for $C_{12}EO_5$ aqueous solution is 32.1 °C, whereas $Si_mC_3EO_{3.2}$ ($m = 25, 14, 5.8$) are insoluble in water and can be considered as silicone oils. The clouding phenomenon of poly(oxyethylene)-type non-ionic surfactant solution is related to the dehydration of the EO chain [55]. As a result, the surfactant layer curvature decreases, the micellar aggregation number increases and eventually, the surfactant is separated from water. As observed in Fig. 17.24, the cloud temperature decreases upon addition of $Si_{5.8}C_3EO_{3.2}$. On the other hand, when $Si_{14}C_3EO_{3.2}$ or $Si_{25}C_3EO_{3.2}$

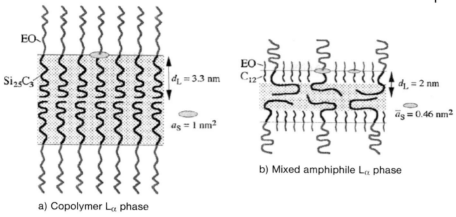

Fig. 17.23 Schematic representation of copolymer L_α phase (a) and L_α phase in the mixed amphiphile system (b) [39].

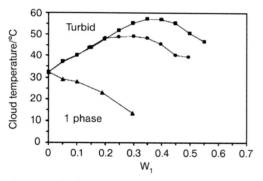

Fig. 17.24 Cloud temperatures for $Si_m C_3 EO_{3.2}$–$C_{12}EO_5$ aqueous systems as a function of the weight fraction of $Si_m C_3 EO_{3.2}$ in the mixture, W_1 at constant $W_S = 0.02$. ■ $m = 25$; ● $m = 14$; ▲ $m = 5.8$ [53].

are added to the $C_{12}EO_5$ aqueous solution, the cloud temperature increases, similar to what is observed in non-ionic surfactant aqueous solution upon addition of long-chain alkanes, such as C_{12} or C_{16}. The hydrophobic volume of $Si_{5.8}C_3EO_{3.2}$ is about 0.84 nm^3 per chain, and it corresponds approximately to the volume of a C_{28} chain. The hydrophobicity of $Si_m C_3 EO_{3.2}$ increases on increasing the hydrophobic chain, and it is then expected that the surfactant layer curvature would be less positive upon addition of longer chain silicone amphiphiles. However, the experimental fact is opposite to this prediction. Soni et al. [56] found a gradual increase in the hydration of the ethylene oxide moiety with a rise in temperature in poly(dimethylsiloxane) graft polyether copolymer micelle. This unusual temperature-dependent hydration was explained in terms of a preferential dehydration from the inner micellar core and core–corona interface, so at elevated temperatures, asymmetric micelles with highly water swollen coronas can be imagined. This replaces the repulsive interactions between the hydrophobic part of the solute and solvent (or water) with the solute–solute or intermicellar attractive interactions.

The apparent hydrodynamic radii ($R_{H\,app}$) of the aggregates formed in the $C_{12}EO_5$–$Si_{25}C_3EO_{3.2}$ system are shown in Fig. 17.25. The concentration dependence of R_{Happ} can be described by Eq. (17.10):

$$R_{H\,app} = \frac{R_H}{1 + A\,C} \tag{17.10}$$

where R_H is the hydrodynamic radius at infinite dilution, A is a constant with a value of 1.56 for the hard-sphere model [57, 58] and C is the concentration of particles (micelles) in the system. In a real situation, A should include the effect of hydrated water in the poly(oxyethylene) chain. Approximately four water molecules are hydrated in one poly(oxyethylene) chain as measured by NMR [59, 60], and dielectric relaxation spectroscopy [61]. Even if hydrated water molecules are taken into account, the difference between R_H and $R_{H\,app}$ is less than 1.5 wt-%. For the following calculations, it is assumed that all surfactants and copolymers are involved in the formation of the mixed spherical micelles, in which the effective cross-sectional areas of surfactant and copolymer are constant. The hydrophobic volumes are also assumed to be constant. a_S for $C_{12}EO_5$ ($a_{S,S}$) is 0.44 nm², and a_S for $Si_{25}C_3EO_{3.2}$ ($a_{S,P}$) is 0.70 nm² (these values correspond to the lamellar phase). The volumes of the hydrophobic part of the copolymer [39] and surfactant [42, 54] are $V_{L,P} = 3.29$ nm³ and $V_{L,S} = 0.357$ nm³, respectively. For spherical micelles, the radius of the hydrophobic part, r, is calculated from the surface area, A_L, and

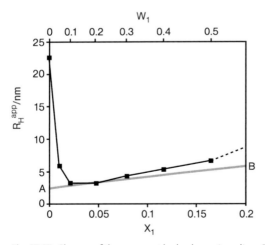

Fig. 17.25 Change of the apparent hydrodynamic radius, $R_{H\,app}$, as a function of the copolymer mole fraction, X_1, in a water–$C_{12}EO_5$–$Si_{25}C_3$–$EO_{3.2}$ system at 25 °C. The line A–B represents values calculated using Eqs. (17.10) to (17.13) [53].

the volume, V_{ML}, of the hydrophobic parts of the mixed micelles as shown in Eqs. (17.11) to (17.13):

$$r = \frac{3V_{ML}}{A_L} \tag{17.11}$$

$$A_L = N_{agg}\left[X_1 a_{S,P} + (1-X_1)a_{S,S}\right] \tag{17.12}$$

$$V_{ML} = N_{agg}\left[X_1 V_{L,P} + (1-X_1)V_{L,S}\right] \tag{17.13}$$

where N_{agg} is the surfactant and copolymer total aggregation number. The change of r as a function of X_1 for this model is shown as the A–B line in Fig. 17.25. In the copolymer-rich region, $W_1 = 0.1$–0.5, the $R_{H\ app}$ value is in relatively good agreement with the radius r calculated from the spherical model, whereas $R_{H\ app}$ deviates from r in the surfactant-rich region ($W_1 < 0.1$). This disagreement can be attributed to a change in the micellar shape. In fact, a temperature-induced structural evolution from spheres to ellipsoids was found for poly(dimethylsiloxane) graft polyether copolymer micelles [55, 62]. Moreover, Bernheim-Groswasser et al. [63] directly observed the micellar shape of 0.5 wt-% $C_{12}EO_5$ aqueous solution by cryo-TEM and reported that 50–100 nm threadlike micelles exist between 18 and 29 °C. All of these results support our hypothesis that rod-like micelles are formed in the surfactant-rich region, and that, upon addition of copolymer, a rod–sphere transition takes place.

Thus far, in previous works [64, 65] it has been considered that the cloud temperature for the non-ionic surfactant decreases upon addition of a hydrophobic amphiphile, because the hydrophobic amphiphile forms a mixed microinterface with the non-ionic surfactant and changes the mixed interface curvature towards negative. However, we have seen that when an amphiphilic block copolymer with a large hydrophobic chain, such as $Si_{25}C_3$ and $Si_{14}C_3$, is mixed with the non-ionic surfactant, the chain tends to make an oil pool in the core of the micelle to minimize the surface area, leading to a change in the curvature towards more positive at the mixed interface. As a result, the cloud temperature for the non-ionic surfactant solution increases upon addition of the amphiphilic block copolymer containing a bulky hydrophobic chain.

17.8
Conclusions and Outlook

It has been shown that PEO-PDMS block copolymers can form a variety of stable self-organizing structures as melts or when dispersed in water, oils and other solvents. The morphology and size of the nanostructures can be controlled by changing the PEO or PDMS chain lengths, or by mixing with low molecular weight surfactants. The peculiar properties of these block copolymers make them suitable as structure-directing agents in both polar and non-polar media. Much research still needs to be done in this field, for example, on the use of PEO-PDMS block copolymers in the preparation of nanostructured organic and inorganic materials.

References

1 R. M. Hill, in *Special Surfactants*, ed. I. D. Robb, Blackie Academic and Professional, London, **1997**.
2 M. Gradzielski, H. Hoffmann, P. Robisch, W. Ulbricht, *Tenside Surf. Deterg.* **1990**, *27*, 366–379.
3 A. Stuermer, C. Thunig, H. Hoffmann, B. Gruening, *Tenside Surf. Deterg.* **1994**, *31*, 90–98.
4 G. L. F. Schmidt, in *Industrial Applications of Surfactants* ed. D. R. Karsa, Royal Society of Chemistry, London, **1973**, p. 24.
5 R. M. Hill, (ed.) *Silicone Surfactants*, Marcel Dekker, New York, Surfactant Sci. Series 86, **1999**.
6 A. Bouridah, F. Dalard, D. Deroo, H. Cheradame, J. F. Le Nest, *Solid State Ionics* **1985**, *15*, 233–240.
7 N. Huesing, B. Launay, J. Bauer, G. Kickelbick, D. Doshi, *J. Sol.-Gel Sci. Technol.* **2003**, *26*, 609–613.
8 A.-W. Xu, *J. Phys. Chem. B* **2002**, *106*, 13161–13164.
9 D. L. Bailey, F. M. O'Connor, US Patent **1958**, 2,834,748.
10 D. C. Steytler, D. L. Sargeant, B. H. Robinson, J. Eastoe, R. K. Heenan, *Langmuir* **1994**, *10*, 2213–2218.
11 B. Luhmann, H. Finkelmann, *Colloid Polym. Sci.* **1987**, *265*, 506–516.
12 T. Iwanaga, Y. Shiogai, H. Kunieda, *Progr. Colloid Polym. Sci.* **1998**, *110*, 225–229.
13 N. Nagatani, K. Fukuda, T. Suzuki, *J. Colloid Interf. Sci.* **2001**, *234*, 337–343.
14 R. Fink, E. J. Beckman, *J. Supercrit. Fluids* **2000**, *18*, 101–110.
15 R. M. Hill, M. He, Z. Lin, H. T, Davis, L. E. Scriven, *Langmuir* **1993**, *9*, 2789–2798.
16 Z. Lin, R. M. Hill, H. T. Davis, L. E. Scriven, Y. Talmon, *Langmuir* **1994**, *10*, 1008–1011.
17 H. Kunieda, H. Taoka, T. Iwanaga, A. Harashima, *Langmuir* **1998**, *14*, 5113–5120.
18 X. Li, R. M. Washenberger, L. E. Scriven, H. T. Davis, *Langmuir* **1999**, *15*, 2278–2289.
19 X. Li, R. M. Washenberger, L. E. Scriven, H. T. Davis, *Langmuir* **1999**, *15*, 2267–2277.
20 J. Yang, G. Wegner, R. Koningsveld, *Colloid Polym. Sci.* **1992**, *270*, 1080–1084.
21 G. Kickelbick, J. Bauer, N. Huesing, in *Silicon Chemistry*, ed. U. Schubert, Wiley-VCH, Weinheim, **2003**, pp. 439–450.
22 F. S. Bates, G. H. Fredrickson, *Phys. Today* **1999**, 32–39.
23 F. J. B. Callejia, Z. Roslaniec, *Block Copolymers*, Marcel Dekker, New York, **2000**.
24 L. Leibler, *Macromolecules* **1980**, *13*, 1602–1617.
25 M. Matsen, M. Schick, *Curr. Opin. Colloid Interf. Sci.* **1996**, *1*, 329.
26 Md. H. Uddin, C. Rodríguez, A. Lopez-Quintela, D. Leisner, C. Solans, J. Esquena, H. Kunieda, *Macromolecules* **2003**, *36*, 1261–1271.
27 H. Kunieda, Md. H. Uddin, M. Horii, H. Furukawa, A. Harashima, *J. Phys. Chem. B* **2001**, *105*, 5419–5426.
28 Md. H. Uddin, Y. Yamashita, H. Furukawa, A. Harashima, H. Kunieda, *Prog. Colloid Polym. Sci.* **2004**, *123*, 269–274.
29 C. Rodríguez, Md. H. Uddin, K. Watanabe, H. Furukawa, A. Harashima, H. Kunieda, *J. Phys. Chem. B* **2002**, *106*, 22–29.
30 D. A. McQuarrie, in *Statistical Thermodynamics*, Harper & Row, New York, **1976**, chap. 13.
31 V. Luzzati, R. Vargas, A. Gulik, P. Mariani, J. M. Seddon, E. Rivas, *Biochemistry* **1992**, *31*, 279-285.
32 J. M. Seddon, *Biochim. Biophys. Acta* **1990**, *1031*, 1–69.
33 P. Alexandridis, U. Olsson, B. Lindman, *Langmuir* **1998**, *14*, 2627–2638.
34 A. K. Khandpur, S. Forster, F. S. Bates, I. W. Hamley, A. J. Ryan, W. Bras, K. Almdal, K. Mortensen, *Macromolecules* **1995**, *28*, 8796–8806.
35 M. Galin, A. Mathis, *Macromolecules* **1981**, *14*, 677–683.

36 Y. Ren, T. P. Lodge, M. A. Hillmyer, *Macromolecules* **2002**, *35*, 3889–3894.
37 M. W. Matsen, *Phys. Rev. Lett.* **1995**, *74*, 4225–4228.
38 F. S. Bates, M. F. Schulz, A. K. Khandpur, S. Forster, J. H. Rosedale, K. Almdal, K. Mortensen, *Faraday Discuss. Chem. Soc.* **1994**, *98*, 7–18.
39 H. Kunieda, Md. H. Uddin, H. Furukawa, A. Harashima, *Macromolecules* **2001**, *34*, 9093–9099.
40 C. Rodríguez, Md. H. Uddin, H. Furukawa, A. Harashima, H. Kunieda, *Prog. Colloid Polym. Sci.* **2001**, *118*, 53–56.
41 K.-L. Huang, K. Shigeta, H. Kunieda *Progr. Colloid Polym. Sci.* **1998**, *110*, 171–174.
42 H. Kunieda, K. Shigeta, K. Ozawa, M. Suzuki, *J. Phys. Chem. B* **1997**, *101*, 7952–7957.
43 K. Shigeta, U. Olsson, H. Kunieda, *Langmuir* **2001**, *17*, 4717–4723.
44 Md. H. Uddin, H. Kunieda, *Progr. Colloid Polym. Sci.* **2004**, *126*, 1–5.
45 K. Watanabe, N. Kanei, H. Kunieda *J. Oleo Sci.* **2002**, *51*, 771–779.
46 K. Konno, A. Kitahara, *J. Colloid Interf. Sci.* **1971**, *35*, 636–642.
47 T. Iwanaga, M. Suzuki, H. Kunieda, *Langmuir* **1998**, *14*, 5775–5781.
48 K. Aramaki, U. Olsson, Y. Yamaguchi, H. Kunieda, *Langmuir* **1999**, *15*, 6266–6232.
49 H. Kunieda, Md. H. Uddin, Y. Yamashita, H. Furukawa, A. Harashima, *J. Oleo Sci.* **2002**, *51*, 113–122.
50 S. A. Madbouly, B. A. Wolf, *J. Chem. Phys.* **2002**, *117*, 7357–7363.
51 Y. Lin, P. Alexandridis, *J. Phys. Chem. B* **2002**, *106*, 12124–12132.
52 M. H. Uddin, D. Morales, H. Kunieda, *J. Colloid Interf. Sci.* **2005**, *285*, 373–381.
53 M. Kaneko, K. Matsuzawa, M. H. Uddin, M. A. López-Quintela, H. Kunieda, *J. Phys. Chem. B* **2004**, *108*, 12736–12743.
54 C. Tanford, *J. Phys. Chem.* **1972**, *76*, 3020–3024.
55 G. Karlström, *J. Phys. Chem.* **1985**, *89*, 4962–4964.
56 S. S. Soni, N. V. Sastry, J. George, H. B. Bohidar, *J. Phys. Chem. B* **2003**, *107*, 5382–5390.
57 P. N. Pusey, R. J. A. Tough, in *Dynamic Light Scattering*, ed. R. Pecora, Plenum Press, New York, **1985**.
58 H. M. Fijnaut, *J. Chem. Phys.* **1981**, *74*, 6857–6863.
59 P. G. Nilsson, B. Lindman, *J. Phys. Chem.* **1983**, *87*, 4756–4761.
60 G. Klose, S. Eienblatter, J. Galle, A. Islamov, U. Dietrich, *Langmuir* **1995**, *11*, 2889–2892.
61 T. Sato, Md. K. Hossain, D. P. Acharya, O. Glatter, A. Chiba, H. Kunieda, *J. Phys. Chem. B* **2004**; *108*, 12927–12939.
62 Y. Lin, P. Alexandridis, *J. Phys. Chem. B* **2002**, *106*, 10845–10853.
63 A. Bernheim-Groswasser, E. Wachtel, Y. Talmon, *Langmuir* **2000**, *16*, 4131–4140.
64 H. Kunieda, M. Horii, M. Koyama, K. Sakamoto, *J. Colloid Interf. Sci.* **2001**, *236*, 78–84.
65 E. J. Kim, D. O. Shah, *J. Phys. Chem. B* **2003**, *107*, 8689-8693.

Subject Index

a

ABA triblock copolymers   193, 358
– hydrogenated PB block   367 ff.
– semicrystalline   380 f.
– thermal behavior   380
– thermoplastic elastomers (TPEs)   369
ABC triblock copolymers
– crystallization–microphase separation   384 ff.
– hydrogenated PB block   367 ff.
– mesostructure   300
– semicrystalline   381 ff.
actuators   112
adhesion of polymer vesicles   57
– PEO-PBD   57
Ag nanoparticles   164
Al_2O_3 302
alignment   106
aluminosilicate mesostructures   311
aluminum-*sec*-butoxide   310
amphiphilic   24, 28, 39, 117
– P*t*BA-*b*-PEO-*b*-P*t*BA   28
– PEO-*b*-PS   28
– PS-*b*-PEO-*b*-PS   28
amphiphilic block copolymers   265, 291 ff., 391 ff.
– end-functionalized homopolymers   172
amphiphilic dendrons   111, 319 ff.
anionic polymerization   14, 259
– PB-*b*-PEO   16
– PEO-*b*-PS   15
– PI-*b*-PAA   16
– PMMA-*b*-PEO   15 f.
– PS-*b*-PAA   16
– PS-*b*-PEO   16
– PS-*b*-PI-*b*-PS   15
anionic promoter   156
anisotropic susceptibility   201
annealing   106, 192 ff., 203

anthracene   207, 209, 216, 218
anti-tumor   78
anticancer   74
architecture   10, 112
atomic force microscopy (AFM)   174, 196, 214, 261 ff., 266, 268, 301, 342

b

BAM® Ionomers   354 f.
– SANS   354
– SAXS   354
benzoic acid (BA)   206 ff., 216
biocompatibility   112, 147
biodegradability   112
biohybrid polymer vesicles   60
biological membranes   118
biomimetic   118, 146
biosensors   87
blends   346, 396
block copolymer–inorganic hybrid nanofibers   247
block copolymers
– containing free metal–ligand side-chains   178 ff.
– containing tris(bpy)   178
– cylindrical brushes   233
– diblock   10
– morphology   191 ff., 361
– nanostructured carbons   257 ff.
– silica composites   270
– templating   292 ff.
– triblock   10
block copolypeptides   123
block ionomers   337 ff.
bovine serum albumine   137
Brij   292
broken lamellar   141
bulk   109, 112
bulk state   164

Block Copolymers in Nanoscience. Massimo Lazzari, Guojun Liu, Sébastien Lecommandoux
Copyright © 2006 WILEY-VCH Verlag GmbH & Co. KGaA, Weinheim
ISBN: 3-527-31309-5

c

cancer 79, 81, 87
carbon
– films 260 ff., 271
– nanoobjects 258, 265
– nanoporous 258, 266 ff.
catalysis 64
catalyst 21, 163 ff.
– PMMA-b-PnBMA 22
cationic polymerization 17
– PIB-b-PVE 18
cells 140
CeO_2 296, 299, 302
ceramic particles 165
ceramic replica 164
– shape retention 163
channels 107
charge mobility 339
chemical cross-linking 237
chemical modification 30
chemical patterning 204, 278
chemical vapor deposition 164
chimeras 118
CMC 93
CMT 93 f.
cohesive energy 391, 400
comb-like copolymer 340
complex architecture
– arm-first 31
– core-first 31
– dendrimers 31
– graft 31
– stars 31
compression 196
confinements 203
confocal fluorescence microscopy 46
controlled radical polymerization 19
– atom transfer radical polymerization (ATRP) 19 f., 259, 267
– MADIX 26
– nitroxide-mediated radical polymerization (NMP) 19, 23, 183, 260
– reversible addition fragmentation chain transfer (RAFT) 19, 25, 97, 260
convergent 173
copolymerization 347 ff.
copolypeptide 124 f.
core 100 f.
core-shell 4
corona 100 f.
coupling reactions 13
– PS-b-PtBA-b-PS 14
– PS-b-PEVE 13
– PS-b-PI-b-PS 14

crew-cut micelles, PS-PAA 102
cryo-TEM 415
cylindrical micelles 158, 163, 237
– $PLLeu_{180}$-b-$PLGlu_{180}$ 123
cylindrical nanoceramic replica 162
cylindrical pores 213

d

2D-chord analysis 343
2D-confinement 213
2D-SAXS 301
– data 318
dendritic structure 111
diagnostics 64
diblock 10
diblock copolypeptides 133
dielectric constant 198
dielectric relaxation spectroscopy 414
dip coating 193
– evaporation induced self assembly 316
1,1-diphenylethylene (DPE) 154
direct methanol fuel cell (DMFC) 339, 358, 360
directional crystallization 209 ff., 214 ff.
disclination 192
dislocation 192
divergent 173
DLS 174, 186
DNA 62
domain orientation switch
– electric field 279
– epitaxial means 278
– rough surface 279
– solvent atmosphere 280
double hexagonal 131
– PS-b-PBLGlu 132
– PS-b-PZLLys 132
double hydrophilic block copolymers 92
– P2VP-PEO 95
drug 82, 85, 98
– carriers 85
– delivery 64, 73 f., 79, 112, 124
– release 83 f.
dual initiators 29
– PS-b-PCL 29
dumbbell-shaped molecules 252
dynamic density functional theory 200

e

electric breakdown 201
electric field 197 ff., 359
electrodeposition 304
electroless plating solutions 249

electron diffraction pattern 207, 217, 220 f.
electronics 64
electrospinning 197
electrostatic interactions 100
– P2VP-PEO 99
elongational flow field 201
encapsulation 64 f.
endosomes 77
energy filtering TEM 313
environmentally responsive surfaces 284
enzymatic degradation 138
enzymes 113
epitaxial crystallization 206, 215, 381
epitaxial self-assembly 204
epitaxy 204 ff., 212, 215
etching 107
etching resists 160
eutectic
– behavior 209
– crystallization 214
– solidification 223
evaporation-induced self-assembly (EISA) 293, 295
extrusion 195 f.

f
γ-Fe$_2$O$_3$ 323
ferrocenylsilane 151
– anionic ring-opening polymerization 152
– photolytic living anionic ROP 156
– ROP 154
fibrillar structures 144
– PBLGlu-*b*-PHF-*b*-PBLGlu 145
fibrils 123, 125, 145
field cooling 324
Flory–Huggins 1
– interaction parameter 369, 395, 398
fluorescence microscopy 245
folate 80 f.
free-radical polymerization 13
fuel cells 304, 337 ff.
functional polymer vesicles 59
– PEG-PPS 60
– PEO-PBD 59
– PEO-PLA 59
– PEO-poly(lactic acid) (PLA) 59
– PMOXA-PDMS-PMOXA 59
– PNIPAAM-PLA 60
fusion and fission of polymer vesicles 58
– HBPO-star-PEO 59
– PS-PAA 59

g
gels 146
– PS-*b*-PLLys 128
giant vesicles 55, 95
Gibbs phase rule 396
GISAXS 263
glass transition temperature T_g 193, 197, 202, 359
(3-glycidyloxypropyl) trimethoxysilane 310
gradient copolymer brushes 284
grafting 345
"grafting from", "grafting to" 283
grain boundary defects 192
graphoepitaxy 203, 205, 212 ff.
grazing incidence small-angle X-ray scattering (GISAXS) 262, 330
group transfer polymerization 16
– PMMA-*b*-PAA 16
"Grubbs" catalyst 178

h
heat-exchange capacity 339
α-helical 120, 126, 128, 139
α-helix 124, 131, 141
hexagonally packed cylinders 166
HfO$_2$ 302
hierarchical order 361
hierarchical structures 129, 146
– ordered 133
– PS$_{52}$-*b*-PBLGlu$_{104}$ 130
– self-organized 361
high-density magnetic media 328
hybrid 4
– PELLys-*b*-PNaLAsp 124
hybrid vesicles 126
hydrodynamic radius 410, 414 f.
hydrogels 112, 124
hydrogen-bonding 110
– P4VP-MSA-PDP 110
– PS-P4VP 110
hydrogenation, synthetic elastomers 367 f.
hydrolysis 107
hydrophobic core 297

i
incompatibility 2
interaction parameter, χ 53
interfaces 275
interfacial interactions 200
interlayer spacing 408
inverse cubic bicontinuous morphology 313

ionic exchange capacity (IEC)   342 f., 348, 351, 359
ionic strength sensitive micellization   98
ionomers
– applications   340
– molecular weight (MW)   344
– SAXS   343
IR spectroscopy   349
γ-iron oxide–aluminosilicates   322

k

Kapton   198, 200
KLE as templates   292 ff., 300 ff.
Kraton   195, 372 f.
Kraton-liquid-b-poly(ethylene oxide) (KLE)   292

l

large amplitude oscillatory shear (LAOS)   197
LCST   93, 95, 97, 99, 101, 103, 105, 121
liquid crystal phases   393 ff.
– bicontinuous cubic phase   394, 399
– cubic phase   394 f., 400 ff.
– hexagonal phase   393, 395, 400 ff.
– HPL phase   399
– lamellar phase   393, 395, 404 ff.
lithium ion charge transport   321
lithographic techniques   204
lithographic templates   275
luminescence intensity of the Ru(II) complex   179
lyotropic liquid crystalline phases   292 ff.
lysosomes   77

m

macromolecular engineering   9
macroscopic order of sol–gel   318
maghemite
– PB_{48}-b-$PLGlu_{56–145}$ 125
– $PELLys_{100}$-b-$PAsp_{30}$ 125 f.
magnetic colloids   65
magnetic field   197, 201
magnetic memory materials   163
magnetic mesoporous materials   322
magnetic nanocomposite   125
magnetic nanowire array   280
mechanical flow fields   196 f.
melting temperature T_m 395, 408
membrane   40, 45, 53, 123, 137, 139
– mechanical properties   54
– PBLGlu-b-PI-b-PBLGlu   138
– PEG-PDMS-PMOXA   63
– PEO-PB   55

– PEO-PBD   56
– PEO-PEE   54, 56
– PHLGln-b-PI-b-PHLGln   138
– PMOXA-PDMS-PMOXA   56, 63
– thickness   53
membrane-electrode assembly (MEA)   337 ff., 343
mesoporous ceramics   324
mesoporous materials   180, 182, 294, 391
– SAXS   298
– SAXS–WAXS   300, 302
mesoporous metal oxides   299 ff.
mesoporous metals, fabrication   304
mesoporous silicas
– as templates   257
– fabrication   291 ff.
mesostructured block copolymer–sol nanoparticle co-assemblies, flow-induced alignment   316
mesostructured inorganic materials   291 ff.
mesostructured polymer–inorganic hybrid materials   309
– formation mechanisms   313
– morphologies   311
– solid-state ^{29}Si NMR   314
mesostructured thin films   328
metal complexes   172
metal oxides, nanoparticles   296
metal–ligand containing block copolymers   186
metal–ligand core   172
metal-catalyzed polymerization   18
– PP-b-PHex   18
– PP-b-POct   18
metal-containing polymers   151
metal-driven self-assembly   169
metallic nanodots   107
metallo-supramolecular
– PS-P2(4)VP-PMAA   101
– PS-P2VP-[Ru]-PEO   101
– PS_{20}-[Ru]-PEO_{70} 175
micelles   4, 29, 40, 73 f., 80, 82 ff., 91, 95 ff., 101, 112, 118 f., 121, 125, 292, 297 ff., 415
– $PAEI_{41}$-b-$PLPhe_{4, 8}$ 122
– PAMPS-PNIPAM-PCEMA   105
– PB-b-PLGlu   121
– PCL-b-PAA   29
– PDEAEMA-PDMAEMA-PEO   98
– PDMAEMA-PMAA   99
– PEG-pGlu   82
– PEG_{113}-b-$PAsp_{18, 78}$ 121
– PEG_{113}-b-$PLLys_{18, 78}$ 121

– PEG$_{75}$-b-PMLGlu$_{32}$ 121
– PI$_{49}$-b-PLLys$_{123}$ 120
– PLL$_{95, 270}$-b-PAsp$_{47-270}$ 120
– PS-b-PZLLys   121
– PS-P4VP   92
– shell cross-linked   265 ff.
microfluidically aligned   162
microphase separation   2
microporosity quantification, SANS   298
miktoarm star copolymer   156
mixed brush   284
morphology   46
– cubic   2, 112
– cylindrical   106 f., 119, 135, 180, 194 f., 197, 199, 202, 206, 211, 214, 216, 261, 352
– – PBLGlu-b-PI-b-PBLGlu   137
– – PMLGlu-b-PB-b-PMLGlu   136 f.
– gyroid   2, 117
– hexagonal   2, 112, 117, 126, 141
– – PELGlu-b-PB-b-PELGlu   137
– hexagonal-in-hexagonal   131
– – PS-b-PBLGlu   127
– hexagonal-in-lamellar   126 f., 131, 135
– – PBLGlu-b-PGly   134
– – PI-b-PZLLys   127
– lamellar   2, 106, 110, 112, 117, 119, 126, 132, 135, 197, 202, 205, 211, 216, 261
– – domains   180
– – PBLGlu-b-PBAN-b-PBLGlu   144
– – phase   294
– – PS-b-PBLGlu   128
– – PZLLys-b-PB-b-PZLLys   137
– – supramolecular   131
– PB-PEO   46
– PEG-PPS   52
– PEO-PBD   52
– PEO-PEE   52
– PEO-PEE-PEO   52
– PI-b-PCEMA   50
– PS-b-PCEMA   50
– PS-PAA   47, 51 f.
– PS-PEO   47, 51
– PS-PI   50
– PS-PIAT   47
– (P4VP)-PS   50
– spherical   102 f., 117, 119, 144, 174, 197, 206, 261
– – PMGlu-b-PB-b-PMGlu   136 f.
– – PZLLys$_{49}$-b-PDMS$_{400}$-b-PZLLys$_{49}$ 139
multi-responsive micellar systems   103
– PBD-PGA   104
– PMAA-PEO   105
– PNIPAM-PAA   103
– PPO-PDEAEMA   103

– P2VP-PEO   105
MW determination   352

n

N-methacryloxysuccinimide (OSu)   184
Nafion®   339 ff., 347, 353, 362
nanocapsules   104
nanoceramics   161
nanocomposites, SEBS   377 f.
nanocylinders   158
nanodevices   4, 76
nanofibers   233 f.
– alignment   248
– chemical reactions   247
– dilute solution viscosity properties   242
– dispersity   247
– end functionalization   251
– fractions   242
– Huggins coefficient   243
– hydrodynamic diameter   244
– intrinsic viscosity   243
– length distribution function   241
– liquid crystalline to disorder transition temperature   246
– persistence length   244
– shear thinning   242
– solution properties   240
– superparamagnetic   248
nanomaterials   4
nanoparticles   266, 299 f.
– metal oxides   296
nanopillars   330
nanopores   107
nanoporous alumina templates   213
nanoporous block copolymer thin films   107
nanoscience   73
nanotubes   158, 164, 233 f.
– backbone modification   247
– chemical reactions   247
– end functionalization   251
NCA polymerization   147
neutral substrate   276
NMR   343, 349, 414
non-equilibrium structures   193
non-uniform cross-linking   237

o

octamethylcyclotetrasiloxane (D$_4$)   396, 406
once-broken rods, PBLGlu$_{53-188}$-b-PB$_{64}$-b-PBLGlu$_{53-188}$ 119
"optical magnetic nanohand"   253
order–disorder transition   109 f.

ordered structures
- cylinders 91
- spheres 91
OsO$_4$-stained 211 f.
oxidative degradation 339
oxygen plasma 212
ozonolysis 107

P

P[VDF-co-HFP]-b-PS 359
palladated pincer complex 182
PAN 258
- synthesis 259 f.
PAN block copolymers, as carbon precursors 258 ff.
PAN-b-PBA-b-PAN 264
patterns 191, 203, 205, 212
PB-b-PEO 294, 302
PB$_{27-119}$-b-PLGlu$_{24-64}$ 120
PBA-b-PAN 266, 268 ff.
- pyrolysis 268
PCEMA-PtBA nanotubes 239
PDMS 152
PDMS chain 403 ff., 411
PE crystals 208, 220
PE-b-PEP-b-PE 207 ff., 216
PE-b-PEP-b-PEO 382
PEO chains 297 ff., 395, 402, 404 ff.
PEO crystallization 382
PEO-b-PAN 267 f., 270
PEO-b-PDMS 392 ff.
- phase diagram 396
- SAXS 397, 404
PEO-b-PPO-b-PEO, silica composites 270
PEO-b-PS 281
peptide folding 129
peptosomes 60
periodic arrays 205
periodicity 191, 203
Perovskites 302
PFP-b-PFS-b-PDMS 153
PFS-b-PBLG 157
PFS-b-PDMAEMA 153, 159
PFS-b-PDMS 152, 158
PFS$_{50}$-b-PDMS$_{300}$ 158
PFS-b-(PI)$_3$ 156
PFS-b-PMMA 154 f., 164
- living free radical polymerization 154
PFS-b-PMVS 153, 161
pH-sensitive 97
pH-sensitive core-shell corona aqueous micelles 177
pH-sensitive micelle 77 ff.

- PEG-PBLA 79
pH-sensitive micellization 95
- PDMAEMA-PDEAEMA 97
- PEO-P(M)AA 96
- PMAA-PEO 96
- P2VP-PDMAEMA 97
- P2VP-PEO 96
- P4VP-PEO 96
phase diagram
- PS-b-PBLGlu 131
- PS-b-PZLLys 131
photoactive 144
photonic crystals 328
PI-b-PFS 153, 159, 161, 164
PI-PtBA-P(CEMA-HEMA)-PGMA 249
PI-PCEMAP-tBA 238
plasma etching 161
platelets 124
Plumber's Nightmare morphology 313
Pluronics 293, 297, 303
polarized optical microscopy 210
poly(acrylic acid)–polyacrylonitrile (PAA-b-PAN) 265, 267
polyampholytic 98
polyarylene 351
polybenzimidazole (PBI) 351
poly(n-butyl acrylate)–polyacrylonitrile (PBA-b-PAN) 260, 262
poly(t-butyl acrylate)–polyacrylonitrile (PtBA-b-PAN) 260
poly(cyclohexylethylene)-b-polyethylene-b-poly(cyclohexylethylene) (PCHE-b-PE-b-PCHE) 381
polydiene-based diblock copolymers 126
- PB-b-PBLGlu 126
- PB-b-PHLGln 126
- PB-b-PZLLys 126
polydiene-based triblock copolymers 134
- PBLGlu-b-PB-b-PBLGlu 134 f.
polyester-based diblock copolymers 133
- PCL-b-PBLGlu 133
- PCL$_{50}$-b-PBLGlu$_{40}$ 133
- PLL-b-PBLGlu 133
polyether-based diblock copolymers 131
- PEG-b-PAla 132
polyether-based triblock copolymers 140
- PBLAsp$_{25}$-b-PEG$_{250}$-b-PBLAsp$_{25}$ 143
- PBLGlu-b-PEG-b-PBLGlu 140 f.
- PMLGlu-b-PTHF-b-PMLGlu 143
poly(ether ketones) (PEK) 351
polyethylene-b-polystyrene-b-poly-(ε-caprolactone) (ESC) 384
poly(ferrocenyl phosphine) 153

polyferrocenylsilane 151
poly(3-hexylthiophene) 50
poly(hydroxyl ethyl methacrylate)-b-PMMA 285
polyisobutylene-b-poly(ethylene oxide) (PIB-b-PEO) 302
poly(isoprene-b-dimethylamino ethyl methacrylate) 325
poly(isoprene-b-ethylene oxide) 310
polymer brush 282
polymer–drug conjugates 75
– PEG-p(Asp-ADR) 76
– PEG-PBLA 76
polymeric phthalocyanine 319
polymerization of terpy functionalized monomers 184
polymerizing pre-formed metal complexes 178
polymersomes 41, 45, 118
poly(methylvinylsiloxane) 153
poly(MMA-b-OSu) 185
polymorphism 408
poly(oxyethylene) chain 414
polypeptide-based copolymer vesicles 60
– PBD-PG 61
– poly(L-leucine) 61
– poly(L-lysine) 61
polypeptides 104, 117 ff.
polysaccharides 118
poly(S-b-OSu) 185
polysiloxane-based triblock copolymers 139
– PBLGlu-b-PDMS-b-PBLGlu 139
polystyrene-based diblock copolymers 127
– PS-b-PZLLys 127
polystyrene-based triblock copolymers 138
– PBLGlu-b-PS-b-PBLGlu 138
– PSar-b-PS-b-PSar 138
polystyrene-b-polybutadiene-b-poly(ε-caprolactone) (SBC) 369
polystyrene-b-poly(2-cinnamoyloxyethyl methacrylate) 235
polystyrene-b-poly(ethylene-co-butylene)-b-polystyrene (SEBS) 367 ff.
polystyrene-b-polyethylene-b-poly-(ε-caprolactone) (SEC) 384 ff.
polystyrene-b-poly(ethylene propylene) (PS-b-PEP) 197
polystyrene-b-poly(ethylene-alt-propylene)-b-polyethylene (PS-b-PEP-b-PE) 216, 218, 222, 381 ff.
– elastic recovery 383 f.

polystyrene-b-poly(ethylene-alt-propylene)-b-polystyrene (PS-b-PEP-b-PS) 384
polystyrene-b-poly(hydroxyethyl methacrylate)-b-poly(methyl methacrylate), PS-b-PHEMA-b-PMMA 200
polystyrene-b-polyisoprene-b-polystyrene (SIS) 372
polystyrene-b-poly(2-vinyl pyridine) (PS-b-P2VP) 177, 212, 277
polystyrene-b-poly(4-vinyl pyridine) (PS-b-P4VP) 180, 271
polystyrene-b-poly(2-vinyl pyridine)-b-poly(t-butyl acrylate) (PS-b-P2VP-b-PtBA) 195
polystyrene-b-poly(2-vinyl pyridine)-b-poly(t-butyl methacrylate) (PS-b-P2VP-b-PtBMA) 280, 284
polystyrene-b-poly(2-vinyl pyridine)-b-poly(ethylene oxide) 177
polysulfones (PSU) 351
poly(tetrafluoroethylene) (PTFE) 212, 345
poly(ureamethylvinyl)silazane 325
poly([vinylidene difluoride-co-hexafluoropropylene]-b-styrene) 359
poly(vinyl pyridine) 180
porous lamellar structure 180
porous material 180
post-polymerization attachment of the metal–ligand 184
potential applications of polymer vesicles 64
– PBD-PEO 65
– PMOXA-PDMS-PMOXA 65
pre-ceramic polymer 325
preferential interactions 199
preferential wetting 205
pressure-sensitive adhesives 372
proteins 45
– PMOXA-PDMS-PMOXA 62
proton exchange membrane (PEM) 339
protonic conductivity 340, 355, 358, 361
PS-b-P4VP/alkylphenol 180
PS-b-P4VP[Zn(DBS)$_2$]$_y$ 180
PS-b-PE 207, 216, 219 ff.
PS-b-PEO 174, 202, 298
PS-b-PFS 152, 164 f., 202, 212 f., 280
PS-b-PI 200, 209, 212, 215
– SAXS 201
PS-b-PMMA 195 f., 198 ff., 203 f., 209 f., 278

PS-g-PFS   154
PS-PCEMA   235, 238
PS-PCEMA-PAA nanotubes   239 f.
PS-PI   237 f.
pyrolysis   163

q

quantum dot arrays   328
quasi-elastic neutron scattering   343

r

Raman spectroscopy   266, 270
redox-activity   163
reverse micelles   394, 396
RIE (reactive ion etcher)   213 f.
ring opening metathesis polymerization (ROMP)   18, 178
rod–coil   127
rod–coil block copolymers   117
rod-like   123
rod-like micelles   95
rods   102 f.
roll-casting   197, 316
ROMP   182
rotating cylinder viscometer   242
Ru(II) tris(bipyridine) metal complex   178
RuO_4-stained   210, 218 f.

s

"sacrificial blocks"   262, 268
SANS
– BAM® Ionomers   354
– microporosity quantification   298
SAXS   180, 326
– BAM® Ionomers   354
– ionomers   343
– mesoporous materials   298
– PEO-b-PDMS   397, 404
– PS-b-PI   201
– SEBS   357
– thin film   199
SAXS–WAXS, mesoporous materials   300, 302
SBS   194, 368
– adhesives, sealants and coatings   372 f.
– asphalt modification   373 ff.
scanning electron microscopy (SEM)   213 f., 382
scanning force microscopy (SFM)   382
"schizophrenic"   104
– PGA-PLys   104
"schizophrenic" micelles   103

SEBS
– adhesives, sealants and coatings   372 f.
– applications   371 ff.
– asphalt modification   373 ff.
– compounds   375 ff.
– grafted maleic anhydride   372 ff.
– medical applications   378 f.
– nanocomposites   377 f.
– SAXS   357
– thermoplastic elastomer gels (TPEGs)   377 f.
secondary structure   140, 144
– PBLGlu-b-PEG-b-PBLGlu   142
– PZLLys-b-PB-b-PZLLys   136
– PZLLys-b-PEG-b-PZLLys   142
segregation   412
self-assembled monolayer (SAM)   204
self-assembly   1, 33, 39, 117
sensors   112
shearing   196
sheet-like structures   122
shell cross-linked nanocylinders and nanotubes   161
shell cross-linked PI_{320}-b-PFS_{53} 162
shell cross-linking reaction   161
silica-type ceramic materials   310
silicone block copolymers   391 ff.
single-walled carbon nanotubes   164
small-angle X-ray scattering (SAXS)   199, 294, 313
smart block copolymers   112
smart materials   118
SnO_2 299
sol–gel precursors   311
solution-casting   349
solvent evaporation   202
sPI   352, 355
spin-coating   193, 197, 204
SQUID (super-conducting quantum interference device)   323
star   3
star-like copolymers   172
– convergent approach   172
– divergent approach   172
stealth   45
stimuli   147
stimuli-responsive   91
stimuli-responsive micelles   100
– PBO-PEO   101
– PIBVE-PMVE   101
– PPO-PEO   101
– PS-P2VP-PEO   101
stimuli-responsive thin films   106
– PS-[Ru]-PEO   107

– PS-P4VP   107
– PS-P4VP-HABA   108
– PS-PMMA   106
– PVP-HABA   107
– P4VP-PDP   107
stimuli-sensitive micellization   92
stimulus   64
– electric field   9
– ionic strength   9
– pH   9
– temperature   9
storage applications   106
β-strand   124, 128, 141
structure-directing   320
structure-directing agents   319
structure-within-structure morphology   180
sulfonated
– polybenzimidazole   353
– poly(ether sulfone)   351
– polyimides (sPI)   347 ff.
sulfonation   344 f., 358
sulfur monochloride   236
"super-surfactant"   252
supercritical CO_2 195, 303
superstructures   146
supramolecular   87
"suprapolymer" chain   241
surface effects   193, 205
surface energy   275
surface regeneration   286
surfactant–block copolymer mixtures   391 ff.
surfactants   292, 392, 410
switch
– PI-b-PLA   28
– PIB-b-PMMA   28
– PNB-b-PMA   27
– PNB-b-PS   27
– PS-b-PDMS-b-PS   28

t

tapping mode AFM   328
TEM specimen staining, iodine   277
temperature-responsive   109
temperature-sensitive micellization   93 ff.
– PBO-PEO   93
– PEO-PNIPAM   94
– PEO-PPO   93
– PEO-PPO-PEO   93
– PEOVE-PMOVE   94
– PEOVE-PMOVE-PEOEOVE   95
– PMVE-PMTEGVE   94

– PMVE-PVA   94
– poly(NIPAM-co-HEMA-lactate) block-PEO   94
templated self-assembly   203
TEMPO   24
– PtBA-b-PDMA   24
– PnBA-b-PS   24
– PS-b-PI   24
terracing   192
tetrablock copolymer   249
thermal gradient   202
thermoplastic   4, 33
thermoplastic elastomers   28
thermotropic phase   395
thickeners   124
thin films   106, 112, 139
TiO_2 296, 299, 302
topographic pattern   204, 214
transmission electron microscopy (TEM)   174, 200, 207 f., 211, 216 ff., 264, 293 ff., 342, 355 ff., 382
transport properties   341
triblock   10
triblock copolymers
– semicrystalline   379 ff.
trifluorostyrene-based ionomers (sPTFS)   354
triphenylene electron donor   319
tubular micelles   234
tubular polymersomes   52
tumors   75, 81, 85 ff.
twisted ribbon   123

u

UCST   93, 103
unimers   97
UV-etching   106

v

V_2O_5 299
vesicles   39 ff., 96, 102 ff., 118, 125, 294, 404
– PB$_{27-119}$-b-PLGlu$_{20-175}$ 119
– PB$_{48}$-b-PLGlu$_{20}$ 122
– PEO$_{272}$-b-PBLGlu$_{38-418}$ 121
– PGA-Plys   105
– PLGlu$_{15}$-b-PLLys$_{15}$ 123
– PLLeu$_{10-75}$-b-PELLys$_{60-200}$ 123
– PNIPAAm$_{203}$-b-PBLGlu$_{39-123}$ 121
– PS$_{258}$-b-PZLLys$_{57}$ 122

w

water-dispersible polymer–Pd hybrid catalytic nanofibers   249

y

Yamakawa–Fujii–Yoshizaki (YFY) theory   243

z

zero-field cooling   324
zigzag   130
zigzag lamellar, PS-*b*-PZLLys   129
Zimm plot   241
zinc dodecylbenzene sulfonate   180
ZrO_2 299, 302
zwitterionic   123